THEORY OF
SPECTROCHEMICAL EXCITATION

Theory of
Spectrochemical Excitation

P. W. J. M. BOUMANS

Laboratory for Analytical Chemistry
University of Amsterdam

HILGER & WATTS LTD
LONDON

ISBN 978-1-4684-8430-4   ISBN 978-1-4684-8428-1 (eBook)
DOI 10.1007/978-1-4684-8428-1

Published by
HILGER & WATTS LTD
98 ST PANCRAS WAY, LONDON, NW1

To

my parents

my wife

and

Christien, Lidwien and Baptist

PREFACE

There are more things in heaven and earth, Horatio,
Than are dreamt of in your philosophy.

Hamlet

There exists a fairly large number of textbooks concerned with spectrochemical analysis. Most of them deal with practical applications and instrumental factors, and provide the reader with the knowledge indispensable for conducting analyses with the help of emission spectra. Practical knowledge and experience are indeed important requisites for successfully exploiting the spectrochemical method in the field of analytical chemistry. As the method is essentially empirical, it is, in principle, a simple one, provided that we succeed in exciting all samples in an identical manner; for then, relative intensities of spectral lines can serve as the 'weights' by which to measure amounts of elements. However, creating the required constancy of excitation conditions is hampered by the very nature of the sample, whose composition profoundly influences the excitation characteristics of the light source. Therefore, spectrochemists are inevitably engaged in all the processes that determine the radiation output of the light source for a given sample.

Dealing with this ensemble of processes, that is, with 'excitation' in the widest sense, is the object of this book (cf. § 1.1). The reader will seek in vain for enumerations of practical rules that would tell him how to tackle a particular analysis problem. What he will find is a detailed and specified exposition on the laws that govern the excitation of samples in the d.c. carbon arc. This should enable him to derive the rules for performing an analysis *rationally* in a particular case. Conclusions pertaining *immediately* and *explicitly* to analytical applications also occur sometimes, for instance, in §§ 7.12, 8.8, and 8.9, where they appear as natural steps in the discussion. Although I have purposely not considered the field of analysis, I have continuously kept in close touch with it.

The text is primarily intended as a help for understanding and interpreting numerous phenomena observed in the spectro-analytical laboratory. It comprises themes that are merely *interesting* to know as well as questions that *ought* to be known. The main line of the work follows my own experience in the field of fundamental spectrochemistry. So, my book, which is not a mere survey of literature of the subject, includes expositions that have not been published before or were contained in my thesis, which is in Dutch and is thus confined to a small circle of readers. New material was obtained as the text progressed; as an example I mention the larger part of Chapter 8, which concerns spectral-line intensity and radial distributions in the arc. Essentially new are experimental results on absolute concentrations and concentration distributions of elements in the arc (§ 9.4). These results were taken from Dr de Galan's thesis, published in 1965. Also new is a recent development in the mathematical treatment of vapour transportation through the arc (§ 9.5).

Obviously I could not cover the entire field of spectrochemical physics comprehensively. I have made special reference to the d.c. carbon arc, primarily because my own experience is chiefly with this light source. A great many of the conclusions given in the text have a *general* scope, however, and indeed apply to any excitation source. At the same time, many of those problems that specifically refer to the d.c. carbon arc in air at atmospheric pressure also occur in related sources. The reader will be able to make the necessary adaptations; for, as a matter of fact, detailed knowledge of the excitation conditions in *one* source, viz. the d.c. carbon arc, will give sufficient insight into *general* fundamentals, and brings specialized literature concerned with other sources easily within reach. The fundamental approach of spectrochemical problems attempted in this work unfortunately involves the risk that the book is doomed to remain 'the voice of one crying in the wilderness', as such a treatment entails that many topics must be depicted in a more complicated and involved manner than spectrochemists have been accustomed to. Let us hope the contrary, however.

For those readers who do not want to read all details in the text at once, I have marked with an asterisk those sections that are the most important on a first reading. In addition, the conclusion summarizes my discussion on matrix effects. To fill any gaps in the literature cited, the list of references at the end of the book is preceded by an enumeration of the most important compilations, namely *Spectrochemical Abstracts, Index to the Literature on Spectrochemical Analysis,* and the reviews in *Analytical Chemistry.*

Not all problems of spectrochemical physics have been solved in this work; I have endeavoured to strike a balance for the d.c. carbon arc. I hope, therefore, that this book will also contribute to further progress in the field of theoretical spectrochemistry, serving as a guide to a field that, although it may seem the reserve of physicists, should be common ground for spectrochemists working and arguing on fundamentals as well. The book has been written first of all for spectrochemists by an author who himself is a chemist too.

P. W. J. M. BOUMANS

AMSTERDAM

March 1966

ACKNOWLEDGEMENTS

At the beginning of this book but at the end of the 'work' I must acknowledge my deep obligation to a number of people who have assisted in bringing the text into the present form.

I thank Mr E. H. S. van Someren for his critical comments upon the manuscript. His criticism induced a beneficial reduction of detail in my final revision of the text.

Grateful acknowledgement I make to the publishers, particularly to Mr Neville Goodman for giving continuous encouragement during the years of writing and to Mr David Tomlinson for the many valuable suggestions made for improving my English version of the text.

I am indebted to Mr Charles H. Corliss of the National Bureau of Standards at Washington D.C. and to Mr K. Schurer of the Physical Laboratory of the State University at Utrecht for reading critically the sections on transition probabilities; to Prof. Dr H. A. Lauwerier and to Mr H. Bavinck of the Mathematical Centre at Amsterdam for their help with the mathematical treatment of transport problems; to Mr R. Gerbatsch of the Institut für Metallphysik und Reinstmetalle at Dresden for providing me with a German translation of a Russian paper; to Mr J. van Leeuwen of the Laboratory for Analytical Chemistry of the University of Amsterdam for his aid with the illustrations; and to my co-operator, Mr F. J. M. J. Maessen, for his assistance with the preparation of the plates and for many useful discussions.

Finally, I feel obliged to all the people who kindly gave me permission to reproduce figures from their work: Prof. Dr H. Brinkman, Mr C. H. Corliss, Prof. Dr G. Ecker, Dr L. de Galan, Prof. Dr G. Herzberg, the late Dr A. M. Kruithof, Prof. Dr H. Maecker, Dr R. L. Roes, Prof. Dr J. A. Smit, Dr J. Sperling, and Dr L. H. M. van Stekelenburg. At the same time I acknowledge Springer-Verlag and D. van Nostrand Company for permitting me to use illustrations from works published by them.

P.W.J.M.B.

CONTENTS

CHAPTER ONE

VARIABLES IN SPECTROCHEMISTRY
AND SPECTRAL-LINE INTENSITY

★ § 1.1 *Introduction*

Matrix effects, influence of third component, inter-element effects

In spectrochemical analysis, emission spectra of atoms and ions (and sometimes of molecules) are used for detecting and determining chemical elements. To ascertain the concentrations of the elements in the samples, 'intensities' of spectral lines are evaluated. The basis of quantitative spectrochemical analysis is a simple, empirical relationship between the content G of an element in the sample and the intensity I of a spectral line in the source of excitation. This relationship is usually expressed by the Scheibe–Lomakin equation

$$I = K \times G^m \tag{1.1}$$

or by an equivalent expression.

The equation is based on the assumption that, in principle, intensity is proportional to concentration; deviations from this proportionality, primarily that caused by self-absorption, are taken into account by the exponent m.

Particularly when photographic recording is employed, equation (1.1) is depicted graphically in logarithmic coordinates, viz.

$$\log I = m \log G + \log K \tag{1.2}$$

The resulting plot of $\log I$ versus $\log G$ is commonly referred to as the working curve. Generally, intensity I is not measured absolutely, in the sense meant at the end of § 1.2, but is taken to be the intensity ratio of the analysis line and a suitable reference line.

In practical applications the analyst deals with the experimental determination of the slope m and the constant $\log K$ of the working curve for each selected analysis line or line pair. Whatever method is chosen for constructing the working curve, each time that such a curve is set up or referred to for analysing samples, the analyst must make the conditions such that the intensity is proportional to the concentration. This involves the fulfilment of the condition that all the factors affecting the intensity of the analytical line or the intensity ratio of the chosen line pair, other than the concentration of the relevant element, be rigorously constant. Considering the different variables in detail and stating accurately what measures must be taken to eliminate their influence on analytical results, so as to have optimum working conditions, is the ideal of spectrochemistry.

There are many variables: first, the great variety of samples; second, the manifoldness of sources of excitation, such as different types of flame, arc, spark, plasma jet and related gas-stabilized source, radio-frequency torch, and hollow-cathode and high-frequency discharge; and third, the diversity of equipment for dispersing the emitted radiation into a spectrum, for recording the spectrum, and for measuring intensities of spectral lines and background. Even if discussion is confined to a particular category of samples, to a specified source of excitation, and to specified equipment, we have to consider a vast aggregate of variables. They can be conveniently classified under the following headings.

1 Preparation of samples and standards.
2 Excitation of specimens.
3 Optics.
4 Recording and measuring.
5 Evaluation.
6 Statistical interpretation of results.

We shall, as has been stated in the Preface, be particularly concerned with the variables involved in the excitation of specimens in the d.c. carbon arc. By excitation we mean the aggregate of processes that determine the radiation output of the source for a given sample. Thus it encompasses:

1 The excitation of particles to higher energy levels, i.e. excitation in the strictest sense, which naturally includes self-absorption.
2 Emission.
3 Ionization.
4 Chemical reactions in the plasma (dissociation and formation of molecular species).
5 Transportation phenomena in the plasma (diffusion, convection, migration of charged particles in the electric field).
6 Volatilization and transportation of material at the electrode.
7 Chemical and physical processes in the electrode (decomposition, reduction, chemical changes, diffusion into the electrode walls, etc.).

Exciting a sample, which usually is either a solid or a liquid, involves its conversion into vapour. The original molecular structure is thereby disrupted, new molecules may be formed, and fractions of the constituents become ionized. The vapours spread out from the electrode tip into and around the arc column. Of each constituent of the sample, a fraction, whose magnitude is not necessarily alike for different elements, enters into the discharge zone. Diffusion and migration under the influence of the electric field cause the vapours to move through the gas in the discharge gap; in addition, the gas as a whole, i.e. the composite of the sample vapours and the atmosphere in which the discharge takes place, tends, because of convection, to flow upwards. The electric current energizes the moving gas. The current is essentially a continuous flow of electrons that gain energy in the electric field. In the arc in air at atmospheric pressure electrons do not travel, on the average, over a distance longer than one mean free path without imparting the gained energy to molecules. For that reason *local thermal equilibrium* (LTE) is reached. In the hot gas, commonly termed 'plasma' because of its partially ionized nature, numerous collisions occur and lead to the excitation and de-excitation of atoms, ions, and molecules. (In fact, all of the processes mentioned above, such as ionization, dissociation, and recombination, are dominated by collisions.) A small fraction of the excited particles gives off excitation energy by spontaneous emission. Some of the quanta produced are intercepted and absorbed on their way through the gaseous layers to the surroundings, whereupon the excitation energy is converted by impact into kinetic energy again. Part of the radiation that finally escapes from the source enters the aperture of the optical equipment and provides the informative signals whose 'intensities' are recorded by the detector.

Roughly, this is the picture of sample excitation. We collect the signals that are emitted by the fractions of the constituents of the specimen, as they rush past the observation window. It does not take much imagination to see that the theoretical relationship between the number of atoms of an element in the sample and the intensity of the recorded radiation is complex and involved. Curiously, it is not surprising that simple relationships between numbers of atoms in the sample and intensities of spectral lines in the source do occur if we succeed in establishing constant excitation conditions. The realization of this ideal is the first task in spectrochemical analysis; unravelling the network of intermediate steps and interfering side reactions belongs to the field of *spectrochemical physics*. The latter, inherently, is closely associated with spectrochemical analysis and physics. I adopt the term to distinguish our field of study from that of analytical applications and from that of profound theoretical physics. In my opinion, the spectrochemical physicist is the scientist who conducts physical investigations from the chemical point of view, endeavouring to enrich the general understanding of the processes encountered in spectrochemical analysis and thus furthering possibly the development of spectrochemical methods and techniques. It is in this frame of mind that I have written the book, attempting to show the roles played by the separate factors that influence the excitation conditions.

The crucial problem is the powerful influence exerted by the sample on the excitation conditions. So, in addition to the content of an element to be determined, the chemical composition as well as the structure and the physical state of the sample are decisive factors that affect in one way or another the intensities of the relevant spectral lines. All these 'internal influences', which are attributed to the sample, are generally known as *matrix effects*. The term suggests that it is mainly the matrix, i.e. the base of the material to be analysed, which must be considered. The nature of the base, and more generally that of the major constituents, is the main source of trouble, though not the only one.

Instead of the term 'matrix effects', the expressions 'influence of third component' or 'inter-element interference' are often employed as alternatives. The term *influence of third component* has its origin in metallurgical analysis, where it was adopted for denoting interferences that became apparent when alloys containing more than two elements, the matrix and the analysis element, were investigated spectrochemically. In common parlance, the terms 'matrix effects', 'influence of third component', and 'inter-element effects' now refer to interference by the chemical and physical properties of the sample with the excitation conditions. Customarily, the aggregate of matrix effects is divided into two broad categories, viz. the effects associated with the mechanism of entry of the material into the discharge gap and those pertaining to the processes that take place in the plasma. In this book we shall deal extensively with the latter aspect and consider the former one briefly (see Chapter 10).

For practical, analytical applications the occurrence of matrix effects evokes the serious problem of accurate standardization. Spectrochemical determinations are commonly based on the comparison of spectral-line intensities in the spectra of samples and in the spectra of standards; hence, it is essential that the elements sought are excited under entirely identical conditions in both the samples and the standards. Ideal standards should have properties similar to those of the samples, the only difference being the concentration of the element to be determined. As selecting and composing standards is often a question of compromise, it is necessary to know which properties the sample must have in common with the standards. Unless this problem is given proper attention, results will be systematically either too high or too low, and they are said to be incorrect or inaccurate.

The accuracy of a determination should be distinguished from the precision. The latter, also termed 'reproducibility', refers to random fluctuations in the results that originate from our inability to conduct exactly all the processes encompassed in an excitation when repeating the experiment with identical specimens.*

Substantial progress in improving the precision and accuracy of spectrochemical analysis resulted from the application of the *internal standard principle*, which was enunciated by Gerlach (1925) and Schweitzer (1927).† Nowadays this principle is rarely used in its original form. (I even wonder how many spectrochemists today still have a notion of performing an analysis using only *fixation points* of a number of homologous line pairs.) We have retained only the original conditions that a reference element (internal standard) and a reference line be used and that the intensity ratio of the analysis and the reference line be insensitive to changes in the excitation conditions. For the lines to meet this requirement, both should be either atom lines or ion lines. Moreover, it is frequently stated that their excitation potentials should be equal. This, however, holds true only if the ionization energies of the internal standard and the analysis element are also alike; otherwise a proper difference in the excitation energy can be an advantage in compensating for the discrepancy between the ionization energies (see § 8.9). In d.c. carbon arc analysis,

---

* As I have restricted the discussion to excitation conditions, all other sources of error, such as those arising from detection and measurement, will not be considered here.

† See also Gerlach and Schweitzer (1930). Some interesting remarks on the original scope of the Gerlach–Schweitzer principle and its adaptation to modern demands have been made by Kaiser (1949).

where fractional distillation is likely to occur, considerable attention must be paid to the volatilization properties of the internal standard and to those of the element to be determined. Altogether, it is difficult to select the reference element and the reference line so that all demands are satisfied. On the other hand, practice has demonstrated that the use of an internal standard, in spite of compromises, frequently leads to an appreciable improvement in precision and accuracy in spectrochemical analysis. In the ideal case, matrix effects should be completely annulled by internal standardization, a situation that is seldom met with, however.

To avoid confusion when discussing matrix effects, we must clearly define our concept of spectral-line intensity. Different aspects of this fundamental concept will be reviewed in the next section. For the present it suffices to distinguish between intensity in *total energy procedures* and that in *steady state methods*. In the former we have the sample, or at least the elements sought, volatilized to completion, and we integrate spectral-line intensities over the entire evaporation time (Slavin, 1938; and Leme, 1938). In the latter a steady (or a quasi-steady) state is created by feeding the sample at a constant rate into the discharge gap;* intensity is integrated then over a short period of arbitrary length, sufficiently long for random fluctuations to cancel out in the recorded mean value. Here intensity pertains to the instantaneous situation. In so far as these instantaneous conditions are concerned, we do not immediately encounter fundamental difficulties when interpreting the influence of the rate of volatilization upon line intensity. We must recognize, of course, that, in steady state observations, intensity depends inherently on the rate of entry of the material into the vapour phase. Consequently, in a powder insufflation method† for instance, the average size of the grains as well as the

---

* For example, powder insufflation methods (see below† for references) and related methods for continuous injection of powdered samples into the excitation source, e.g. the powder sifting technique (see Czakow, 1960, 1962; Czakow and Minczewski, 1960, 1962a; Czakow and Grzelak, 1964; and Plško, 1964) and the tape method (see Danielsson, Sundkvist, and Lundgren, 1959; Danielsson, Lundgren, and Sundkvist, 1959; Zeeman and Coetzer, 1961; and Roubault, de la Roche, and Govindaraju, 1962–63).

Further, methods employing solutions that are sprayed or otherwise fed into the plasma (see e.g. the reviews on solution techniques by Young, 1962; and Matherny, 1962; and the articles by Baer and Hodge, 1960; Bass and Soulati, 1963; and Gegus, 1964a and b).

Interesting recent developments are various kinds of plasma jets and gas-stabilized arc sources (see Margoshes and Scribner, 1959, 1963; Margoshes, 1960, 1965; Korolev and Vainshtein, 1959; Korolev and Kvaratskheli, 1961; Vainshtein, Korolev, and Savinova, 1961; Owen, 1961, 1962; Vainshtein, 1962; Scribner and Margoshes, 1962; Jahn, 1962, 1963; Neeb and Gebauhr, 1962; Grechikhin and Shimanovich, 1962; Grechikhin, 1963; Mitteldorf and Landon, 1963; Kranz, 1963, 1964, 1965; Webb and Wildy, 1963; Adcock and Plumtree, 1964; Collins and Pearson, 1964; Raisen, Carrigan, Razicunas, Loseke, and Grove, 1964, 1965; Lincoln and Kohler, 1964; Sirois, 1964; Greenfield, Jones, and Berry, 1964; Serin and Ashton, 1964; and Kaiser, Laqua, and Schirrmeister, 1965) as well as radio-frequency torches (see van Calker, 1962; Dunken, Mikkeleit, and Kniesche, 1962; van Calker and Tappe, 1963; Tappe and van Calker, 1963; Jecht and Kessler, 1963, 1964; Mavrodineanu and Hughes, 1963, 1964; West and Hume, 1964; Greenfield, Jones, and Berry, 1964; and Wendt and Fassel, 1965).

See also references on gas-stabilized arcs quoted in § 2.2.

† See e.g. Rusanov and Khitrov (1958), Raikhbaum and Luzhnova (1959), Rusanov and Batova (1961, 1962), Kibisov and Kubasova (1961), Kibisov and Antropov (1962), Raikhbaum, Malykh, and Luzhnova (1963), Rusanov and Vorob'ev (1964), and Kessler, Jecht, and Zottmann (1965).

chemical structure of the specimens are likely to be serious matrix effects. Hence accuracy tends to be poor if standards prepared by simply grinding oxides with a base are referred to for analysing samples where the elements sought are embedded in the crystal lattice of that base.

A different situation arises when samples are arced to completion; now it seems to be less clear what part is played by the rate of entry of the material. Evidently the rate of volatilization acts upon the instantaneous concentrations of the vapours. Apart from influencing the degree of self-absorption of spectral lines, the vapour concentrations affect the temperature and the electron pressure, including the spatial distributions of these arc parameters. A quantitative account of the transportation of the vapours through the plasma should now complete the picture (see particularly Chapter 9). It is not certain whether this holds true for all cases. When judging the situation we must bear in mind, however, that in many investigations no attempts have been made to consider in a really quantitative fashion all the obvious changes that happen in the plasma, which, admittedly, are involved. Unfortunately, incompleteness, lack of knowledge of fundamentals, and lack of satisfactorily founded experimental evidence, have led to the idea that all those phenomena that cannot be explained by simple reasoning must be collected in the rag-bag of volatilization defects. Let us, however, not anticipate more detailed discussions that are to come later. I have endeavoured to assess the present state of our knowledge of matrix effects as far as d.c. carbon arc spectrochemical analysis is concerned.

Dilution of samples with suitable admixtures occupies an important place in d.c. arc powder analysis. Admixtures are in use for various reasons, e.g. they can improve the general burning properties of the arc, they can act as a flux or as a 'carrier' (see §§ 9.3 and 10.4), and by initiating thermochemical changes they can affect the volatilization rates of the sample constituents (§ 10.4; and the end of § 7.12); above all, they are the *diluents* in which the sample is embedded, chosen so as to dominate the characteristics of the arc. When the latter effect is especially intended, the admixture is frequently called a *spectrochemical buffer*.

There may be disagreement about the use of this term. In the most restricted sense, the buffer is an agent whose presence in the arc defines the temperature so that the latter does not depend on the composition of the sample. For an admixture to satisfy this condition, the ionization energy of its major constituent element should be low.

Common parlance has enlarged the scope of the concept of spectrochemical buffers. For example, Holdt (1962) states a buffer to be a 'substance that is added to a sample in order to minimize the influence of sample composition and characteristics on spectral-line intensities.' A similar definition is proposed by the A.S.T.M. (1964, p. 117): 'a spectrochemical buffer is a substance which by its addition or presence tends to minimize the effects of one or more of the elements on the emission of other elements.' These definitions reflect perfectly the meaning now generally attached to the concept of a spectrochemical buffer: it is an admixture that should act as a buffer against matrix effects. Thus its function is more comprehensive than in the original sense. Nevertheless, it remains true that the buffer must possess, among other properties, the capacity of guaranteeing a definite temperature in the arc and should consequently have a low ionization energy, unless dilution of the samples is carried extremely far or analysis is restricted to test substances that do not contain elements of low ionization energy.

In practical applications the efficacy of an admixture is judged primarily according to its influence upon analytical results (e.g. Makoto Nakano, 1962); moreover, the problem of establishing the most favourable operating conditions is for the greater part tackled empirically. Indeed, in selecting an appropriate admixture much information on the behaviour of different substances in the arc can only be collected by experiment, since available data allow only rough predictions. The empirical approach has thus introduced a great variety of diluent materials.*

All admixtures must improve the precision and the accuracy of spectro-analytical results and hence they should reduce matrix effects in one way or another. If we consider that the concept of the buffer has acquired the wide scope mentioned above, we are not surprised to find in the literature a collection of widely differing substances all designated 'spectrochemical buffers'. Thus, for example, graphite and carbon, Li_2CO_3, LiF, SiO_2, Cu, CuO, ZnO, Ga_2O_3, AgCl, GeO_2, Ge, $CaCO_3$, NaCl, Al_2O_3, and others, are listed among the spectrochemical buffers, though their role is not always strictly that of stabilizing the temperature. We must note that a particular admixture is often used only with samples of restricted diversity. Accordingly, in the one case the necessity for preventing a drop in temperature may be unimportant, compared to securing a smooth burn; in the other, the analyst wants to prevent a considerable drop in the temperature in view of the high alkali content of the specimens. The latter, specific buffer action, which will be considered in detail in § 7.12, is a vital factor in the processes that dominate the excitation conditions in the plasma.

Buffering with an element of low ionization energy results in a reduction of the source temperature and in turn influences the *limits of detection*† of elements. It will be seen in §§ 8.8 and 8.9 that moderating the temperature must often lead to an improvement in the detection limits. We should also realize that spectral-line intensities and the intensity and the structure of the background (§§ 7.10 and 8.9) must be considered when the influence of the excitation conditions on the limit of detection is discussed.

To enable an analyst to decide upon the most favourable circumstances for analysis, the analytical problem should be specified in concrete form. General rules can be given only for general problems. However, specific rules for specific

---

* See, for example, Ahrens and Taylor (1961, p. 46 ff.), and Kemp (1958), who has compiled a survey in tabular form of some fifty routine methods and listed among other things the dilution procedures.

† In accordance with the terminology that has become customary, particularly in the German literature, and is recommended by the A.S.T.M. (1964, p. 117), we prefer to avoid the term 'sensitivity' for denoting the limit of detection. The term 'sensitivity', whenever needed, will be used to indicate the degree of change in the intensity of the analysis line in relation to a change in the concentration of the element, i.e. dI/dC.

problems are implied in an exposition of the general laws that govern the excitation of samples. Therefore, the reader who is willing to take cognizance of the matter explained in the present work may find some answer to his own particular questions, even if they are not dealt with explicitly in the text.

★ § 1.2 *Intensity of a spectral line*

Every spectrochemist may be assumed to have his own idea about the concept of spectral-line intensity; therefore it is not mere padding if I make clear what my concept is. The reader will, of course, admit that in adopting terms and definitions my choice and preference is never indisputable.

Before introducing the radiometric quantities* that are of interest to us, I note the following about the light source that will concern us. The source to be considered chiefly is a vertical, free-burning d.c. arc; its inner portion, the core, is more or less cylindrically symmetrical, whereas the enveloping flame has a parabolic profile (see § 2.2). On the whole, in a cross-section perpendicular to the arc axis, circular symmetry is found in the ideal steady arc. We must realize that this source is a volume source. An optical system, however, views it as a plane source and observes 'intensities' that are integrated along the line of sight. A convenient approach in defining the radiometric quantities of interest to us is to consider first the common concepts for a plane source and to connect these subsequently to *emission per unit volume*, which is the link between observed intensities and the number of radiating particles.

Radiant flux: The primary concept is that of the *radiant flux* (Φ) of a source, that is, the amount of radiant energy emitted into space in unit time. Usually, we consider only a fraction of the total flux, namely the flux within a relatively small solid angle in the direction of observation, for example, the flux that reaches the dispersing medium or the detecting system.

In spectroscopy there is seldom any interest in radiation that is not dispersed into a spectrum; hence, our concept of flux needs further definition. We introduce the term *spectral flux* (Φ_λ) to denote the flux at a certain wavelength (λ) per unit of wavelength. Of course, quantities like $\Phi_\lambda \Delta\lambda$ and

$$\Phi_{\lambda_1, \lambda_2} = \int_{\lambda_1}^{\lambda_2} \Phi_\lambda \, d\lambda \tag{1.3}$$

only have practical meaning and represent finite amounts of energy flowing through a surface in unit time.

---

* Intensity and related quantities are often introduced in photometric terms (e.g. by Jenkins and White, 1957; and by Sawyer, 1951), sometimes as radiometric concepts (e.g. by Bauer, 1962). Strictly speaking, we shall be concerned here with radiometry since our field extends beyond the visible range of the spectrum.

A spectral line as emitted by the source is not a quantum of sharply defined frequency, but, owing to natural line broadening, collisional broadening (including Stark effect) and Doppler broadening (see §§ 3.8 and 5.7), encompasses a small range of wavelengths. Here we shall define the flux at the wavelength of a spectral line as *spectral flux integrated over the profile of the line*, viz.

$$\Phi(\lambda) = \int_{\text{profile}} \Phi_\lambda \, d\lambda \qquad (1.4)$$

Normally, it is not necessary to specify the wavelength; thus, Φ alone will symbolize the integral function (1.4), unless otherwise stated. This holds equally well for all quantities derived from Φ.

Strictly speaking, a spectral-line flux (or intensity, as defined below) should be measured in compliance with its definition (1.4), that is by integrating over the line profile. Indeed, this is often done in physics, for instance, when transition probabilities are determined (§ 1.3). Examples are given by Terpstra and Smit (1958), and by van Hengstum and Smit (1956). In spectrochemistry, the time-consuming procedure of evaluating intensities by profile integration is highly inconvenient; therefore, the peak value of the line profile is taken to be representative of the intensity.* This method of measuring rests upon the assumption that the profile, as received by the detector, is very nearly rectangular. This assumption holds reasonably well if the spectrograph slit is made sufficiently wide.

In conclusion, the starting-point of our considerations is the radiant flux (Φ) emitted within the small wavelength range of a spectral line, determined experimentally either accurately by integration over the line profile, or approximately by measuring the peak value. The dimension of this quantity is energy per unit time; it is conveniently expressed as ergs.sec^{-1}.

Radiance and intensity: The flux flowing from a point source Q to a surface of area dA at the (mean) distance a from Q is proportional to the solid angle ($d\omega$) subtended by dA (Fig. 1.1). The solid angle can be written as

$$d\omega = \frac{dA \cos \varphi}{a^2} \qquad (1.5)$$

where φ is the angle between the normal on the surface and the mean ray in the pencil being considered. Thus we have

$$d\Phi = I_r \, d\omega \equiv I_r \frac{dA \cos \varphi}{a^2} \qquad (1.6)$$

---

* In photographic photometry a procedure has been developed that uses effective line widths as a measure of spectral-line intensity. The principles of this method, as well as the theory of the optical and photographic line profile, have been extensively discussed in a series of noteworthy articles by Junkes and Salpeter (1958, 1959, 1961, 1963). In addition to the references cited by these authors we mention Rubeška and Polej (1962, 1964a and b), and Preobrazhenskii (1962).

The constant of proportionality I_r has the dimension of flux per unit solid angle. It is often denoted as radiant intensity. Curiously, this concept is not a very practical one for our purpose, since it is convenient only if the source can be regarded as a point source. Particularly in our case, where, as will be seen later, we take an interest in the inhomogeneous spatial distribution of the emission in the arc discharge, quantities defined on the basis of point source approximations are of little value.

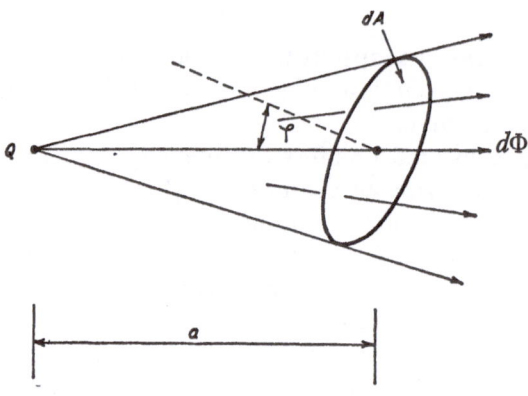

FIG. 1.1. Illustration of the geometrical relationships involved in the definition of the radiant flux emanating from a point source Q to a surface of area dA at the (mean) distance a from Q.

Therefore, let the source be a small disk of area dS radiating to the surface dA (Fig. 1.2). We suppose the elements of area, dS and dA, to be small in comparison with their (mean) distance a, so that the solid angle at which dA is seen from any point of dS can be regarded as having the constant value of $d\omega$. The total flux from dS to dA is expressed then by

$$d\Phi = B\frac{dS \cos\varphi\, dA \cos\vartheta}{a^2} \tag{1.7}$$

where B is a constant. The flux is seen to be proportional to the projected area of dS, that is the apparent area of dS viewed from dA along the direction of observation. This direction is represented by the path followed by the 'mean ray' from dS to dA (see the lower part of Fig. 1.2). When writing down the flux in the form of equation (1.7) we have tacitly assumed that the source is a Lambert radiator, which means that B is independent of the direction at which radiation is emitted.

The constant B is the fundamental quantity that characterizes the radiation properties of the source. In photometry, B is known as brightness or luminance. The radiometric equivalent is radiance (sometimes called steradiancy).* According

---

* Clark Jones (1963) proposes the term 'sterance' qualified, as appropriate, by either luminous or radiant. The terms 'luminous sterance' and 'radiant sterance' are thus constructed in exactly the same manner as, for example, luminous intensity and radiant intensity. A similar alternative is proposed for illuminance and irradiance, which would be called then luminous incidence and radiant incidence. The main advantages of the proposal are that it facilitates generalization and allows for further qualification, e.g. energic incidence.

to equation (1.7), B is the rate at which radiation energy is sent out from unit area (projected in the line of sight) per unit solid angle. It can be expressed as ergs.sec^{-1}.sr^{-1}.cm^{-2}, where sr is the abbreviation of steradian.

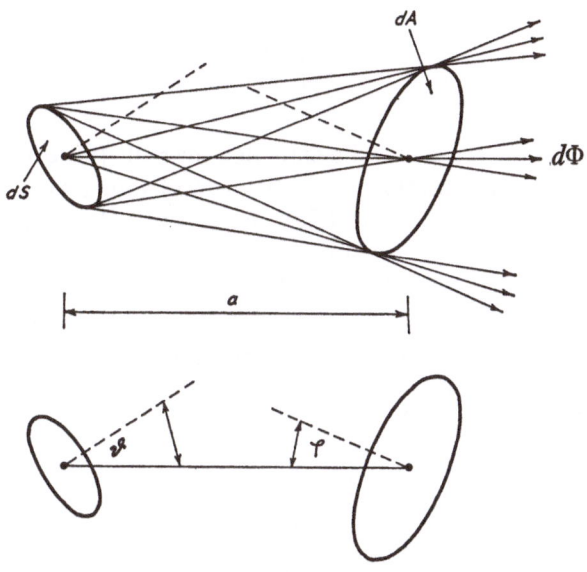

FIG. 1.2. Illustration of the geometrical relationships involved in the definition of the radiant flux emanating from a small disk of area dS to a surface of area dA at the (mean) distance a from dS.

It is convenient to identify *absolute intensity* with the concept of radiance; therefore, throughout this work

$$I_{abs} = B \qquad (1.8)$$

The symbol I will be used also to denote *relative intensity*. Only where confusion might arise are distinctive labels or primes added. Needless to say, relative intensity will occur more often in our equations than the absolute quantity. Below we shall see what concepts are covered by the term 'relative intensity'; inherently, the properties of the optics and the detecting system then enter into the discussion.

Previously we shall examine the relation between the intensity as defined by (1.8) and (1.7), i.e., the radiance of the surface of a volume source, on the one hand, and the rate of generation of radiation in the interior of that source, on the other. An adequate link is to introduce the concept of emissive power per unit volume per unit solid angle, or simply the *emission per unit volume per unit solid angle*, for which the symbol J will be adopted here. We define it as the total flux divided by 4π, emanating in all directions from a unit volume of the source. It is conveniently expressed as ergs.sec^{-1}.sr^{-1}.cm^{-3}.

For a homogeneously radiating gaseous layer of thickness d (measured in the line of sight), intensity I and emissive power J are interrelated by

$$I = J.d \qquad (1.9)$$

In general we have a relationship of the type

$$I = \int_{-\infty}^{+\infty} J(y)\,dy \qquad (1.10)$$

where the integration is actually to be extended over the (finite) depth of the source along the direction of observation y (see below and Fig. 1.3).

We note that equations (1.9) and (1.10) hold true only if the self-absorption is negligible, since J represents a quantity that is directly and exclusively related to the number of radiative transitions per sec in unit volume, whereas in I the modifications of the radiant energy flow upon passing from the field point to the observer through the gaseous layer are also considered.

If the number of transitions per sec per cm^3 leading to emission of a light quantum $h\nu$ is equal to N, we have the relations

$$J = \frac{1}{4\pi}Nh\nu \qquad (1.11)$$

and

$$I = \frac{d}{4\pi}Nh\nu \qquad (1.12)$$

for the case of uniform distribution of emission along the line of observation. Generally, geometrical relationships between I and J are more involved. To make the situation clear for the arc, Fig. 1.3 shows two parallel planes, the xz and the XZ plane. The arc is represented by a paraboloid of revolution along the z axis of the rectangular xyz coordinate system. The y axis is supposed to be the direction along which the arc is projected on to the XZ plane.

The orthogonal geometric projection of Fig. 1.3 serves as a very good approximation of an optical projection in emission spectroscopy, since solid angles occurring in the common optical arrangements are small. For that reason we shall treat the present problem entirely in terms of the geometrical projection. All equations derived for this mathematical model can be immediately transferred to its optical analogue. The only adaptation needed is the mutual conversion of units of length in source and image if the optical magnification is not equal to unity, in contrast with that of the orthogonal geometric projection.

To avoid possible misunderstanding we mention the important theorem of optics about the invariance of radiance, which states that the radiance in a point of an image is the same as that of its conjugate point in the source, of course disregarding reflection and absorption losses in the optics. For orthogonal geometric projections, such as considered in Fig. 1.3, this means simply that the distance between the source and the plane of projection is immaterial; therefore, the radiance in a point of that plane always equals the radiance in the corresponding point at the surface of the source. Remember that radiance is flux emitted per unit apparent area and per unit solid angle and that it must not be confused with irradiance, i.e. flux (incident) per unit area.

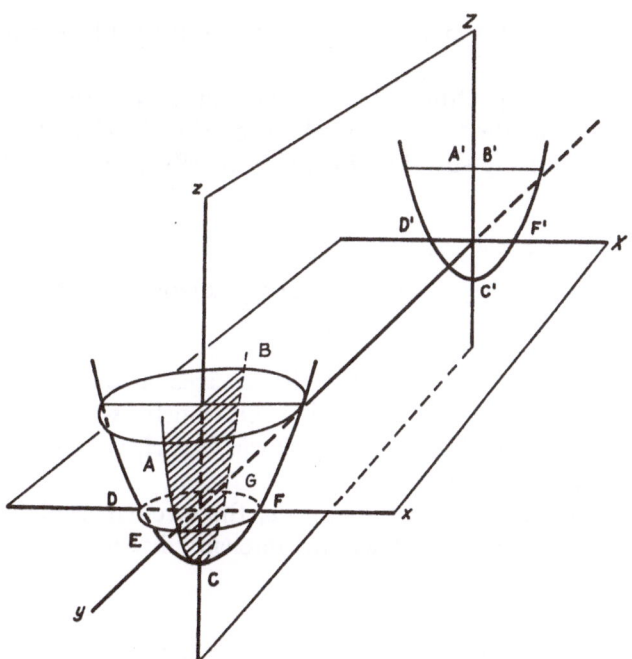

FIG. 1.3. Illustration of the geometrical relationships to be considered when an arc is projected on the slit of a spectrograph. The arc, represented by a paraboloid of revolution in the rectangular xyz coordinate system, is projected along the direction of the y axis on to the XZ plane.

The slit of the spectrograph is assumed to coincide either with the Z axis or with the X axis. The latter exemplifies edge-on projection of a circular slice of plasma (DEFG). In practice, edge-on spectra of a vertical arc are commonly taken by adding two correctly positioned plane mirrors or an inverting prism to the focusing system, so as to rotate the image of the arc through 90°.

If solid angles are small, the orthogonal geometric projection presented in the figure serves as a good approximation of actual optical projections.

Returning to Fig. 1.3 and denoting the emission per unit volume per unit solid angle in an arbitrary point of the source as $J(x, y, z)$, we represent the radiance or intensity $I(x, z)$ at the surface of the source, observed in the direction of the y axis, as

$$I(x, z) = \int_{-\infty}^{+\infty} J(x, y, z)\, dy \qquad (1.13)$$

which, for reasons of symmetry, is conveniently written as

$$I(x, z) = 2 \int_0^\infty J(x, y, z)\, dy \tag{1.14}$$

$I(x, z)$ is the intensity in a point (x, z) in an arbitrary plane parallel to the xz plane.

We now visualize the spectrograph slit as coinciding with the Z axis in the plane of projection XZ. Thus the vertical, parabolic section ABC of the arc is projected as the line (A'B')C' on the slit. The intensity distribution along the slit is given by

$$I_{x=0}(z) = 2 \int_0^\infty J(0, y, z)\, dy \tag{1.15}$$

So, for example, the point (A'B') of the slit receives all the radiation emitted on the line AB in the direction of observation. Obviously, when the arc is actually focused on the slit, the 'line' has some thickness, $\Delta x \Delta z$, which holds equally well for the 'point' (A'B') of the slit.

If in Fig. 1.3 the slit is along the X axis, the circular section DEFG of the arc is projected as the line D'F'. Then, the intensity distribution in the x direction at the height z_1 can be written as

$$I_{z=z_1}(x) = 2 \int_0^\infty J(x, y, z_1)\, dy \tag{1.16}$$

Usually, the dependence on z is not explicity expressed in the equations; thus we write instead of (1.16):

$$I(x) = 2 \int_0^\infty J(x, y)\, dy \tag{1.17}$$

Equation (1.17) is commonly used to correlate the emission distribution in a horizontal slice through the source and the radiance from that slice in edge-on projection. For sources with circular symmetry, as considered here, the function $J(x, y)$ can be replaced by $J(r)$, which expresses the dependence of J on the distance (r) to the z axis. The functions $J(x, y)$ and $J(r)$ are mutually convertible by the use of the relation $r^2 = x^2 + y^2$. Equation (1.17) serves as a fundamental relationship in the treatment of radial distributions (§§ 6.12 to 6.14).

For taking edge-on spectra of an arc it is highly inconvenient to mount the spectrograph vertically so that the slit is horizontal.* Positioning the arc horizontally is not an

* To save space this arrangement is sometimes used in commercial direct-reading equipment.

acceptable alternative either, except when proper means for arc stabilization are applied and the character of the arc does not depend on its spatial orientation (cf. § 2.2). However, if two properly mounted plane mirrors or an inverting prism are added to the focusing system, imaging a vertical arc horizontally on a vertical slit is easily achieved.

Concluding the mathematical formulation of geometrical relationships between I and J we give the expression for the intensity at the slit when the arc is imaged on the collimator lens of the spectrograph. Under proper conditions the slit is evenly illuminated along its entire length. For the (averaged) intensity we have

$$I = \frac{2}{\Delta x \Delta z} \int_{0}^{\infty} \int_{x_1-\Delta x}^{x_1+\Delta x} \int_{z_1-\Delta z}^{z_1+\Delta z} J(x, y, z) \, dy \, dx \, dz \qquad (1.18)$$

where $x_1 \pm \Delta x$ and $z_1 \pm \Delta z$ define the effective area of the source.

So far we have been concerned with absolute radiometric quantities, representing definite amounts of radiation energy. Normally in spectrochemical analysis absolute intensities of spectral lines are not used. In physics and spectrochemical physics they are used much more frequently. An interesting application for studying the transport mechanism of metal vapours in the d.c. carbon arc will be dealt with in § 9.4.

An absolute intensity measurement involves comparison of the signals produced at the detector successively by a spectral line emanating from the source under investigation and by the radiation emitted at the same wavelength by a standardized source, whose spectral characteristic is known (e.g. a tungsten-strip lamp or the positive crater of a carbon arc*). An absolute intensity calibration must always pertain to the same piece of optical equipment and to the same detector, since the properties of the optics and the detector enter jointly into the calibration graphs. Therefore, we must remember that a spectral sensitivity characteristic of an 'emulsion' includes the spectral properties of the optics.

Given the optics and the detector, the need for absolute intensity determinations does not arise in normal spectrochemical work. It suffices to measure some kind of relative intensity, usually either of the following alternatives.

1 We compare the radiances (intensities) of a source in a definite wavelength region (spectral line) for different concentrations of an element that generates this radiation under constant excitation conditions; in other words, we consider only the variation of the intensity of a spectral line with the content of the relevant element. The relationship expressing this variation is the working curve. It is established with the aid of standard samples of known composition. In fact, we assume for one of the standards an arbitrary value of the spectral-line intensity, i.e. the signal as recorded by the detector, which the other numerical values are referred to. The relative intensities thus defined are proportional to the radiance of the source (at the given wavelength interval), but the proportionality constant,

* See MacPherson (1940, 1941), Euler (1956/57), and Null and Lozier (1962).

which depends entirely on the optics and the detector, remains unknown. It is self-evident that the proportionality constant must be kept rigorously constant if the working curve is to be relied upon for subsequent analyses. This implies that all instrumental factors should be constant. If the working curves from the one set of equipment are to be used with another set, the pertinent proportionality constants (light conductance, *Lichtleitwert*) must be compared. The problem has been recently treated theoretically by Kaiser (1964).

2 We compare the intensities of two spectral lines that are simultaneously emitted from the source, the one line pertaining to the analysis element, the other to the reference element (internal standard). To establish a working curve we use a series of reference samples with constant concentration of internal standard and varying content of analysis element. Under ideal conditions, that is with lines having similar profiles and differing only very little in wavelength, relative intensity, viz. the intensity ratio of the line pair, does not depend on instrumental factors. It is proportional to the radiance of the source at the wavelength of the analysis line; the proportionality constant is the unknown, but constant reciprocal radiance at the wavelength of the internal standard line. Needless to say, the ideal situation is not met with very frequently, so that at least the spectral characteristics of the optics and the detector enter into the calibrations. Again, as long as these factors remain constant, they need not be known for analytical work. Note, however, that a photographic or photoelectric intensity ratio of two lines differing markedly in wavelength contains an unknown constant and therefore does not represent the *true* intensity ratio. This must be borne in mind, for example, when source temperatures and electron concentrations are derived from intensity ratios of spectral lines (see §§ 6.4, 6.5, and 7.6).

It is customary to denote spectrochemical procedures based on the type of working curves defined in (1) as absolute methods, and those using an internal standard (2) as relative methods. True absoluteness is reached only, however, if reference standards of known composition can be totally abandoned. Such a procedure, which cannot claim utmost accuracy and is not a preferable alternative to existing methods, has been shown by de Galan (1965a, 1966b) to be theoretically feasible (see also § 9.4).

Finally, we should point out that the concept of intensity is normally modified to mean an amount of radiation energy received by the detector during a definite time. Thus, when a photographic plate is employed, intensity is radiant flux incident per unit area, that is the irradiance H, integrated over the exposure time t_e. It is convenient to denote this integrated irradiance as the *exposure E*; accordingly

$$E = \int_{t_e} H \, dt \qquad (1.19)$$

Given the optical system between source and photoplate, H is proportional to the radiance (intensity) of the source for a given wavelength region.

An unfavourable property of the photographic emulsion is the *reciprocity-law failure*. The variables H and t are not entirely equivalent, so that the effect produced by irradiating an emulsion is not uniquely determined by the product of H and t, but depends also on the magnitudes of H and t separately. Therefore, if the irradiance is constant, we must express E as

$$E = Ht_e{}^p \qquad (1.20)$$

where p is the *Schwarzschildt exponent*, that usually differs from unity. In virtue of (1.20) equation (1.19) must be replaced then by

$$E = \int_{t_e} Hpt^{p-1}\, dt \qquad (1.21)$$

An example will be considered in conjunction with volatilization effects in the d.c. arc (§ 10.2).

In this work instrumental factors will not be considered and we shall be concerned only with the influence of the source characteristics upon intensities of spectral lines and background. The subject will be discussed in terms of the emission J per unit volume per unit solid angle in the interior of the source, and in terms of the radiance or intensity I at the outer surface of the source. Where experimental intensity values are given, the data in fact represent exposures E. In general, J, I, and E are considered on a relative scale only. E values that must be mutually compared refer to equal exposure times.

§ 1.3   *Transition probability for spontaneous emission*

According to quantum mechanics, spectral lines originate from transitions between energy levels of atoms or ions. The frequency v of the light quantum that is emitted when a transition from the higher state q to the lower state p takes place is related to the energy difference between the two states by the Bohr frequency rule:

$$hv = E_q - E_p \qquad (1.22)$$

where h is Planck's constant and E is energy.

The quantum theoretical treatment of transitions between quantum states in terms of probabilities was introduced by Einstein (1917). If we consider an optically-thin source, the radiative transition $q \to p$ happens spontaneously. According to Einstein, the number of spontaneous transitions taking place in unit time is governed by a law similar to that of radioactive disintegration; thus, if the number of atoms in the excited level q at a given time is N_q, we have

$$-\frac{dN_q}{dt} = A_{qp}.N_q \qquad (1.23)$$

which is equivalent to saying that the number of atoms leaving the state q per sec by spontaneous emission is proportional to the number N_q present in that state. The constant A_{qp} (expressed as sec^{-1}) is called *transition probability*; it is also referred to as the Einstein coefficient of spontaneous emission.

If the upper level q is involved in several transitions, e.g. $q \to p_1$, $q \to p_2$, ..., equation (1.23) should read

$$-\frac{dN_q}{dt} = N_q \cdot \sum_p A_{qp} \qquad (1.24)$$

where, instead of the transition probability of a single transition, the sum of the probabilities of all the possible transitions occurs. Equation (1.23) evidently holds true if only one transition from the level q can take place.

The quantity

$$\gamma_q = \sum_p A_{qp} \qquad (1.25)$$

is designated as the *damping constant*; its reciprocal is the *natural mean lifetime* τ_q of the relevant state. For a further discussion of the subject the reader is referred to, for example, Unsöld (1955, pp. 278 ff.).

Summarizing, we must clearly distinguish between the probability of a transition, mostly referred to as the transition probability of a spectral line, and the damping constant of a level. Only the former will be of great interest to us, as it is immediately related to intensities of spectral lines.

In view of our previous definition (§ 1.2), the absolute intensity I of an atom line is given by the expression

$$I_{qp} = \frac{d}{4\pi} A_{qp} n_{aq} h\nu_{qp} \qquad (1.26)$$

where d is the depth of the source (cm), A_{qp} the transition probability (sec^{-1}), n_{aq} the concentration or density of excited neutral atoms in the level q (number per cm^3), h the Planck constant (ergs.sec), and ν_{qp} the frequency of the spectral line emitted in the transition $q \to p$ (sec^{-1}).

The transition probability A_{qp} is a line constant just like the frequency; as follows from (1.26), it determines the intrinsic intensity of the line. In spectrochemical analysis, where we are only concerned with relative intensities, numerical values of transition probabilities are not of special interest. In spectrochemical physics, however, the need for reliable A-values of atom and ion lines is ever increasing, as will become apparent in the course of this text.

The present position of our knowledge of atomic transition probabilities has been recently reviewed by Wiese (1963). Interesting comments are also found in King's paper (1963) and in a review by Foster (1964). In June 1960, a Data Center on atomic constants was established in the Atomic Physics Division of the National Bureau of Standards (N.B.S.); here, all the literature on atomic transition probabilities is collected, catalogued, and studied. The first definite result is an imposing bibliography by Glennon and

Wiese (1962). It contains, besides references on individual elements in the main body, a table of conversion factors and general references to tables of numerical values, to previous literature compilations, to review articles and general comments, and to literature on fundamental relationships and basic concepts, on experimental or theoretical methods for determining transition probabilities, and on environmental influences.

Transition probabilities or related quantities, such as oscillator strengths (see § 1.4), which are readily converted into A-values, can be determined either by theoretical calculations or by experimental procedures. The basic principles of the most popular methods have been lucidly summarized and commented upon by Wiese (1963). A more extensive treatment of various principles is found, e.g. in Unsöld's work (1955, pp. 269–370). Discussion here will be confined to a few notes and comments. References will be found in Wiese (1963) or in Glennon and Wiese (1962). Also, estimates of the accuracy of the results obtained by various methods are given by Wiese.

Two methods are used for calculating transition probabilities theoretically: firstly, the procedure based on the self-consistent field approximation developed by Hartree and improved by Fock, Biermann, Treffzt, and others; secondly, the Coulomb approximation by Bates and Darmgaard.

Among the experimental methods we mention the following.

1 *Measurements in emission:* This method is based on the measurement of the intensities of spectral lines that are emitted from plasmas, particularly from various types of arc, whose conditions should be accurately known. The reverse procedure is a familiar technique in spectrochemical physics now, i.e. determining the excitation conditions in arc or spark plasmas from the intensities of spectral lines with known transition probabilities. We shall deal with it extensively throughout this work (see particularly Chapters 6 and 7).

The emission method originates from the Physical Laboratory of the State University at Utrecht; the names of Ornstein, Brinkman, Smit, *et al.* are associated with it. The consistency of numerical values of the transition probability found at various temperatures and field strengths confirmed Ornstein and Brinkman's hypotheses about thermal equilibrium (to be more exact, local thermal equilibrium) prevailing in the arc column (see Chapter 5).

A survey of the work done in this field in the Utrecht Laboratory up to 1946 has been given by Smit (1946). Among the papers on the subject published since then we mention those by Voorhoeve (1946), van Stekelenburg and Smit (1948), van Hengstum (1955), van Hengstum and Smit (1956), Terpstra (1956), and Terpstra and Smit (1958).

Most temperature determinations at Utrecht University have been carried out with the aid of smoothed intensity profiles of CN bands, especially the 3883 Å and 4216 Å sequences (see § 6.11). Somers (1954, pp. 23 to 48) points out that an inaccurate value for the ratio of vibrational transition probabilities of the second and first band of the 4216 Å sequence was at first used. In a critical discussion he concludes that the ratio should be decreased by several per cent. Therefore, those atomic transition probabilities that are based on the erroneous ratio of vibrational A-values require revision. The whole question of temperature measurement in the arc with the aid of CN bands is still being thoroughly investigated at Utrecht University and we shall not attempt to give the exact corrections for published atomic A-values. There is one exception, however, that must not escape our attention here, namely the relative transition probability of the atom lines Zn I 3076 and Zn I 3072. In view of the use of these lines for temperature measurements in the arc, the choice of the best A-values deserves discussion.

The relative transition probability of the zinc lines just mentioned is commonly taken from Schuttevaer (1942) or from Schuttevaer and Smit (1943). According to these authors we have

$$(gA)_{3072} : (gA)_{3076} = 176 \cdot 5 : 0 \cdot 244 = 723$$

This value requires the Somers correction and should read

$$(gA)_{3072} : (gA)_{3076} = 380$$

Corrected values for different Zn I lines have been summarized by van Hengstum (1955, p. 52). His object was only to *compare* transition probabilities of Zn I, Cd I, and Hg I lines and he rounded off too rigorously the relative value of the resonance-intercombination line Zn I 3076, so that the ratio 450 instead of 380 is deduced from his table. On the basis of temperature determinations from resolved rotational lines of CN bands, Schurer of the Physical Laboratory at Utrecht has recently checked the appropriateness of the value 380 (personal communication).

An independent indication that the Somers correction leads to better numerical values for transition probabilities is obtained by comparing the work of Penkin and Red'ko (1960) with the works of Schuttevaer and Smit (1943) and of van Hengstum and Smit (1956). Penkin and Red'ko determined the relative oscillator strengths of Zn I and Cd I lines using Rozhdestvenskii's 'hook' method, while Schuttevaer and Smit, and van Hengstum and Smit made measurements using the emission method. For the Cd I lines Penkin and Red'ko obtained agreement within the limits of experimental error; for the Zn I lines they noted a systematic deviation between Schuttevaer and Smit's results and their own values. This discrepancy is removed if we apply the Somers correction to the values reported for Zn I. (To van Hengstum's results for the Cd I lines the correction has been applied.) However, a systematic difference at wavelengths below 3000 Å, for Cd I and Zn I, now becomes evident.

Among the institutes at which the emission method is used for determining transition probabilities we mention Kiel University, where Lochte-Holtgreven, Maecker, and collaborators developed the high-current arc and several related types of stabilized arc source, and also Michigan University, where shock tubes have been used (Laporte and Wilkerson).

A laudable effort made by Addink (1959) to derive transition probabilities from spectro-chemical calibration constants has yielded results that are unfortunately of little value, since various concepts of his theoretical treatment of ionization and mass transport are essentially erroneous. Various errors have offset one another to some extent so that reasonable agreement with literature values was obtained in a number of instances.

A large-scale investigation involving the determination of transition probabilities of 25 000 lines has been made by Corliss and Bozman (1962) at the N.B.S.

The procedure followed by them incorporates many noteworthy features and offers so many problems of general interest for further discussion that it appears advantageous to consider it separately (see § 1.5). Corliss and Bozman's work has demonstrated, better than a literature survey could, the poor reliability of available data on atomic transition probabilities.

Finally I mention the measurements of A- or f-values in emission by Eberhagen (1955b), Eicke (1962), and Köstlin (1964) at Göttingen, those by Margoshes and Scribner (1963) at N.B.S., and those by Hefferlin and Gearhart (1964) at Collegedale.

2 *Measurements in absorption:* Closely related to the emission method is the absorption method, which measures the absorption of radiation that occurs when light from a source giving a continuous spectrum passes through an absorption tube containing atomic vapour of the element to be investigated. The method has been elaborated by R. B. King, A. S. King and their co-workers at the California Institute of Technology. An important improvement in the method was achieved by using an atomic beam as the absorber and condensing the atoms on a microbalance built into the chamber (see Kopfermann and Wessel, 1951; King, 1963; and Lawrence, Link, and King, 1965).

3 *Measurements from the anomalous dispersion at the edges of spectral lines:* The method of measuring oscillator strengths (f-values) from the anomalous dispersion at the edges of spectral lines has been developed by Rozhdestvenskii. It involves determining the index of refraction in the neighbourhood of the spectral line, which is done interferometrically with the 'hook' method. The procedure is applied particularly in the U.S.S.R. A review of this work was published by Penkin (1964).*

4 *Lifetime determinations:* The determination of lifetimes of excited atomic states has long been attempted. In the last decade the method has regained popularity with the application of fast counting techniques developed for nuclear physics. Atoms are excited by radiation or electron impact in short bursts and the subsequent depopulation of excited levels is observed by studying the time decay of the emitted radiation. The method can only be applied in gases at very low pressure since lifetimes do not depend exclusively on the transition probability for spontaneous emission. (It must be remembered that *natural* mean lifetime defined as reciprocal damping constant [equation (1.25)] is not identical with lifetime.)

Recent papers on lifetime determinations are those by Demtröder (1962), by Hulpke, Paul, and Paul (1964), and by Lurio (1964).

---

* Other valuable sources of information, particularly for those unable to read Russian, are *Optical Transition Probabilities, A Collection of Russian Articles, 1924–1960,* and *Optical Transition Probabilities, A Representative Collection of Russian Articles, 1932–1962.* (Translated from the Russian and published for the National Science Foundation, Washington, D.C., and the Department of Commerce, U.S.A., by the Israel Program for Scientific Translations. Available from the Office of Technical Services, U.S. Department of Commerce, Washington 25, D.C.)

2

§ 1.4   *Transition probability and related quantities*

In the preceding section we defined the *transition probability for spontaneous emission*. In the same way we introduce the *transition probability of absorption* B_{pq}, also known as the Einstein coefficient of absorption. It is the constant of proportionality that relates the number of absorption processes p $\rightarrow$ q occurring per sec to the number (N_p) of atoms in the lower level p and to the density (ρ_ν) of the radiation of frequency ν_{qp}:

$$\frac{dN_q}{dt} = B_{pq}.N_p.\rho_\nu \qquad (1.27)$$

With respect to intensity, radiation density is expressed as

$$\rho_\nu = \frac{4\pi}{c} I_\nu \qquad (1.28)$$

where c is the velocity of light.

A third quantity to be defined is the *transition probability for induced emission* B_{qp}. A light quantum of frequency ν_{qp} can cause an excited atom to emit a quantum of identical frequency and direction as the arriving one. The number of this type of process in unit time is given by

$$-\frac{dN_q}{dt} = B_{qp}.N_q.\rho_\nu \qquad (1.29)$$

where N_q is the number of atoms in the excited state q and ρ_ν is the density of radiation of frequency ν_{qp}.

In a state of complete thermodynamic equilibrium we have in consequence of detailed balancing (see § 5.5)

$$(A_{qp} + B_{qp}.\rho_\nu).N_q = B_{pq}.\rho_\nu.N_p \qquad (1.30)$$

which follows from (1.23), (1.27), and (1.29) if we require that $dN_q/dt = 0$, that is the number of atoms arriving in the state q per sec should equal the number leaving that state per sec.

By applying the Boltzmann equation (§ 6.1) and the Rayleigh-Jeans radiation formula, the following interrelations can be derived (see Unsöld, 1955, p. 277):

$$g_q.B_{qp} = g_p.B_{pq} \qquad (1.31)$$

and

$$A_{qp} = \frac{8\pi h\nu^3}{c^3} B_{qp} = \frac{8\pi h\nu^3}{c^3} \frac{g_p}{g_q} B_{pq} \qquad (1.32)$$

where h is the Planck constant, c is the velocity of light, ν is the frequency ν_{qp} of the radiation involved in the transition q $\rightarrow$ p, and vice versa, and g_p and g_q are the statistical weights of lower and upper level, respectively. The *statistical weight* of a particular state is the probability of populating a state under identical conditions. A state with total angular momentum defined by the inner quantum number commonly known as J has statistical weight $(2J+1)$. J-values of energy levels of atoms and ions can be found in Moore's tables *Atomic Energy Levels* (1949, 1952, and 1958).

By inserting the numerical values of the constants (see Glennon and Wiese, 1962) we obtain from equation (1.32):

$$B_{qp} = 6 \cdot 01 \lambda^3 A_{qp} \tag{1.33}$$

and

$$B_{pq} = 6 \cdot 01 \lambda^3 \frac{g_q}{g_p} A_{qp} \tag{1.34}$$

where λ is expressed in ångström units.

Finally, we must consider a quantity that is often, probably even more frequently, used instead of transition probability, namely the (absorption) oscillator strength, f_{pq}, introduced by Ladenburg (1921) to connect the classical and quantum theoretical treatment of atomic absorption. By definition the *oscillator strength* f_{pq} for the transition p → q is the number of classical oscillators that produce the same absorption effect as one atom in the state p.

The following useful interrelations are deduced and discussed by Unsöld (1955, pp. 278 ff.):

$$f_{pq} = \frac{g_q}{g_p} A_{qp} \frac{mc^3}{8\pi^2 e^2 \nu^2} = \frac{g_q}{g_p} \frac{A_{qp}}{3\gamma_{cl}} \tag{1.35}$$

where

$$\gamma_{cl} = \frac{8\pi^2 e^2 \nu^2}{3mc^3} \tag{1.36}$$

is the classical damping constant of an oscillator of frequency ν. The quantities m, e, and c are the mass of an electron, the electronic charge, and the velocity of light respectively.

The numerical conversion of transition probability A_{qp} into oscillator strength f_{pq} is made by applying the equation (Glennon and Wiese, 1962):

$$f_{pq} = 1 \cdot 4992 \cdot 10^{-16} \lambda^2 \frac{g_q}{g_p} A_{qp} \tag{1.37}$$

The reverse conversion is accomplished with

$$A_{qp} = \frac{6 \cdot 6702 \cdot 10^{15}}{\lambda^2} \frac{g_p}{g_q} f_{pq} \tag{1.38}$$

where the wavelength λ is expressed in ångström units.

TABLE 1.1

Spectrum	$\lambda(\text{Å})$	Transition	g_p	g_q	f_{pq} (line)	$g_p f_{pq}$ (line)	$\Sigma g_p f_{pq}$ (multiplet)	f_{pq} (multiplet)	$g_q A_{qp} \cdot 10^{-8}$ (line)	$A_{qp} \cdot 10^{-8}$ (line)	$(\Sigma) A_{qp} \cdot 10^{-8}$ (level)	Reference
Mg I	2852	$3S_0-3P_1$	1	3	1·11	1·11	—	—	9·12	3·04	3·04	Demtröder (1962)
Ca I	4227	$4S_0-4P_1$	1	3	1·6	1·6	—	—	6·0	2·0	2·0	Odintsov (1963)
Na I	5889	$3S_{1/2}-3P_{3/2}$	2	4	0·654	1·308	1·964	0·98	2·52	0·63	0·63	Demtröder (1962)
	5896	$3S_{1/2}-3P_{1/2}$	2	2	0·328	0·656			1·26	0·63	0·63	
Ag I	3280	$5S_{1/2}-5P_{3/2}$	2	4	0·506	1·012	1·506	0·75	6·28	1·57	1·57	Penkin and Slavenas (1963)
	3382	$5S_{1/2}-5P_{1/2}$	2	2	0·247	0·494			2·88	1·44	1·44	
Ga I	4033	$4P_{1/2}-5S_{1/2}$	2	2	0·0878	0·176	0·518	0·086	0·72	0·36	1·01	Demtröder (1962)
	4172	$4P_{3/2}-5S_{1/2}$	4	2	0·0854	0·342			1·31	0·654		
Ga I	4033	$4P_{1/2}-5S_{1/2}$	2	2	0·129	0·258	0·798	0·13	1·06	0·53	1·56	Penkin and Shabanova (1963)
	4172	$4P_{3/2}-5S_{1/2}$	4	2	0·135	0·540			2·06	1·03		
Mg II	2795	$3S_{1/2}-3P_{3/2}$	2	4	0·60	1·20	1·80	0·90	10·4	2·6	2·6	Bierman and Lübeck (1948), cf. Allen (1955)
	2803	$3S_{1/2}-3P_{1/2}$	2	2	0·30	0·60			5·2	2·6	2·6	
Ba II	4524	$6P_{1/2}-7S_{1/2}$	2	2	0·28	0·56	1·80	0·30	1·8	0·9	2·6	Eicke (1962)
	4900	$6P_{3/2}-7S_{1/2}$	4	2	0·31	1·24			3·4	1·7		

Examples of numerical values of transition probabilities and oscillator strengths that have been determined according to different methods (lifetime measurements, Rozhdestvenskii's hook method, measurements in absorption, measurements in emission, and self-consistent field calculation).

For the sake of illustration Table 1.1 summarizes transition probabilities and related quantities of a few spectral lines. The absolute values shown in the table have been determined according to different methods: lifetime measurements (Demtröder, 1962), measurements in absorption (Odintsov, 1963), Rozhestvenskii's hook method (Penkin and Shabanova, 1963; Penkin and Slavenas, 1963), self-consistent field calculation (Biermann and Lübeck, 1948), and measurements in emission (Eicke, 1962). The f-value entered into column nine, that is f_{pq} for the multiplet, is defined as

$$(f_{pq})_{\text{multiplet}} = \frac{\Sigma g_p (f_{pq})_{\text{line}}}{\Sigma g_p} \qquad (1.39)$$

It must be noted that most papers on transition probabilities or oscillator strengths give *relative A-* or *f*-values, which are easier to obtain than absolute ones. Often a set of relative values is then adjusted as a whole to an absolute scale with the aid of literature data. Furthermore, it has become customary to list *gf*- or *gA*-values, i.e. *weighted oscillator strengths* or transition probabilities.

The relative intensities of the spectral lines in multiplets show various regularities which are described by so-called sum rules. We shall not enter into this matter here and refer the reader to works on spectroscopic theory and atomic structure, e.g. White (1934), Herzberg (1944), Richtmyer and Kennard (1947), Unsöld (1955), and Kuhn (1962).

§ 1.5 *Transition probability measurements at the N.B.S.*

To provide relative intensity values on a uniform scale, Meggers, Corliss, and Scribner (1961a and b) measured the relative intensities of 39 000 spectral lines with wavelengths between 2000 and 9000 Å for the seventy chemical elements commonly determined in spectrochemistry. The source of excitation was a free-burning 10-amp, 220-V d.c. arc between solid copper electrodes formed by pelleting pure copper powder, mixed with 0·1 atomic per cent of one of the seventy elements. Details of the procedure together with a historical sketch about intensity measurements in general and those in the present investigation in particular are given in the original papers referred to above.

The fact that uniform intensity data had become available must have been a challenge for the workers concerned with it, so that further derivations, originally not intended, were made; a logic consequence of saying '*I*' is to follow it by '*gA*' and '*gf*'. Had the goal been initially to determine transition probabilities, the investigators would most probably have chosen a different way of taking the spectra, especially as regards the portion of the arc falling within the aperture of the spectrograph. With the procedure adopted, in which all the radiation of the short arc contributed to the spectra, the intensities were averaged over a source of maximum spatial inhomogeneity of excitation. This, of course, has entailed troubles that the authors had to overcome subsequently. On the other hand, there is much reason to doubt whether the source inhomogeneity is the primary cause of a noticeable scatter of the values of temperature and electron concentration that has been reported for the N.B.S. copper arc. Before entering more thoroughly into this matter, let us trace the main line of argument followed by Corliss and his co-workers for deriving absolute transition probabilities from relative intensities of spectral lines (see Corliss, 1962a, b, and c; and Corliss and Bozman, 1962).

In general, when determining absolute transition probabilities by an emission method, we are concerned with

1 The temperature of the source.

2 The relative numbers of atoms and ions of each element, i.e. the individual degrees of ionization.

3 The absolute numbers of either atoms or ions of each element.

To obtain the necessary data for the copper arc, Corliss and Bozman applied spectroscopic methods, requiring accurate knowledge of the transition probabilities of suitable sets of spectral lines. The authors selected these basic data from published values, and, notably, used a large number of sets to determine the excitation conditions (temperature and electron concentration) in the copper arc. Thus, their results are 'world averaged' and the scatter of individual values clearly reflects disagreement among the different sets of basic data taken from the literature.

For the temperature determination an extensive search of the literature disclosed thirty-one sets of data suitable for the purpose. The data came from five laboratories, viz. in Leningrad, Utrecht, Pasadena (California), Moscow, and Amsterdam. They encompassed the first spectra of the following elements: Ag, Ba (2 sets), Ca (2 sets), Cd (2 sets), Co (2 sets), Cr, Fe (4 sets), Ga, Hg, In, K, Mg, Mn, Ni, Sc, Sr, Ti (3 sets), Tl, V (2 sets), and Zn (2 sets).

By plotting the log ratio of intensity $\times \lambda^3$ to the gf-value or the log ratio of intensity $\times \lambda$ to the gA-value versus upper level, straight lines were obtained, thus confirming the assumption of a Boltzmann distribution of energy-state population, and the appropriateness of an effective temperature (see § 6.4). From the slope of the lines temperatures were derived. The numerical values thus found with the thirty-one sets of independent data showed a standard deviation of 600°K and a mean value of 5100°K.

It is not plausible that the temperature of the copper arc with a single element added in the ratio of 1 atom of element to 1000 atoms of copper should vary perceptibly from one element to another (cf. § 7.9). A slight influence of the electrode composition would begin to appear at a ratio of 1 atom of element to 100 atoms of copper, as seems to be the concentration level used in the experiments conducted by Kessler and Dobeneck (1963), who comment upon the N.B.S. measurements. Therefore Corliss suggests with good reason that the variations arise from systematic errors in the various sets of transition probabilities. For example, a difference of 790°K is found between the temperature based on the two sets of gf-values for V I, obtained from Leningrad and from Pasadena. Since in both temperature determinations the same intensities are involved, the discrepancy must arise from systematic errors in one or both sets of gf-values.

Individual temperature values are scattered about the mean value according to a Gaussian distribution function, and no correlation exists between the temperature found and the ionization potential of the thermometric species, as is shown by Fig. 1.4. Such a correlation is expected when uniform sets of *gf*-values are employed (see §§ 6.16 and particularly 8.6). In fact, the systematic differences between the various sets of basic data have become random in the present experiment and conceal the systematic dependence of the temperature on the ionization potential of the relevant element, which should

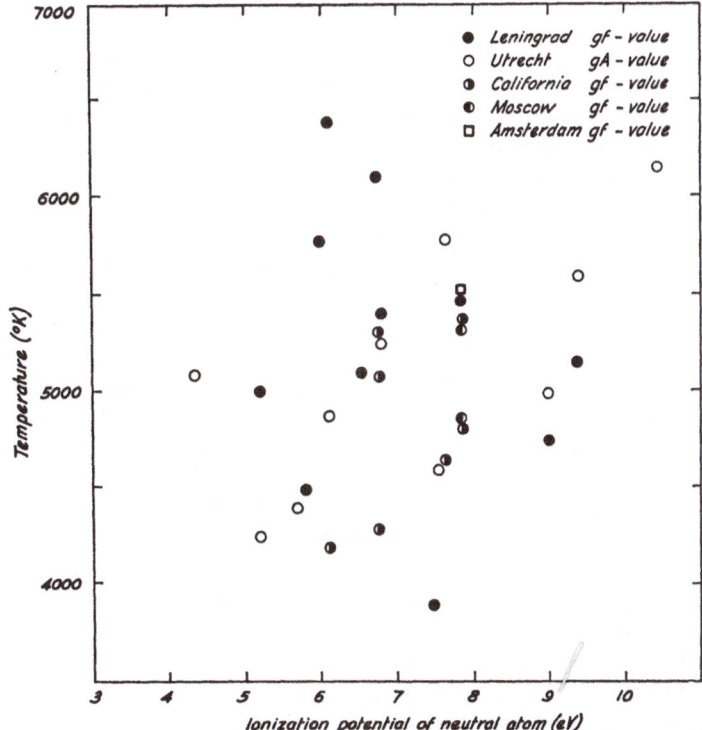

FIG. 1.4. Temperature for the N.B.S copper arc, measured on the basis of various sets of *gf*- or *gA*-values, plotted against the ionization potential of the thermometric element used. The figure shows clearly that the large scatter of temperature values, caused by systematic errors in the *gf*- or *gA*-values, conceals any functional relationship between the temperature and the ionization potential of the thermometric species, that would be expected for the inhomogeneous source under investigation.

occur as a natural consequence of source inhomogeneity (not as a result of a matrix effect). As will be discussed more extensively in §§ 6.12 and 6.16 and in Chapter 8, each atomic or ionic species emits predominantly from that zone in the discharge where the temperature, with regard to excitation and ionization, is at an optimum for that particular species.

Corliss adopted the mean value of all individual temperature determinations as the effective excitation temperature for all elements. This is probably the best procedure to follow; for, by employing numerous independent sets of *gf*-values, systematic errors in these basic data are averaged out to some extent, while, by the simultaneous use of elements of widely differing ionization potential as temperature indicators, some kind of average over the different emission zones of the discharge is obtained. The adoption of the standard deviation of the mean, i.e. $\hat{\sigma}/\sqrt{n} = 110°K$, as a measure for the uncertainty of the effective temperature appears, however, to be too favourable a judgment, since the effective excitation temperature incontestibly varies with the ionization potential of the element under consideration.

Once the effective temperature had been estimated, the next step in deriving transition probabilities was to ascertain the relative numbers of atoms and ions for each element, in other words, to determine the various degrees of ionization. As will be discussed in Chapter 7, the electron concentration must be known if the calculation is to be carried out using Saha's ionization relationship. When measuring the electron concentration spectroscopically following the procedure described in § 7.6, Corliss encountered an analogous and even more serious difficulty than when determining the temperature, namely the choice of pairs of arc and spark lines with known reliable transition probabilities. There turned out to be eleven elements suitable for the method. The average value of the electron concentration (n_e) was found to be $2\cdot4 . 10^{14}$ per cm^3; the scatter of the individual values was large: on the basis of the calculated average deviation in $\log n_e$ the range of n_e was from $0\cdot9 . 10^{14}$ to $6 . 10^{14}$ per cm^3. So, the uncertainty, again originating in the uncertainty of the absolute gf-values, is high. Here, obviously, similar comments to those made about the determination and application of the effective temperature hold. Since the spatial distribution of the electron concentration and the temperature had to be neglected because of lack of data, it must be considered to be the main source of error when determining the various degrees of ionization.

On the whole, we should not be too pessimistic. The straight lines that Corliss obtained when plotting the log ratio of intensity $\times \lambda^3$ to the gf-value versus the upper level for determining the temperature fit remarkably well; also the plot of Saha's relationship and its fit to experimental points is striking (Corliss, 1962b, Fig. 1; the figure is reproduced and discussed in § 7.6). The location of the points with respect to the slope of the line, fixed by the temperature of $5100°K$, indicates that this effective temperature is appropriate for calculating the ionization as well as the excitation.

The last step of the calculation was the normalization of the uniform relative scale of transition probabilities to the absolute scale. Here one becomes concerned with the very critical and vital problem of spectrochemical physics, a quantitative approach to the determination of the concentrations (and the concentration distributions) of atoms and ions in the arc. We shall postpone a discussion of this matter until Chapter 9.

In summary, Corliss and Bozman (1962) brought out an imposing collection of absolute A- and f-values for 25 000 lines of seventy elements. The values are on a uniform scale.* Perhaps the scale is too uniform since the systematic variation

---

* As a result of a calibration error, the gf- and gA-values listed for lines with $\lambda < 2500$ Å must be corrected by about a factor of 10 (Margoshes, personal communication).

in the excitation conditions from one element to another is neglected. For many purposes the great uniformity of the N.B.S. data will be an important gain, the more so as the very procedure which led to those data revealed the lack of uniformity of f-values published in the literature. Although comparatively large errors in the absolute f-values of the order of a factor of 2 are to be expected, the set of N.B.S. data should prove valuable for applications where the highest accuracy is not needed.

CHAPTER TWO

GENERAL DISCUSSION OF
THE D.C. CARBON ARC

★ § 2.1   *Introductory note*

In this work we are concerned with the carbon arc in air at atmospheric pressure. By the term 'carbon' we mean primarily the *element* of which the electrode material is composed, and do not particularly refer to the 'amorphous' allotrope carbon.

The general characteristics of the carbon arc vary little for carbon or graphite electrodes. However, we should not conclude that it makes no difference whether we use carbon or graphite electrodes for a particular analysis. These materials differ markedly in thermal and electrical conductivity, and consequently, spectrochemical specimens volatilize differently according to whether the supporting electrode is of carbon or graphite. The question will be examined more closely in § 10.3. For our general discussion of the carbon arc and the processes that occur in the plasma, we need not specify the electrode material beyond naming the element composing it.

We shall consider the carbon arc chiefly as a source of excitation for spectrochemical analysis. This means that we shall concentrate, not on the discharge character of the arc, but on those aspects that pertain to light emission and occur when samples are introduced. For that reason, the expositions brought out in this book often have a much wider scope than the field of the carbon arc alone, and can be extended, *mutatis mutandis*, to other sources of excitation.

★ § 2.2   *The free-burning vertical arc*

Our discussion will be chiefly restricted to the *free-burning arc*, that is the arc with free convection; it must be distinguished from arc discharges with suppressed or forced convection, in which the arc is stabilized by peripheral cooling, e.g. a tangential air blast.* Convection in the free-burning arc is caused by the temperature

---

* A review of literature concerning the physical aspects of gas-stabilized, wall-stabilized, and free-burning arcs has been produced by Olsen (1963) (see also Finkelnburg and Maecker, 1956). The gas-stabilized arc has found extensive application in spectrochemical analysis. A well-known device is the Stallwood-jet (see Stallwood, 1954; Hoens and Smit, 1957; Margoshes, 1960; the review articles by Champ, 1962, Snooks, 1962, and Shaw, 1962, contain additional references). The Stallwood-jet when combined with a chamber is a convenient expedient for producing a discharge in a controlled atmosphere (see the end of § 5.6 for further references). In this connection we also mention different types of plasma jets and related gas-stabilized sources (see note on p. 5 for references). [*continued on facing page*]

difference between the arc and the surrounding medium. When an arc burns horizontally, the upward flow of gas produces the curved shape of the discharge from which its familiar name has been derived. The ideal vertical arc is symmetrical about an axis that coincides with the axis of the electrodes.

The arc discharge consists of the following parts: the *core* or *column*, the *flame*, fringe, mantle, aura, or envelope, and the layers immediately adjacent to the electrodes. These regions merge more or less gradually into one another; consequently their boundaries cannot be defined sharply (see § 6.15). In common parlance, the core or column in the most restricted sense, as defined below, and the areas in the immediate vicinities of the electrodes are often taken together as the 'column'; in this sense the column extends from pole to pole. It carries most of the current and accordingly the temperature is here at a maximum. In the absence of added substances, the column of the carbon arc in air shows the well-known violet colour caused by cyanogen band emission. The cross-sectional area or diameter of the region through which the current flows is largest in the middle of the gap and decreases toward the electrodes. If the gap width exceeds a certain minimum value (see § 3.1), the middle part of the discharge becomes uniform and takes the shape of a small cylinder. A further increase of electrode separation only causes the uniform part to lengthen without any change in the regions adjacent to the electrodes. To be exact, the term 'core' or 'column' must be confined to the uniform, cylindrical region of the arc, which in the terminology of gas discharges is known as the positive column. In arcs between pure carbon electrodes, i.e. when no material has been added, the properties of the column (e.g. electric field strength, current density, and temperature) are practically constant along the entire length; moreover, they are almost unaffected by an increase in the width of the gap, provided the current strength is kept constant.

The uniform, cylindrical region, which will be referred to hereafter as the core or column (in the most restricted sense), is for spectrochemical analysis by far the most important part of the arc. The immediate vicinities of the electrodes still deserve attention because it must be recognized that the excitation conditions in these gas layers may differ greatly from those in the column. These differences can be understood by considering the energy balance qualitatively.

The energy loss from the column is primarily by convection and by thermal conduction to the surrounding medium;* from the gas layers in the immediate vicinity of the poles, however, there is at the same time an appreciable heat loss by conduction to the electrodes; this is clearly so, since the temperature of the

---

An example of the wall-stabilized arc is the cascade arc referred to at the end of § 3.5. Its application to spectrochemical analysis has been described by Riemann (1964b, 1965). Another type of wall-stabilized arc was used by Svoboda (1962).

The graphite ring surrounding the cathode and the upper part of the arc column that I used for stabilizing purposes (see § 6.7) forms a simple device for producing a wall-stabilizing effect on the discharge, without much expense or complication.

* Diffusion of dissociated molecules to the boundaries may be regarded as a special case of radial conduction (see § 4.3).

electrodes is low, namely $\leq 4000°K$, in comparison with the temperature of the adjacent gas layers, viz. 5000–7000°K. The maintaining of the high temperature of the gas at the electrodes can be explained by assuming that the power dissipated here (per cm) is enhanced as compared with the energy supply (per cm) in the column. In fact, the electric field strength and the current density are considerably higher in the neighbourhood of the electrodes than in the column, while the arc section is noticeably reduced. We shall not attempt to dwell upon the complicated theories of the mechanism of the arc as a whole. For detailed discussion the reader is referred to appropriate texts that will be mentioned in § 4.1. For our purpose it will be adequate to realize that the greater part of a "long" carbon arc consists of a cylindrical column which is homogeneous throughout its length.

Radially, the column passes over rather sharply into the fringe, also termed 'flame', 'mantle', 'aura', or 'envelope'. In this region, current density and electric power dissipation are negligible. The high temperature of 3000–4000°K in the flamy fringe is due to thermal conduction and convection from the column. The temperature gradient in the flame is rather large (see § 6.15).

The shape of the flame, in contrast with that of the column, is not cylindrical: the diameter increases with the distance from the lower electrode; the profile is more or less parabolic. According to Smit (1950, p. 39) this can be explained in the following way. Let us consider a thin horizontal gas layer just above the lower electrode. The layer as a whole will migrate upwards with a convection velocity of about 100 cm/sec (see § 2.3). Ascending in the arc the layer remains in a horizontal position and preserves a flat shape. When it has just passed over the lower electrode spot, a circular section in the centre forming part of the column has gained a very high temperature, whereas the surrounding zones still consist of unheated air. Now, from the centre heat is conducted radially to the outer zones; accordingly, as the layer has travelled a greater distance, more heat has been transported to the outer regions. In the central part of the layer the high temperature is maintained, because power dissipation here is virtually constant. The continued production of heat in the centre of the migrating gas layer and the radial conduction account for the gradual broadening of the flame, i.e. a zone where temperature is intermediate between room temperature and the temperature prevailing in the column. Strangely enough, the temperature in the column does not increase with the distance travelled by the gas layer. This is closely related to the rise of the electrical and thermal conductivity of the arc gas with temperature. The increase is markedly rapid in the temperature range of the column. It must be recalled that the governing feature of the thermal conductivity of air is the molecular dissociation, particularly that of nitrogen (see § 4.3).

Although in some respects the transition between column and flame may be regarded as abrupt, the boundary of the column cannot be defined exactly. Useful criteria are the gradients of the electron concentration and the temperature (see §§ 3.8 and 6.15). Fairly rough information can be obtained experimentally from optical projections or photographs (see, for example, Hörmann, 1935; and Mason, 1948). Spectroscopically, the diameter of the column can be determined from 'edge-on' spectra (see §§ 6.13 and 6.14). Indirect information is deduced from observed values of the mean electron concentration, the current strength, and the electric field strength [see equation (3.33)].

When a substance is evaporated in the carbon arc, the vapour spreads over column and flame. In the column, the vapour becomes ionized to a degree consistent with the temperature and electron pressure that ensue, and with the ionization potential of the relevant element. In the flame, ionization may be ignored; here, on the other hand, formation of molecules and radicals must be considered. Electron concentration and temperature change more or less gradually when we pass from the centre of the column to the flame; in other words, excitation conditions show a radial variation which accounts for the observed radial emission distribution of atom and ion lines and molecular bands. For example, elements of low ionization potential ionize almost completely in the hot core and emit their atom lines mainly from the flame, where ionization is negligible and temperature is still sufficiently elevated for the relatively low levels of these atoms to become excited. The familiar 'hollow flames' that are observed when a small quantity of an alkali salt is introduced into the carbon arc have been investigated already by Lenard (1903, 1905; cf. § 6.12 and Chapter 8).

In the cooler fringe, ground-state and low-level atoms predominate and, consequently, since radiation emitted by the column must pass this fringe, it may be greatly affected by self-absorption (see § 5.7).

Up to the present we have restricted our attention to the arc; but the physical conditions of the electrodes, particularly that of the supporting lower electrode, should not be overlooked in spectrochemical analysis. The temperature of the electrode spots is assumed to be the boiling or sublimation point of carbon. The numerical value of the anode spot temperature has been established to be 4000°K. The temperature drops very rapidly with increasing distance from the spot. Graphite and carbon differ markedly in thermal conductivity and for this reason they behave differently. (Various problems connected with the properties of the electrodes will be dealt with more thoroughly in § 10.3.)

★ § 2.3   *The convection velocity in the arc*

The speed of convection upwards in the arc is of the order of magnitude of 100 cm/sec. Van Stekelenburg (1943, 1946) deduced the convection velocity in carbon arcs of 3 cm length from the velocity of small carbon particles carried along in the arc. In this way, he measured the convection velocity (v_c) as a function of the height above the lower electrode and as a function of the electric current strength (i). Both functions show a positive slope. For short arcs (electrode separation ≤ 1 cm), as commonly employed in spectrochemical analysis, the longitudinal variation of the convection velocity may be ignored and mean values for the gap as a whole can serve as a satisfactory approximation. From the graphs published by van Stekelenburg the following numerical values of the convection velocity can be taken: $v_c = 80$ cm/sec at $i = 4$ amp, $v_c = 100$ cm/sec at $i = 6$ amp, $v_c = 120$ cm/sec at $i = 8$ amp, and $v_c = 140$ cm/sec at $i = 10$ amp.

Polarity appeared to have little influence on the speed of the gas flow; similarly, the effect of elements of low ionization potential, which exert a powerful influence on the arc temperature (see § 7.8), proved to be insignificant.

The investigations quoted were carried out in the vicinity of the arc axis. There is evidence, however, that in the column the radial gradient of the speed of convection is small (van Stekelenburg, 1943, p. 55).

Hagenah (1950) determined the streaming field in a free-burning 10-amp d.c. carbon arc. His Fig. 4 showing streaming lines and curves of equal gas flow velocity illustrates the vertical and radial change of convection velocity in column and fringe, and the impeding effect of the electrodes on the streaming field. The absolute values of the convection velocity found by Hagenah are somewhat higher than those given by van Stekelenburg.

Recently Gol'dfarb and Il'ina (1961) reported the value $v_c = 70$ cm/sec for the convection velocity in a 5-amp d.c. carbon arc of 8 mm gap width. It agrees fairly well with van Stekelenburg's results. Finally, we should mention the experimental study of convection velocity made by Suits (1939a).

CHAPTER THREE

ELECTRIC PARAMETERS
OF THE D.C. CARBON ARC

★ § 3.1 *Basic electric properties*

The basic electric parameters of an arc discharge are the *current strength* (i) and the *voltage drop* across the gap (V). Both quantities can be measured without difficulty. In general, a relationship between V and i such as shown graphically in Fig. 3.1 is found; clearly Ohm's law is not obeyed: for a given value of the electrode distance (l) the arc voltage decreases as the current rises. The voltage-current curves lie higher for larger l. With carbon electrodes the voltage suddenly drops as the current exceeds a certain minimum value, which is accompanied by acoustic hissing (*Zischen*) of the arc.*

The kind of relationship plotted in Fig. 3.1 is referred to as a negative or *descending characteristic*. It means that an arc cannot be run without an additional resistance connected in series to limit the current; for, a sudden increase of the current would result in a decrease of the voltage, which again would bring about another rise of the current, etc., a steady burn thus being impossible. The available voltage must be always higher than that which actually exists between the electrodes; thus, random fluctuations of the current strength are automatically controlled by the series resistance.

The following picture taken from Finkelnburg and Maecker (1956, p. 264 ff.) gives a good idea of the various factors to be considered. In Fig. 3.2, curve B is the negative characteristic of a free-burning carbon arc; the straight lines R_1, R_2, and R_3 represent the voltage–current relationships of three series resistances obeying Ohm's law ($R_1 < R_2 < R_3$). A characteristic of the circuit as a whole is obtained by adding that of the arc alone and that of the series resistance; thus, with R_1, R_2, and R_3 respectively, we find the curves K_1, K_2, and K_3.

---

* Hissing is an anode phenomenon when observed with carbon arcs. It is associated with the increase of the anode spot with increasing current, on the one hand, and the limited area of the anode surface, on the other. The anode becomes overloaded at a certain current strength, the numerical value of which depends, among other factors, on the cross-sectional area of the anode. The acoustic noise is caused by the escape of fine gas jets from the electrode.

The phenomenon of hissing has been studied by Bräuer (1919), Schmick and Seeliger (1928), Weizel and Fassbender (1940, 1943), Schluge and Finkelnburg (1944), Höcker and Bez (1955), Bez and Höcker (1956), Lochte-Holtgreven (1956a), and recently by Eather (1963) and Baker (1964); see also Ecker (1961, p. 49), Finkelnburg and Maecker (1956, p. 425 ff.), and Weizel and Rompe (1949, pp. 103–104).

We now consider the horizontal broken line V_a representing the total available voltage. Necessarily, for a steady state we have

$$V_a = V_B + V_R \qquad (3.1)$$

where V_B is the arc voltage and V_R the potential drop across the series resistance. The current strength at which the arc burns is found as the abscissa of the intersection point of the total characteristic K and the horizontal line V_a. The ordinates

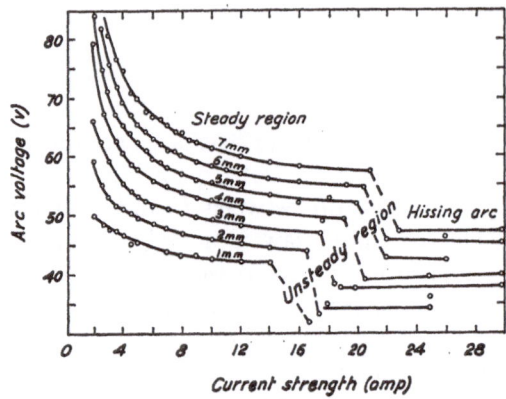

FIG. 3.1. Voltage–current characteristics of a low current arc at various electrode separations.

From W. Finkelnburg and H. Maecker, 'Elektrische Bögen und thermisches Plasma' in *Handbuch der Physik*, Vol. 22, pp. 254–444 (Springer, Berlin, Göttingen, Heidelberg, 1956).

of B and R at this current strength give the arc voltage and the voltage drop across the series resistance. Obviously, the value of the latter should be such that the total characteristic K shows a positive slope in the neighbourhood of the working point

$$\frac{dV_B}{di} + \frac{dV_R}{di} > 0 \qquad (3.2)$$

Equation (3.2) is the condition for a steady state enunciated by Kaufmann (1900). Evidently, smoothing out random fluctuations takes place more rapidly as the total characteristic runs steeper in the vicinity of the working point, or the greater the series resistance the higher the stability of arc burn. (In spectrochemical applications, economy of power input does not play a role of any importance.) Fig. 3.2 also demonstrates that for a given total voltage V_a the numerical value of the series resistance has an upper limit (R_2 in Fig. 3.2); from the point of contact of the curve K_2 and the line V_a we find the minimum current strength at which the arc can be run for a given value of V_a.

The various conditions are also illustrated by the diagram of Fig. 3.3. Here the resistance lines R_1, R_2, and R_3 are shown as differences of total available voltage and voltage drop across the resistance; the ordinates of the lines thus represent the voltage available for the arc. The point of intersection of a resistance line and the arc characteristic now gives the working point for a steady arc, whereas the point of contact of R_2 and B yields the maximum value of the series resistance and the minimum value of the current strength for the given total available voltage.

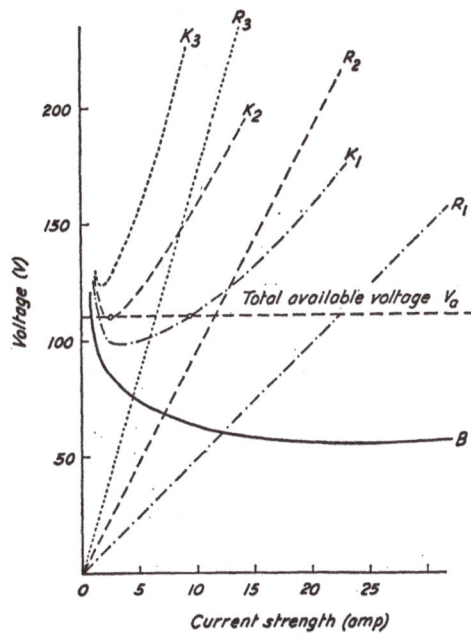

FIG. 3.2. Stabilizing effect of series resistances of different magnitude on free-burning arcs with falling characteristic.

B: arc characteristic.

R_1, R_2, R_3: voltage–current relationships of series resistances obeying Ohm's law.

K_1, K_2, K_3: characteristics of complete circuit, that is arc plus series resistance, $B+R_1$, $B+R_2$, and $B+R_3$.

The current at which the arc burns is found from the intersection point of the characteristic K and the horizontal line V_a representing the total available voltage.

From W. Finkelnburg and H. Maecker, 'Elektrische Bögen und thermisches Plasma' in *Handbuch der Physik*, Vol. 22, pp. 254–444 (Springer, Berlin, Göttingen, Heidelberg, 1956).

Fig. 3.2 elucidates the stability condition [equation (3.2)]; Fig. 3.3 shows clearly how the intersection points of the various resistance lines with the arc characteristic shift when the total available voltage is altered. It can also be used to illustrate the rise of the current strength observed when a substance of low ionization potential is evaporated into the carbon arc. Introducing such a substance causes a change of the characteristic. Let us assume that curve B' now applies. The point of intersection with R_1 thus shifts to higher current strength. In order to restore the original value we have to raise the value of the series resistance up to R_1'.

The significance of the series resistance for the steadiness of the current is confirmed by the following arguments (cf. Zaidel, Kaliteevskii, Lipis, and Chaika, 1960, p. 219). From Ohm's law,

$$i = \frac{V_a}{R_B + R_S} \tag{3.3}$$

where V_a is the voltage available from the power source, R_B the resistance of the

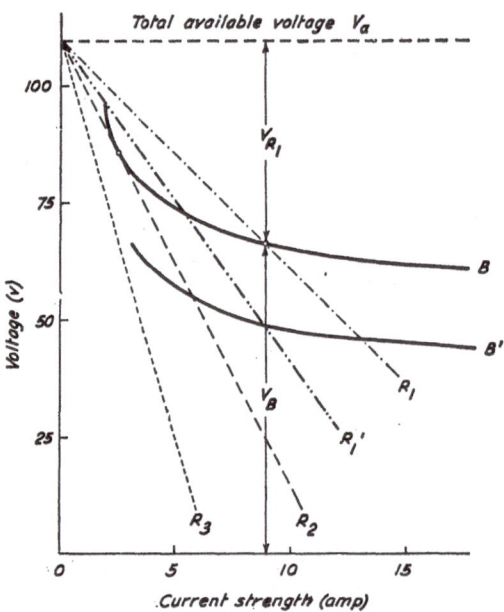

FIG. 3.3. Stabilizing effect of series resistances of different magnitude on free-burning arcs with falling characteristic.

B: characteristic of carbon arc.

B': arc characteristic after introducing metal of low ionization potential.

R_1, R_2, R_3: series resistance characteristics represented as difference of total available voltage and voltage drop across resistance.

The current strength at which the arc burns is found from the intersection point of the arc characteristic B (or B') and the resistance line.

With slight modification from W. Finkelnburg and H. Maecker 'Elektrische Bögen und thermisches Plasma' in *Handbuch der Physik*, Vol. 22, pp. 254–444 (Springer, Berlin, Göttingen, Heidelberg, 1956).

arc, and R_S the series resistance, we find for a small change dR_B in the arc resistance:

$$\frac{di}{i} = \frac{dR_B}{R_B + R_S} = \frac{i}{V_a} dR_B \tag{3.4}$$

The relative changes in the current strength decrease when the series resistance is raised; hence the stability of the arc is enhanced. However, when R_S is increased, the supply voltage V_a should be raised in order to maintain the necessary current strength. Thus the value of the series resistance is limited by the available voltage, and it must be adapted to the current strength needed.

Let us consider an example and evaluate the magnitude of the change in the current strength for an arc between pure carbon electrodes burning with a current strength of 5 amp. At a gap width of several millimetres the resistance of the arc is about 10 Ω, while random variations of it are of the order of 2 to 5 Ω. When the arc is fed by a voltage of 110 V, we have

$$\frac{\Delta i}{i} \approx 0\cdot14$$

if we consider that the change in the resistance R_B is 3 Ω on the average. The series resistance R_S is computed to be 12 Ω. When the same arc is fed from a source having a voltage of 250 V, we find

$$\frac{\Delta i}{i} \approx 0\cdot06$$

and $R_S = 40$ Ω.

Experience has demonstrated that good stability for carbon arcs is secured if the total available voltage is about twice the voltage consumed by the arc. In the spectrochemical field Belyakov-Bodin and Mandelshtam (1944) measured the reproducibility of the intensity ratio Cu 5106/Cu 5153 using a 110-V and a 220-V d.c. source. In the latter case, they obtained remarkably better results than in the former. The variability of the intensity ratio of the lines chosen in this experiment is a good criterion for the stability of excitation, since the excitation potentials of the lines, 3·80 and 6·16 eV respectively, differ by more than 2 eV, so that the intensity ratio is reasonably sensitive to temperature fluctuations (see also §§ 6.5, 6.9, 6.17, and 7.13).

For a given current strength the arc voltage rises with electrode separation (l). Fig. 3.4 gives an example of the kind of curves that are obtained when the voltage of a carbon arc, through which the current is kept constant, is plotted against the gap width. The slope of the curve is constant for arc lengths exceeding about 0·5 cm; for smaller lengths the slope increases rapidly as the arc shortens. The linear portion of the curve in Fig. 3.4 is correlated with the lengthening of the uniform column (cf. § 2.2). Owing to the linearity between voltage and length, it is assumed that the potential gradient in the column of a carbon arc is constant, i.e. independent of the arc length. Probe measurements confirm the appropriateness of this hypothesis (cf. Mason, 1948; and Aarts, 1952, p. 8).*

* The merits of probes for measuring potential distributions in the arc have been discussed by Finkelnburg and Maecker (1956, pp. 279–280), by Müller and Finkelnburg (1956), by Müller (1958), and by Ringler (1962). A general discussion of probe measurements in discharges is given by Francis (1956).

The potential gradient in the uniform column is the longitudinal or axial field strength E:

$$E = \frac{dV}{dl} \qquad (3.5)$$

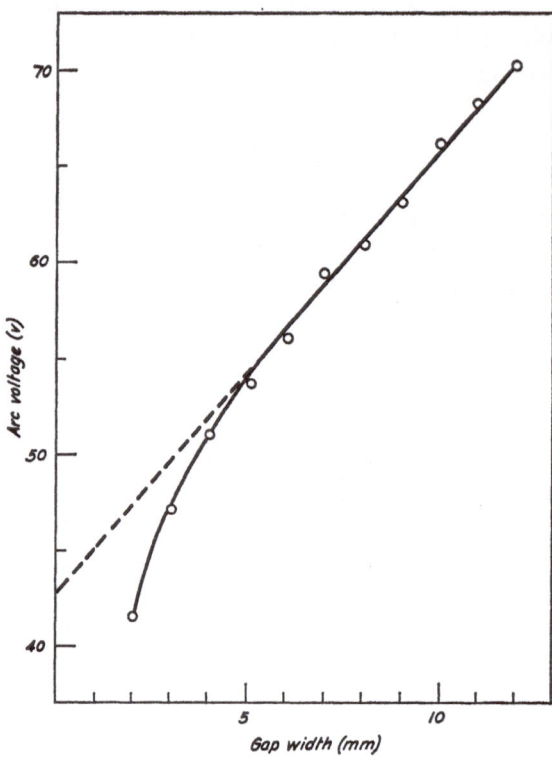

FIG. 3.4. Plot of arc voltage against gap width for an arc between pure graphite electrodes in air at atmospheric pressure.

Current strength = 10 amp	Electric field strength = 23 V/cm
Diameter of anode = 4 mm	Sum of anode fall and cathode fall = 43 V
Diameter of cathode = 4 mm	

In this work it will be briefly designated as the *field strength*. Note that this quantity is the *voltage gradient*. It is not simply calculated as the quotient of the total potential drop (V) and the electrode distance (l).

In addition to the longitudinal field, an arc possesses a radial field, which is mainly due to the (small) positive space charge in the column. The radial field counteracts the tendency of electrons and ions to separate as a result of difference in diffusion rate (see § 3.10).

It is convenient to write for the linear part of the curve in Fig. 3.4

$$V = El + V_A + V_C \qquad (3.6)$$

where V_A and V_C are potential drops occurring near the electrodes, *anode fall* and *cathode fall* respectively. These potential drops are related to the existence of space charge zones at the electrodes and to the contraction of the discharge in front of them (cf. § 2.2).

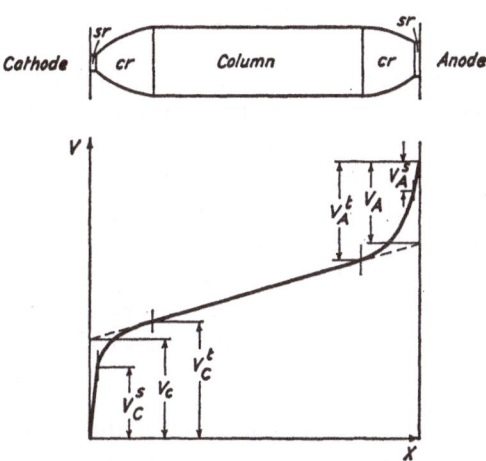

FIG. 3.5. Schematical representation of the various definitions of the cathode fall and the anode fall.

cr = contraction region sr = space charge region

From G. Ecker 'Electrode Components of the Arc Discharge' in *Ergebnisse der exakten Natur-wissenschaften*, Vol. 33, pp. 1–104 (Springer, Berlin, Göttingen, Heidelberg, 1961).

The complex and involved problem of the electrode components of the arc discharge has been recently reviewed *in extenso* by Ecker (1961) (see also Weizel and Rompe, 1949; and Finkelnburg and Maecker, 1956). Here we only note that in discharge theory the terms 'cathode and anode fall' are used in different ways, as is illustrated by Fig. 3.5. The figure shows schematically how the voltage in the arc changes as a function of the distance from the cathode surface. The potential drop in the column is given by the linear section. The curved portions in front of the cathode and anode correspond to the alteration of the field caused by contraction. The very steep voltage drops in the vicinity of the electrodes belong to the space charge zones.

The following quantities should be noted:

1 $V_C{}^t$, the total voltage over the cathode discharge region.

2 V_C, the difference of the total voltage $V_C{}^t$ and the voltage the column would require if it could continue unaltered to the cathode surface.

3 $V_C{}^s$, the cathode fall in the most restricted sense, i.e. a potential drop across the space charge zone that extends only over a distance of several times the mean free path. $V_C{}^t$ and V_C cover broader regions of the order of 1 mm.

Similar definitions apply for the anode voltage drops ($V_A{}^t$, V_A, and $V_A{}^s$).

Which of the three quantities is actually measured depends on the conditions of the experiment. For example, from the plot of the arc voltage as a function of the arc length (Fig. 3.4) we deduce the sum $V_A + V_C$ when extrapolating to the electrode distance zero.

In this sense the terms 'anode fall' and 'cathode fall' will be employed throughout this text.

The descending characteristic of the arc discharge can be expressed by the experimental equation of Ayrton (1902)

$$V = a + \frac{c}{i} + \left(b + \frac{d}{i}\right)l \tag{3.7}$$

where a, b, c, and d represent constants that are independent of i and l; their numerical values depend on the properties of the electrodes, on the pressure of the surrounding medium, and on the composition of the arc gas.

Actually equation (3.7) does not hold rigorously, so that the validity of particular experimental values of the constants is always restricted to a definite range of i and l. This, however, is not the only reason for discrepancies among the values reported in the literature. As the numerical values of a, b, c, and d depend on the properties of the arc gas, it is essential that grade, dimensions, and purity of electrodes are included when stating operating conditions. Especially in older papers, those details have often been omitted.

Various numerical results about the characteristic of the carbon arc in air have been critically summarized by Aarts (1952, p. 128 ff.), purposely in order to compare the results reported for the normal carbon arc with those obtained in arcs stabilized by an air blast. Numerical values of the field strength and the anode and cathode falls will be discussed in the next section.

We should recognize that equations (3.6) and (3.7) state that both the field strength (E) and the combined potential drops at the electrodes ($V_A + V_C$) behave similarly with respect to their dependence on the current; thus when plotting the field strength versus the current strength we also obtain a descending characteristic.

★ § 3.2 *The electric field strength*

The theoretical interest of the field strength may be summarized as follows. In the *power dissipation* (*Ei*), the quantity enters into the energy balance equations of the column (see Chapter 4). Along with the mobilities of electrons and ions, it determines the drift velocities of the charged particles in the column; thus it plays an important part when the conditions for thermal equilibrium are to be considered (see § 5.4) and in discussions regarding the transport of metal vapour through the arc (see Chapter 9).

In spectrochemical analysis the field strength is normally not a convenient parameter, because it cannot be readily measured. It is usually deduced from a plot of the total potential difference across the gap against the arc length; when this relation is determined, the current strength should be kept constant. The method applies only if the column is cylindrical (see Busz–Peuckert and Finkelnburg, 1955). An accurate measurement of the gap width may offer serious difficulties.

Unambiguous results can be obtained only if the physical state in the column, notably the composition of the arc gas, does not vary within a series of measurements. With a carbon arc in air alone, no special problems will arise, since the conditions in the column remain virtually constant as the arc length is varied. When a substance is evaporated from the lower electrode, the composition of the arc gas may change with time, so that successive observations do not apply to the same arc and the correlation between arc voltage and gap width is poor. If, however, changes of the gas composition are mainly variations of the concentration of a major constituent of the sample, the field strength remains constant, whereas the anode fall and the cathode fall change. These changes can be allowed for when the field strength is determined from a plot of the total arc voltage against the arc length (see below).

It may happen that the curve relating V and l does not become linear at all, since the field strength is not necessarily constant along the entire column when the arc contains metal vapour. An inhomogeneous distribution of added vapour tends to destroy the homogeneity of the longitudinal field. It will occur particularly if the lower electrode is the cathode.

Among the early measurements of the field strength in arcs containing metal vapour, those of Hörmann (1935) ought to be mentioned. To assure a high degree of accuracy, he projected the arc simultaneously in two mutually perpendicular directions perpendicular to the arc axis; the images were photographed along with the deflection of a voltmeter.

Suits (1939b) devised a method for directly observing the field strength in an arc. The method involves the measurement of the periodic changes the arc voltage exhibits when the arc length is varied periodically with a constant amplitude. The procedure is not entirely free from objections since the conditions of the arc gas may be affected by the high-frequency vibration of the electrode (J. A. Smit, personal communication; see also King, 1962).*

At normal current strengths (4–12 amp), the field strength in the carbon arc in air alone ranges from 20 to 40 V/cm. The sum of the anode fall and the cathode fall is of the same magnitude or slightly higher. According to Finkelnburg and Maecker (1956, p. 279) the cathode fall is about 10 V, whereas the anode fall varies between 0 and over 30 V. In the range of current strength between 0·5 and 30 amp the field strength for the arc in air is lower than that for the arc in nitrogen, which can be attributed to the formation of NO (King, 1962).

I carried out some experiments on arcs between graphite electrodes (Johnson and Matthey, graphite grade 3B) in air at a current strength of 10 amp. The arc length was estimated from an optical projection. The voltage was measured across the electrode

---

* Of investigations concerned with the field strength in carbon arcs we quote further, for air, those of Ayrton (1902), Mannkopff (1943), and Aarts (1952); for air, nitrogen, argon, and carbon dioxide, those of Kohn and Guckel (1924); for hydrogen, water vapour, helium, argon, air, nitrogen, carbon dioxide, and mercury, those of Suits (1934, 1939b); for xenon, those of Bauer and Schulz (1954); for nitrogen at various pressures, those of Somers (1954, pp. 8–22), and Somers and Smit (1956); see further Maecker (1951), and Finkelnburg and Maecker (1956).

clamps using an electronic strip-chart recorder. With electrodes having a diameter of 4 mm, burning in air alone, the arc was allowed to lengthen continuously with time, while in the recorder diagram the points were indicated where the electrode distance equalled an integral number of millimetres. With larger electrodes the rate of consumption is too small for this procedure to be practical; discontinuous variation of the gap width is more convenient then. This applies also when an alkali metal salt is evaporated into the discharge zone.

Fig. 3.4 shows a plot of the arc voltage versus the electrode separation for a graphite arc in air. Each point represents the mean of 3 to 6 single observations. For the arc in air without added material I determined the field strength E to be about 25 V/cm, independent of the electrode shapes. By contrast, the sum of the anode and the cathode fall varied markedly with the type of electrode used. For example, if an anode of 4 mm diameter was substituted by one of 6 mm diameter, an increase of $V_A + V_C$ from 46 to 56 V was found. The cathode had a diameter of 4 mm in both cases. The presence of a bore in the anode did not affect $V_A + V_C$.

When elements of low ionization potential enter into the arc, the field strength and the potential drops near the electrodes are markedly reduced. An idea of the magnitude of the effect is obtained from the following data. For a carbon arc at 10 amp with steady volatilization of NaCl from the anode, I found $E = 11$ V/cm, $V_A + V_C = 22$ V; when KCl was evaporated the values $E = 13$ V/cm and $V_A + V_C = 17$ V were obtained. In these experiments a steady burn was secured by means of 'furnace electrodes' and 'rings' (see § 6.7).

Hörmann (1935) reported somewhat lower values for the field strength in his sodium and potassium arcs, namely $E = 9$ V/cm and 7 V/cm respectively. For the sum of anode and cathode falls the values of 19 and 16 V are derived from his Fig. 5. He employed a free-burning carbon arc with a current strength of 10 amp; the metal was fed from the cathode.

Eberhagen (1955a) determined the field strength in an air-stabilized, horizontal arc fed with strontium vapour. He found it to vary, depending on the strontium content, only between 10·5 and 11·0 V/cm. The value of $V_A + V_C$, however, was strongly affected by the metal vapour concentration and changed from 26 to 69 V.

De Galan (1965a) measured the field strength in 10-amp d.c. arcs, whose temperatures were also determined (cf. § 9.4). He arced metal salt + graphite mixtures from electrodes similar to that illustrated in Fig. 7.14, which were made the anode. He observed the field strength E and the sum of the potential drops $V_A + V_C$ during the evaporation of the specimens. To that end the electrode separation was increased stepwise from 0·8 to 2·0 cm, while the arc current was kept constant at 10 amp; immediately after a run the original electrode distance of 0·8 cm was restored, after which a new run was commenced. Thus, an overall change of the arc voltage during a run could be allowed for when computing the field strength as the slope of the linear plot of voltage versus arc length. Each series yielded one value of E, and for each arcing five series could be run. Experiments were conducted with a graphite anode and a graphite cathode as well as with a graphite anode and a carbon cathode. Results are summarized in Table 3.1.

It is interesting to note that the field strength proved to be constant during the entire burn of a specimen and that results for a given specimen can be reproduced well from one arcing to another. From the results in Table 3.1 it can be seen that the field strength is somewhat enhanced when graphite is replaced by carbon. The results also demonstrate a correlation between the average temperature in the relevant arcs and the field strength. I note, however, that changes of the temperature observed during the arcing of a specimen were not associated with perceptible

TABLE 3.1

Specimen		Temperature (°K)	Field strength (V/cm)	
			Graphite anode and graphite cathode	Graphite anode and carbon cathode
KCl+C	(1+1)	—	10·6	—
KF+C	(1+1)	5100	—	10·5
LiF+C	(1+4)	5600	12·1	12·7
Al$_2$O$_3$+C	(1+3)	6000	15·0	15·9

Electric field strength and temperature in arcs with metal vapour (after de Galan, 1965a). Specimens composed of a metallic compound and graphite were arced from anodes of the type shown in Fig. 7.14 (graphite grade U-1). The cathode was either graphite grade U-1 or carbon (preform C-105-U, Ultra Carbon Corporation). The temperature was measured with the aid of the atom-atom line pair Zn 3076/3282 (see § 6.6).

changes of the field strength. On the basis of the correlation shown in Table 3.1, a temperature variation as large as 200°K, which occurred with the Al$_2$O$_3$ and KF specimens, should give rise to a change in the field strength by 1 V/cm. Such a change would have been readily detected; in fact, it was not found.

In contrast with the field strength, the arc voltage did change during the evaporation of a specimen. This change must be attributed to a variation of the sum of the anode fall and the cathode fall, which, in turn, is most probably caused by changes in the metal vapour concentration in the arc. De Galan also demonstrated that the electrical properties of the arc were not disturbed by the process of measurement, i.e. by varying the arc length from 0·8 to 2·0 cm. To that end he compared voltage-time curves calculated for $l = 1·0$ cm, from mean values of the field strength and observed values of $V_A + V_C$, with voltage-time curves recorded at a constant electrode separation of 1·0 cm. Agreement was excellent.

§ 3.3 *The arc voltage as a spectrochemical parameter*

In spectrochemistry, one has always been aware of the existence of some kind of correlation between the total voltage across the arc and the mean temperature in the column. Detailed empirical studies have been reported by Rusanov (1945),

Semenova (1946), Hegemann and Schöntag (1953), Golling (1957), Boumans 1961, 1962), Doerffel and Geyer (1964), and Mellichamp (1965).

It can be understood qualitatively that arc voltage and arc temperature are correlated; for, if the current is kept constant, the field strength and the potential drops near the electrodes are reduced when a substance of low ionization potential enters into the arc. The power dissipated decreases and hence a fall of temperature is to be expected. This explanation is only a rough approximation, since the change of the thermal conductivity of the arc gas with temperature is not allowed for; moreover, a constant diameter of the discharge is postulated. Nevertheless, the picture is suitable for qualitative considerations (see also § 3.9).

A quantitative theoretical approach to the problem is involved and reaches far beyond the scope of the spectrochemical field. After all, a theoretical treatment requires the mechanisms of the column and of the regions of the anode and cathode falls to be examined separately (see Chapter 4, where a brief introduction on arc theories, particularly the theory of the energy balance of the column, is given). For spectrochemical purposes it is less troublesome and more promising if an empirical approach to the problem is made. Then, by fixing the operating conditions (current strength, electrode separation, polarity, grade and shape of electrodes) we make the arc voltage depend entirely upon the plasma composition in the discharge zone. So, the voltage may either provide qualitative information on the passage of material into the arc, or be related quantitatively with an excitation parameter, e.g. the temperature, which also depends on the plasma composition.

In a detailed study that I made of electron pressure and temperature in carbon arcs containing various metal vapours, spectra for optical measurements were taken at the same time as the voltage was recorded. Profitably, I postpone a thorough discussion of the results and their interpretation until § 7.14. For the present, suffice it to report that a non-linear relation between the median values of the voltage and the temperature was established. For the low temperature range, i.e. in arcs containing alkali metal vapour, the slope of the curve $V = f(T)$ is markedly reduced in comparison with the value found for the high temperature range (see Fig. 7.28).

On the whole, as will be considered more extensively in § 7.14, we must exercise some caution and avoid interpreting small variations of the arc voltage, say of the order of ± 5 V, in terms of temperature variations. In my experience, temperature often turns out to be constant, although the gap voltage changes. To understand this paradox we must clearly bear in mind that the arc voltage is the resultant of contributions by the uniform column (field strength times arc length) and the sum of anode and cathode falls [see equation (3.6)]. A change of the vapour composition tends to affect primarily the excitation conditions and the potential drops in the vicinity of the electrodes, while leaving the uniform column and the field strength unaffected. Thus, if spectroscopic observation is restricted to the column, changes in the excitation conditions are not disclosed, even though the arc voltage may vary considerably.

Conversely, I found also substantial changes of the electron pressure at constant voltage and temperature (see § 7.14). Evidently, constant voltage does not fully establish invariant excitation conditions. Yet, I believe the voltage to be a very useful expedient in spectrochemical analysis. When it is used for monitoring the excitation conditions and for following the entry of material into the discharge zone, the worker's personal knowledge and experience should be the guiding principle (see further § 7.14).

It must be realized that empirical relations between voltage and temperature hold quantitatively only for closely defined operating conditions, among which current strength, electrode shapes, gap width, and the grade of the carbon or graphite have to be specified.*

The positive slope of the voltage-temperature curves at present under consideration does not imply that long arcs are hotter than short ones. Indeed, arc voltage also increases with electrode separation; this rise, however, is related to the extension of the homogeneous column: every millimetre added to its length requires a corresponding increase in the total power input; the power dissipation per millimetre, which determines the temperature, remains virtually unchanged. When the arc is shortened below the critical value, the field strength increases sharply; therefore, the temperatures found in very short arcs will be higher than in long ones.

§ 3.4   *Electrical conductivity, electron mean free path, electron mobility, and effective cross-section of neutral particles for elastic collisions with electrons*

The current flow in the arc per unit area of cross-section, i.e. the *current density* j, is given by

$$j(r) = \sigma(r)E \qquad (3.8)$$

where σ is the *electrical conductivity* and E the longitudinal field strength; the functions $j(r)$ and $\sigma(r)$ express the radial dependence of j and σ upon the distance (r) to the arc axis. Numerical values of j are expressed as amp/cm^2, of σ as ohm^{-1}/cm, and of E as V/cm.

Owing to their small mass and consequently high mobility, electrons carry more than 99 per cent of the total current in the arc; therefore electrical conductivity can be written as

$$\sigma(r) = e\mu_e n_e(r) \qquad (3.9)$$

where e is the electronic charge ($= 1.6 . 10^{-19}$ C), n_e the electron concentration (number per cm^3), and μ_e a quantity called *mobility*. It requires ample discussion.†

---

* The influence of the gap width is easily understood, provided it can be assumed that the field strength is independent of the arc length; thus the arc length should exceed the minimum value for the homogeneous column to develop; moreover, the metal vapour density should be independent of the gap width. If those conditions are fulfilled, an increase of the electrode separation will only cause shifts of the individual voltage-temperature points to higher voltages. The numerical values of the shifts will increase gradually with the voltage.

† See also, for example, von Engel (1955, Chapter 4), Unsöld (1955, pp. 594–595), Weizel and Rompe (1949, p. 10 ff.), and Compton and Langmuir (1930).

Let us consider a partially ionized gas in thermal equilibrium, which means (see Chapter 5) that the average kinetic energy of all kinds of particles, i.e. molecules, atoms, ions, and electrons, is the same, while for each separate kind of particle the velocities are equally distributed in all directions about an average value. If a uniform electric field is applied, the swarm of electrons in the gas will be bodily moved in the direction of the field. The average speed with which the centre of the swarm migrates in the field direction is called *drift velocity*. The drift velocity of electrons in a field of unit strength is the *electron mobility* (μ_e), which is expressed as cm/sec per V/cm or as cm^2 V^{-1} sec^{-1}.

Thermal equilibrium also implies that the electron concentration is maintained at a constant level by thermal ionization (see § 3.8 and Chapter 7); consequently, there will be a continuous stream of electrons passing through the gas. Ions will travel in the opposite direction, their drift velocity being much lower, however.

If the field is relatively weak, that is when the drift velocity of the electrons is well below the average random velocity, the action of the field will not interfere with the state of thermal equilibrium: an electron, having moved along a path whose average length equals a mean free path, collides with a gas atom or molecule and rebounds at random. The energy gained in the field direction is transferred for the greater part to gas atoms and molecules, whence the drift velocity of electrons after an impact can be regarded as zero and their velocity distribution is hardly affected by the action of the field.

The picture given above applies to the column of the carbon arc in air at atmospheric pressure. The conditions of the gas mixture in this column can be considered as a state of thermal equilibrium (see Chapter 5). The high temperature is produced and maintained by a continuous flow of electrons that energizes the gas. The processes near the electrodes are more complex and involved (see, e.g. Ecker, 1961; and Finkelnburg and Maecker, 1956).

In the carbon arc in air, electron mobility is of the order of $3 . 10^4$ cm^2 V^{-1} sec^{-1} at 6000°K and 1 atm (see § 3.6). Assuming a field strength of 25 V/cm (see § 3.2) we find the drift velocity to be $7 \cdot 5 . 10^5$ cm/sec. The average random velocity at 6000°K evaluated as the root mean square velocity

$$\bar{v}_e = \sqrt{\left(\frac{3kT}{m_e}\right)} \tag{3.10}$$

is $5 \cdot 2 . 10^7$ cm/sec; evidently this value is well above the drift velocity.

According to the kinetic theory of gases (see for example Jeans, 1925; and Kennard, 1938; cf. Maecker, 1951) electron mobility is given by the relation

$$\mu_e = \alpha \frac{e\lambda_e}{m_e \bar{v}_e} \tag{3.11}$$

or, in virtue of (3.10), by

$$\mu_e = \alpha \frac{e\lambda_e}{\sqrt{(3kTm_e)}} \tag{3.12}$$

In equations (3.10) to (3.12), λ_e is the mean free path for electrons, m_e the electronic mass, e the electronic charge, k the Boltzmann constant, and T the absolute temperature. The constant α is an uncertainty factor of unit order of magnitude ($0.75 < \alpha < 1.38$). Its numerical value depends on the method of averaging. The main difficulty in estimating μ_e is not the numerical value of α, but the proper value of λ_e.

For a single gas the mean free path for electrons can be written as

$$\lambda_e = \frac{1}{n^0 Q^0 + n_i Q_i} \qquad (3.13)$$

where n is the particle density or concentration (number per cm^3) and Q the *effective cross-section* (cm^2) for elastic collisions with electrons; the labels o and i refer to neutral and singly-ionized particles respectively.

For a gas mixture $n^0 Q^0$ should be replaced by

$$\sum n_k{}^0 Q_k{}^0 = n_1{}^0 Q_1{}^0 + n_2{}^0 Q_2{}^0 + \ldots n_j{}^0 Q_j{}^0 \qquad (3.14)$$

the subscripts $1, 2, \ldots j$, denoting the various gas components. A similar expression should replace $n_i Q_i$ if higher ionization stages are to be taken into account. By application of (3.13) and (3.14) the problem of evaluating μ_e has been reduced to the determination of effective cross-sections.*

Calculating the mean free path λ_e according to (3.13) and (3.14) for a gas mixture of known composition would be a very simple problem if all the cross-sections, $Q_1{}^0, Q_2{}^0, \ldots Q_j{}^0$, and Q_i, were known; however, in assigning appropriate values many difficulties are met with, the most serious one being the dependence of cross-sections on electron energy; this phenomenon is commonly known as the Ramsauer effect.† In the energy range of the normal low-current arc,‡ available data indicate

---

* If two particles approach each other closely enough, interaction leading to energy transfer may occur. The simplest form of interaction is the exchange of kinetic energy (elastic collisions, elastic scattering). Kinetic energy can also be converted into excitation energy or ionization or dissociation energy (collisions of the first kind). Conversely, excitation energy from one particle can be passed to another and can thus reappear as kinetic or another form of energy (collisions of the second kind). The distance to which two particles must approach each other for a particular process to occur depends in general on the type of process under consideration. The probability that a projectile particle travelling a distance dx collides with a target particle so that a particular process takes place is

$$P\,dx = \frac{\text{target area}}{\text{total area}} = \frac{nQ\,A dx}{A} = nQ\,dx \qquad (1)$$

The probability that the projectile particle covering a distance of 1 cm makes a collision effecting the relevant process is therefore

$$P = nQ \qquad (2)$$

P is the total target area or cross-section per unit volume; it is expressed as cm^2/cm^3 or as cm^{-1}. The quantity Q is the effective cross-section of *one particle* for the particular process; it is expressed as cm^2.

† See, for example, Kollath (1930), Compton and Langmuir (1930, p. 221 ff.), Landolt–Börnstein (1950, pp. 327–328), and von Engel (1955, Chapter 3).

‡ At 5000°K the thermal energy $\frac{3}{2}kT$ ($= 1.035 \cdot 10^{-12}$ ergs) corresponds to $1.035 \cdot 10^{-12} \times 6.2422 \cdot 10^{11} = 0.646$ eV.

reasonably good agreement between Ramsauer cross-sections and the so-called kinetic theory cross-sections for elastic scattering of electrons (see § 3.5).

The kinetic theory cross-section of gas molecules for elastic encounters with electrons is related to the kinetic theory 'radius' of the relevant gas molecules by

$$Q = \pi \rho^2 \tag{3.15}$$

In the literature, values are given for the kinetic theory diameter (2ρ) or for the mean free path (λ) of molecules and atoms in their own gas.* The mean free path λ is related to λ_e (see, for example, Compton and Langmuir, 1930, p.207 ff.; and Jeans, 1925) by

$$\lambda_e = 4\lambda\sqrt{2} \tag{3.16}$$

where $\lambda_e = 1/nQ$.

The dependence of cross-sections on temperature is roughly expressed by the relation

$$Q_T{}^2 = Q_\infty{}^2\left(1 + \frac{C}{T}\right) \tag{3.17}$$

where C is the Sutherland constant in degrees Kelvin, Q_T the cross-section at temperature T, and Q_∞ the cross-section at very high temperature. For most gases C is in the range 100–1000°K (see Landolt–Börnstein, 1950).

§ 3.5 *Effective cross-sections of neutral particles for elastic collisions with electrons: Numerical data and calculations*

For instructional purposes we shall now consider some numerical data, which at the same time may show the order of magnitude of the various quantities involved. After Landolt–Börnstein, for molecular nitrogen we have $2\rho_\infty = 3\cdot22.10^{-8}$ cm. Since the Sutherland constant $C = 105°$K, we can ignore the Sutherland correction [equation (3.17)] in the temperature range at present under consideration, that is 4000–6000°K. Thus the elastic collision cross-section $(Q = \pi\rho^2)$ in this range is found to be $8\cdot2.10^{-16}$ cm^2.

The total cross-section per cm^3 (nQ) depends on temperature and pressure, as it contains the *particle density n*. We have

$$p = \frac{RT}{V} \equiv \frac{NkT}{V} \equiv nkT \tag{3.18}$$

where p is the pressure, T the absolute temperature, V the volume of a gram molecule, and n the particle density or concentration; the constants R, N, and k are respectively the gas constant, Avogadro's number, and Boltzmann's constant.

---

* Absolute values of λ quoted in the literature vary to some extent according to the method of measurement (see, for example, Jeans, 1925, pp. 326–333; and von Engel, 1955, pp. 26). Landolt–Börnstein's tables (1950, pp. 325 and 369) list kinetic theory diameters (2ρ) for various atomic and molecular gases; the values are based on viscosity measurements and critical data.

Inserting numerical values for the constants and rearranging (3.18) we obtain

$$n = 7 \cdot 340 \cdot 10^{21} \frac{p}{T} \qquad (3.19)$$

if n is expressed as number per cm^3, p as atm, and T as °K. Table 3.2 shows numerical values of n at various pressures and temperatures. A more extended conversion table has been added as an appendix.

TABLE 3.2

T(°K)	p	n(cm^{-3})
273	1 mm Hg	$3 \cdot 54 \cdot 10^{16}$
298	1 mm Hg	$3 \cdot 24 \cdot 10^{16}$
273	1 atm	$2 \cdot 69 \cdot 10^{19}$
298	1 atm	$2 \cdot 46 \cdot 10^{19}$
3000	1 atm	$2 \cdot 45 \cdot 10^{18}$
4000	1 atm	$1 \cdot 84 \cdot 10^{18}$
5000	1 atm	$1 \cdot 47 \cdot 10^{18}$
5500	1 atm	$1 \cdot 33 \cdot 10^{18}$
6000	1 atm	$1 \cdot 22 \cdot 10^{18}$
6500	1 atm	$1 \cdot 13 \cdot 10^{18}$

Particle density or concentration (n) at various temperatures and pressures.

Table 3.3. summarizes kinetic theory values of Q, nQ, λ, and λ_e for nitrogen at various temperatures and $p = 1$ atm. The data have been calculated formally with the use of equations (3.15) and (3.16). We have thus neglected the dissociation of nitrogen, the Ramsauer effect and the influence of ionization (see below and § 3.6).* The data in the table give an idea of the magnitude of the quantities as calculated according to kinetic theory. The order of magnitude is in good agreement with experimental and theoretical values of Ramsauer cross-sections for elastic scattering of electrons, if the scarce and uncertain data in the range of electron energies below 1 eV warrant such a statement.†

In view of various uncertainties it appears not necessary to take the dissociation of nitrogen into account, nor to allow for the presence of other components (O_2, C_2, CO, CN, NO, O, and C), since Landolt–Börnstein's data for O_2, CO, and NO suggest no essential differences for those gases in comparison with nitrogen. Alkali metal vapours, however, might be of importance, as their Ramsauer cross-sections (Q) are of the order of $200 \cdot 10^{-16}$–$400 \cdot 10^{-16}$ cm^2 in the low electron

* It should be realized that the variation of nQ, λ and λ_e (Table 3.3) is merely due to the variation of the density n with temperature. When considering literature data on cross-sections we must be aware of the distinction made between cross-sections of one particle (Q) and cross-sections per unit volume (nQ). Q-values depend only slightly on temperature [equation (3.17)] in contrast with nQ-values. Literature data of the latter usually refer to a pressure of 1 mm Hg and $T = 273$°K or $T = 298$°K. To adapt these data to the temperature range of the normal carbon arc (say 5000–6000°K) and a pressure of 1 atm a factor of roughly 40 applies.

† See Kollath (1930), Fisq (1936, 1937), Landolt–Börnstein (1950, p. 374), and Köstlin (1964).

range.* In the type of arc at present under consideration, alkali metal vapour density may become of the order of $5 . 10^{15}$ per cm^3 (see § 7.9), thus making nQ of the order of 100–200 cm^{-1}, which contribution is not entirely negligible (cf. Table 3.3).

TABLE 3.3

$T(°K)$	$Q(cm^2)$	$nQ\,(cm^{-1})$	$\lambda(cm)$	$\lambda_e(cm)$
4000	$8 \cdot 2 . 10^{-16}$	1500	$1 \cdot 18 . 10^{-4}$	$0 \cdot 67 . 10^{-3}$
5000	$8 \cdot 2 . 10^{-16}$	1200	$1 \cdot 48 . 10^{-4}$	$0 \cdot 84 . 10^{-3}$
6000	$8 \cdot 2 . 10^{-16}$	1000	$1 \cdot 77 . 10^{-4}$	$1 \cdot 00 . 10^{-3}$
6500	$8 \cdot 2 . 10^{-16}$	925	$1 \cdot 92 . 10^{-4}$	$1 \cdot 09 . 10^{-3}$

Kinetic theory values for molecular nitrogen. Q is the effective cross-section per particle, nQ the effective cross-section per cm^3, λ the mean free path, and λ_e the mean free path for elastic collisions with electrons. The pressure is 1 atm.

The data discussed hitherto are all based on measurements at low pressure and temperature, from which extrapolations to the temperature range of the arc have been made. Maecker (1950, 1951) reversed the problem and calculated effective cross-sections for air particles from measurements in the arc, assuming the validity of the Elenbaas–Heller equation (see Chapter 4). Later investigations of Maecker, Peters, and Schenk (1955) in a high-current arc with axial temperature $T = 11\,000°K$ yielded the value $Q = 20 . 10^{-16}\,cm^2$ for atomic C, O, and N. According to a private communication by Maecker (see also Maecker, 1962) the value $Q = 5 . 10^{-16}\,cm^2$ with an uncertainty of a factor of about 2 has been deduced for nitrogen from measurements in a cylindrical cascade arc in nitrogen.† Altogether it will not be too harmful if the kinetic theory values summarized in Table 3.3 are regarded as being near the truth.

§ 3.6   *Effective cross-sections of charged particles for elastic collisions with electrons: Numerical data for the electron mobility*

In the preceding sections we have discussed the contribution of neutral particles to the mean free path for electrons. We shall now investigate the second term in the denominator of equation (3.13), namely the contribution of ions to the mean free path. Owing to Coulomb interaction, effective cross-sections of ions for electron encounters are large: Q_i may exceed Q^0 by several orders of magnitude; therefore a theoretical treatment has to discern between ions of difference charge,

* See Brode (1929), Kollath (1930, p. 1003), Landolt–Börnstein (1950, p. 327), Mirlin, Pikus, and Yur'ev (1962), Perel, Englander, and Bederson (1962), and Garrett and Mann (1964).

† Applications of the wall-stabilized cascade arc have been reported by Maecker (1956, 1960, 1962), Burhorn (1959), and Riemann (1963, 1964a and b, 1965).

while difference in 'size' can be ignored. The temperature range to be considered here is such that we can restrict our discussion to singly-ionized particles.*

The cross-section of singly-charged ions for electron encounters (Gvosdover cross-section) can be written as

$$Q_i = \frac{1}{\gamma} \frac{4\pi}{9} \frac{e^4}{(kT)^2} \qquad (3.20)$$

where e, k, and T are the electronic charge ($= 4 \cdot 8 . 10^{-10}$ e.s.u.), the Boltzmann constant, and the absolute temperature respectively. The quantity γ is a function of unit order of magnitude.

The basic form of equation (3.20) is understood roughly if we consider an electron passing a singly-charged ion at a distance ρ_i; the electron will experience a substantial deflection when its potential energy e^2/ρ_i becomes of the order of its average kinetic energy, viz.

$$\frac{e^2}{\rho_i} \approx \frac{3}{2} kT \qquad (3.21)$$

The collision parameter ρ_i defines the effective cross-section according to $Q_i = \pi \rho_i^2$; thus we obtain an equation of the form of (3.20).

Maecker, Peters, and Schenk (1955) concluded from experiments that the following expression is appropriate in gas discharges

$$Q_i = \frac{e^4}{(kT)^2} \ln \frac{kT}{e^2 n_i^{\frac{1}{3}}} \qquad (3.22)$$

where n_i is the ion concentration; in the arc n_i can be put equal to the electron concentration n_e (see below and § 3.8). For singly-charged ions equation (3.22) is virtually equal to the theoretical one, if the proper function for γ is inserted. Numerical values of Q_i according to (3.22) are about half as large as those calculated with the original Gvosdover formula.

Inserting the numerical values for the constants in equation (3.22) we obtain

$$Q_i = \frac{6429}{T^2} . 10^{-9} \log 598 \frac{T}{n_e^{\frac{1}{3}}} \qquad (3.23)$$

where n_i has been replaced by n_e (expressed in cm^{-3}).

To obtain an idea about the contribution of Q_i to electron mobility in carbon arcs with metal vapour, I made some calculations based on the results of an investigation I carried out to study the influences of metal vapours on the temperature and the electron concentration in the arc. I shall describe it in detail in §§ 7.7 to

---

* Gvosdover (1937) was the first to consider the problem. Since then Schulz (1947a), Cohen, Spitzer, and Routly (1950), Spitzer and Härm (1953), and Maecker, Peters, and Schenk (1955) have written papers about Gvosdover cross-sections. The most important features of the theory have been summarized by Unsöld (1955, pp. 594–598), and by Finkelnburg and Maecker (1956, pp. 348–349, 378–380).

3

TABLE 3.4

T	n^0	n_e	Q_i	n_iQ_i	n^0Q^0	$1/\lambda_e$	λ_e	μ_e	$(\mu_e)_{\text{kin.th.}}$
°K	cm⁻³ ×10¹⁸	cm⁻³ ×10¹⁵	cm² ×10⁻¹⁴	cm⁻¹	cm⁻¹	cm⁻¹	cm ×10⁻⁴	cm²V⁻¹sec⁻¹ ×10⁴	cm²V⁻¹sec⁻¹ ×10⁴
5200	1·41	3·8₅	30·9	1180	1155	2335	4·3	1·5₅	3·1₅
5400	1·36	1·9₅	31·1	615	1115	1730	5·8	2·0₅	3·2
5600	1·31	1·0₅	31·1	325	1075	1400	7·1	2·5	3·2₅
5800	1·26₅	0·8₅	29·9	250	1040	1290	7·8	2·6₅	3·3
6000	1·22₅	0·7	28·7	200	1005	1205	8·3	2·8	3·3₅
6200	1·18₅	0·6	27·5	165	970	1135	8·8	2·9	3·4

Calculation of the electron mobility in a metal vapour graphite arc for which a definite relationship (Fig. 7.6) between the electron concentration and the temperature was established.

T = absolute temperature, n^0 = concentration of neutral particles ($\approx$ total concentration), n_e = electron concentration ($= n_i$ = concentration of singly-charged ions), Q_i = effective cross-section of singly-charged ions for elastic electron encounters = Gvosdover cross-section calculated according to equation (3.23), n_iQ_i = effective cross-section per cm³ of singly-charged ions for elastic electron encounters, n^0Q^0 = effective cross-section per cm³ of neutral particles for elastic electron encounters, $1/\lambda_e = n^0Q^0 + n_iQ_i$ [cf. equation (3.13)], λ_e = mean free path of electron: $1/\lambda_e = n^0Q^0 + n_iQ_i$ [cf. equation (3.13)], λ_e = reciprocal mean free path of electron, μ_e = electron mobility calculated with equation (3.24) from the λ_e values in the preceding column, $(\mu_e)_{\text{kin.th.}}$ = electron mobility calculated with equation (3.24) from $\lambda_e = 1/n^0Q^0$.

7.10. I note here that the variations of the excitation parameters (temperature, electron concentration, cross-sectional area of the column, voltage drop, and field strength) were produced by the introduction of the metal. Under those conditions a negative correlation between the (effective) electron pressure p_e and the (effective) temperature was found (see particularly Fig. 7.6). Although this correlation is not generally valid, it can serve as a convenient basis for examining the extent to which changes of the electron concentration, such as disclosed in an actual experiment, may affect the electron mobility.

Results of the calculation of Q_i, λ_e, and μ_e are summarized in a self-explanatory way in Table 3.4. For Q^0, the kinetic theory cross-section, $8 \cdot 2 \cdot 10^{-16}$ cm^2, according to § 3.5 has been taken. The constant α in equation (3.12) for computing μ_e was put equal to unity; the practical form of this equation then reads

$$\mu_e = 2 \cdot 61 \cdot 10^9 \frac{\lambda_e}{\sqrt{T}} \qquad (3.24)$$

The data in Table 3.4 show that Gvosdover cross-sections play an essential role in the type of arc at present under consideration and also in all types that bear a close resemblance to it.

★ § 3.7 *Ion mobility and coefficient of diffusion in the arc*

For spectrochemical inference, particularly for exposing the mechanism of transport of metal vapour through the arc, knowledge of numerical values of *ion mobility* is essential (Chapter 9). Theoretical considerations on ion mobility as well as experimental data (see, e.g. von Engel, 1955; and Meek and Craggs, 1953) always refer to gases at *low* pressures and temperatures, thus to conditions that differ markedly from those at present under consideration, i.e. a pressure of 1 atm and temperatures in the range 4000–7000°K. In the field of high-temperature discharges, electron mobility has attracted more attention than ion mobility; for more than 99 per cent of the current is carried by the electrons, owing to their small mass and consequently high mobility; as a result ions do not contribute perceptibly to the electrical conductivity, which reduces the interest of ion mobility in any physical arc theory. By contrast, in spectrochemical physics we are primarily interested in ion mobility, in view of the transport of metal vapours, and only secondarily in electron mobility.

We define ion mobility (μ_i), as we did electron mobility (§ 3.4), to be the drift velocity of ions in a field of unit strength. We express it as cm/sec per V/cm or as cm^2 V^{-1}sec^{-1}. To arrive at plausible numerical values for μ_i, valid at the conditions prevailing in the arc plasma, the safest way, in my opinion, is to relate μ_i with the coefficient of diffusion, by the well-known kinetic theory formula*

$$\frac{\mu_i}{D_i} = \frac{e}{kT} \qquad (3.25)$$

* See, for example, Kennard (1938) and von Engel (1955, p. 121).

where e is the electronic charge ($= 1 \cdot 6 . 10^{-19}$ C), k the Boltzmann constant ($= 1 \cdot 37 . 10^{-23}$ J°K^{-1}), and T the absolute temperature. Inserting the numerical values of the constants we have the practical form

$$\frac{\mu_i}{D_i} = \frac{1 \cdot 17 . 10^4}{T} \qquad (3.26)$$

where μ_i is expressed as cm^2 V^{-1} sec^{-1} and D_i as cm^2/sec. Thus, $\mu_i = 2 D_i$ at $T = 5850$°K.

According to the rigid-sphere model of simple kinetic theory, the coefficient of self-diffusion, that is the interdiffusion of particles of the same mass and size, is given by (cf. Hirschfelder, Curtiss, and Bird, 1954):

$$D = 2 \cdot 63 . 10^{-3} T^{\frac{3}{2}} \frac{\sqrt{M}}{P \rho^2} \qquad (3.27)$$

where T is the absolute temperature (in °K), M the molecular weight (in atomic units), P the gaseous pressure (in atm), ρ the molecular diameter (in Å), and D the diffusion coefficient (in cm^2/sec).

Similarly, for a mixture of two chemical species we have the binary diffusion coefficient

$$D_{12} = 2 \cdot 63 . 10^{-3} T^{\frac{3}{2}} \frac{\sqrt{[(M_1 + M_2)/2 M_1 M_2]}}{P \rho_{12}^2} \qquad (3.28)$$

with $\rho_{12} = \frac{1}{2}(\rho_1 + \rho_2)$.

Now we must ask to what extent equations (3.27) and (3.28) remain valid in the high-temperature region of the arc. As we conclude from a review by Liley (1959), little experimental work has been reported for even moderately elevated temperatures. Recently, Walker and Westenberg* measured molecular diffusion in the temperature range 300–1100°K, using a point-source technique (cf. § 9.5). In addition, from their experimental results they calculated diffusion coefficients for temperatures beyond the experimental region explored. As a guide, some of Walker and Westenberg's results have been summarized in Table 3.5. Predicted values for the high-temperature diffusion coefficient of He in N_2 have been checked experimentally up to about 1700°K by Ember, Ferron, and Wohl (1962, 1964).

Let us attempt to make some extrapolations to the conditions in the arc plasma so that we can estimate the order of magnitude of diffusion coefficients of metal atoms and ions. Considering the value $D_{12} = 13 \cdot 5$ cm^2/sec for diffusion of CH_4 into O_2 at 3000°K, we find $D_{12} = 33 \cdot 9$ cm^2/sec at 5000°K and $D_{12} = 47 \cdot 0$ cm^2/sec at 6000°K, when, in accordance with the results by Walker and Westenberg for He—A and He—N_2, proportionality of D_{12} with $T^{1 \cdot 8}$ is assumed. We correct these values for the difference in molecular weight of the arc plasma and oxygen. Computing for the partly dissociated plasma, consisting chiefly of N_2, N, O_2, O, and

---

* See Walker and Westenberg (1958a and b, 1959, 1960), Westenberg and Walker (1959), and Walker, Monchick, Westenberg, and Favin (1961).

CO (§ 11.1), the mean molecular weight of 24, we obtain $D_{12} = 35.7$ cm^2/sec at 5000°K and $D_{12} = 49.5$ cm^2/sec at 6000°K, if, in agreement with (3.28), D_{12} is supposed to be proportional to $\sqrt{(1/M_1 + 1/M_2)}$, and variations of the molecular diameter ρ_{12} are ignored.

Fig. 3.6 depicts the dependence of D_{12} upon M_1 according to (3.28) if $M_2 = 24$ and the molecular diameter ρ_{12} is supposed to be constant. When M_1 is in the range 16 to 50, the graph may be expected to represent the right magnitude of D_{12}. It is difficult, however, to predict how large departures will be outside this region.

TABLE 3.5

Gas pair	D_{12} $T = 1000°K$	D_{12} $T = 3000°K$	Gas pair	D_{12} $T = 1000°K$	D_{12} $T = 3000°K$
He—N$_2$	5.67	35.8	CH$_4$—O$_2$	1.98	13.5
CO$_2$—N$_2$	1.46	9.6	H$_2$—O$_2$	7.02	48.4
He—A	6.23	41.0	CO—O$_2$	1.75	11.2
CO$_2$—O$_2$	1.43	9.9	H$_2$O—O$_2$	2.21	13.8

Binary diffusion coefficient D_{12} (in cm^2/sec) for various gas pairs according to Walker, Monchick, Westenberg, and Favin (1961) (see also Walker and Westenberg, 1960).

The original paper lists 4 to 6 values for each system according to the potential energy function used for predicting them. In the present table the averages of four values are given.

To summarize, in the arc plasma at $T = 5000°K$, the coefficient of diffusion (D_{12}) of atoms having an atomic weight between 16 and 50 can be assumed to vary from 36 to 27 cm^2/sec, and at $T = 6000°K$ from 49 to 38 cm^2/sec. For heavier atoms, D_{12} can be expected to decrease only slightly more, namely for the heaviest atom to 24 cm^2/sec at 5000°K and to 33 cm^2/sec at 6000°K, if solely a change of mass is taken into consideration. For lighter atoms, D_{12} is likely to increase more rapidly. Considering the mere effect of mass we would predict for lithium atoms, a value of 48 cm^2/sec at 5000°K and 66 cm^2/sec at 6000°K.

Numerical values of diffusion coefficients thus derived from Walker and Westenberg's data appear to be somewhat higher than those deduced from the data of older literature. Among the early measurements of diffusion coefficients in flames we mention those by Wilson (1912), Davis (1924), Symon (1925), Ginsel (1933), and Ginsel and Ornstein (1933). Results for lithium, sodium, and potassium have been summarized and discussed by Vendrik (1949, p. 33). Van der Held and Miesowicz (1937) concluded on theoretical and experimental grounds that Ginsel's value for sodium, $D_{12} = 3.2$ cm^2/sec at 2120°K, was the most reliable. From this value we compute on the basis of proportionality with $T^{1.8}$ a value of 15 cm^2/sec at 5000°K, which is lower by a factor of 2 than the result derived from Walker and Westenberg's data.

A similar conclusion applies to the estimates of diffusion coefficients made by Raikh-baum and Malykh (1960) from mean times spent by atoms in the arc. Their model, involving a cylindrical diffusion zone with diffusion in the radial direction only, is an oversimplified one (cf. Chapter 9).

Somewhat better agreement exists with diffusion coefficients calculated by Gol'dfarb and Il'ina (1961) (see also Il'ina and Gol'dfarb, 1962) according to formulae given by Arnold (1930). They compute, for instance, the diffusion coefficient for lithium to be .42 cm²/sec in air at 5000°K and normal pressure; for sodium they find $D_{12} = 23$ cm²/sec and for barium $D_{12} = 12$ cm²/sec. These values seem to have been checked by measurements of mean transit times in the arc, although here too the model underlying the calculations

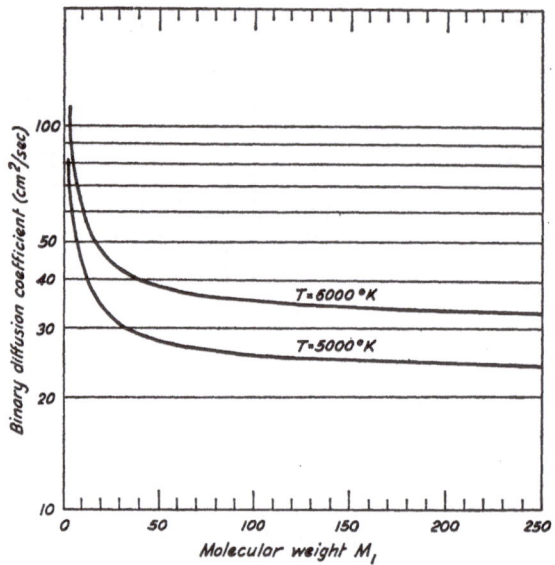

FIG. 3.6. Binary diffusion coefficient (cm²/sec) plotted against molecular weight M_1 of diffusing species. The gas into which diffusion takes place is assigned the molecular weight $M_2 = 24$. This value agrees closely with the mean molecular weight of the arc plasma in the temperature range 5000–6000°K. The curves represent the relationship

$$D_{12} :: \sqrt{\left(\frac{M_1 + M_2}{2M_1 M_2}\right)}$$

Changes of the molecular diameter of species one according to equation (3.28) are not considered.

Basic values of the plots are $D_{12} = 33\cdot9$ cm²/sec at 5000°K and $D_{12} = 47\cdot0$ cm²/sec at 6000°K, both for $M_1 = 16$. These values have been derived from data by Walker and Westenberg (see text).

(spherical diffusion from a point source, neglecting the influence of the electrical field) is the weak point. The values of D_{12} computed by Gol'dfarb and Il'ina are seen to be lower than those derived from Walker and Westenberg's data, viz. 48, 32, and 24 cm²/sec for lithium, sodium, and barium, respectively. Particularly for barium the discrepancy is large.

The high values for the diffusion coefficient according to Westenberg and Walker agree well with results for sodium that Snelleman (1965, p. 102) obtained from measurements in the flame, namely $D = 9\cdot9$ cm²/sec at 2440°K. From this value we compute $D = 36$ cm²/sec at 5000°K, if again we assume that D increases proportional with $T^{1\cdot8}$.

For further calculations we shall consider the values derived from Walker and Westenberg's data to be the most reliable at the moment. Accordingly, we take for a species with an atomic weight of 50 the values $D = 27.5$ cm²/sec at 5000°K and 38 cm²/sec at 6000°K (see Fig. 3.6), which values are for neutral atoms.

To find the order of magnitude of ion mobility μ_i in the arc, we suppose the diffusion coefficients of atoms (D_a) and singly-charged ions (D_i) to be equal.* Using equation (3.26) we obtain $\mu_i \approx 64$ cm² V⁻¹ sec⁻¹ at 5000° K and $\mu_i \approx 74$ cm² V⁻¹ sec⁻¹ at 6000°K, so that a value of 70 would represent the order of magnitude of μ_i for the temperature range under consideration. For light ions we must expect higher values; so, considering the mere effect of mass (Fig. 3.6), we compute the mobility of lithium ions to be enhanced by a factor of 1.7 over the value of 70 mentioned previously. For heavy ions, mass is not likely to entail appreciable changes.

§ 3.8 *Electron density, column radius, electron pressure, and total degree of ionization in the arc column*

In equation (3.9), which defines the electrical conductivity, we introduced the quantity *electron density* or electron concentration n_e, i.e. the number of electrons per cm³ in the discharge column. Although electron density is essentially connected with electrical conductivity, we are primarily concerned with it in spectrochemistry because it is a vital parameter for dealing with ionization equilibria and relative intensities of atom and ion lines. So, in this text, we shall often refer to it, particularly in Chapters 7 and 8. It was only recently that the part played by electron density in excitation theory was recognized in spectrochemistry, although the literature concerning the physical aspects of the subject has used the concept ever since the arc mechanism has been considered as thermal (Compton, 1923; see Chapter 5).

In the arc column, net space charge is negligible compared with the total charge of the free electrons in the plasma. The main reason for this is that the charge carrier density is so high that extremely large forces would be needed to pull out even a small fraction of either the positive or the negative charges (see Rompe and Steenbeck, 1939; Smit, 1950, p. 40; and Finkelnburg and Maecker, 1956, p. 312). Thus, in the normal carbon arc, where singly-charged ions only have to be taken into consideration, we have

$$n_e = \sum_j n_{ij} \tag{3.29}$$

* I do not know whether assuming $D_a = D_i$ is justified, neither shall I attempt to approach this problem theoretically. I only point out here that we found a discrepancy between experimental and theoretical results for the transport of elements in the arc when assuming $D_a = D_i$. Should, on the contrary, D_i be larger than D_a by a factor of about 2, excellent agreement will be obtained. As the transport model may have shortcomings, we must not attach too much value to such an agreement, although it must be mentioned here (see § 9.5).

If D_a and D_i are different, we can compute the diffusion coefficient of a species from

$$D = (1 - \alpha)D_a + \alpha D_i$$

where α is the degree of ionization of that species.

where the label j is used to distinguish the densities (n_i) of different ionic species.

To conclude, in the arc column we have *quasi-neutrality*; moreover, the densities of electrons and ions are unambiguously related to the gas temperature and consequently to the gas composition, since all charged particles originate from the thermal ionization of the gas mixture in the gap between the electrodes. It is true that the cathode emits electrons, but the anode absorbs in unit time the same number as the cathode supplies. The flow of charge carriers heats the gas and maintains the elevated temperature. The presence of the charged particles results immediately from the high temperature; paradoxically, it is not caused directly by the electric current. If the gas were heated externally in a furnace to the same temperature, thermal ionization would give rise to similar densities of electrons and ions.

The electron concentration can be measured spectroscopically from the intensity ratio of an ion-atom line pair of one element (see § 7.6). The method is indirect, as it requires an independent measurement of the source temperature for computing the electron density from a combined Saha–Boltzmann equation containing the ratio of spectral-line intensities mentioned above. Although this method fits common spectrochemical conditions very well and is readily applicable, it has the serious drawback of being indirect, which may weigh heavily upon precision and accuracy of results, especially if spatial resolution of non-uniform sources is dealt with.

Direct methods for determining the electron density have become of great importance in plasma physics. They are based on the interferometric measurement of optical dispersion, on micro-wave techniques, or on the measurement of line profiles (Stark widths). In the last method, experimental Stark profiles are compared with theoretical profiles. Considerable progress in this field has been made after the old Holtsmark theory had been extended to include the effects of both ions and electrons on Stark broadening.*

Although the procedure for deriving electron densities from Stark profiles does not seem to be well suited for the low temperature arc, which is principally considered in this work, it can be expected to gain in popularity in spectrochemical physics for studying discharges in inert gas atmospheres (see, e.g. Kitayeva and co-workers, *loc. cit.*), plasma jets, and spark discharges (see, e.g. Bardócz, Vörös, and Vanyek, 1961, 1962). Stark widths are almost solely determined by the electron density and are nearly independent of the temperature. Therefore, electron density can be calculated from measured line profiles even if the temperature is only approximately known, which is a decisive advantage over the Saha method,

---

* See Kolb and Griem (1958), Griem, Kolb, and Shen (1959). Recently, Griem published a review of the articles on this subject by himself and his co-workers (Griem, 1964a).

Of the literature on this subject I cite further Gol'dfarb and Il'ina (1964), Hansen (1964), Hill (1964), Roder and Stampa (1964), Griem (1962), Lochte–Holtgreven (1963, 1960, 1958, 1956b), Kitayeva and Sobolev (1962), Kitayeva, Obukhov–Denisov, and Sobolev (1962), Kitayeva, Kolesnikov, Obukhov–Denisov, and Sobolev (1962), Fucks, Bohn, Heinrich, and Platz (1962), Wiese, Paquette, and Solarski (1962), Bergstedt, Ferguson, Schlüter, and Wulff (1962), Mandelshtam and Mazing (1960), Finkelnburg and Maecker (1956, p. 374), Unsöld (1955), and Maecker (1953).

where errors in the temperature measurement have great influence on the resulting values of n_e (cf. § 7.6). Finally, the Stark profile method also applies if the system is not in local thermal equilibrium.

As has been expressed in writing down equation (3.9), electron density is a function $n_e(r)$ of the radial coordinate r. Determining the radial variation of n_e experimentally is involved, as it requires the use of space-resolved information (see §§ 6.12 to 6.16, 7.17, and Chapter 8). When the Saha method is applied, we must determine the radial temperature distribution $T(r)$ and the radial distribution of the ion-atom line pair intensity ratio. The function $n_e(r)$ that results finally cannot claim high precision and accuracy, because the fairly large errors in the primary, space-resolved data manifest in amplified form. Nevertheless, the procedure is feasible and it has yielded interesting results (see the end of § 7.6 and §§ 7.17 and 9.4).

As the work involved in finding $n_e(r)$ is laborious, we might often content ourselves with estimates of the mean electron density ($\bar{n}_e$) in the column. The bearing of such a mean value in spectroscopy should be carefully recognized. Mean temperatures and mean electron concentrations, as determined by spectroscopic methods, are so-called effective or population-averaged quantities, which must be distinguished from volume-averaged ones (see § 6.16). Therefore, the results of measuring T and $\bar{n}_e$ depend on the type of element and on the kind of spectral line used. The tenor of the concepts of effective temperature and effective electron concentration will be elucidated by detailed examples in Chapter 8. It will also become evident then that numerical values of the effective electron density, observed spectroscopically, do not coincide exactly with the volume-averaged values dealt with in simplified equations describing the current transport.

In virtue of (3.8) and (3.9) we write for the current strength through an arc

$$i = \int_0^R 2\pi r j(r)\, dr \equiv E \int_0^R 2\pi r \sigma(r)\, dr \qquad (3.30)$$

or

$$i = E e \mu_e \int_0^R 2\pi r n_e(r)\, dr \qquad (3.31)$$

where $j(r)$ is the current density, $\sigma(r)$ the electrical conductivity, e the electronic charge, E the electric field strength, and μ_e the electron mobility, which, for the sake of convenience, has been supposed to be independent of r.

If we define formally

$$\bar{n}_e = \frac{1}{\pi R^2} \int_0^R 2\pi r n_e(r)\, dr \qquad (3.32)$$

the current equation reduces to

$$i = Ee\mu_e\bar{n}_e\pi R^2 \tag{3.33}$$

In these equations we have introduced the radius R of the column as a formal quantity. This concept cannot be defined entirely satisfactorily, because the arc does not have well-defined boundaries. Thus, different criteria, leading to different numerical values of R, are resorted to. From a spectroscopical point of view, the dimensions of the luminous region or the steepest fall in the radial temperature distribution appear to be attractive as a basis for defining the arc radius. Although there are objections to using these criteria, they can sometimes serve as useful approximations for discussing emission phenomena (§ 6.15). Obviously, in conjunction with the current transport [equation (3.33)] a definition of R in terms of the radial distribution of electron concentration, or, still better, in terms of the radial change of electrical conductivity, is most adequate.

Such a rigorous treatment has been given recently by Kolesnikov and Sobolev (1962) for the column of an argon arc. Experimentally they found the shape of the radial distribution of electrical conductivity $\sigma(r)$ to be the same for current strengths ranging from 1 to 200 amp. Consequently, all the radial distributions $\sigma(r)$, which by their nature are approximated by Gaussian curves, could be normalized to the same half width, the peaks being made equal to unity. The column radius now was conveniently taken to be the quantity $R_{0.5}$, determined by the condition $\sigma(R_{0.5}) = 0 \cdot 5 \cdot \sigma(0)$. So, by the introduction of the *half width radius*, ambiguity in using equations such as (3.30) to (3.33) can be widely removed.

In Chapter 8 we shall delimit the arc core in an analogous way to relate spectroscopically measured, effective values of the electron concentration to the type of mean value defined by equation (3.32). As the calculations in Chapter 8 pertain to a model with smooth distributions, $n_e = n_e(r)$, we could venture to use the core radius $R_{0.1}$, defined by $n_e(R_{0.1}) = 0 \cdot 1 \cdot n_e(0)$. If a measured radial distribution is dealt with, the half width radius is to be preferred, in view of the experimental difficulty of determining a radial profile accurately in the outer wings.

At any rate, we must clearly understand that the absolute value of $\bar{n}_e$ defined by (3.32) or (3.33) depends largely on the definition of the column radius. Conversely, when we insert a spectroscopic, effective value of $\bar{n}_e$ into equation (3.33) to compute R, we obtain an *effective* value of this radius.

For qualitative considerations equation (3.33) is often a convenient tool, though not more than that. By referring to this equation we are able to explain some general regularities. Yet, we should bear in mind that a rigorous approach must be made in terms of radial distributions; equation (3.33) allows only rough conclusions if we discuss interrelations between i, E, μ_e, $\bar{n}_e$, R, and T (cf. § 3.9).

Numerical values of electron density are often advantageously converted into partial pressure values by means of the relation [cf. (3.19)]:

$$p_e = 1 \cdot 3625 \cdot 10^{-22} \, Tn_e \tag{3.34}$$

where p_e is expressed as atm, n_e as number per cm^3, and T as °K. An abridged table for converting partial pressures into densities and vice versa has been added as an appendix.

The use of partial pressure is particularly profitable when the arc burns at atmospheric pressure (and atm is used as a unit of pressure), because, in the range of the low-current arc, the numerical value of p_e then virtually equals the *total degree of ionization* ($\bar{\alpha}$) of the arc gas (cf. §§ 7.15 and 7.16):

$$p_e = \frac{\bar{\alpha}}{1 + \bar{\alpha}} P \approx \bar{\alpha} \qquad (3.35)$$

since $p_e \ll P$ and the total pressure $P = 1$ atm.

★ § 3.9 *Mutual dependence of electron concentration and temperature*
Their relation with arc gas composition

Electron concentration is a fundamental concept in thermal arc theory: the arc gas is heated by the electric current flow to such an extent that the resultant thermal ionization produces the 'required' electrical conductivity (cf. § 5.3). To arrive at an understanding of the mutual dependence of electron concentration and temperature, as implied in this statement, we first consider a composite gas of fixed composition. The system is heated at constant pressure in an enclosure by means of an external source. If the elemental gas composition is known, we are able to evaluate the concentrations of all components, i.e. neutral particles, ions, and electrons, as functions of the temperature. For that purpose we use the equation for quasi-neutrality [cf. (3.29)]

$$n_e = \sum_j n_{ij} \qquad (3.36)$$

and the set of equations that govern the dissociation and ionization equilibria involved. Of course, numerical values of all relevant constants, viz. dissociation and ionization constants, and partition functions, must be available. Such calculations[*] will be dealt with more fully in § 7.8 and in Chapter 11.

With reference to Fig. 11.1 we show in Fig. 3.7 the relation between electron density n_e and the temperature, as computed by Roes (1962) for a gas whose composition corresponds with the plasma of a low-current carbon arc in air at atmospheric pressure. The diagram also shows the concentrations of the different positive charge carriers as functions of the temperature. Other examples of thermal ionization of gases appear in § 7.8 (Figs. 7.8 and 7.9).

In conclusion, for a gas mixture of given elemental composition and constant pressure, the concentration or the partial pressure of the electrons is uniquely

[*] Calculations have been made by Brinkman (1937) for hydrogen and nitrogen, by Höcker (1946), Huldt (1948a and b), and Burkhardt (1948) for air, by Finkelnburg and Maecker (1956, p. 350), Wienecke (1956), and Burhorn (1959) for nitrogen (temperature range $T = 1000-35\,000°K$), and by Roes (1962) for nitrogen and for air with atomic carbon vapour and C_2, CN, and CO (temperature range $T = 5000-6600°K$). Some typical results taken from Roes' work are given in Chapter 11.

determined by the temperature. The function relating n_e (or p_e) and T shows a marked positive slope.

Let us now visualize the following model of an arc (cf. § 4.4). The column is a cylinder of fixed radius (R), uniformly filled with a gas mixture of given composition and pressure; thus, radial distribution functions can be entirely ignored. Since the composition and the pressure of the gas are fixed, the electron concentration solely depends on the temperature, e.g. in the manner depicted in Fig. 3.7.

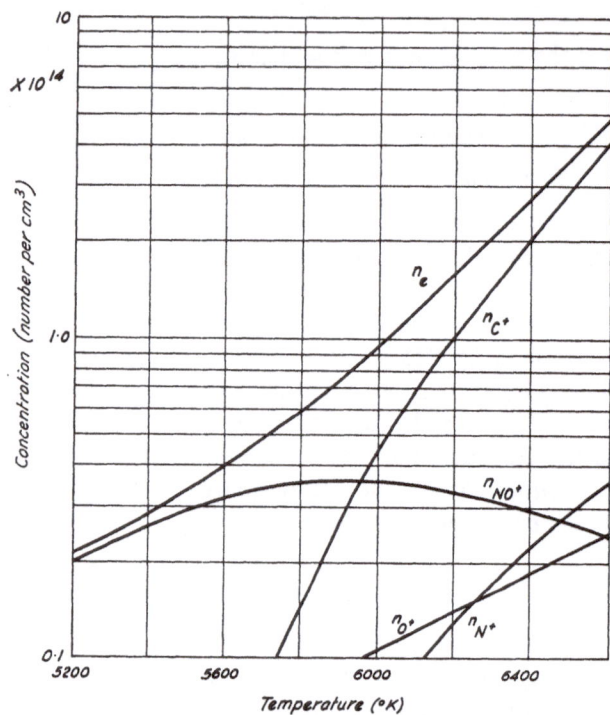

Fig. 3.7. Dependence of electron concentration n_e upon temperature in the plasma of a low-current carbon arc in air (after Roes, 1962, p. 110). The diagram shows also the concentrations of the different positive charge carriers (C^+, NO^+, N^+, and O^+) as functions of the temperature (cf. § 11.1).

This interrelation between temperature and electron density entails that our idealized arc, given its current strength, takes up a definite temperature and consequently a definite electron concentration. This is easily grasped by a qualitative consideration of the energy balance.

The electric power that is dissipated per cm in the column is

$$W = Ei \equiv \frac{i^2}{\sigma} \equiv \frac{i^2}{e\mu_e n_e \pi R^2} \tag{3.37}$$

where E is the field strength, i the current strength, σ the electrical conductivity,

e the electronic charge, μ_e the electron mobility, n_e the electron concentration, and R the column radius. For given current strength and column radius, power dissipation is seen to be a decreasing function of n_e. In view of the positive correlation between n_e and T, power dissipation is a decreasing function of T too. (Changes of μ_e with temperature can be neglected in comparison with those of n_e.)

In summary, a rise of temperature causes n_e to increase, which, in turn, enhances the electrical conductivity of the gas and consequently reduces the power dissipation, provided that the current strength and the column radius are fixed.

On the other hand, the heat produced by the electric current flows away by thermal conduction to the surroundings. In the temperature range considered here, that is $T = 5000\text{–}7000°\text{K}$, the thermal conductivities of nitrogen and air are increasing functions of the temperature (see § 4.3). Therefore, thermal conduction (S) and power dissipation (W) can be represented schematically by the solid curves shown in Fig. 3.8. The intersection point defines the temperature (T_1) at which the power supply equals the energy loss by thermal conduction: hence the arc takes up precisely that temperature T_1.

The nature of the functions $W(T)$ and $S(T)$ is decisive for the temperature that is established. Both functions depend principally on the gas composition; there is, however, an important difference in the effect of minor constituents of the plasma upon $W(T)$ and $S(T)$. In the arc in air, thermal conduction is governed by the dissociation of nitrogen (§ 4.3). As a result, $S(T)$ is virtually independent of the presence of added metal vapours, since their proportion in the plasma does not exceed a few per cent (cf. § 7.9); the nitrogen content of the arc gas is thus practically unaffected by the introduction of a substance. In contrast with $S(T)$, the power dissipation $W(T)$ is very sensitive to small additions of extraneous elements to the arc plasma, as the relationship between n_e and T is markedly altered by the presence of even a slight amount of an element of low ionization potential (cf. §§ 7.8 and 7.9). So, in our schematical representation of Fig. 3.8, the broken curve will depict the power dissipation when an element of low ionization potential has been added. As $S(T)$ is unchanged, the arc is forced to burn at the lower temperature T_2 instead of T_1.

According to the foregoing discussion, the temperature of the arc gas is completely determined if the composition of the gas, the diameter of the column, the current strength, (and the pressure of the surrounding medium) are all fixed: too high a temperature is not feasible with the corresponding low power input and high thermal conductivity; too low a temperature cannot exist as it is automatically raised in consequence of high power supply and low thermal conductivity (Fig. 3.8). This is equivalent to saying that the gas must have that precise temperature at which the power dissipation, in balance with thermal conduction, is adequate for maintaining that temperature.

For adapting our simplified arc model to the actual free-burning arc, we must first drop the premise that the column radius be fixed. It then enters as a variable into our equations; therefore, given the composition of the arc gas, the arc 'selects' proper values of T, n_e, and R. Those variables all are increasing functions of the current strength. We will not go into details of the complications that originate from the additional variability of R. Suffice it to say that these problems can be dealt with rigorously in terms of radial distributions; this involves the use of differential equations that must be tackled numerically (cf. §§ 4.1 and 4.2). Also,

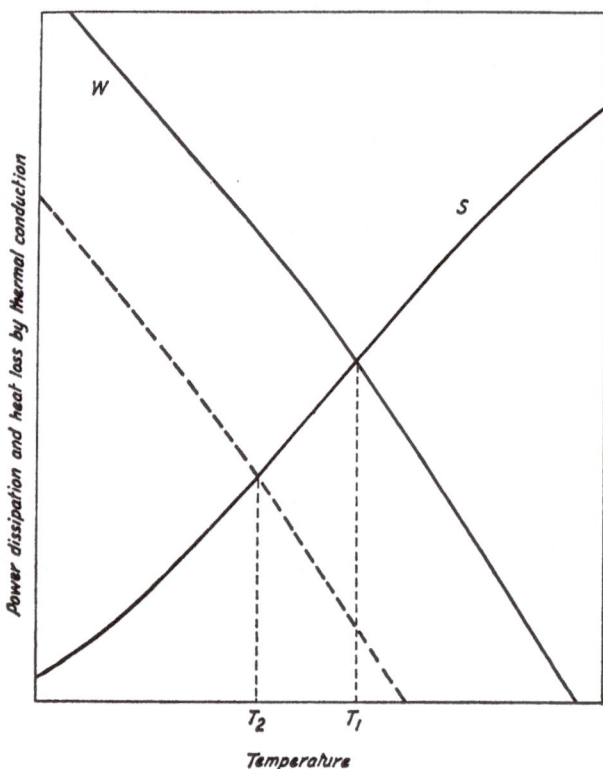

FIG. 3.8. Schematical representation of the power dissipation (W) and the thermal conductivity (S) for a carbon arc. The arc takes up the temperature (T_1) at which the power supplied equals the energy loss by thermal conduction. The addition of a substance of low ionization potential produces a marked change of the intrinsic electrical conductivity of the gas, whereas it leaves the intrinsic thermal conductivity unaffected. The repercussion of the altered dependence of power dissipation upon temperature (broken curve) causes the temperature to drop to the value T_2, where again the power supplied is balanced by the energy loss by thermal conduction.

in terms of the simplified model, Steenbeck's minimum principle is sometimes useful (cf. § 4.4).

If we content ourselves with a rather rough picture, such as has been outlined, conclusions have a qualitative nature only; however, where in spectrochemistry a free-burning carbon arc in air at atmospheric pressure in a very restricted range of current strengths is exploited, appreciable variations of the arc radius are not likely to occur, so that considerations in terms of an approximate model, taking R to be constant, can be made without distorting the truth greatly. Of course, quantitative details cannot be included in such a discussion.

An important feature of the normal carbon arc, that was recognized as early as 1932 by Ornstein, Brinkman, and Beunes (1932), is the restricted range of values the electron concentration can assume. In view of the self-consistency of thermal ionization in the arc plasma and the actual behaviour of the arc radius, n_e ranges from about 10^{14} to several times 10^{15} per cm^3 in the normal low-current arc, which is equivalent to an interval of some 10^{-4} to some 10^{-3} atm for the electron pressure p_e, when T is in the range 4000–7000°K.

The limited range of electron concentration in the carbon arc is an adequate argument for explaining the influence exerted upon the temperature by a substance of low ionization potential (e.g. Rollwagen, 1939; and Smit, 1950, p. 53). Evidently, the arguments are of the same nature as those brought out above: we must expect the arc gas to assume the temperature at which the 'required value' of the electron concentration is reached. When a vapour component of low ionization potential is present, this state of equilibrium will be reached at a lower temperature than in the absence of that substance. In § 7.8, it will be shown mathematically how a definite value of the temperature follows if for a gas mixture of given composition the electron density n_e be constant.*

In fact, the electron density is only constant within an order of magnitude. This variability does not weaken, however, the line of argument used for explaining the lowering of the temperature effected by a substance of low ionization potential (§ 7.8). Some information on the changes of the electron concentration in carbon arcs with metal vapour is furnished by my spectroscopic study of the correlations between electron density, temperature, and metal vapour concentration, which will be considered in detail in §§ 7.7 to 7.10. For an arc with constant current strength and electrode separation, I observed a more or less systematic variation of the electron concentration with temperature, if the changes of the temperature were produced by evaporating successively various metal salts from the lower electrode, thus by altering the plasma composition. Under the conditions of the experiment, I found that the electron concentration tends to *increase* when an element of low ionization potential volatilizes into the arc column and simultaneously brings about a decrease in the temperature. The enhancement of the electron concentration was closely associated with an increase in the vapour concentration of the metal carrying the discharge. The rise of electron density accompanied by a fall in the temperature does not contradict the previous statement that for a gas of given composition ionization increases with temperature, since for the system to which the negative correlation between n_e and T applies, variation of the plasma composition is an essential feature.

---

* We note that the temperature fall caused by the entry of alkali metal vapour is not ascribed to cooling down as a result of high energy loss by emission of resonance lines. The amount of energy given off from the column by radiation to the surroundings is, in contrast with that emitted by the electrodes, a negligible fraction in the energy balance (see §§ 5.3).

§ 3.10   *The radial electrical field*
Ambipolar diffusion

Owing to the radial drop in concentration of charged particles in the arc column, electrons and ions diffuse outwards, and since electrons and ions have different diffusion coefficients, the former tend to move ahead of the latter. The process does not lead to demixing, however, since even a small displacement of the positive and negative charge centres produces an electrical field which retards the movement of the electrons and accelerates that of the ions. This field is called the radial electrical field. The motion of electrons and ions can be treated as if the assemblage of charges diffused with a common velocity. The process is referred to as *ambipolar diffusion*.

The coefficient of ambipolar diffusion D_{amb} is a mean diffusion coefficient averaged in the ratio of the mobilities. It is computed to be (see e.g. Weizel and Rompe, 1949, p. 14; Maecker, 1951, p. 317; and von Engel, 1955, p. 122):

$$D_{amb} = 2D_i \tag{3.38}$$

in other words, the outward motion of the electrons causes the ions to diffuse with a velocity that is twice that of the ions alone.

CHAPTER FOUR

REMARKS AND REFERENCES
REGARDING ARC THEORY

★ § 4.1 *General remarks and references*

Although many questions considered in this work might be regarded as touching the field of arc theory, I wish to restrict the term 'theory of the arc' to those discussions that offer a complete or, at least, far-reaching account of the inter-relations between the arc parameters. In this chapter a basic equation of arc theory used for arcs having cylindrical symmetry will be merely outlined (§ 4.2); moreover, some attention is drawn to the problem of thermal conduction (§ 4.3) and to the 'channel model' (§ 4.4). The main purpose is to give the interested reader a few references—and the inevitable chain of sub-references—that may serve as a starting-point for an elaborate and laborious study.*

A glance at the literature on arc theories gives an impression of such a great complexity that spectrochemists—myself not excluded—might easily become confused or even discouraged. Not only is our knowledge of physics and mathematics, in general, fairly limited, but even if we succeed in seizing the quintessence of a physical discussion of arc theory, we are apt to be unsatisfied, because the discussion does not fully answer our spectrochemical problems or perhaps does not answer them at all. This observation must not be taken as a reproach against the admirable work done in physics: it is a natural consequence of the different objectives of physicists and spectrochemists; besides, the range of conditions explored by each group of scientists is entirely different. Physicists have studied, for example, arc discharges in various atmospheres (H_2, N_2, CO_2, Hg, inert gases, air, or steam), have investigated different types of arc in wide ranges of current strengths (up to 1500 amp), temperatures (over 50 000°K) and pressures (0·01 to 1000 atm), and have examined horizontal and vertical arc discharges with free, forced or suppressed convection. Although, on the other hand, spectrochemists use different types of arc, they commonly restrict observations to the free-burning

---

* Among the numerous texts or papers discussing the subject we cite above all Finkelnburg and Maecker (1956), Weizel and Rompe (1949), and Elenbaas (1951); moreover: Elenbaas (1934), Heller (1935), Brinkman (1937), Mannkopff (1943), Höcker and Finkelnburg (1946), Maecker (1951, 1959, 1960, 1962), Maecker and Peters (1956), Champion (1952, 1953, 1956), King (1954, 1956, 1957), Burhorn (1959), Ecker (1961), Roes (1962), Schmitz and Patt (1962, 1963), Schmitz, Patt, and Uhlenbusch (1963), Olsen (1963), Uhlenbusch (1964), and Krinberg (1964). To understand the theory developed by Maecker and co-workers, a thorough knowledge of the kinetic theory of gases and of the thermodynamics of irreversible processes is essential. References can be found in Finkelnburg and Maecker's article (1956).

vertical arc in air at atmospheric pressure and low current strength. Their range of external conditions might thus be very limited compared with that explored by physicists; however, they are confronted with a train of internal variations caused by the volatilization of samples, the influence of metal vapours upon the excitation conditions, and the transport of elements through the arc. Therefore, in spite of the limitations set to physical variables in spectrochemistry, a great variety of problems is met with, the more so as spectrochemists are inclined—and forced— to think of precision and accuracy in terms of percentage rather than in terms of orders of magnitude. We must admit that the percentage level is rather easily attained if all conditions except the concentration of an element sought are kept constant; finding the influence of changes of excitation conditions upon spectral-line intensi- ties if, on the contrary, the concentration of that particular element is the only invariant, offers a great deal more trouble. As spectrochemical physics is still developing, we cannot predict if the need for introducing arc theories will arise. The day may come when spectrochemists apply these theories as they now build upon those results of physical investigations that are primarily dealt with through- out this text. Yet this prospect is not so likely to be a reasonable one; as Maecker (1951) points out, recognition of the nature of the physical processes proceeding in the arc is not the greatest problem; knowledge of only a few elementary mechanisms suffices to explain the various types of arcs. The main difficulty is the quantitative incorporation of all relevant physical elements into a significant and manageable mathematical model; moreover, available data on the properties of arc gases at high temperatures are scarce and often speculative.

Altogether, the progress of modern science coupled with the possibilities of electronic computers might solve a great many problems that are now still looked upon as complex and involved. Which spectrochemist will be able, however, to follow the track of physicists without being regarded as a physicist too, thus leaving spectrochemistry once more entangled with its own problems? Admittedly, the task of spectrochemical physics is a critical one, since half knowledge is perilous, full knowledge is considered as involved and perhaps apparently irrelevant, whereas *no* knowledge once more means mere empiricism. For the time being we must content ourselves in spectrochemistry with partial explanations that yet might take us forward, since the main purpose is not to establish a com- plete theory of the arc but to provide some insight into those phenomena that affect the precision and accuracy of our spectrochemical determinations. Thus we might be able to find, partly by theoretical reasoning, partly by experimental ways, the most favourable and soundest conditions for exciting our samples. We should be cautious, however, in absolutizing our theoretical considerations, because they are necessarily incomplete, simplified, and adapted to our analytical- chemical 'brains', development, and education.

§ 4.2 *Energy balance: Elenbaas–Heller equation*

A fundamental concept of arc theory is the energy balance equation that ex- presses equality of energy supply and energy loss per unit time per unit volume.

If the energy loss by radiation can be neglected (see § 5.3) in comparison with losses by thermal conduction, we have for a steady state:

$$\text{div}(\kappa \operatorname{grad} T) + jE = 0 \tag{4.1}$$

where E is the electrical field strength, j the current density, T the absolute temperature, and κ the thermal conductivity. The meaning of κ becomes apparent from the following definition: the rate at which heat crosses an area of 1 cm^2 perpendicular to the direction x of the temperature gradient dT/dx is $\kappa . dT/dx$. Numerical values of κ will be expressed as ergs . $\text{sec}^{-1} \text{cm}^{-1} \text{degree}^{-1}$.

Equation (4.1) is known as the Elenbaas–Heller equation (Elenbaas, 1934; Heller, 1935). If discussion is restricted to a cylindrically symmetrical arc, the Elenbaas–Heller equation can be put into the form

$$-\frac{1}{r}\frac{\partial}{\partial r}(r\kappa \operatorname{grad}_r T) = jE \tag{4.2}$$

where r is the radial coordinate in a cylindrical coordinate system with axis z ($r = 0$). Since cylindrical symmetry is assumed, any additional equation expressing dependence on z vanishes. The equation holds true in good approximation for an arbitrary part of the column of the normal arc, although in principle free convection involves certain objections (see below).

By using equation (3.8) equation (4.2) can be rewritten as

$$-\frac{1}{r}\frac{\partial}{\partial r}(r\kappa \operatorname{grad}_r T) = \sigma E^2 \tag{4.3}$$

where σ is the electrical conductivity (see § 3.4).

Multiplying by 2π and integrating we obtain

$$-2\pi r \kappa \frac{dT}{dr} = 2\pi E^2 \int \sigma r \, dr \tag{4.4}$$

where the boundary condition, $dT/dr = 0$, and $E = $ constant have been used. The right-hand side of (4.4) is the electrical energy dissipation in a cylinder of radius r and unit height; the left-hand side is the heat leaving this cylinder per sec.

An additional boundary condition is $T = T_R$ for $r = R$. As Maecker (1951) shows, the volume of a cylindrically symmetrical arc should be bounded by a tube of radius R and temperature T_R in order that a steady state will be reached. To a first approximation, the column of a free-burning, vertical arc can be regarded as cylindrical, convection now occupying the role of the walls of the discharge vessel.

If E is given, the Elenbaas–Heller equation, combined with the boundary conditions stated, should yield the radial distribution of temperature. Solving the equation can be done only numerically, as the functions κ and σ depend in a complicated manner on temperature, particularly when the arc burns in a molecular gas—or in a mixture of such gases—where dissociation (see Chapter 11 and § 4.3) and ionization (see Chapter 7) equilibria entail additional difficulties. So, even the apparently simple case of an arc in pure nitrogen is in fact extremely involved. It serves no useful purpose to enter into the details of mathematical solution methods, for which we refer the interested reader to the literature cited in § 4.1.

§ 4.3   *Thermal conduction in the arc*

The larger part of the energy dissipated from the column is carried away by radial heat conduction; only a small percentage is converted into radiation (see § 5.3). Outside the central core, convection becomes more and more important for heat transport to the surroundings (see Hagenah, 1950).

The thermal conductivity has been defined in § 4.2. Accordingly, the rate at which heat crosses an area of 1 cm^2 perpendicular to the direction x of the temperature gradient dT/dx is $\kappa.dT/dx$. Numerical values of κ are expressed as ergs.sec^{-1} cm^{-1} degree^{-1}.

The total thermal conductivity of a gas is made up of several components. Which of these have to be taken into account in a particular case depends on the temperature interval under consideration.*

According to Meixner (1952) the total thermal conductivity κ is composed of

1 The normal classical conductivity due to the motions of atoms, molecules, ions, and electrons.
2 The conductivity due to the transport of energy of reaction, viz. energy of dissociation and ionization.
3 The conductivity due to thermal diffusion.

If the temperature is in a range between 5000 and 7000°K, the governing feature of the thermal conductivity of air and nitrogen is the dissociation of nitrogen. Atoms of the partly dissociated gas diffuse outwards from the axis to the boundaries and recombine in the cooler fringe, yielding up the energy of dissociation. Molecules travel in the opposite direction and are dissociated at the higher temperature, a process that consumes energy. The complete cycle, which is associated with the radial transport of dissociation energy, contributes by far the greatest portion (κ_d) to the total thermal conductivity. As dissociation increases with temperature, the contribution κ_d shows a marked dependence on temperature and passes through a maximum at about 7000°K. Older calculations (e.g. that by Brinkman, 1937) yield the maximum in the vicinity of 5000°K, as they are based on the value of 7·38 eV for the dissociation energy of nitrogen. Nowadays the higher value of 9·76 eV is generally accepted. With the latter one there is also good agreement between theory and experiment when the transition from the low-current to the high-current arc is considered (King, 1954, 1956, 1957).

Fig. 4.1 shows the total thermal conductivity κ of nitrogen and air according to various calculations. In addition to the conductivity term due to the dissociation

* See King (1954, 1956, 1957), Pelzer (1960, 1962), Roes (1962), Burhorn (1959), Wienecke (1956), Finkelnburg and Maecker (1956, p. 346 ff.), Meixner (1952), Höcker and Finkelnburg (1946), Schulz (1947b), and Höcker and Schulz (1949).

of nitrogen, the portion contributed by the classical conductivity has been taken into account. Formulae for computing the partial conductivities due to the motions of atoms (κ_a), molecules (κ_m), ions (κ_i), and electrons (κ_e) are given by Finkelnburg and Maecker (1956, pp. 346–348) and by Wienecke (1956).

Fig. 4.2 shows the total conductivity and its components κ_d, κ_a, and κ_m according to Roes' calculations; the components κ_i and κ_e are negligible in the temperature range at present under consideration. This holds true equally well for the conductivity due to thermal diffusion and for that due to transport of ionization energy (Roes, 1962, pp. 70 and 73).

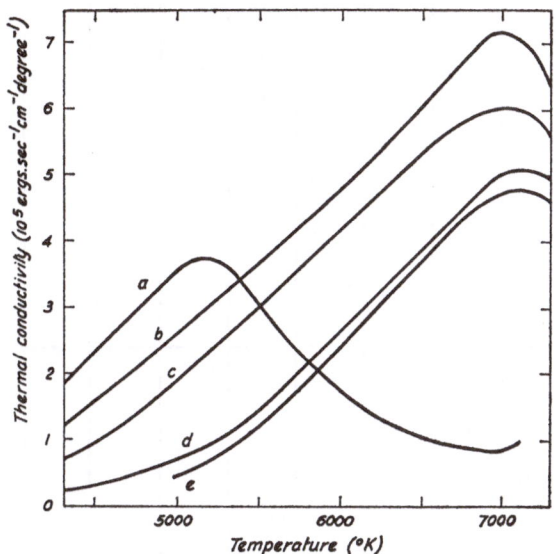

FIG. 4.1. Comparisons of total thermal conductivity of nitrogen and air according to various calculations (after Roes, 1962, p. 65).

    (a) κ of nitrogen after Brinkman (1937), based on the value of 7·38 eV for the dissociation potential (V_d) of nitrogen.

    (b) κ of air with 30 per cent carbon for the high-current arc after Wienecke (1956).

    (c) κ of nitrogen after Finkelnburg and Maecker (1956, p. 352, Fig. 41), based on $V_d(N_2)$ = 9·762 eV.

    (d) κ of nitrogen after King (1956), based on $V_d(N_2)$ = 9·762 eV.

    (e) κ of nitrogen after Roes (1962), based on $V_d(N_2)$ = 9·762 eV.

Fig. 4.3 presents the total thermal conductivity κ and its components for a mixture of air and carbon vapour, such as should be found at about 0·5 cm above the anode in a 5-amp carbon arc. In computing the components the following assumptions and estimates have been made. Firstly, the number of carbon atoms in the gas (both as free and as bound carbon) is estimated to be $1·6 . 10^{17}$ per cm^3, which is equivalent to saying that at 5000°K the number of carbon atoms amounts to about 1/11 of that of the nitrogen atoms (also taken as the total of free and bound nitrogen). The total carbon concentration is supposed to be independent of the temperature. The estimate of $1\cdot6 \ .\ 10^{17}$ carbon atoms per cm³ rests on mass transport considerations of the type outlined in § 9.5; carbon is assumed to enter into the arc mainly in the form of CO (see Chapter 11). Secondly, the classical thermal

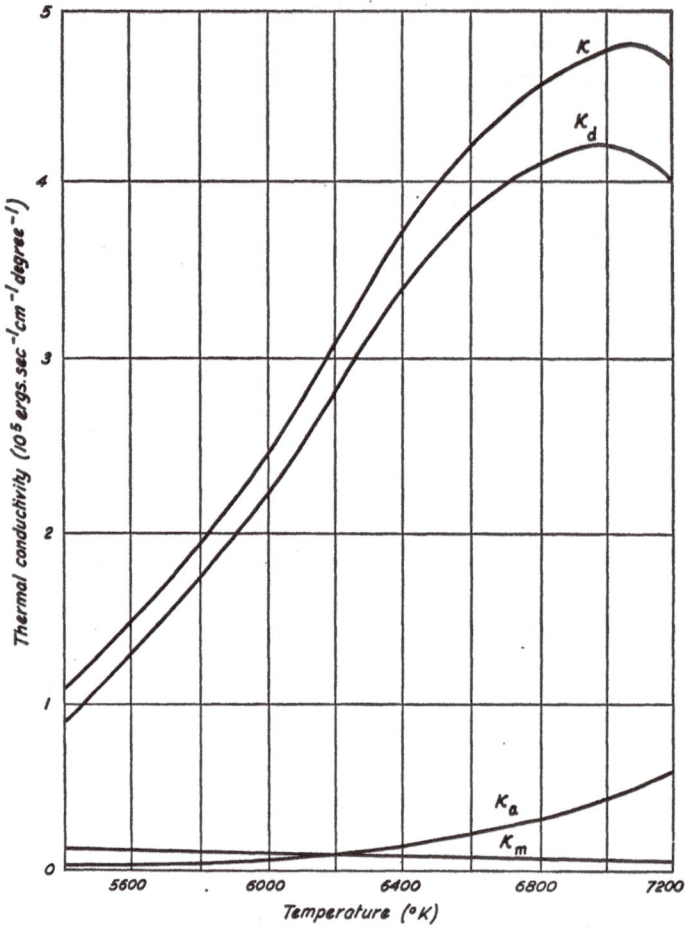

FIG. 4.2. Total thermal conductivity (κ) of nitrogen plotted as a function of the temperature (after Roes, 1962, p. 72). In the temperature range considered here, κ is made up of

    1 The classical conductivity due to the motions of atoms (κ_a) and molecules (κ_m).

    2 The conductivity associated with transport of dissociation energy (κ_d).

conductivity has been reasonably assigned the same value as that of pure nitrogen, since the contributions of atoms and molecules separately (κ_a and κ_m) do not differ greatly. Furthermore, in addition to κ_d for nitrogen, the conductivity due to the transport of dissociation energy by CO had to be allowed for. Similar contributions by O_2, C_2, NO, and CN can be ignored here.

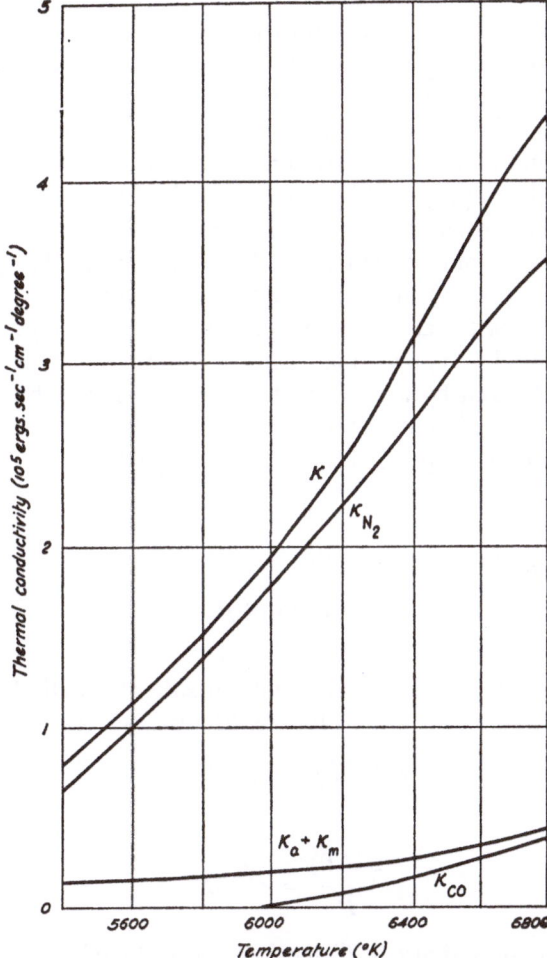

FIG. 4.3. Total thermal conductivity (κ) of a mixture consisting of air, C, C_2, CO, and CN (after Roes, 1962, p. 108). The total conductivity is composed of

1 $\kappa_a + \kappa_m$, the classical conductivity due to the motions of atoms and molecules, which is taken to be equal to that of pure nitrogen.

2 $\kappa_d(N_2)$ and $\kappa_d(CO)$, the conductivities due to the diffusion of nitrogen and of carbon monoxide (transport of dissociation energy).

The number of carbon atoms (total of free and bound carbon) is estimated to be $1/11$ of that of the total number of nitrogen atoms, when $T = 5000°K$ (see text).

Altogether, as somewhat uncertain assumptions may underlie the calculations of thermal conductivities, results should not be taken too rigorously. Yet, the general behaviour of nitrogen and air must be fairly well approximated by the curves shown in Figs. 4.2 and 4.3.

§ 4.4   *The channel model and the Steenbeck minimum principle*

The conditions of a cylindrically symmetrical arc can be completely determined from the Elenbaas–Heller equation (4.2) and the current transport equation (3.8), provided that the relevant functions, such as electrical and thermal conductivity, are known functions of the temperature. They are of a complicated nature, and so a solution of the equations is to be attempted by numerical procedures. Approximate methods have been proposed in order to retain the surveyability of functional relationships between separate arc parameters. Most of these approximations maintain a sufficient number of equations for determining the system completely with the given boundary conditions. A well-known exception, however, is the so-called 'channel model of the arc', where one equation is missing.

In the channel model we approximate the arc column by a homogeneous cylinder of radius R, characterized by a uniform temperature T and by an electrical conductivity $\sigma(T)$, that is uniquely determined by Saha's relationship.* The current strength (i) is given then by [cf. (3.8), (3.9), and (3.34)]

$$i = \sigma(T)E\pi R^2 \tag{4.5}$$

and the power dissipation by [cf. (3.38)]

$$W = Ei \equiv \sigma(T)E^2\pi R^2 \tag{4.6}$$

where E is the electric field strength.

Assuming a radial temperature gradient outside the cylinder and neglecting energy losses by radiation, we obtain for the Elenbaas–Heller equation

$$2\pi RS(T) = \sigma(T)E^2\pi R^2 \tag{4.7}$$

or

$$\frac{2S(T)}{R} = \sigma(T)E^2 \tag{4.8}$$

where the thermal conductivity $S(T)$ has been introduced.

Equations (4.5) and (4.8) enable us to compute the temperature as a function of R, if the field strength E is given and the functions $S(T)$ and $\sigma(T)$ are known. The essential difference between this method and that with the rigorous treatment is that the temperature is not found as a function of the radial coordinate r, a solution that inherently allows the exact calculation of the current strength, the half width radius, and other parameters. In spite of the shortcoming of the channel model to yield only a function $T = f(R)$ instead of definite values of T and R, it has often proved to be a useful approximation.

Steenbeck (1932) (cf. von Engel and Steenbeck, 1934) extended the channel model and postulated that the arc parameters tend to assume values which cause the electric field strength to be at a minimum. Thus, an additional condition completes the set of equations of the model. There is the contradiction, however, that

---

* Note that § 3.9 was based in principle on the use of the channel model.

in the exact procedure no additional equations are needed, so that the minimum principle seems to be superfluous.

Steenbeck's postulate has been successfully applied in arc theory; naturally, it has been also the object of much discussion (see, e.g., Maecker, 1951, p. 331; and Ecker, 1961, references 168, 481, 483, 500, 545, and 626). Recently it has regained interest (see Steenbeck, 1960/61; Kischl and Wilhelm, 1960, 1960/61; and particularly Peters, 1962, 1956).

CHAPTER FIVE

THERMAL EQUILIBRIUM IN THE CARBON ARC

★ § 5.1 *Introduction*

Although the arc is the oldest example of an electric discharge that can be maintained for an arbitrary long time, the properties of chief interest for spectrochemistry bear a closer relation to the essential features of a furnace than to discharge phenomena. This fact, at first sight, is curious and will be examined in the following sections.

The electric current flow is of vital importance for maintaining the elevated temperature in the arc column; it is a minor point, however, in a discussion of excitation, ionization, and dissociation phenomena. Since the arc is a source in (partial) thermal equilibrium, those processes can be treated in a similar way as if the gas mixture were contained in a furnace at the same temperature; therefore, excitation is described by Boltzmann's distribution law, ionization by Saha's relationship, and dissociation by the general equation for chemical equilibria; moreover, in the range of temperature and current strength at present under consideration, electron pressure can vary only between definite limits, viz. 10^{-4} to 10^{-3} atm, whence the ionization equilibria of vapour components bear a close resemblance to dissociation equilibria of weak acids in a buffered aqueous solution.

★ § 5.2 *The concept of temperature*

If an arc in air at atmospheric pressure is the source of excitation, the temperature is the gas temperature. The conditions of the arc plasma can be regarded as a state of thermal equilibrium; therefore all definitions of temperature give the same numerical values. In non-equilibrium sources, different definitions will, in general, lead to different values of the parameter.

The various temperatures each describe a distinct aspect of the physical state of the system. According to Mohler (1941) and Lochte-Holtgreven (1955) the conditions in an atomic gas can be characterized by four different kinds of temperature, viz.

1 The *electron temperature*, which is determined by the kinetic energy of the electrons.

2 The *gas temperature*, which is defined by the kinetic energy of the neutral atoms.

3 The *excitation temperature*, which describes the population of the various energy levels.

4 The *ionization temperature*, which governs the ionization equilibria.

Summarizing, in an atomic gas we recognize translation, excitation, and ionization temperatures. In molecular gases, temperatures assigned to dissociation equilibria and to the rotational and vibrational states of the molecules should be added.

In a discharge the numerical differences among the various temperatures depend mainly on the strength of the electric field and on the gaseous pressure, actually on the ratio p/E (see Smit, 1946). At reduced pressure and/or in a strong field the electron temperature is high, whereas the gas temperature is low. As the field strength is attenuated and/or the pressure is raised, electron temperature decreases and gas temperature rises; finally all temperatures become numerically equal when the system has reached a state of thermal equilibrium. This holds true in a carbon arc in air at atmospheric pressure (see §§ 5.3 to 5.6). As for the spark, Mandelstham (1957) has shown that excitation and ionization temperatures are equal to the electron temperature $T_e(= 40\,000–50\,000°K)$; thus Boltzmann's law and Saha's relationship apply to the spark if electron temperature is used as a parameter. The thermodynamic equilibrium of spark plasmas has been treated by Krempl (1962).

★ § 5.3   *Thermal equilibrium in the arc*

The existence of *thermal equilibrium* in the column of the carbon arc was first assumed by Compton (1923). According to Compton's theory the arc gas is heated by the electric current to such a degree that the resulting thermal ionization suffices to afford the required electric conductivity of the gas mixture. The correctness of the statement has been justified theoretically and experimentally by several investigators. The theoretical consideration of Ornstein and Brinkman (1934) still deserves study; their vision should be supplemented, of course, by recent discussions of the subject, e.g. that by Finkelnburg and Maecker (1956). For instructional purposes the principal points of the theory will be considered here in some detail.

For a gas to be in complete thermal equilibrium, it is required that the various gas components, i.e. neutral particles, charge carriers, and photons, are in equilibrium mutually and with respect to the surroundings. Complete thermal equilibrium prevails in an enclosure whose walls and interior have a uniform temperature with respect to radiation and internal energy. It holds also for a gaseous body of such dimensions that numerous energy exchange processes, particularly absorption and emission of radiation, take place. Thermal equilibrium is interfered with by those processes that tend to affect the energy flow to or from the system in one direction only.

A gaseous system in complete thermal equilibrium is characterized by the following conditions.

1 The velocity distribution of all kinds of free particles (molecules, atoms, ions, and electrons) in all energy levels satisfies Maxwell's equation.

2 For each separate kind of particle the relative population of energy levels conforms to Boltzmann's distribution law (see §§ 5.5 and 6.1).

3 Ionization of atoms, molecules, and radicals is described with Saha's equation (see § 7.2) and dissociation of molecules and radicals with the general equation for chemical equilibria (see § 11.1).

4 Radiation density is consistent with Planck's law.

In the arc column equilibrium between light quanta and material particles is not reached, owing to the absence of walls with the same temperature as the arc gas; consequently the radiation density does not satisfy Planck's law (see also § 5.5). The energy loss from the arc by radiation, however, amounts to only a few per cent* of the total energy supply. For this reason a state of partial thermal equilibrium can exist, and so the conditions 1 to 3 are closely met. In order that this state of partial equilibrium will ensue, a Maxwellian velocity distribution for all kinds of particles is imperative.

In an arc, we always have a radial decline of temperature that fundamentally interferes with a state of thermal equilibrium; however, if the change of temperature along one mean free path is small in comparison with the mean temperature in this region, influence of the temperature gradient on equilibrium conditions is negligible. In an arc at atmospheric pressure the mean free path for the various types of process is small and so the temperature fall is not likely to cause any difficulties. The same holds true for the density gradient. It has become customary to consider each element of volume separately and to denote the equilibrium conditions in a non-homogeneous source as *local thermal equilibrium* (LTE).

★ § 5.4   *Electron temperature and gas temperature in the arc*

Whether electrons have a Maxwellian velocity distribution corresponding to the gas temperature depends on the energy exchange between electrons and neutral particles. The energy gained by the electrons from the electric field in unit time per unit volume is $en_e v_d E$.† In the presence of heavy particles, electrons are scattered and their velocity in the field direction is converted into random velocity, so that the kinetic energy they have acquired in the field appears as a rise of the electron temperature. A uniform electron temperature is established after only a few collisions, which according to Weizel and Rompe (1949) (cf. Finkelnburg and Maecker, 1956) requires 10^{-12} to 10^{-17} sec.

---

* In the normal carbon arc in air at atmospheric pressure Brinkman (1937, p. 61) found the following values for the energy emitted by a part of the column (thus excluding the electrodes; see § 10.3): 0·8 per cent of the total power input at a current strength of 5 amp, 1·4 per cent at 10 amp, and 3·0 per cent at 14 amp.

† e = electronic charge, n_e = electron density, v_d = drift velocity (= $\mu_e E$), μ_e = electron mobility, E = electric field strength.

The time necessary for reaching equilibrium between electron temperature and gas temperature is much larger, since an electron when colliding elastically with a heavy particle loses only a very small amount of energy, which is on the average a fraction $2m_e/m_h$ of its excess energy, i.e. the energy gained in the field (m_e = mass of electron, m_h = mass of heavy particle).

An idea of the order of magnitude of various quantities involved is obtained if we make an estimate of the time required for complete passing on of excess electron energy, e.g. in the case of an arc in nitrogen at 6000°K and 1 atm, neglecting dissociation. We find the required number of collisions to be $\frac{1}{2}m_h/m_e = 2 \cdot 5 \cdot 10^4$. The root mean square velocity of the electrons at 6000°K has been calculated previously as $\bar{v}_e = 5 \cdot 2 \cdot 10^7$ cm/sec (§ 3.4). Taking the value $\lambda_e = 10^{-3}$ cm for the mean free path (Table 3.3) we find the collision period of electrons with heavy particles, viz.

$$\tau_{eh} = \lambda_e/v_e \qquad (5.1)$$

to be about $2 \cdot 10^{-11}$ sec. Accordingly, complete energy exchange takes about $5 \cdot 10^{-7}$ sec. During this period, however, electrons continuously acquire new energy in the electric field, so that the system will approach a state of equilibrium in which the electron temperature T_e is so far above the gas temperature T_g that the energy gained by the electrons in unit time per unit volume equals the energy transferred to the gas as a result of the temperature difference; therefore

$$en_e\mu_e E^2 = \tfrac{3}{2}k(T_e - T_g)\frac{2m_e}{m_h}\frac{n_e}{\tau_{eh}} \qquad (5.2)$$

Substituting (5.1) and inserting equations (3.10) and (3.11) for v_e and μ_e we obtain as a condition for thermal equilibrium

$$\frac{\Delta T}{T} = \frac{T_e - T_g}{T_e} = \frac{m_h}{4m_e}\frac{(e\lambda_e E)^2}{(\tfrac{3}{2}kT_e)^2} \ll 1 \qquad (5.3)$$

if at the same time we demand that the relative temperature difference is small compared with unity. Obviously, the condition for thermal equilibrium is not satisfied when the energy ($e\lambda_e E$) gained by an electron from the field upon travelling along one mean free path is small in comparison with its thermal energy ($\tfrac{3}{2}kT_e$): the unfavourable exchange mechanism expressed in the ratio m_h/m_e should also be taken into account.

Let us consider the numerical value of the left-hand side of the inequality (5.3) when inserting the following data for an arc in nitrogen at 6000°K and 1 atm: $m_h/m_e = 5 \cdot 10^4$, $\lambda_e = 10^{-3}$ cm, $E = 25$ V/cm, $T_e = 6000°$K, and $e = 1 \cdot 6 \cdot 10^{-19}$ C, $k = 1 \cdot 38 \cdot 10^{-23}$ J°K^{-1}. We find the value 13, so that, strangely enough, the condition for thermal equilibrium appears to be far from fulfilled, although we believe that a low-current arc in nitrogen, or the normal carbon arc in air at atmospheric pressure, is a thermal source.* The reason for the discrepancy is that equation (5.3) is based on energy transfer by elastic collisions only; actually the arc in nitrogen or air *is* a thermal source owing to the molecular nature of the medium, which aids electrons in giving up their

---

* A check made by Lochte–Holtgreven (1958) on the validity of the equilibrium condition (5.3) for a free-burning arc in air at atmospheric pressure with $T = 6000°$K, $E = 20$ V/cm, and $\lambda_e = 1 \cdot 3 \cdot 10^{-5}$ cm, yielding $T = 6°$K, is misleading, because the chosen value of λ_e is incorrect.

excess energy by inelastic impact leading to excitation of vibrational levels of molecules and radicals.*

In an excellent article by Gurevich and Podmoshenskii (1963) the energy balance for the electrons has been considered and estimates have been made of the proportion of the energy dissipated through elastic and inelastic collisions with heavy particles, and through diffusion and conduction to the colder regions of the plasma. The authors show that for molecular gases the energy lost by excitation of vibrational levels contributes by far the largest fraction; it exceeds that of the energy transferred in elastic collisions by about two orders of magnitude.

In inert gases, which do not have vibrational degrees of freedom, thermal equilibrium will be consequently less easily established, unless at high current strength and correspondingly low field strength and high electron density (see also § 5.6). According to Gurevich and Podmoshenskii (1963) only a small amount (about 1 per cent) of molecular impurities, such as N_2 or CO, in argon suffices for local thermal equilibrium to be reached.

In general, however, assuming local thermal equilibrium will be admitted only if it has been proved experimentally, as holds true for the column of the carbon arc in air at atmospheric pressure. Of course, the assumption is equally well, or even better, justified when metal vapours are introduced into the arc, since the presence of those vapours reduces the field strength and the mean free path for electrons (see §§ 3.2, 3.5 and 3.6); those effects are counteracted, but not at all compensated for by the simultaneous fall of temperature. In arcs between metal electrodes in air at atmospheric pressure, in carbon arcs in air at reduced pressure, and in arcs in inert gas atmosphere, departures from thermal equilibrium might be found (see § 5.6). Similarly, in the cathode fall region of the normal carbon arc and in electric sparks equality of electron temperature and gas temperature does not apply.

The experimental determination of discrepancies between electron temperature and gas temperature has been the subject of various studies. Usually electron temperature is identified with excitation temperature as derived from the intensities of spectral lines (see § 6.4). At known pressure the gas temperature can be determined from the gas density which is found from the velocity of sound, from the absorption of α-rays or X-rays, or from the intensity of radioactive radiation (see Finkelnburg and Maecker, 1956, pp. 376–377). The gas temperature follows also from the intensity distribution in rotational bands of molecules (see §§ 6.4 and 6.11) or from the Doppler line width of properly chosen spectral lines (see Burhorn, 1955; Finkelnburg and Maecker, 1956, p. 372; and Lochte-Holtgreven, 1958).

Mannkopff (1932, 1933) and Witte (1934) compared electron temperature with gas temperature by observing the rate at which the intensities of spectral lines emitted from the column decrease after the current has been switched off. Experiments carried out for

---

* Mandelshtam (1962) brings out arguments that show the validity of local thermal equilibrium to depend largely on the magnitude of the electron concentration in the plasma. It seems to me that these arguments apply particularly to inert gases and other such plasmas, but do not pertain to the molecular gas in the normal arc in air or nitrogen (cf. §§ 5.5 and 5.6). Mandelshtam's vagueness in this respect is confusing and might lead too easily to the discovery of 'anomalous matrix effects' in spectrochemistry.

the normal carbon arc in air at atmospheric pressure show that spectral-line intensities fall off at a rate corresponding to the cooling of the molecular gas (intensity half life period for resonance lines about 10^{-3} sec) and not at the rate at which the electron gas would cool ($\approx 10^{-7}$ sec) if it had possessed initially a much higher temperature than the molecular gas. Besides, the experiments furnished evidence for departures of the thermal mechanism in the vicinity of the cathode. An improved version of the methods used by Mannkopff and Witte was applied by Gurevich and Podmoshenskii (1963) to measure the differences between electron and gas temperature in arcs in air, argon, and mercury, at different pressures. For the arc in air at atmospheric pressure they confirmed that the electron temperature does not exceed the gas temperature by more than 0·5 per cent.

★ § 5.5 *Excitation and ionization temperatures*
Excitation and 'de-excitation'

If a gaseous body in an enclosure is in thermal equilibrium with the walls, the population of energy levels for each separate kind of particle follows a Boltzmann distribution, namely

$$\frac{n_q}{n_0} = \frac{g_q}{g_0} e^{-\epsilon_q/kT} \tag{5.4}$$

where n_q is the density or concentration of particles in the excited state q, n_0 is the density of ground state atoms, g_q and g_0 are the *statistical weights** of the corresponding levels, ϵ_q is the excitation energy of the state q, k is the Boltzmann constant, and T the absolute temperature.

Under those conditions we have 'detailed balancing', i.e. in equilibrium the total number of particles leaving a certain quantum state per sec equals the number arriving in that state per sec, and the number leaving by a particular path equals the number arriving by the reverse path. In order to apply the principle of detailed balancing let us consider three kinds of process leading to excitation and 'de-excitation' of atoms, namely

1a Collisions with neutral particles in which atoms are excited from a low level to a high level (collisions of the first kind).

1b The converse of (1a): collisions in which excited states are destroyed and no radiation is emitted (collisions of the second kind).

2a An analogous process as pictured under (1a) but occurring by collisions with electrons.

2b The converse of (2a).

3a Excitation of atoms by absorption of light quanta (cf. § 5.7).

3b Return of atoms from a high level to a low level by spontaneous or induced emission (cf. §§ 1.3 and 1.4).

For a diluted gas (density = n particles per cm³) mixed with a second gas

---

* The statistical weight of a particular state is the probability of populating a state under identical conditions. A state with total angular momentum defined by the inner quantum number commonly known as J has a statistical weight $(2J + 1)$. The J-values of the various energy levels of atoms and ions have been included in Moore's tables *Atomic Energy Levels* (see Moore, 1949, 1952, 1958).

(density $= N$ particles per cm^3, $n \ll N$), the following equations express detailed balancing:

$$\alpha N n_0 = \beta N n_q \qquad (5.5)$$

$$\alpha_e n_e n_0 = \beta_e n_e n_q \qquad (5.6)$$

$$B' \rho_v n_0 = (A + B \rho_v) n_q \qquad (5.7)$$

where N, n_0, and n_q denote the quantities defined above, n_e is the electron density, ρ_v is the density of radiation of frequency $v = \epsilon_q/h$; A, B, and B' are the Einstein coefficients (probabilities) of spontaneous emission, induced emission, and absorption (see §§ 1.3 and 1.4). Each of the quantities, α, β, α_e, and β_e, is a function of the effective cross-section for the relevant process (see § 3.4) and of the velocity distribution of the exciting species (see e.g. Finkelnburg and Maecker, 1956; and Ornstein and Brinkman, 1934). Obviously, in writing down equations (5.5) to (5.7) it has been assumed that only excitation from the ground level occurs. This restriction, as well as that made above with respect to the nature of the gas mixture, is not essential for the validity of the conclusions arrived at below (see Ornstein and Brinkman, 1934).

If the system under consideration is in thermal equilibrium, both neutral particles and electrons have a Maxwellian velocity distribution with temperature T; then it follows as a matter of course that

$$\frac{n_q}{n_0} = \frac{\alpha}{\beta} = \frac{\alpha_e}{\beta_e} = \frac{B'}{A/\rho_v + B} = \frac{g_q}{g_0} e^{-\epsilon_q/kT} \qquad (5.8)$$

i.e. the population of energy levels is a Boltzmann distribution.

In consequence of detailed balancing, the Boltzmann distribution of energy levels is maintained by *each* pair of processes, $1a$ and $1b$, $2a$ and $2b$, and $3a$ and $3b$, separately. So it must be clearly understood that collisions of the first and the second kind alone can establish this distribution without emission and absorption taking place.

It would seem difficult to picture a heated gaseous body without emission; however, we should realize that the plasma of the normal carbon arc in air at atmospheric pressure is rather like that imaginary source. In § 5.3 we pointed out the low radiation efficiency of the arc gas, which implies that, in spite of the absence of an enclosure with the same temperature as the plasma, local thermal equilibrium can be reached. Since walls are lacking, detailed balancing [equations (5.5) to (5.7)] is no longer appropriate; instead we have a steady state characterized by the equation

$$\alpha N n_0 + \alpha_e n_e n_0 + B' \rho_v n_0 = \beta N n_q + \beta_e n_e n_q + (A + B \rho_v) n_q \qquad (5.9)$$

or by

$$\frac{n_q}{n_0} = \frac{\alpha N + \alpha_e n_e + B' \rho_v}{\beta N + \beta_e n_e + A + B \rho_v} \qquad (5.10)$$

Ornstein and Brinkman (1934) argue that in a carbon arc in air at atmospheric pressure, $\alpha_e n_e + B'\rho_\nu$ can be neglected compared with αN, and likewise $\beta_e n_e + A + B\rho_\nu$ in comparison with βN. Thus, remembering (5.8) we arrive again at the Boltzmann distribution. When we cannot ignore the functions for electron processes, the final conclusion, i.e. the validity of Boltzmann's law, is not affected if the velocity distribution of electrons agrees with Maxwell's equation. Provided the mean free path for electrons (λ_e) and the field strength (E) are small, this condition is met with, so that electrons are not likely to play a separate part in excitation, particularly if an arc burns in a molecular gas.

Since the emission and absorption terms in (5.10) do not completely vanish, the population of any excited state is in principle less than follows from the Boltzmann law. The magnitude of the deviations in the normal carbon arc is, in general, beyond the precision and accuracy of our measurements.

It cannot be over-emphasized that the validity of Boltzmann's distribution law in the normal carbon arc in air at atmospheric pressure rests partly on the low radiation efficiency of this light source. Therefore emission and absorption of radiation are of minor importance; excited states are chiefly produced and destroyed by collisions.

In order to underline the plausibility of this statement, without giving laborious details, we compare the order of magnitude of transition probabilities of spectral lines with that of the collision frequency (f_c), i.e. the average number of collisions one particle makes with other particles in unit time,

$$f_c = v/\lambda \qquad (5.11)$$

where v is the random velocity and λ the mean free path. The collision frequency is the reverse of the collision period (τ_c) defined by (5.1) for the special case of electrons colliding with heavy particles. For molecular nitrogen at 6000°K and 1 atm, we find the root mean square velocity to be $2 \cdot 3 \cdot 10^5$ cm/sec [equation (3.10)], if dissociation is ignored. In virtue of $\lambda = 1 \cdot 77 \cdot 10^{-4}$ (Table 3.3) the collision frequency is of the order of 10^9 sec^{-1}. For excited particles, f_c will be still higher.

The gas mixture in the carbon arc in air at atmospheric pressure consists chiefly of diatomic molecules (N_2, CO, NO) and radicals (CN), whereas added vapours contribute only a negligible fraction to the total composition (see § 7.9). Atoms of substances volatilized from the electrode cavity will thus mainly collide with diatomic molecules, which possess rotational and vibrational degrees of freedom and are able to accept widely differing amounts of energy. When an excited atom and a molecule collide, the excitation energy of the atom is easily passed on to the molecule; therefore the molecular nature of the arc gas involves high efficiency for collisions of the second kind in which excitation energy of atoms is converted into another form of energy but not into radiation. For this reason the frequency at which excited states of atoms are destroyed is practically equal to the collision frequency ($\approx 10^9$ sec^{-1}). Comparison of this value with the magnitude of transition probabilities (10^7 to 10^8 sec^{-1} for most levels, cf. Table 1.1 and Corliss and Bozman, 1962) shows that the greater part of excited states will be lost in collisions of the second kind, whereas only a small fraction of the atoms involved is likely to

4

give up excitation energy by emission of light quanta. The predominance of collisions in producing and destroying excited states also implies that metastable levels are of no importance in the arc.

Self-absorption causes no complications regarding the state of thermal equilibrium; on the contrary, errors introduced by neglecting radiation losses in equation (5.10) are compensated for when self-absorption occurs. In the extreme case of equilibrium between radiation and matter the Boltzmann distribution holds exactly. According to Smit (1950, p. 92; 1946) the radiation density at almost all wavelengths is much smaller than would follow from Planck's law; only in the centre of spectral lines with a high degree of self-absorption Planck's upper limit is approximately attained (see also § 5.7).

The foregoing discussion does not imply that self-absorption can be neglected when studying spectral-line intensities: obviously the proportionality between intensity of emitted radiation leaving the source, on the one hand, and number of excited atoms, on the other, is interfered with by self-absorption (see § 5.7).

Experimentally the conception of local thermal equilibrium in the arc is justified by numerous measurements of transition probabilities (see § 1.3) of excited states of atoms and ions: by assuming a population of energy levels according to Boltzmann's law the values of transition probabilities found proved to be practically independent of temperature and electric field strength, i.e. for the normal low-current carbon arc in air at atmospheric pressure. Many investigations in this field have been done by Ornstein and his co-workers in the Physical Laboratory of the State University at Utrecht. The work up to 1946 has been reviewed by Smit (1946) (cf. § 1.3).

The appropriateness of Saha's relationship, also treated theoretically by Ornstein and Brinkman (1934), has been demonstrated by Kruithof (1943a) and by Kruithof and Smit (1944). Thus it can be assumed that the ionization temperature is equal to the gas temperature.

The production of ions in the arc is mainly by collisions with thermal electrons and rarely by absorption of light quanta (photo-ionization), evidently because radiation density is low. The reverse path is for the greater part recombination in the presence of a third body, three-body collision (*Dreierstossrekombination*). The third body can be an electron, an atom, an ion, or a molecule either excited or in the ground state. The energy liberated in the recombination process is shared by three bodies and reappears as kinetic energy, or as excitation or dissociation energy. Occasionally recombination occurs in two-body encounters; in this type of recombination process the energy evolved is emitted (see also § 7.11).

Because photo-ionization is lacking, the latter process might be the cause of a deficiency in the balance such that Saha's relationship cannot hold rigorously (see Finkelnburg and Maecker, 1956, pp. 308–309; and Unsöld, 1955, p. 652, ff.); elementary processes involved in ionization and recombination are treated by von Engel (1955).

★ § 5.6 *Conclusions and further remarks*

Summarizing our previous discussion we conclude that in the carbon arc in air at atmospheric pressure the electric current heats the gas in the arc column, whereas the resulting high temperature governs excitation, ionization, and dissociation of atoms and molecules.

Electrons play a role of secondary importance.* They have a Maxwellian velocity distribution corresponding to the gas temperature; electrons cannot acquire high velocities owing to excitation of vibrational levels of molecules. Radiation is emitted only by a small fraction of the excited particles. The molecular nature of the arc gas is indispensable in order that local thermal equilibrium be reached (cf. Gurevich and Podmoshenskii, 1963).

The population of the various energy levels satisfies Boltzmann's distribution law, the ionization equilibria are governed by Saha's relationship, and the dissociation equilibria are described by the general equation for chemical equilibria.

Spectrochemical literature does not always give evidence of a right understanding of the essential features of the thermal mechanism of the arc discharge, for example several papers quoted by Ahrens and Taylor (1961, pp. 126–127), among which is Strock's discussion (Strock, 1953, 1954) on the significance of collisions of the second kind in d.c. arc spectrochemical analysis; it conflicts with the picture given above.†

It must be noted that in discharges at reduced pressure or in atomic gases the thermal mechanism is no longer valid. Under these conditions interactions of energy levels differing from those described in the preceding section (i.e. interactions leading to the Boltzmann distribution) are feasible (see for example von Engel, 1955, p. 257; and Fowler, 1956).

Discharges in inert gas atmospheres, which are coming into use in ever-increasing number for spectrochemical analysis,‡ are not necessarily thermal, because the efficiency for collisions of the second kind is low in inert gases. Recently, Kolesnikov and Sobolev (1963)¶ made an experimental study of thermal equilibrium in arcs burning in argon or helium atmospheres. They determined temperatures (1) from the rotational lines of molecular bands, (2) from relative line

In contrast with Ornstein and Brinkman (1934), Mannkopff (1932, 1933) and Witte (1934) believe that excitation and ionization in the arc are chiefly caused by electrons. The conception does not interfere with the final conclusion regarding the existence of thermal equilibrium; for the difference between electron temperature and gas temperature can be ignored as was demonstrated by Mannkopff and Witte.

† Hurwitz (1959) was able to show that the deviations of the working curves from a slope of unity, upon which Strock's arguments are based, could be satisfactorily explained by the experimental errors inherent in the method or by self-absorption.

‡ See for example, Mochalov and Raff (1956), Andermann and Kemp (1959), Shaw, Wickremasinghe, and Weber (1960), Majkowski and Schreiber (1960), Farský (1960), Annell and Helz (1960, 1961), Morrison and Rupp (1961), Curtis (1962), Margoshes and Scribner (1964), Wang and Cave (1964), Margoshes (1965), and references on the Stallwood-jet and the plasma jet cited in notes on pp. 5 and 30.

Many interesting observations on d.c. arcs in inert gas atmospheres have been reported by Vallee, Reimer and Loofbourow (1950), Vallee and Peattie (1952), Vallee and Adelstein (1952), Adelstein and Vallee (1954), Vallee and Baker (1956), Vallee and Thiers (1956), and Thiers and Vallee (1957, review); the interpretation of the facts observed is sometimes in dispute, however.

¶ See also Sobolev, Kolesnikov, and Kitayeva (1960), Kitayeva, Obukhov–Denisov, and Sobolev (1962), and Shvangiradze, Oganezov, and Chikhladze (1962).

intensities, (3) from absolute line intensities, and (4) from the ionization. The experiments led to the conclusion that thermal equilibrium is not reached in these arcs when the current strength is below 15 amp; for higher current strength the four temperatures proved to be numerically equal, so that thermal equilibrium can be assumed.

We remind ourselves of the observation made by Gurevich and Podmoshenskii (1963) that even a small quantity of molecular impurities (about 1 per cent) suffices for establishing local thermal equilibrium in a low-current (5-amp) argon arc. Therefore, the exact composition of the inert gas atmosphere in which an arc burns is of crucial importance for the behaviour of the discharge with respect to local thermal equilibrium.

Summarizing, departures from the thermal mechanism may occur in the cathode fall region of the normal carbon arc in air at atmospheric pressure, in arcs between metal electrodes, in arcs at reduced pressure and in arc discharges taking place in inert gas atmospheres. Deviations from the thermal mechanism will be first noticeable as departures from Boltzmann's distribution law for high energy levels, from Saha's relationship for elements of high ionization potential, and from the dissociation equation for molecules of high dissociation energy.

★ § 5.7   *Self-absorption of spectral lines*

Radiation generated in the interior of a source is subject to absorption on its passage to the outside. The absorption occurs by atoms or molecules of the same kind as those causing the emission. The phenomenon is called *self-absorption*.

An atom or molecule is capable of absorbing a particular light quantum if it populates the lower level of the relevant transition. By the absorption process it is excited to the corresponding upper level. In a thermal source, excitation energy is predominantly converted by collisions of the second kind into kinetic energy; only a few de-excitations happen by spontaneous emission (see § 5.5). Consequently, an absorbed light quantum has only a small chance of being re-radiated, so that self-absorption weakens the intensity of spectral lines and destroys the proportionality between intensity of radiation and concentration of emitting particles.

To a first approximation absorption of radiation by atomic vapours can be treated similarly to that in solutions, that is, in terms of the Lambert–Beer law. However, a rigorous theoretical treatment of atomic absorption is more complex and involved, since we must consider the intensity distribution within one spectral line, i.e. the line profile.

When discussing line profiles, one conveniently distinguishes between the *physical profile* and the *instrumental profile*. The former is the shape the line has in virtue of the physical conditions in the source of excitation, that is before interference from the spectrograph; the latter profile is the pattern that results from the influence of the spectrographic equipment on a uniform intensity distribution.

Superposing the instrumental line contour on the physical line contour gives the finally observed optical profile.*

The physical shape of a line is controlled by three major factors.

1 *Natural broadening*, which has its origin in the finite lifetime of an excited state. Its contribution to the half width of a line is normally of the order of 0·001 Å only and can be ignored in comparison with the other contributions.

2 *Doppler broadening*, due to the random motion of the emitting atoms. The Doppler half width is proportional to $\lambda\sqrt{(T/M)}$, where λ is the wavelength, T the absolute temperature, and M the molecular weight of the emitting species. In the temperature range of the normal arc Doppler half widths are of the order of several hundreds of an ångström.

3 *Collisional broadening*, resulting from the perturbation of energy levels by impact (Lorentz broadening) and by the presence of an electric field of surrounding charges (statistical broadening, Holtsmark broadening). Collisional broadening usually gives line widths much larger than those due to the other causes.

In recent years the theory of statistical line broadening has become of utmost interest in the field of plasma physics for determining electron density from Stark profiles (see references in § 3.8).

The natural and Lorentz profiles can be represented mathematically by a dispersion function $1/(1+x^2)$; the Doppler profile follows a Gaussian distribution e^{-x^2}. Both combine to form an intensity distribution which will be given by a Voigt function (see e.g. Junkes and Salpeter, 1961, p. 472).

The intensity distribution within a spectral line emitted by a point source can be represented as a function $I_0 . P_E(\nu)$, where I_0 is the total intensity of the line for no absorption and $P_E(\nu)$ is taken to be a normalized function expressing the shape of the emission line. The intensity distribution $I(\nu)$ that results when the radiation has travelled through a vapour layer containing n_A atoms per cm^3 capable of absorbing that radiation can be written as

$$I(\nu) = I_0 . P_E(\nu) \exp - \left[p \frac{P_A(\nu)}{P_A(\nu_0)} \right] \tag{5.12}$$

where ν_0 is the frequency in the centre of the line, $P_A(\nu)$ the absorption profile, and p an absorption parameter, defined as

$$p = (h\nu_0 B/c) P_A(\nu_0) \int n_A(x) \, dx \tag{5.13}$$

---

* The contribution of instrumental factors to the optical profile, including the part played by the photographic emulsion, has been discussed at length by Junkes and Salpeter (1961, 1963).

An extensive bibliography on physical line broadening exists. We mention only a few sources, viz. Mitchell and Zemansky (1961), Cowan and Dieke (1948), Unsöld (1955), Sobolev (1957), Ch'en and Takeo (1957), Breene (1961), Bartels and Zwicker (1962), Zwicker (1962), Pingel and Sichmeier (1962), Fishman (1962), Fishman, Shaimanov, and Ilin (1963), Preobrazhenskii (1959, 1963), Grechikhin (1962), Gorbacheva and Preobrazhenskii (1963), and the articles on the Stark profile quoted on p. 60.

The main line of the exposition given here is based on the article by Cowan and Dieke.

Here h and c are the Planck constant and the velocity of light respectively, and B is the Einstein coefficient of absorption. According to § 1.4, B is proportional to the probability (A) and to the oscillator strength (f) of the transition. We have

$$A :: \nu_0{}^3 B :: \nu_0{}^2 f \tag{5.14}$$

From equations (5.12) and (5.13) the following conclusions are drawn.

1 Absorption is strongest at the peak of the line; hence self-absorption tends to flatten the profile.

2 If the absorption parameter becomes > 1, there is a dip in the centre of the line. This phenomenon is termed 'self-reversal'; it is only an extreme case of self-absorption.

3 The absolute number of absorbing particles determines the relative attenuation of spectral-line intensities.

4 When we consider different spectral lines of one element which have upper levels of nearly the same energy and likewise have lower levels of nearly equal energy (e.g. lines due to transitions between two multiplet terms in an atom having fairly small multiplet splitting), then the larger the transition probability of a line the more strongly it is absorbed. In other words the most intense line of the multiplet is the most affected by self-absorption.

5 Finally we should consider that in a light source with a mixture of elements and with given excitation characteristics, self-absorption is primarily determined by the intensity I_0 of the line. It does not matter, for example, whether a line appears relatively strong because it is an intrinsically strong line (large transition probability) with a relatively small concentration of the element, or whether the line is intrinsically weak but appears strong owing to the large concentration of the element in the source.

Formula (5.12) pertains to a source where the emitting and absorbing atoms are spatially separated. An instructive example conforming to this model is furnished by the recently developed atomic absorption spectrophotometry: radiation emanating from a source of constant intensity (e.g. a hollow-cathode discharge tube) is absorbed by atomized vapour sprayed in a flame that is located at some distance from the primary source.

The pendant of the model with complete spatial separation of emission and absorption is the source where emission and absorption take place in one and the same volume. If the conditions of excitation are uniform throughout, the function $I(\nu)$ can be evaluated (see Cowan and Dieke, 1948). It must be considered then that the nearer the origin of emitted radiation to the observer the thinner the absorbing layer. Obviously, the fraction of the emitted radiation that is absorbed in this case is smaller than in the former one. With increasing concentration of the emitting and absorbing particles the intensity in the centre of the line asymptotically approaches the intensity of blackbody radiation at the same temperature and wavelength (Planck's upper limit). Self-reversal does not occur.

If the part of the light source with uniform temperature is surrounded by a zone of lower temperature, which still contains free atoms, self-reversal is feasible. In the outer layer the ratio of the numbers of absorbing and emitting particles is higher than in the core of the source and consequently, in the envelope absorption is much less compensated for by emission than in the inner high-temperature region. In addition, the profiles of the lines emitted from inner and outer zones differ.

Such a vapour layer of relatively low temperature, which envelops the light source, is the typical example for demonstrating the occurrence of self-reversal. Hence, it has become customary to refer to self-reversal as self-absorption by a vapour layer of low temperature rather than as an extreme case of self-absorption.

In an are the actual situation is complicated in view of the radial distributions of temperature and emission. To a first approximation we can assume the core to contain absorbing as well as emitting atoms with constant excitation throughout, the fringe being a layer with absorbing atoms only. Thus it should be described in terms of a model that is a combination of the preceding ones. Still, the sudden change between excited and unexcited portions makes it a poor approximation of the true situation.

Further remarks on self-absorption can be found in § 10.2 (see also Ovechkin, 1964).

CHAPTER SIX

EXCITATION PHENOMENA AND
TEMPERATURE MEASUREMENT

★ **§ 6.1** *Mathematical expressions for Boltzmann's distribution law*

In § 5.5 Boltzmann's law was formulated as a relation between temperature and the ratio n_q/n_0 of the concentrations of atoms in the excited state q and in the ground state o. We shall now find an expression giving the dependence of the ratio n_q/n on temperature. It is the more important one because the total concentration n occurring here is an independent parameter.

To derive the equation we write instead of (5.4)

$$n_0 : n_1 : n_2 : \ldots n_j =$$

$$g_0 : g_1 \exp(-\epsilon_1/kT) : g_2 \exp(-\epsilon_2/kT) : \ldots g_j \exp(-\epsilon_j/kT) \qquad (6.1)$$

where the energy of the levels 1, 2, ..., j, has been taken from the ground level zero. Now, since

$$\sum_{m=0}^{m=j} n_m = n \qquad (6.2)$$

we have*

$$\frac{n_q}{n} = \frac{g_q \exp(-\epsilon_q/kT)}{\sum_{0}^{j} g_m \exp(-\epsilon_m/kT)} \qquad (6.3)$$

The quantity in the denominator at the right-hand side of (6.3), viz.

$$Z = \sum_{0}^{j} g_m \exp(-\epsilon_m/kT) \qquad (6.4)$$

---

* In terms of probabilities, this result can be derived as follows. The quantity
$$P_q = C g_q \exp(-\epsilon_q/kT) \qquad (1)$$
is the probability of finding an atom in the state q; obviously, the value of the constant C is such that

$$\sum_{0}^{j} P_q = 1 \qquad (2)$$

whence

$$C = \frac{1}{\sum_{0}^{j} g_m \exp(-\epsilon_m/kT)} \qquad (3)$$

is called the *partition function* (or the sum over states) of the particular atom. The summation will include all possible energy levels of the relevant particle (see § 6.2 for further comments).

To distinguish the quantities n and Z of neutral atoms and singly-charged ions, the subscripts a and i will be used throughout; a superscript $+$ will be added to distinguish g for ions from that for atoms. Accordingly for atoms we write

$$n_{aq} = n_a \frac{g_q}{Z_a} \exp(-\epsilon_q/kT) \qquad (6.5)$$

A similar equation holds for ions.

The practical form of expression (6.5) is

$$n_{aq} = n_a \frac{g_q}{Z_a} 10^{-5040\, V_q/T} \qquad (6.6)$$

where V_q is the excitation potential expressed in electron volts.*

Equation (6.6) is often conveniently employed in its logarithmic form, i.e.

$$\log n_{aq} = \log n_a + \log g_q - \frac{5040}{T} V_q - \log Z_a \qquad (6.7)$$

§ 6.2 *Partition functions*

The partition function of an atom or ion as defined by (6.4) is computed by adding the terms in the right-hand side of the equation. This addition when performed for an unperturbed atom involves an infinite number of terms and thus the difficulty of a divergent series arises. Theoretical considerations indicate, however, that, in view of perturbations and interactions within the plasma, the high terms in the neighbourhood of the series limit must be discarded. We shall not engage in theoretical discussions about the precise location of the cut-off level,† as this level is in most instances beyond what we should regard as the practical limit for our purpose.

Numerical values of partition functions have been published by Claas (1949, 1950), Boumans (1961, 1962), Corliss (1962b and c), and Drawin and Felenbok (1965). I, being initially unacquainted with Claas' work, computed the partition functions Z_a and Z_i of the neutral atoms and singly-charged ions of fifty-four elements for $T = 5000$ and $6000°K$ (Table 6.1). They may serve as a check on Claas' values and on those reported by Corliss. There is serious disagreement between Claas' values and mine for La I, Nb I, Nb II, and Re I; small discrepancies occur for Au I, Mo I, Mo II, P II, Pt I, and Ru I. The few serious discrepancies are most probably due to revisions of numerical values of energy

---

* See Appendices 2 and 9 for the conversion of units.

† The question of divergence of partition functions is closely related to the lowering of the ionization potential within the plasma. See Unsöld (1948, 1955, p. 84), Claas (1949, 1951), Ecker and Weizel (1956), Theimer (1957), Griem (1962, 1964b), and Lochte–Holtgreven (1963).

levels or term classifications. My calculations are based on the data in Moore's tables, *Atomic Energy Levels* (Moore, 1949, 1952, 1958). For practical reasons I allowed for levels up to 26 000 cm^{-1} or 3·3 eV when $T = 5000°$K, and up to 31 000 cm^{-1} or 3·9 eV when $T = 6000°$K, except for Cs I, where I stopped adding at the level of 30 000 cm^{-1}. Corliss' summation was only for all levels up to 14 000 cm^{-1}; for that reason his results tend to be somewhat lower than mine, though the discrepancy is immaterial.

If we compare 'complete' partition functions with more or less obvious approximations, we notice the following. The statistical weight (g_0) of the ground level

TABLE 6.1

Element	Z_a 5000°K	Z_a 6000°K	Z_i 5000°K	Z_i 6000°K	Element	Z_a 5000°K	Z_a 6000°K	Z_i 5000°K	Z_i 6000°K
Ag	2·00	2·01	1·00	1·00	Ni	30·8	32·6	10·8	12·4
Al	5·9	5·9	1·00	1·00	Os	19·7	24·3	17·8	20·6
As	4·5	4·8	5·9	6·5	P	4·4	4·7	8·6	9·0
Au	2·43	2·68	1·14	1·26	Pb	1·55	1·90	2·08	2·14
B	6·00	6·00	1·00	1·00	Pd	2·92	3·84	7·4	7·8
Ba	2·53	3·64	4·2	4·8	Pt	19·8	21·1	9·7	11·2
Be	1·02	1·04	2·00	2·00	Rb	2·2	2·62	1·00	1·00
Bi	4·2	4·4	1·10	1·20	Re	7·5	9·3	7·6	8·5
C	9·2	9·4	5·9	5·9	Rh	26·5	30·4	15·4	17·1
Ca	1·17	1·38	2·21	2·41	Ru	33·9	41·1	23·7	27·6
Cd	1·00	1·00	2·00	2·00	Sb	4·7	5·2	3·3	4·0
Co	33·5	38·0	29·6	33·4	Sc	11·9	14·0	22·7	25·1
Cr	10·3	12·6	7·1	8·3	Se	7·5	7·9	4·2	4·4
Cs	2·5	3·1	1·00	1·00	Si	9·5	9·8	5·7	5·7
Cu	2·32	2·60	1·02	1·07	Sn	5·15	5·8	3·2	3·4
Fe	27·7	31·6	42·6	47·6	Sr	1·23	1·56	2·16	2·32
Ga	5·1	5·3	1·00	1·00	Ta	17·0	22·5	22·3	28·2
Ge	7·5	8·1	4·4	4·6	Te	6·3	6·7	4·4	4·7
Hf	13·8	16·6	12·4	15·5	Th	19·0*	—	34·3*	—
Hg	1·00	1·00	2·00	2·00	Ti	29·4	36·0	55·4	61·2
In	4·1	4·4	1·00	1·00	Tl	2·42	2·62	1·00	1·00
Ir	20·9*	—	14*	—	U	54*	—	80*	—
K	2·18	2·44	1·00	1·00	V	47·6	55·9	43·2	49·9
La	25·7	32·6	29·4	33·8	W	12·7	16·5	13·8	18·1
Li	2·09	2·20	1·00	1·00	Y	11·7	14·3	15·8	18·1
Mg	1·02	1·04	2·00	2·00	Zn	1·00	1·00	2·00	2·00
Mn	6·4	6·9	7·7	8·4	Zr	33·9	43·0	45·2	52·9
Mo	8·8	11·2	7·6	9·5					
Na	2·05	2·11	1·00	1·00					
Nb	53·3	63·9	43·0	52·2					

Partition functions for the neutral atoms (Z_a) and the singly-charged ions (Z_i) of fifty-seven elements of spectrochemical interest, according to calculations by Boumans (1961, 1962) and Corliss (1962b and c). Corliss' values are marked with an asterisk.

is a good approximation only for atoms and ions having a simple level diagram (see Fig. 6.1). For complex atoms the sum of the statistical weights of the ground term components gives better agreement, although it may be too high an estimate of the partition function. Very good results are obtained if we calculate Z following equation (6.4), while allowing for a few terms only, say those with an exponential factor > 0.1. This corresponds to a cut-off level of 9700 cm^{-1} or 1.2 eV when $T = 6000°K$. The number of terms to be taken into account then ranges from one to fifteen and the work involved in the calculation is not too elaborate.

In most applications of Boltzmann's distribution law and Saha's ionization relationship in spectrochemistry, partition functions are entirely omitted. In some instances the partition function is taken to be constant, and equal to the statistical weight of the ground level of the atom or ion in question; then, according to equation (6.5), the population of a particular energy level depends exponentially on the temperature if the concentration n_a is fixed. Actually, the exponential function is smoothed out to some extent because the partition function increases with temperature. This rise is rather sharp for a number of elements. Let us consider the physical reason for it.

Boltzmann's law can be written either in the form (5.4), viz.

$$n_q = n_0 \frac{g_q}{g_0} \exp(-\epsilon_q/kT) \tag{6.8}$$

or in the form (6.3), viz.

$$n_q = n \frac{g_q}{Z} \exp(-\epsilon_q/kT) \tag{6.9}$$

From the former equation it could appear that n_q is a truly exponential function of T; however, we must bear in mind that it is reasonable to take as the independent parameter the concentration (n) of all the atoms of the relevant type and not the concentration (n_0) of the ground-state atoms. Therefore, if n is independently fixed, n_0 is a function of the temperature. When the temperature rises, the population of the ground state is depleted in favour of the population of excited states; other low levels also become less populated in favour of high levels, as will be brought out below (see also the end of § 6.3).

From equation (6.8) we conclude that the rise of n_q is not truly exponential, because the ground-state population (n_0) declines with increasing temperature; from equation (6.9) it is explained by considering the increase of the partition function Z, which results from an increase in the population of excited states.

In this connection it will be instructive to examine the proportions of ground-state atoms and excited ones a little more closely. From equations (6.8) and (6.9) we obtain

$$\frac{n_0}{n} = \frac{g_0}{Z} \tag{6.10}$$

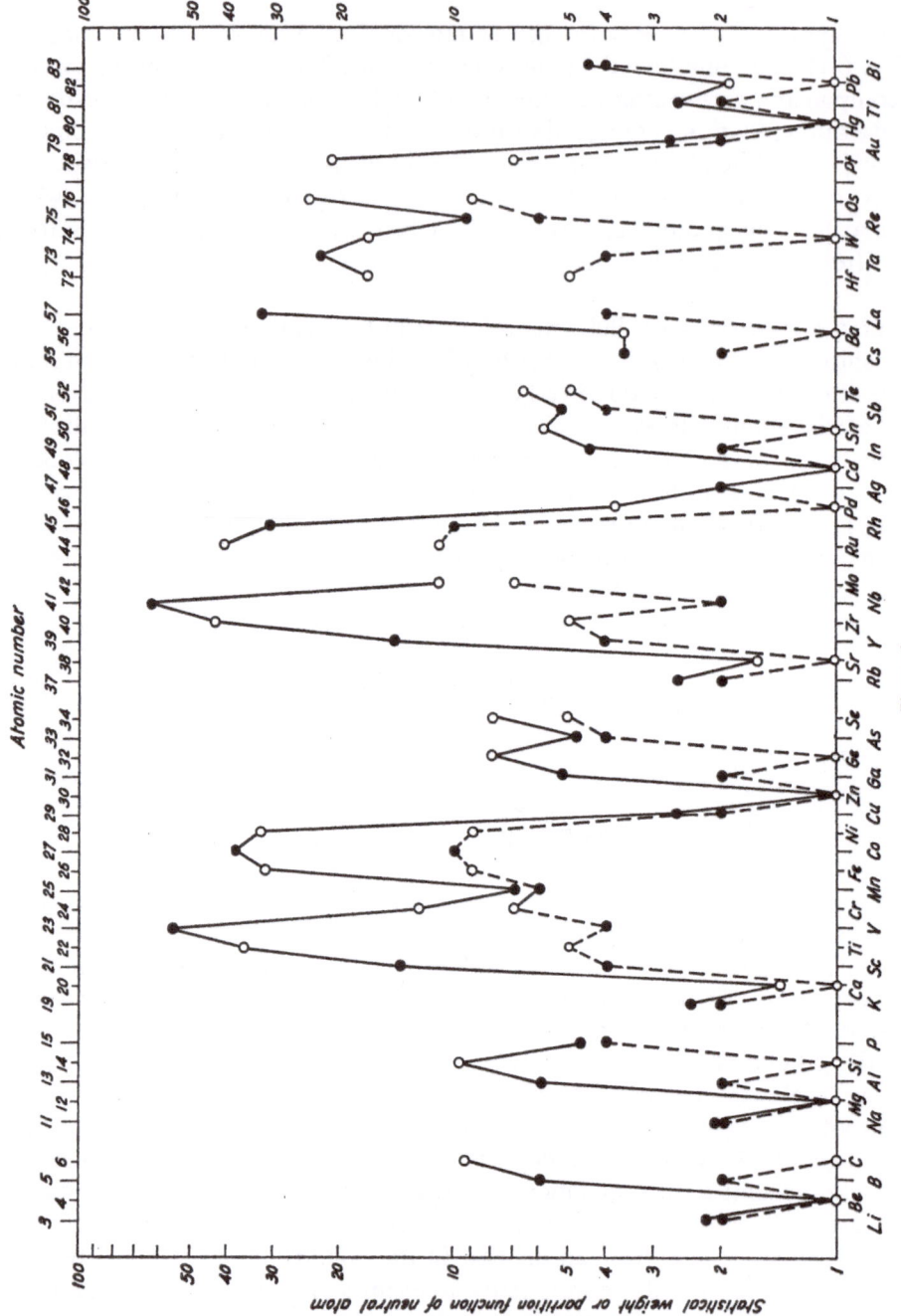

FIG. 6.1. Statistical weight of the ground level (broken lines) and partition function (continuous lines) of the neutral atom for $T = 6000\,^{\circ}K$ plotted as functions of the atomic number of the elements listed in Table 6.1.

and

$$1 - \frac{n_0}{n} = 1 - \frac{g_0}{Z} \qquad (6.11)$$

The expressions give the proportion of ground-state atoms and the proportion of excited atoms to the total number of the relevant atoms per cm³. When Z does not differ markedly from g_0, the concentration of ground-state atoms is practically equal to the total concentration, which is so for Ag, Be, Mg, Cd, Zn, and Hg.

For those atoms and ions that possess a dense energy level diagram, depletion of the ground-state population may be appreciable in the temperature range of the normal carbon arc; this holds particularly for transition elements.

To illustrate this point Fig. 6.2 shows the proportions n_0/n for Ag, Al, Fe, Zr, W, and Nb for $T = 5000$ and $6000°K$. Curiously, the ground level of Al, and similarly of B, Ga, In, Tl, C, Si, Ge, Sn, and Pb, is depleted to a large extent, although these elements possess relatively simple level diagrams. The behaviour is accounted for by the ground term being a doublet or a triplet term; thus the population of the component above the ground level competes easily with the population of the ground state. If we evaluate the proportion of atoms in the ground term according to

$$\frac{\sum\limits_{0} n_m}{n} = \frac{\sum\limits_{0} g_m \exp(-\epsilon_m/kT)}{Z} \qquad (6.12)$$

its numerical value will turn out to be practically equal to unity for all the elements just mentioned.

For complex atoms, which in general have a multiplet ground term (except for Cr, Mn, Mo, Pd, and Re), the proportion (6.12) is likewise significantly larger than the proportion n_0/n, but it does not approach unity (see Fig. 6.2).

To get an idea of the proportion of the populations of excited states, it is not rational to consider the ground level as the only non-excited state; apart from many low levels being metastable and thus having no importance for emission, permitted transitions from low levels to the ground state involve infra-red lines only. If we take a level of $1·2$ eV above the ground state as an arbitrary limit between low and high levels, we obtain for Fe, Zr, W, and Nb the proportions indicated in Fig. 6.2.

Summarizing, we have seen by considering partition functions, that for a number of elements the population of the ground state cannot be regarded as constant; for these elements the continued depletion of the ground state with temperature causes the populations of high levels to rise less sharply with temperature than would be expected if the population of the ground level were virtually constant. The extent to which the increase of the population n_q with temperature, for a fixed total concentration n, departs from the truly exponential form depends on the relative rate at which the partition function Z rises with temperature.

Remembering that $n_0/n = g_0/Z$ we see from Fig. 6.2 that the extent of the departures will be in the succession Ag $\approx$ Al $\ll$ Fe < Nb < Zr < W.

In order that partition functions can be conveniently taken into account when applying Boltzmann's relation, I examined the possibility of using simple corrections for excitation potentials; it will be discussed in the following section.

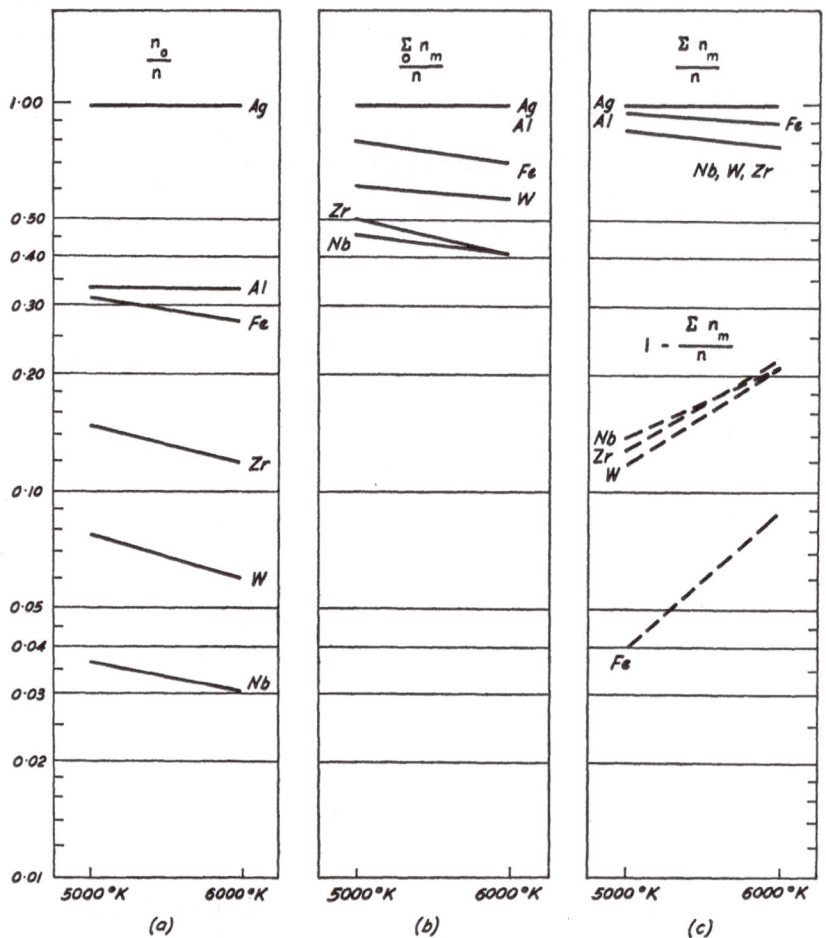

FIG. 6.2. Illustration of the interrelation between the partition function and the proportions of the ground level population, of the ground term population, and of the low-level population, according to equations (6.10) to (6.12), for the neutral atom of Ag, Al, Fe, Zr, W, and Nb, at 5000 and 6000°K.

    (a) Proportion of ground level population (n_0/n).

    (b) Proportion of ground term population ($\Sigma_0 n_m/n$).

    (c) Upper part: proportion of population of levels having an excitation potential < 1·2 eV ($\Sigma n_m/n$). Lower part: proportion of population of levels having an excitation potential > 1·2 eV ($1 - \Sigma n_m/n$).

The straight lines have been drawn for convenience only; they do not truly represent functional relationships.

A similar, approximate correction to ionization potentials applies if allowance for partition functions is to be made for in Saha's ionization relationship (see § 7.3).

§ 6.3 *Correction of excitation potentials*
Apparent values

The dependence of partition functions on temperature can be approximately taken into account by simply applying a correction to the numerical values of the

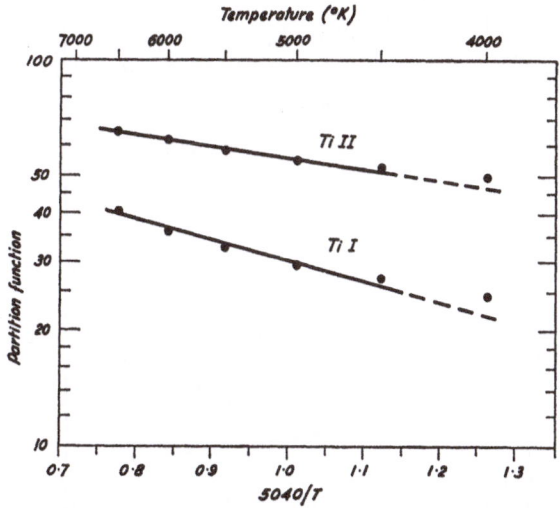

FIG. 6.3. Partition functions Z_a and Z_i of titanium plotted against $5040/T$ to demonstrate the approximate validity of a linear relationship between log Z and $1/T$, in the temperature range 4500–6500°K.

excitation potentials of energy levels. This correction can be introduced if we assume that the partition function Z can be written in the form

$$Z = Z' \exp(-\Delta\epsilon/kT) \qquad (6.13)$$

where k is the Boltzmann constant and T the temperature; Z' and $\Delta\epsilon$ are supposed to be constants for a particular type of atom or ion. We must not expect that this hypothesis will hold for large ranges of temperature; in restricted intervals, however, equation (6.13) proves to be a sufficiently accurate approximation. I first demonstrated this for Ti I and Ti II in the temperature range 4500–6500°K. In Fig. 6.3, Z_a and Z_i, according to van Stekelenburg (1943, p. 73), have been plotted logarithmically against $5040/T$. The plots show that the relationship between log Z and $5040/T$ is reasonably well approximated by the linear function

$$\log Z = \log Z' - \frac{5040}{T}\zeta \qquad (6.14)$$

which is the practical, logarithmic form of (6.13). The constant ζ is equivilent to $\Delta\epsilon$ in (6.13) and is expressed in electron volts. From the slope of the curves in Fig. 6.3, it follows that $\zeta_a = 0.53$ eV and $\zeta_i = 0.29$ eV.

On the basis of the Z-values published by Claas (1951) for 4930, 5140, 5370, 6170, 6450, and 7370°K, the ζ-hypothesis turned out to be appropriate in a

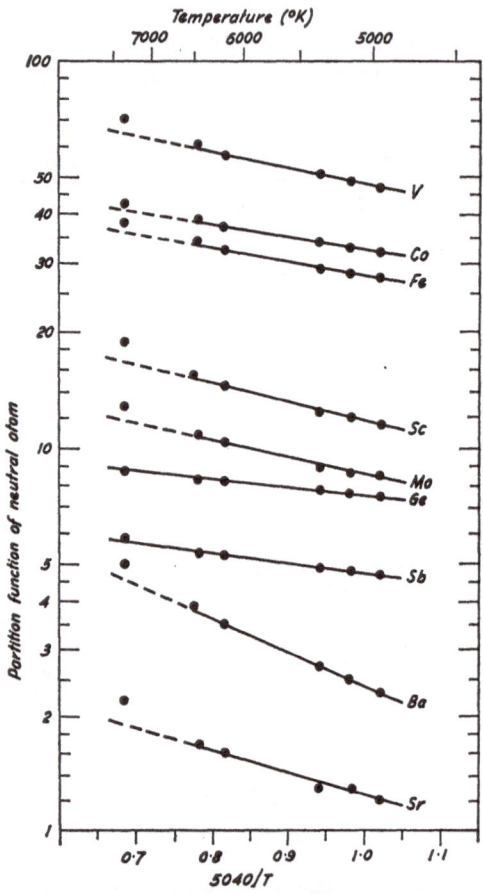

FIG. 6.4. Partition function of the neutral atom of various elements plotted as a function of $5040/T$. The examples show a linear relationship between log Z and $1/T$ to be appropriate in the temperature range 4900–6500°K.

temperature range between 5000 and 6500°K for all relevant elements, with the exception of Ca I. In the majority of cases the ζ-values calculated for that range are no longer valid beyond 6500°K. A few examples are shown in Fig. 6.4. The straight lines have been drawn through the points corresponding to $T = 4940°$K and $T = 6170°$K.

Table 6.2 gives the ζ-values calculated from the partition functions for $T = 5000$ and 6000°K listed in Table 6.1.

For the alkali metals and for calcium, strontium, barium, and lanthanum application of the ζ-correction is critical. Those elements having a low ionization potential will be ionized to a high degree in the arc at temperatures beyond 5000°K. As will be explained in detail in § 6.12 and Chapter 8, atom lines of these elements are mainly emitted from zones with temperatures less than 5000°K; consequently, the ζ-correction does not apply rigorously, since it represents a mean value for the temperature range 4500–6500°K. Moreover, the values of the partition functions for the alkali metals and barium are incomplete (Claas, 1949, 1951).

Conclusion: In the Boltzmann equation, (6.7), the dependence of the partition function on temperature can be taken into account, to a first approximation, by applying a correction ζ to the excitation potential V_q, viz.

$$\log n_q = \log n + \log g_q - \log Z' - \frac{5040}{T}(V_q - \zeta) \tag{6.15}$$

in other words, by replacing the excitation potential V_q by an *apparent value* $\bar{V}_q$, defined as

$$\bar{V}_q = V_q - \zeta \tag{6.16}$$

The quantity ζ is defined by equations (6.13) and (6.14). Numerical values of ζ have been tabulated in Table 6.2 for the neutral atoms and the singly-charged ions of fifty-four elements. Accordingly, the ζ-correction is fairly large ($\geqslant$ 0·4 eV) for Ba, Ca, Cr, Cs, Hf, La, Mo, Nb, Os, Pb, Pd, Rb, Re, Ru, Sc, Sr, Ta, Ti, V,

TABLE 6.2

Element	ζ_a	ζ_i	Element	ζ_a	ζ_i	Element	ζ_a	ζ_i
Ag	—	—	Hf	0·45	0·60	Re	0·55	0·30
Al	—	—	Hg	—	—	Rh	0·35	0·25
As	0·15	0·25	In	0·15	—	Ru	0·50	0·40
Au	0·25	0·25	K	(0·30)	—	Sb	0·25	0·50
B	—	—	La	(0·60)	0·35	Sc	0·45	0·25
Ba	(0·95)	0·35	Li	(0·10)	—	Se	0·10	0·10
Be	0·05	—	Mg	0·05	—	Si	0·10	—
Bi	0·10	0·20	Mn	0·20	0·20	Sn	0·30	0·15
C	0·05	—	Mo	0·60	0·60	Sr	(0·60)	0·15
Ca	(0·45)	0·20	Na	(0·10)	—	Ta	0·70	0·60
Cd	—	—	Nb	0·45	0·50	Te	0·15	0·15
Co	0·30	0·30	Ni	0·15	0·35	Ti	0·50	0·25
Cr	0·50	0·40	Os	0·55	0·40	Tl	0·20	—
Cs	(0·60)	—	P	0·15	0·10	V	0·40	0·40
Cu	0·30	0·10	Pb	0·55	0·10	W	0·70	0·70
Fe	0·35	0·30	Pd	0·70	0·10	Y	0·50	0·35
Ga	0·10	—	Pt	0·15	0·35	Zn	—	—
Ge	0·20	0·10	Rb	(0·45)	—	Zr	0·60	0·40

ζ-corrections for excitation potentials of atom and ion lines between 4500 and 6500°K as derived from the partition functions for $T = 5000$ and 6000°K listed in Table 6.1 (cf. Boumans, 1961, 1962).

W, Y, and Zr. Yet I believe that it is too small to entail great practical interest in spectrochemical analysis, for example, in selecting line pairs. Surely, there is more justification in taking into account the ζ-correction than omitting partition functions at all; as a matter of fact, however, in spectrochemical analysis, so many compromises are necessary that I doubt whether a perfection like the ζ-correction will produce perceptible improvements in analytical results. From a theoretical point of view, it is often very convenient to use the ζ-correction formally, particularly in general discussions, such as produced in Chapter 8; the main advantage is that we are precise in our statements, though we use simplified formulae.

At first sight it may appear strange that equation (6.16) predicts *negative* excitation potentials for low levels. This result, however, is completely consistent with the picture given in § 6.2 concerning the continued depletion of low levels with increasing temperature in favour of high levels; clearly, the Boltzmann equation (6.15) predicts a *decrease* of the population with temperature if the apparent excitation potential ($\bar{V}_q$) becomes negative. Therefore, between 4500 and 6500°K, the ζ-values also represent quantities that indicate the approximate level values at which the decrease of populations with temperature changes into an increase.

★ § 6.4   *The measurement of arc temperature*

Methods for determining the temperature of the arc have been summarized by several authors.* In the spectrochemical field spectroscopic methods have gained the greatest popularity. Discussion here will be chiefly restricted to those methods that are commonly used for measuring the temperature in a low-current arc in air.

As we must distinguish between different kinds of temperature (see § 5.2), so a particular method gives information about one kind of temperature only, the gas temperature, the electron temperature, the excitation temperature, or the ionization temperature. If we want to examine whether local thermal equilibrium is established or not, we must use different methods involving different definitions of temperature; coincidence of the numerical temperature values then implies that thermal equilibrium prevails. An interesting example of non-coinciding values is provided by the investigation of arcs burning in inert atmospheres (helium and argon) carried out by Sobolev, Kolesnikov, and Kitayeva (1960) (see also § 5.6).

Observation of spectral-line intensities leads primarily to the temperature of excitation (and ionization). If the procedure is applied to rotational bands of molecules, the gas temperature is found, even when the excitation is not thermal. This has been demonstrated by experiment and can be understood theoretically if we consider that excitation by electron impact changes only the electron configuration of a molecule, whereas the angular momentum of the rotation is

---

* See, for example, Brinkman (1937, pp. 10–28), Suits (1941), Smit (1946, 1950, pp. 90–132), Maecker (1951), von Engel (1955, pp. 235–238), Finkelnburg and Maecker (1956, pp. 369–374), Lochte-Holtgreven (1958), and Griem (1962, 1964b).

practically unaffected. So, provided that the moments of inertia of the excited molecules and the ground-state molecules do not differ appreciably, the concentrations of the excited molecules and the ground-state molecules in their corresponding rotational levels are proportional to each other. Therefore, the temperature of the non-excited gas is derived from the distribution of rotational energy among the excited molecules.*

The temperature that follows from the intensities of atom or ion lines is not necessarily equal to the gas temperature; for example, as has been noticed in § 5.2, the excitation temperature in the spark equals the electron temperature, which differs appreciably from the temperature of the gas. In the arc in air at atmospheric pressure, all definitions of temperature lead to the same numerical values (Chapter 5).

The methods for measuring the temperature that will be considered here in some detail are all based on the assumption that the populations of atoms, ions, or molecules of the thermometric species at the different energy levels follow a Boltzmann distribution. To arrive at an understanding of the underlying principles, we shall first write down the equation for the absolute intensity of a spectral line involved in the transition from an upper level q to a lower level p. Substituting (6.5) into (1.26) and omitting the index a since this is a generalization, we obtain

$$I_{qp} = \frac{d}{4\pi} A_{qp} h\nu_{qp} n \frac{g_q \exp(-\epsilon_q/kT)}{Z} \tag{6.17}$$

Taking logarithms and rearranging we have

$$\ln \frac{I_{qp}}{g_q A_{qp} \nu_{qp}} = \ln \frac{d}{4\pi} h + \ln \frac{n}{Z} - \frac{\epsilon_q}{kT} \tag{6.18}$$

Inserting relative intensity I'_{qp} and relative transition probability A'_{qp} we write

$$\ln \frac{I'_{qp}}{g_q A'_{qp} \nu_{qp}} = \ln \frac{n}{Z} - \frac{\epsilon_q}{kT} \tag{6.19}$$

Applying (6.19) to a group of spectral lines emitted by atoms, ions, or molecules of one and the same kind, we see that $\ln I'_{qp}/g_q A'_{qp} \nu_{qp}$ is a linear function of ϵ_q because n and Z are now constants, that is, n and Z have the same values for all the lines involved. Consequently, if for a series of spectral lines of a particular type of atom, ion, or molecule, $\ln I'_{qp}/g_q A'_{qp} \nu_{qp}$ or $\ln I'_{qp} \lambda_{qp}/g_q A'_{qp}$ is plotted against the corresponding excitation energy ϵ_q, a straight line with slope proportional to $1/T$ will result, provided self-absorption is negligible.

* See van Wijk (1932), Oldenberg (1934), and Brinkman (1937, p. 14). Though in many experiments with electric or thermal excitation the rotation actually yields the correct gas temperature, abnormal rotation is not to be a priori excluded (Oldenberg, 1934).

The gas temperature can be also evaluated from the Doppler broadening of spectral lines (Burhorn, 1955; and Lochte-Holtgreven, 1958), from the absorption of X-rays or α-rays, and by measuring the gas density by interferometric methods, Schlieren methods, the measurement of the velocity of sound, or the measurement of radioactivity (for references, see Lochte-Holtgreven, 1958; and Finkelnburg and Maecker, 1956, pp. 369–374).

Obviously, if this principle is applied, the relative transition probabilities of the lines must be known. Theory gives simple expressions only for the relative transition probabilities of the rotational lines of one molecular band. Hence, it is understandable that the first spectroscopic temperature determinations in the arc have been made with molecular rotational lines (see § 6.11). Fig. 6.5 shows experimental plots of equation (6.19) for rotational lines of AlO and CN according to Brinkman (1937).

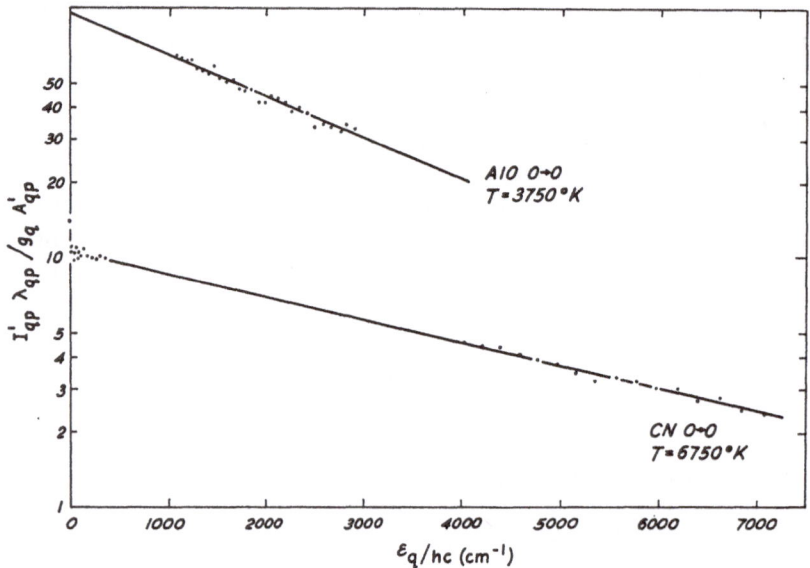

FIG. 6.5. For illustrating the principle of temperature determination from relative intensities of spectral lines with known relative transition probability [equation (6.19)] the figure shows experimental plots for rotational lines of AlO and CN (according to Brinkman, 1937, p. 13). The temperature values inserted in the figure have been derived from the slope of the lines (cf. Fig. 6.15).

Temperature measurements using molecular bands have provided new ways for determining relative transition probabilities of atom and ion lines. As we have already seen in §§ 1.3 and 5.3, investigations in this field, which involve the application of equation (6.19) in the reverse direction, have demonstrated the validity of Boltzmann's distribution law in the arc. Nowadays, absolute or relative transition probabilities of spectral lines of a variety of atoms and ions determined according to various procedures are known (see §§ 1.3 to 1.5). Thus intensity ratios of atom or ion lines can be used for temperature determinations in thermal sources. These procedures, which are less laborious than those involving the use of molecular bands, find application in ever-increasing number in the study of spectrochemical problems. A remarkable work in the field of temperature measurement in the low current arc is the investigation undertaken by Corliss and co-workers at the N.B.S. (see § 1.5). Here we present a few examples taken from Corliss' work to demonstrate the measurement of temperature from a plot of equation (6.19) when atom lines are used (see Figs. 6.6 and 6.7).

In order that high precision be attained in determining temperature from linear plots of the type shown, it is required that spectral lines of widely differing excitation energy are employed. In general, this will involve the use of lines scattered in a wide range of wavelengths, which necessitates the time-consuming operation of determining the spectral sensitivity curve of the emulsion (including the optics). It must be remembered, however, that an estimate of the slope of a straight line, based on experimental points at either extremity, is made with higher precision

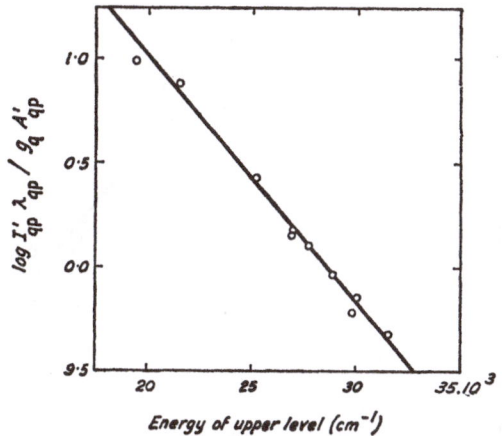

FIG. 6.6. Illustration of the principle of determining the temperature from relative intensities of spectral lines with known relative transition probability [equation (6.19)]. The log ratio of intensity $\times \lambda$ to gA-value for lines of Ti I in the N.B.S. copper arc plotted versus energy of the upper level (gA-values according to van Stekelenburg, 1943; and van Stekelenburg and Smit, 1948).

An arc temperature of 5260°K is derived from the slope of the line of best fit. After Corliss (1961a).

than an estimate based on points scattered along the line. Unless systematic errors, revealed by the curvature of the line, occur, it is thus advantageous to proceed with experimental points at the extremities of the line. For equation (6.19), curvature may be caused by departures of the Boltzmann distribution, by self-absorption of the lines, or by the spatial inhomogeneity of the light source. The latter factors, self-absorption and source inhomogeneity, require careful consideration (see §§ 6.8 and 6.16, and Chapter 8).

Let us for the present assume that neither of those complications is likely to arise; then temperature measurement can be based on the relative intensity determination of only two spectral lines, preferably differing little in wavelength so that the spectral sensitivity calibration of the emulsion can be omitted. This procedure, which has now become very common indeed in spectrochemical physics, will be discussed more fully in the following section.

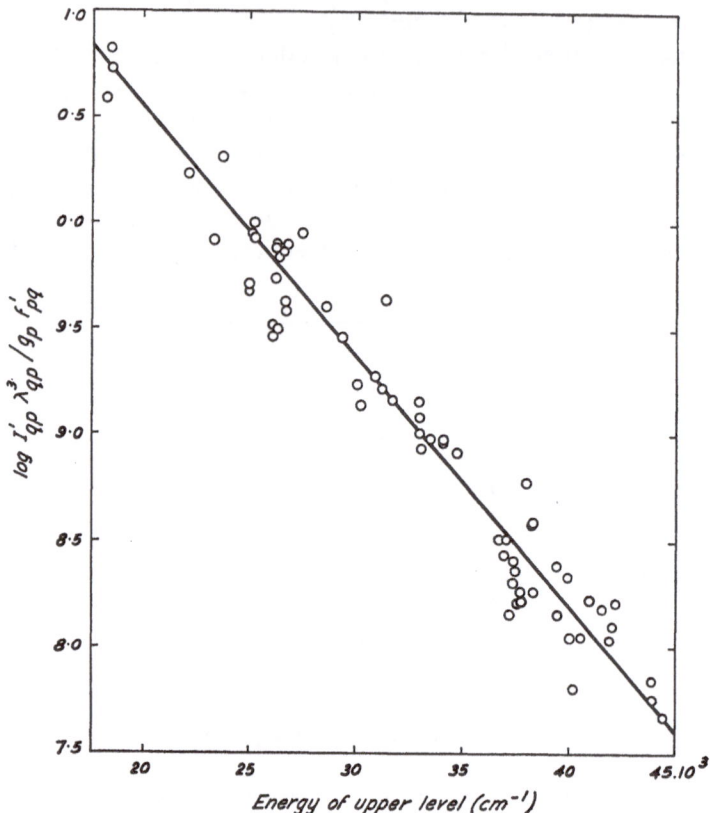

FIG. 6.7. Illustration of the principle of determining the temperature from relative intensities of spectral lines with known relative transition probability [equation (6.19)]. The log ratio of intensity × λ^3 to gf-value for multiplets of V I in the N.B.S. copper arc plotted versus the energy of the upper level (gf-values according to King, 1947).

An arc temperature of 5310°K is derived from the slope of the line of best fit. After Corliss (1961a).

★ §6.5 *Temperature measurement from the intensity ratio of two lines*

The fundamental equation for temperature measurement according to the two-line procedure is easily derived from (6.17). If the lines are labelled a and b, and the subscripts p and q designating lower and upper levels are omitted for the sake of convenience, we obtain

$$\frac{I_a}{I_b} = \frac{(gA)_a}{(gA)_b} \frac{\nu_a}{\nu_b} \exp[-(\epsilon_a - \epsilon_b)/kT] \tag{6.20}$$

Obviously, both lines should belong either to the spectrum emitted by the neutral atom or to the spectrum emitted by the ion of an element.* This does not imply that an ion–atom line pair cannot be used for temperature determinations; it must be borne in mind, however, that an ion–atom line intensity ratio *alone* is not an immediate measure of the temperature. As will be explained in § 7.13, the temperature can be estimated from this ratio when simultaneously an independent determination of the electron pressure is made. The reverse procedure, that is determining the electron pressure from an ion–atom line intensity ratio and a simultaneous measurement of the temperature from an atom–atom or an ion–ion line pair, is more often followed (see § 7.6).

Of course, an ion–atom line pair may be used simply as a *fixation pair* to obtain an idea of the stability of excitation conditions. The interpretation of the numerical results of such measurements is more troublesome, however. For further comments see § 7.13.

Rearranging equation (6.20) and remembering (6.6) we obtain the following practical expression

$$T = \frac{5040(V_a - V_b)}{\log(gA)_a/(gA)_b - \log \lambda_a/\lambda_b - \log I_a/I_b} \tag{6.21}$$

where V is the excitation potential (eV), A the (relative) transition probability, g the statistical weight, λ the wavelength, and I the (relative) intensity; the subscripts a and b refer to the lines a and b respectively.

Now we shall examine what properties the thermometric species and the line pair should have to secure a high degree of precision and accuracy.

1 The thermometric species should be an element of high ionization potential. This is desirable for two reasons. Firstly, to prevent the source temperature from dropping when the thermometric element is introduced. As will be explained more fully in § 7.8, elements of low ionization potential exert a marked influence on the arc temperature; thus, the measurement would disturb the very excitation conditions that we want to evaluate. Secondly, in view of the interpretation of the spectroscopic temperature measurement. This is a common complication resulting from the inhomogeneous, radial intensity distribution in the arc (see §§ 6.12 and 6.16, and Chapter 8).

2 The volatility of the thermometric element should neither be too high nor be too low if it is to be evaporated without difficulty under widely differing conditions.

3 The difference of the excitation potentials V_a and V_b should be large, as follows from the basic equation that relates the relative error (dT/T) of the temperature determination to the relative error (dI/I) of measuring the intensity

---

* Clearly, the use of spectral lines of different elements is not practical for temperature determinations in the normal arc, since the ratio of the concentrations in the source should be known. The procedure is feasible, however, in closely controlled inert gas atmospheres when 'demixing effects' and departures from the Boltzmann distribution do not entail complications (see Frie, 1963a; Maecker, 1962; Frie and Maecker, 1961; and Lochte-Holtgreven, 1963).

ratio (I_a/I_b), viz.

$$\frac{dT}{T} = \frac{T}{5040(V_a - V_b)} 0 \cdot 434 \frac{dI}{I} \tag{6.22}$$

The equation is readily derived by differentiating T in equation (6.21) with respect to I_a/I_b, which is a convenient approximation of the rigorous treatment in terms of a Taylor series (cf. § 7.5).*

In most instances when $V_a - V_b$ is large, the ratio I_a/I_b becomes very small, and thus the determination of I_a/I_b will be liable to large errors.

4 In order that a large value of $V_a - V_b$ is compatible with high precision in determining I_a/I_b, the ratio $(gA)_a/(gA)_b$ should be large so that extreme values of I_a/I_b can be avoided.

5 As has been mentioned in § 6.4, the lines to be used should differ little in wavelength, not only because the time-consuming sensitivity calibration can then be omitted, but also to obtain enhanced precision.

6 It is self-evident that the relative transition probabilities of the lines should be accurately known. The magnitude of the systematic error in T, caused by a slight systematic error in the relative gA-value, can be estimated using the following equation

$$dT = \frac{T^2}{5040(V_a - V_b)} d \log \frac{(gA)_a}{(gA)_b} \tag{6.23}$$

which is obtained by differentiating T in equation (6.21) with respect to $\log(gA)_a/(gA)_b$. Equation (6.23) can be used to evaluate (small) corrections for those temperatures values that have been based on an erroneous value of the relative transition probability (see also § 6.6). As revisions of gA-values are always occurring, it is desirable in reports on temperature determinations that the gA-values used and their origin are included.

§ 6.6 *Zinc as a thermometric species*

In the last few years several workers† have used zinc as a thermometric species. The element has an ionization potential as high as $9 \cdot 39$ eV; so, it causes no difficulties regarding point one discussed in the previous section (see also § 7.8). A less favourable property is its high volatility; for this reason special precautions must be taken. If, for example, we wish to study the variation of temperature during the volatilization of a sample, it will be difficult to achieve a steady and prolonged volatilization of the thermometric species when it is added simply as sponge or oxide powder to the specimen under investigation. It must be emphasized that a proper adjustment of the zinc concentration *in the arc* is imperative; notably, too high a concentration in the arc should be avoided, because self-absorption

* A further discussion involving the use of equation (6.22) is given in § 7.13.

† Dikhoff (1957), Vukanović (1960, 1964), Frisque (1960), Boumans (1961, 1962), Boumans and Rouws (1965), and de Galan (1965a, 1966a and b).

of the line Zn 3076 is likely to produce considerable errors in the resulting temperature values (see § 6.8). Although by choosing the operating conditions carefully temperature measurements with zinc can be accomplished with normal electrode forms, particularly if the line pair Zn 3076/3282 is used, it may be desirable to employ special shapes when extensive investigations are intended (see § 6.7).

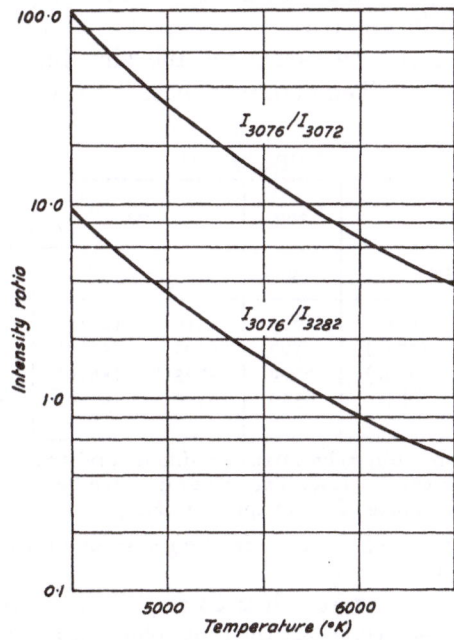

FIG. 6.8. Intensity ratio of the atom–atom line pairs Zn 3076/3072 and Zn 3076/3282 plotted as a function of the temperature [basis: $(gA)_{3072}$: $(gA)_{3076} = 380$].

The lines Zn 3072·06 and Zn 3075·90 represent an almost ideal line pair for temperature determinations; their excitation potentials of 8·08 and 4·01 eV respectively differ largely, whereas the difference of wavelength is negligibly small. With the aid of equation (6.22) we can estimate that the coefficient of variation of T will be in the range of 0·5 to 1·3 per cent if the corresponding coefficient of the intensity ratio I_a/I_b varies between 5 and 10 per cent and if T is in the range of 5000–6500°K. Thus the determination error, expressed as standard deviation, will range from 25 to 85°K.

The best value of the ratio of gA-values is 380.* Arguments for this choice have been given on p. 20. For evaluating temperatures from measured values of

* The advantageous, exceptionally large ratio of gA-values occurs here, since Zn 3076 is a resonance-intercombination line, which has naturally a very low absolute transition probability. The use of such a line entails a danger as well, however, because the low level of the line is the ground level. So, interfering self-absorption must be taken into account, as will be discussed more fully in § 6.8.

the log intensity ratio

$$\log I_{3076}/I_{3072} = \Delta Y_{3072}^{3076} \tag{6.24}$$

the following numerical expression applies

$$T = \frac{20\,510}{2{\cdot}580 + \Delta Y_{3072}^{3076}} \tag{6.25}$$

if the relative gA-value 380 is taken as the basis. The graphical representation of equation (6.25) is shown in Fig. 6.8.

In the literature, different gA-values for the zinc lines have been used for temperature measurement.* Temperatures based on various values are compared in Table 6.3.

TABLE 6.3

$(gA)_{\text{rel}}$	260		380	450		723	
$\gamma \cdot 10^6$	$+8{\cdot}04$		0	$-3{\cdot}6$		$-13{\cdot}6$	
$T(^\circ\text{K})$	5210	(5210)	5000	4910	(4910)	4680	(4660)
	5755	(5745)	5500	5395	(5390)	5115	(5090)
	6305	(6290)	6000	5875	(5870)	5545	(5510)
	6860	(6840)	6500	6350	(6350)	5970	(5925)

> Comparison of temperature values based on different relative gA-values for the atom lines Zn 3072 and Zn 3076. The values in parentheses have been calculated with the approximated form of equation (6.26).

For evaluating the influence of a change in the gA-value on the temperature, the following formulae are useful.

Let T_1 and T_2 be the temperature values corresponding to two different values of $(gA)_{\text{rel}}$, i.e. the ratio $(gA)_{3072} : (gA)_{3076}$; then the following relation exists

$$T_2 = \frac{T_1}{1 - \gamma T_1} \approx T_1(1 + \gamma T_1) \tag{6.26}$$

where

$$\gamma = \frac{[\log(gA)_{\text{rel}}]_1 - [\log(gA)_{\text{rel}}]_2}{5040(V_a - V_b)} \tag{6.27}$$

The validity of the second equality in (6.26) depends on the numerical value of γ. The approximation is particularly convenient in formulae that are derived from the basic equation (6.21), e.g. that for the standard deviation

$$\hat{\sigma}_{T_2} = (1 + 2\gamma T_1)\,\hat{\sigma}_{T_1} \tag{6.28}$$

A disadvantage of the line pair Zn 3076/3072 for measuring relatively low temperatures, say below 5500°K, is the fairly large value the intensity ratio then

---

* Dikhoff (1957) employed the value 260, which probably has its origin in an erroneous application of the Somers correction to Schuttevaer's value of 723. The latter has been the basis of Vukanović's work (Vukanović, 1960, 1964). I initially used the value 450 (Boumans, 1961, 1962); later, I brought all results into line with the best value of 380 (Boumans and Rouws, 1965). This applies to the results presented in this work.

assumes (see Fig. 6.8). To avoid determining large intensity ratios from corres-pondingly diverging densities on the photographic plate, I used a rotating two-step sector with a step ratio of 10 at the uniformly illuminated spectrograph slit. This device, however, can only be used to weaken the intensity of the stronger line (Zn 3076) and not to enhance that of the weaker one (Zn 3072), which must be done by increasing the concentration of the zinc atoms in the discharge zone. In my experience, it is difficult to adjust this concentration so that Zn 3072 contrasts sufficiently with the background, while at the same time Zn 3076 is not seriously interfered with by self-absorption. For that reason I recommend the line pair Zn 3076/3282. The gA-value of Zn 3282 is larger by almost a factor of 6 than that of Zn 3072, so that the intensity ratio I_{3076}/I_{3282} is much more favourable than I_{3076}/I_{3072} (see Fig. 6.8). As the excitation potential of Zn 3282 is high too, namely 7·75 eV, precision will be about the same as when using Zn 3072 [see formula (6.22)]. The only disadvantage is the difference of wavelength between Zn 3076 and Zn 3282. For the emulsions and optical arrangements we used, there was not a great variation of sensitivity in the relatively short range between 3076 and 3282 Å, though it was not entirely negligible either.*

If a sensitivity calibration of the emulsion used is not available, I recommend that an 'intra-laboratory' gA-value for the line Zn 3282 with respect to Zn 3076 should be determined; this is easily accomplished once the temperature has been measured with the aid of Zn 3076/3072.

The practical formula for evaluating T with the aid of the pair Zn 3076/3282 reads

$$T = \frac{18\,850}{3\cdot258 + \Delta Y_{3282}^{3076}} \qquad (6.29)$$

where

$$\Delta Y_{3282}^{3076} = \log I_{3076}/I_{3282} \qquad (6.30)$$

i.e. the true intensity ratio of the relevant lines. A graph of equation (6.29) is shown in Fig. 6.8.†

§ 6.7 *Some details regarding the operating conditions used in an investigation of graphite arcs with metal vapour added*

I carried out simultaneous measurements of electron pressure and temperature in graphite arcs with metal vapour added. A zinc atom–atom line pair was used

---

* For instance, de Galan (1965a) found a relative sensitivity difference of about 12 per cent when using Ilford N 40 process plates and mirror optics for spectrograph slit illumination. Note that the influence of the optics is normally part of the data on the spectral sensitivity variation of an emulsion. Particularly when mirrors are used, the optics may contribute substantially. If this sensitivity correction of +12 per cent to the photographic intensity ratio of I_{3076}/I_{3282} is omitted, a true temperature of 5500°K will be found to be 5595°K, that is about 1·7 per cent too high.

† The value of 3·258 in (6.29) is based on de Galan's measurement of the ratio of gA-values (de Galan, 1965a):

$$(gA)_{3282} : (gA)_{3076} = 1810$$

where a value of 380 for the relative gA-value of the lines Zn 3072 and Zn 3076 has been taken as the basis.

to determine the temperature; magnesium ion–atom line pairs served for obtaining information on the electron pressure (see § 7.6 for a discussion of relevant theory and §§ 7.7 to 7.9 for results). Widely varying excitation conditions were achieved by introducing into the arc elements of differing excitation potential as major constituents of the electrode filling. In view of line interference, the number of elements that can be used for this purpose is restricted; from those to be considered the following have been chosen for our experiments: K, Na, Li, In, Tl, Al, Pb, Ni, Cu, Sb, and C. They were added as KCl, NaCl, LiCl, LiF, In_2O_3, Tl_2O_3, Al_2O_3, PbO, Ni, CuO, Sb_2O_4, and graphite powder. The thermometric and 'manometric' species were ZnO and MgO respectively, except with KCl filling when zinc sponge and magnesium chloride were used so that the elements volatilized more readily.

For producing a regular and prolonged supply of the matrix element and both zinc and magnesium, we employed a type of furnace electrode. Its shape and dimensions are given in Fig. 6.9.* The electrode consists of two tight fitting parts, cap and plug. The short tip, 2 mm in length, on top of the cap promotes the smoothness of the discharge; in the early stages of the burning period the arc is fixed on this tip and usually remains so when the distillation of the filling sets in.

The main factors governing the rate of volatilization of zinc are the distribution of the oxide (or the metal) in the furnace, and the temperature distribution in the electrode. The latter depends, with a given grade of electrode material (in our case, graphite Johnson-Matthey grade 3B), chiefly on the temperature in the arc column; hence, the electrode filling should be adjusted according to the temperature range that is anticipated. Preliminary experiments are indispensable, the more so as it must be ascertained that self-absorption does not interfere.

By correctly filling the furnace electrode we secured a steady volatilization of the matrix, the thermometric element, zinc, and the manometric element, magnesium. In addition, we used a simple method to prevent arc wandering, which I shall refer to as ring-stabilization.† For the experiments at present under consideration we observed only the radiation emanating from the arc through the area W

---

* The electrode resembles those described by Dikhoff (1957), Eichhoff and Addink (1960), and those used in double-arc procedures. See Shaw, Joensuu, and Ahrens (1950); Wedepohl (1953); Ahrens and Taylor (1961, p. 133), Schroll, Brandenstein, Janda, and Rockenbauer (1960), and Hahn-Weinheimer (1962).

† This method for obtaining a stable burn has been reported earlier (Boumans, 1957). The device (see Fig. 6.10) consists of a graphite ring that surrounds the upper electrode (the cathode). The ring has an external diameter of 10–12 mm, an internal diameter of 6 mm, and a height of 10 mm, and is clamped so that it is well isolated from the cathode; the latter is flat-ended and has a diameter of 4 mm. The ring, which envelops the upper part of the arc over a length of 3 mm, exerts a wall-stabilizing effect on the discharge and greatly diminishes arc wandering and climbing. The cathode is not consumed during arcing and can be kept in a fixed position, so that readjusting the anode suffices to maintain a constant electrode separation. The device does not require other accessories than an additional electrode holder with special clamps, whose tips are conveniently made of graphite of spectroscopic quality. Bars, square in cross-section, are a suitable raw material from which tips can be cut.

Although the stabilizing effect of the ring when observed visually is striking, we found in recent experiments, curiously enough, that preventing arc wandering by this means does, in general, not greatly improve the precision of spectrochemical analyses.

indicated in Fig. 6.10; thus the immediate vicinities of the anode and the ring were avoided. This selection of radiation was achieved optically by imaging the arc with unit magnification on a diaphragm, which in turn was magnified twenty-one times and was focused on the collimator lens of a Hilger large quartz spectrograph.

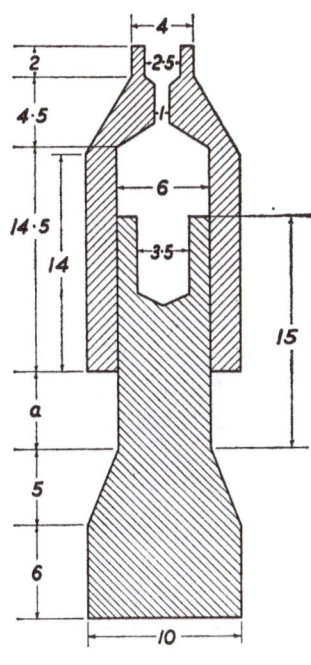

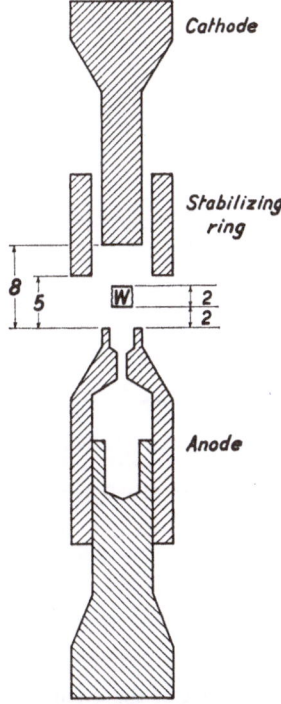

FIG. 6.9. A furnace electrode, which allows regular and prolonged volatilization of a matrix element and of both zinc and magnesium, for determining the temperature and the electron pressure in the arc column under varying conditions. Dimensions are expressed in millimetres. The dimension a is adjustable, so that the total depth of the cavity can be varied.

FIG. 6.10. Cross-section of the electrode assembly, with a stabilizing graphite ring surrounding the cathode and the upper portion of the arc. The dimensions are in millimetres. The outer diameter of the ring is 10–12 mm, the internal diameter, 6 mm, and the height, 10 mm.

For the experiments, only the radiation emanating from the arc through the indicated area (W) was observed.

§ 6.8 Detection of interfering self-absorption of the line Zn 3076

If the line Zn 3076 exhibits marked self-absorption, the temperatures appear to be too high. Errors of this kind can be detected by varying the instantaneous concentration of the zinc atoms in the arc; however, no simple means exist for measuring the concentration of a particular element in the arc. On the other hand, the concentration or the absolute quantity of zinc in the electrode is a poor criterion for establishing the zinc particle density in the arc, since the speed of volatilization depends on the arc temperature—and thus on the nature of the matrix—and on the distribution of the zinc in the electrode cavity.

In our experiments the problem of detecting interference by self-absorption was handled in the following way. By altering the depth of the variable cavity in the furnace electrode (Fig. 6.9) and by varying the distribution of zinc oxide

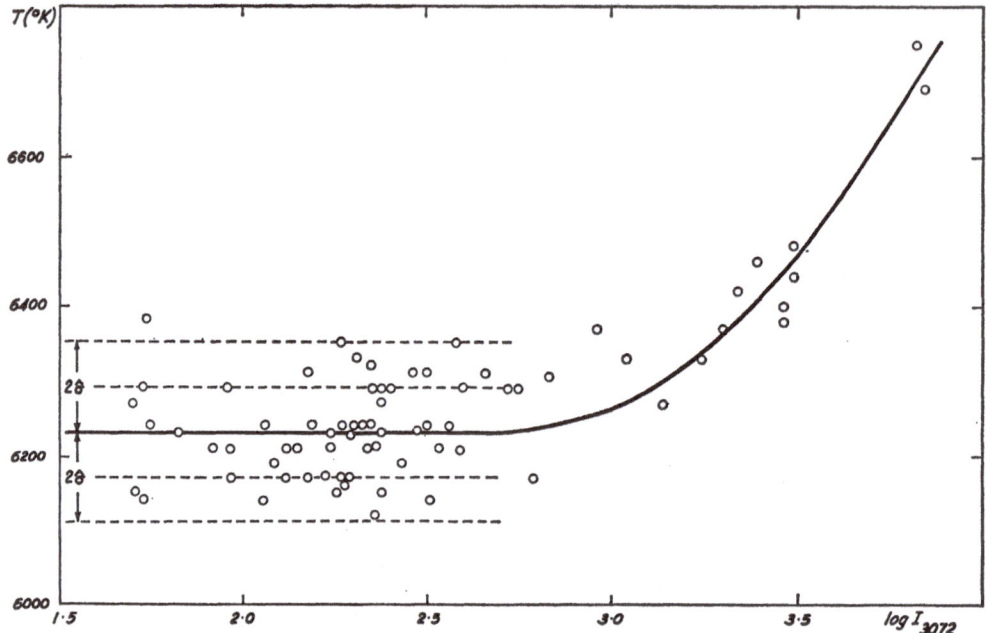

Fig. 6.11. Plot of the evaluated temperature of a graphite arc without added material versus the logarithm of the relative intensity of Zn 3072. As there was plenty of evidence for believing that all temperature values came from the same population, interfering self-absorption could be revealed for those observations that show a correlation between the temperature and the intensity of a zinc line.

Therefore, only the temperature values corresponding with values of $\log I_{3072}$ smaller than 2·7 must be used to estimate the temperature of the graphite arc under consideration. The result is $\bar{T} = 6230°K$, with the standard deviation $\hat{\sigma}_T = 60°K$ (fifty-six observations).

powder within the electrode, we were able to vary the zinc concentration in the arc, whereas the concentration of the matrix element could be kept at a level that was virtually constant. So, we had good reason to suppose that the temperature did not vary greatly during the evaporation of a matrix. Information about the temperature was obtained from moving camera studies and was arranged into groups according to matrix (see § 7.7). If we assume that all observations falling within one group are an estimate of the same temperature and we plot the temperature values within one group against the intensity of a zinc line (e.g. Zn 3072), there should be no correlation between them when self-absorption is absent; if,

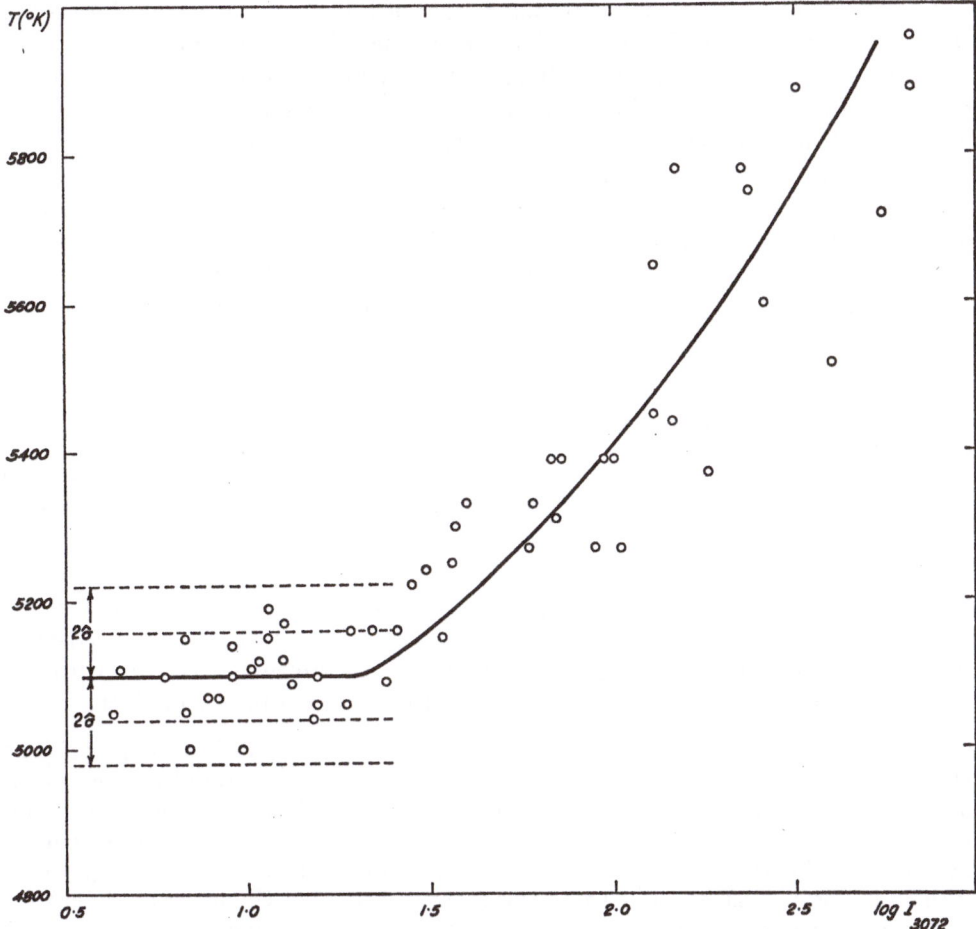

FIG. 6.12. A plot, similar to that of Fig. 6.11, for a graphite arc to which potassium vapour has been added. Here the critical value of log I_{3072} turns out to be 1·3–1·4 instead of 2·7, which is explained by the reduced temperature of the potassium vapour arc and the inherently increased ratio of the number of ground state atoms and excited atoms. From twenty-five observations having log $I_{3072} < 1·4$, it follows that $\bar{T} = 5100°$K and $\hat{\sigma}_T = 60°$K.

on the contrary, the temperature values depend systematically on the intensity, there is strong evidence that interfering self-absorption occurs.

Fig. 6.11 plots the temperature observed for a graphite arc without added material against the logarithm of the relative intensity of Zn 3072. We conclude that reliable temperature values are obtained from those observations where log $I_{3072} < 2·7$. From fifty-six individual measurements satisfying this condition we found the mean and the standard deviation to be $\bar{T} = 6230°$K and $\hat{\sigma}_T = 60°$K.

The plot of Fig. 6.12, which is similar to that of Fig. 6.11, is for a graphite arc to which potassium vapour has been added. Here the critical value of $\log I_{3072}$ turns out to be 1·3 to 1·4. The mean and the standard deviation of the results considered are $\bar{T} = 5100°K$ and $\hat{\sigma}_T = 60°K$ (twenty-five observations) respectively.

Evidently, for the arc with potassium vapour the maximum value of $\log I_{3072}$ compatible with reliable temperature measurements is smaller by a factor of 20 to 25 in comparison with that for the arc in air alone.* Therefore, the temperature determination is cumbersome when we use the line pair Zn 3076/3072 in the conditions prevailing in our potassium vapour arc, notably because the background intensity might reach the same level as that in the arc in air alone (see § 7.10).

§ 6.9   *Copper as a thermometric species*

For determining arc temperatures by the two-line method copper can be used instead of zinc, though this thermometric species is, in several respects, not so satisfactory. Firstly, the ionization potential (7.72 eV) is lower and the energy difference between the upper levels of the lines to be measured (2·4 eV) is smaller than that for zinc; secondly, the thermometric lines are in the visible part of the spectrum (Table 6.4), which might be a disadvantage; and thirdly, the C_2 (Swan) bands interfere with Cu I 5106 and Cu I 5153. An advantage is that with the copper lines lower temperatures can be measured than with the zinc lines.

Relative transition probabilities were originally measured by van Lingen (1936a and b) His values have been corrected by Huldt (1948b) after a suggestion by Schuttevaer, de Bont, and van den Broek (1943). Huldt, moreover, doubted van Lingen's value for the line Cu 5106 and determined it to be 9·2 instead of 18·8. Results of measurements by van den Bold (1945, p. 78) and by van den Bold and Smit (1946) showed agreement with van Lingen's values (after some trivial errors had been eliminated). These were also more or less substantiated by considering the Somers correction (see §§ 1.3 and 6.11) and by recent measurements by Schurer (1964, personal communication); only for the line Cu 5106 was a large discrepancy with van Lingen's value found (14 instead of 18·8).

Relative gA-values according to three authors have been summarized in Table 6.4. For the pairs Cu 5219/5782, Cu 5153/5782, Cu 5219/5700, and Cu 5153/5700 (where the lines Cu 5218 and Cu 5220 are considered to be one line Cu 5219), the extremes differ by 20 per cent. Designating temperatures based on the higher relative gA-values (Huldt) as T_1 and those based on the lower (Schurer) as T_2 we obtain with (6.26) and (6.27) for T_2 respectively 5170, 5710, 6250, and 6795°K if T_1 is taken to be 5000, 5500, 6000, and 6500°K. It is interesting to note that the use of new gA-values after Schurer improves agreement between results for temperature measurements in a 10-amp d.c. carbon arc by a spectroscopic method

---

* This value agrees fairly well with that of the ratio (= 28) of the Boltzmann exponential factors for the line Zn 3072 with $T = 6230°K$ and 5100°K, if the population of the ground state on the threshold for measurements free from self-absorption is assumed to be equal in both cases.

with copper lines, on the one hand, and by shock waves, on the other (Hess, Kloss, Rademacher, and Seliger, 1962a and b).

The different gA-values for the line pairs Cu 5219/5106 and Cu 5153/5106 give rise to much larger differences of temperature: van Lingen temperatures of 5000 and 6000°K respectively read 4330 and 5060°K on Huldt's scale and 4790 and 5700°K on Schurer's.

TABLE 6.4

Author	Line pair					
	5219/5106	5153/5106	5219/5782	5153/5782	5219/5700	5153/5700
van Lingen	55·9	27·9	209	104	1051	525
Huldt	131	65·2	243	121	1206	600
Schurer	71·4	35·7	200	100	1000	500

Ratio of gA-values for copper line pairs according to van Lingen (1936a and b), Huldt (1948b), and Schurer (1964).

§ 6.10 The determination of temperature from CN bands
Introduction: general discussion of band spectra

Temperature determinations using molecular bands are of considerable historical interest, as they have provided the key to unravelling the thermal mechanism of the arc and to the establishment of a great many values of atomic transition probabilities according to the emission method (see §§ 1.3, 5.3, and 6.4). Measuring the temperature from band spectra has not gained popularity in the spectrochemical field, nor is likely to do so in the future, as it is a laborious method, which does not have special advantages over methods using atom or ion lines with known transition probabilities (see §§ 6.4 to 6.6). It is mainly for historical reasons that we shall consider in § 6.11 the principles underlying the temperature determination from CN bands. As an introduction, a few fundamental facts about molecular states and molecular spectra will be summarized in this section.* This discussion may be useful also for those readers who want to arrive at a better understanding of the structure of the bands that appear in the spectra of air and other molecular gases. More information about band emission will be found in Chapter 11.

A molecular state is denoted by its *electronic state* (designated by the number n), by its *vibrational state* (designated by the number v), and by its *rotational state* (designated by the number j). The numbers v and j also represent quantum numbers.

The total energy of a state (n, v, j) is, to a very good approximation, the sum of the component parts, viz.

$$\epsilon_t = \epsilon_{el} + \epsilon_{vib} + \epsilon_{rot} \qquad (6.31)$$

The distance between electronic levels and their position with respect to the ground state are of the same order of magnitude as in atoms, that is 1 to 10 eV. The distance between successive vibrational levels of an electronic state is, in

---

* Cf. Herzberg (1950) and Smit (1950).

5

general, smaller; for CN it amounts to 0·25 eV. Again, the distance between rotational levels of a vibrational level of an electronic state is still smaller (except for large j); for CN we have about 0·005 eV between $j = 10$ and $j = 11$, and about 0·05 eV between $j = 100$ and $j = 101$. Thus the graphical representation of molecular levels is a scheme of electronic levels with vibrational levels superimposed on the electronic states, and rotational levels on the vibrational states. The vibrational levels of an electronic state are nearly equidistant, whereas the distance between rotational levels of a vibrational state increases, to a first approximation, proportionally with the quantum number j.

A transition between rotational levels gives rise to a rotational line; if the level belongs to one and the same electronic state, the line is an infra-red line; if the transition between rotational levels involves also a transition between electronic states, the line is in the ultra-violet or visible range of the spectrum. Transitions of this kind, which here are the only ones of interest, will be designated by the notation n', v', $j' \to n''$, v'', j''. All lines originating from transitions between rotational levels that belong to two distinct vibrational states form a band, which is denoted by n', $v' \to n''$, v'', where n', v', n'', and v'' have definite values, whereas j' and j'', in general, are different for the different lines of the band.

Only those combinations of j' and j'' that are permitted by the selection rules occur, viz.

$$\Delta_j \equiv j' - j'' = \pm 1 \qquad (6.32)$$

and

$$\Delta j \equiv j' - j'' = 0 \qquad (6.33)$$

The transition with $\Delta j = 0$ is forbidden if the angular momenta of the upper and lower electronic states are equal, e.g. the violet CN bands. To summarize, we have to expect two or three series of lines within a band:

1 A series with $j'' = j' + 1$, called the negative branch or P branch.
2 A series with $j'' = j' - 1$, called the positive branch or R branch.
3 A series with $j'' = j'$, called the zero branch or Q branch.

The totality of the transitions between two different electronic states of a molecule is a band system, which is denoted by $n' \to n''$. Any combination of a v'-value (0, 1, 2, . . .) with a v''-value (0, 1, 2, . . .) represents a band of the system since $\Delta v = v' - v''$ is not restricted by selection rules. Bands with equal Δv form a sequence and lie close together in the spectrum because the vibrational quanta in the upper and lower states have similar magnitudes. A system thus consists of a series of sequences, namely $\Delta v = 0$, -1, $+1$, -2, $+2$, etc. Fig. 6.13 shows parts of the level diagram of CN.

The rotational line of a band that corresponds to a transition between levels with $j' = 0$ and $j'' = 0$ is called the *zero line* or null line of the band. If, at the same time, $v' = 0$ and $v'' = 0$, we have the zero line of the system. It is convenient to express the wave-number (σ) of a line with respect to the wave-number of the zero line (of a band). To a first approximation we have

$$\Delta\sigma_P \equiv \sigma_{\text{line}} - \sigma_{\text{zero line}} = -Bj + Cj^2 \qquad (6.34)$$

for a P line,

$$\Delta\sigma_Q \equiv \sigma_{\text{line}} - \sigma_{\text{zero line}} = +Cj + Cj^2 \qquad (6.35)$$

for a Q line, and

$$\Delta\sigma_R \equiv \sigma_{\text{line}} - \sigma_{\text{zero line}} = +B(j+1) + C(j+1)^2 \qquad (6.36)$$

for an R line. Here $B = B_v' + B_v''$, $C = B_v' - B_v''$, and $j = j''$, that is the number

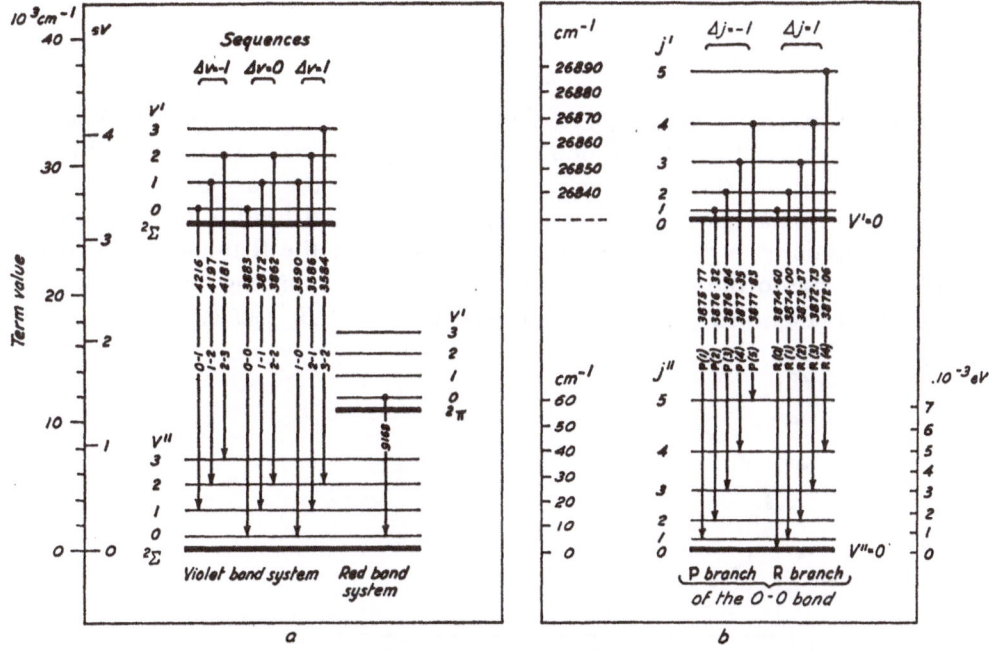

FIG. 6.13. Parts of the energy level diagram of CN (after Smit, 1950, p. 135).
(a) Electronic and vibrational energy levels involved in the violet and the red band system. For each electronic state four vibrational levels are shown. Also, transitions for the violet system and wavelengths of bands grouped in sequences ($v'-v''$ = constant) have been given. Strictly speaking, the transitions shown in the figure correspond to the Q line with $j = 0$ (that actually does not exist in the violet system); the wavelengths inserted are those of the band heads.
(b) Rotational levels of the $0 \to 0$ band. For each electronic vibrational state six rotational levels with transitions and wavelengths of rotational lines grouped in branches ($j'-j''$ = constant) are shown.

of the line. The constants B_v' and B_v'' are the (first) rotational constants of the upper and lower level, according to

$$\frac{\epsilon(j)}{hc} = B_v j(j+1) - D_v j^2 (j+1)^2 + F_v j^3 (j+1)^3 + \ldots \qquad (6.37)$$

where h is the Planck constant and c the velocity of light. In deriving the equations (6.34) to (6.36) the higher terms of the series (6.37) have been neglected.

The dependence of the wave-number of a line on the quantum number j according to equations (6.34) to (6.36) can be represented graphically by a parabola. This representation, which was also suggested by older experimental formulae, was first used by Fortrat, and the parabola is accordingly called a Fortrat parabola. Fig. 6.14 gives the curve for the CN band 3883 as an example. The formula of the parabola as a whole is conveniently written as

$$\sigma = \sigma_0 + Bm + Cm^2 \qquad (6.38)$$

where

$$m = -j \qquad (6.39)$$

for the P branch, and

$$m = j + 1 \qquad (6.40)$$

for the R branch ($\sigma_0 = \sigma_{\text{zero line}}$). As there is no Q branch for the violet CN bands, the smallest value of j in the P branch is $j = 1$, thus $m = -1$. In the R branch $j = 0$, thus $m = +1$ (cf. Fig. 6.13). No line appears in the position $\sigma = \sigma_0$.

From the Fortrat diagram we see clearly how a band head is formed. Owing to the quadratic term in the equations either the P or the R branch turns back, which gives rise to the characteristic band head corresponding to the vertex of the parabola (Fig. 6.14). For CN bands we have $B_v' > B_v''$; so Cj^2 is positive and

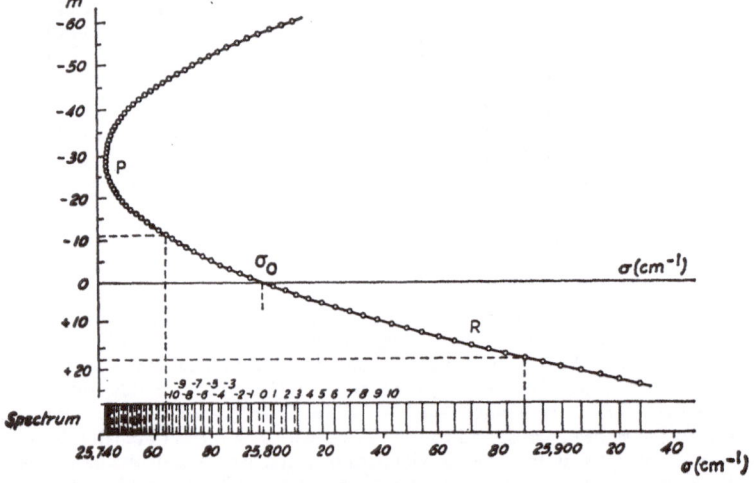

FIG. 6.14. The Fortrat parabola [equation (6.38)] of the CN band 3883. In the figure the intersections of the horizontal lines, having $m = 0, \pm 1, \pm 2, \ldots$, with the parabola are indicated by small circles. The abscissae of the intersections give the wave-numbers (σ) of the lines.

The schematic spectrum below is drawn to the same scale as the Fortrat parabola. The relation between the curve and the spectrum is indicated by broken lines for two points ($m = -11$ and $m + 18$). No line is observed at $m = 0$ (null line gap).

The lines of the 'returning' P branch ($j > j_{\text{head}}$) alternate regularly with those of the 'going' P branch ($j < j_{\text{head}}$) and the R branch, except for large j.

From Herzberg's *Spectra of Diatomic Molecules* (D. van Nostrand Company, Inc., Princeton, New Jersey, 1950).

$-Bj$ negative [equation (6.34)]. Therefore $\Delta\sigma_P$ first decreases with increasing j and then increases. The head lies on the long-wavelength side of the zero line and the band is said to be shaded (or degraded) towards the violet. In the case of AlO, for example, the head appears in the R branch and the band is shaded towards the red.

In the spectrum of the violet CN bands the lines of the 'returning' P branch ($j > j_{head}$) alternate regularly with that of the 'going' P branch ($j < j_{head}$) and the R branch. Small departures from this regularity are observed at high j, where D_v' and D_v'' in (6.37) are to be taken into account.

§ 6.11   *The determination of temperature from CN bands*
Transition probabilities of rotational lines
Temperature measurement from smoothed intensity profiles

The first spectroscopic temperature determinations in the arc were made by Ornstein and van Wijk (1930), and Ornstein and Brinkman (1931a) using molecular rotational lines of CN and AlO.* The method involves the application of equation (6.19) for the rotational lines of one molecular band, that is for constant n', v' and n'', v''. Simple expressions for the relative transition probabilities of these lines can be derived from theory. For example, the following one holds for the relative transition probability $A_P(j)$ of the P line of a $^2\Sigma \to {}^2\Sigma$ transition (e.g. the violet system of CN):

$$A_{P(j)} = \frac{j}{2j-1}\nu^3 U_{n',v' \to n'', v''} \qquad (6.41)$$

where j is the number of the line involved in the transition n', v', j' ($= j-1$) $\to n'$, v', j'' ($= j$) and ν the frequency of the line.

For the R line with number j, and thus for the transition n', v', j' ($= j+1$) $\to n'$, v', j'' ($= j$), we have

$$A_{R(j)} = \frac{j}{2j+3}\nu^3 U_{n',v' \to n'',v''} \qquad (6.42)$$

The constant U has the same value for all the lines of one band.

The relative intensity of a P line with number $(j+1)$ is given by

$$I_{P(j+1)} = A_{P(j+1)}N_{n',v',j}h\nu \qquad (6.43)$$

where h is the Planck constant and ν the frequency of the line.

The population $N_{n',v',j}$ of the upper level is related to the total number of

---

* Similar measurements have been made with C_2 bands by ter Horst en Krijgsman (1934), with CN bands by Gray (1935), and Lochte-Holtgreven and Maecker (1937), with AlO bands by P. Coheur (1942), F. P. Coheur (1942), and Voorhoeve (1946).

molecules N by the Boltzmann equation

$$N_{n',v',j} = N\frac{g_j}{Z}\exp[-\epsilon(j)/kT] \qquad (6.44)$$

where

$$g_j = 2(2j+1) \qquad (6.45)$$

is the statistical weight of the upper level, Z the partition function of the molecule, k the Boltzmann constant, T the absolute temperature, and $\epsilon(j)$ the rotational energy.

Using the first approximation for $\epsilon(j)$ according to (6.37) and inserting (6.41), (6.44), and (6.45) in (6.43), we obtain for the relative intensity of rotational lines of one band, after rearranging,

$$\ln\frac{I_{P(j+1)}}{j+1}\nu^{-4} = \text{constant} - \frac{hc}{kT}\,B_v j(j+1) \qquad (6.46)$$

which is the equivalent of (6.19). A plot of log ratio of intensity $\times$ ν^{-4} to $(j+1)$ versus $j(j+1)$ yields a straight line; from its slope the temperature can be determined.

Analogously, we have for the R lines of one band

$$\ln\frac{I_{R(j-1)}}{j}\nu^{-4} = \text{constant} - \frac{hc}{kT}\,B_v j(j+1) \qquad (6.47)$$

Fig. 6.15 shows examples of empirical graphs of (6.46) and (6.47) applied to the $0 \to 0$ and $0 \to 1$ CN bands. Here the traditional method involving the use of all the lines of one band has been applied. As we remarked at the end of § 6.4, a recommendable procedure for determining the slope of a straight line is one in which the experimental points lying preferably at the extremities of the line are used. Smit (1950, pp. 132–173) discusses the temperature measurement from rotational lines in detail and shows primarily which P and R lines of the $0 \to 1$ CN band are well suited for measurement; such a set of lines consists of low level R lines and high level P lines, in compliance with the recommendation noticed above.

Clearly, the intensity measurements of rotational lines requires a grating spectrograph or a very good prism instrument with a narrow slit. The procedure might thus be less convenient for common practice, the more so because the rotational lines have doublet structure; the splitting increases with the number of the line and is much larger for high j (i.e. P lines) than for low j (i.e. R lines). The doublet structure can affect the results, whence the more laborious procedure involving the measurement of line profiles with subsequent integration—instead of simply determining peak intensities—must be followed to attain the required accuracy (cf. Brinkman, 1937, pp. 32–37). For those reasons procedures for determining the temperature from CN bands with the aid of an ordinary prism spectrograph with a moderately narrow slit have been developed. Much work in this field has been done in the Physical Laboratory of the Utrecht University. The objections (i.e. small luminosity, occurrence of ghosts, astigmatism, and high cost) against the use of a grating spectrograph, stated in the early Utrecht papers published during World War II, may be less stringent nowadays, as perfect stigmatic instruments with blazed gratings (e.g. the Ebert-type spectrographs) are now within the reach of all. There remains, however, the inconvenience of the work involved when measuring the temperature from resolved CN bands.

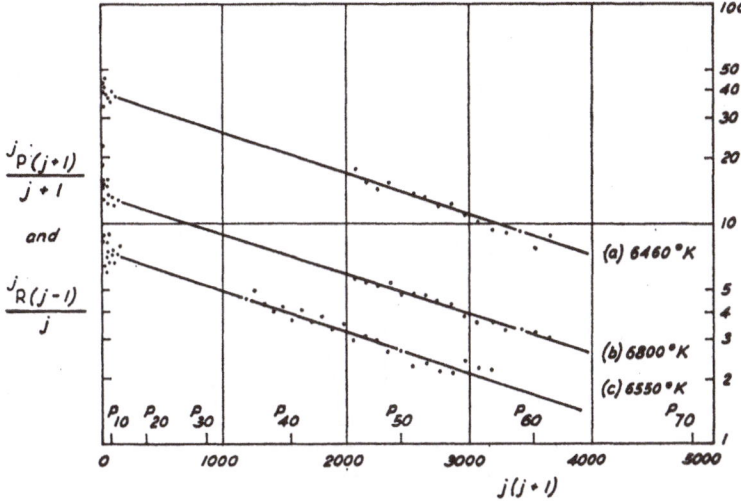

FIG. 6.15. Empirical plots of equations (6.46) and (6.47) for the o → o and o → 1 CN bands, 3883 and 4216, respectively (according to Brinkman, 1937, p. 34). Temperatures determined from the slope of the lines are indicated on the right.

If a spectrum of the CN bands is obtained using the normal type of spectrograph with a moderately narrow slit (i.e. several times the critical width), the rotational structure fades, because the broadened rotational lines partially overlap. This smoothed intensity profile of the bands offers several possibilities for evaluating the temperature; as the intensity distribution among the rotational lines of a band depends on the temperature, so too must the entire intensity profile of the unresolved band; in addition, the ratio of the total intensities of different bands depends on the temperature.

The first step in developing a procedure for measuring the temperature from unresolved bands is the choice of a criterion, that is a pair of heights or areas in the profile whose ratio Θ is readily measurable, is sufficiently sensitive to temperature variations, and is hardly affected by other variables, e.g. the slit width. Secondly, the relationship between the ratio Θ and the temperature T, and thus the calibration curve, must be established. The calibration procedure is for the greater part theoretical.*

* Calculations of smoothed profiles of CN bands based on theoretical rotational A-values and empirical vibrational A-values after Ornstein and Brinkman (1931a) have been undertaken by Smit–Miessen and Spier (1942), by Spier and Smit–Miessen (1942), and by Spier and Smit (1942) (see also Smit, 1946). The subject is treated extensively by Smit (1950, pp. 174–276). These authors considered the CN band pairs o → o and 1 → 1 (that is 3883/3871), and o → 1 and 1 → 2 (that is 4216/4197). Calculations of this type have been also carried out previously, e.g. by Brinkman (1937), though on a smaller scale.

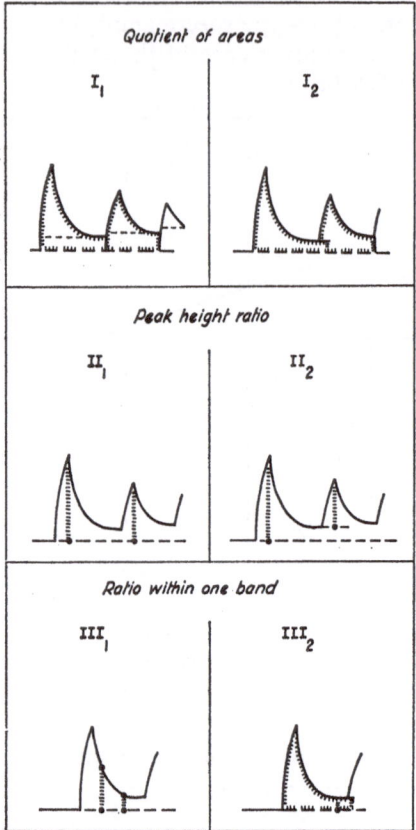

FIG. 6.16. Schematic representation of temperature criteria involving intensity ratios in the first band pair of a CN sequence (after Smit, 1950, p. 219; cf. Smit, 1946).

I_1 Quotient of areas under two band profiles in traditional form, that is area delimitation at the band heads.
I_2 Quotient of areas under two band profiles in traditional form, with modified area delimitation however.
II_1 Total peak height ratio.
II_2 Reduced peak height ratio.
III_1 Slope of the back of a band.
III_2 Tail-to-area ratio.

The different criteria illustrated schematically in Fig. 6.16 are the following:

1 The quotient of the areas under two band profiles, either traditional or in modified form.

2 The peak height ratio or the reduced peak height ratio.

3 The slope of the back of a band or the tail-to-area ratio. In this case the uncertainty caused by an inaccuracy in the ratio of vibrational A-values is avoided.

For details of the calculations and results the reader is referred to the original papers. We note, however, that the published calibration curves require revision, as we have indicated in our discussion on the best gA-values of zinc lines (§ 1.3). This correction, pointed out by Somers (1954, pp. 23–48), is required in view of a systematic error of about 6 per cent in the ratio of the vibrational A-values of the o → 1 and 1 → 2 CN bands (4216/4197), on which the calibrations were based.

§ 6.12 *Radial distributions in the arc*

Introduction

Hitherto we have not considered the spatial structure of the arc and the resulting complications for, e.g., temperature measurements. Here and in the subsequent sections we shall be especially concerned with radial distributions, that is with the radial decline of the temperature and its relation with the radial distribution of the emission. Axial distributions will be discussed elsewhere (§§ 6.17 and 9.4). We shall deal with the problem primarily from a phenomenological point of view, leaving the causal aspects, e.g. how the radial temperature distribution is achieved, to the field of arc theory (see Chapter 4). Taking for granted cylindrical symmetry of the arc column—or at least rotational symmetry—and a radial decrease of the temperature from the centre outwards, we shall consider the following questions.

1. What spatial spectral emission distribution results from the radial variation of the temperature, in conjunction with ionization phenomena, the radial variation of electron density, and the radial distributions of particle concentrations? (See this section and particularly Chapter 8.)

2. How do we visualize a radial emission distribution and in what manner is the average intensity that we actually observe in a spectral line generated from it? (See § 6.13.)

3. What is the reverse way for the special case in which the arc is imaged horizontally on the spectrograph slit, that is how do we derive a radial emission distribution $J(r)$ from the observed variation of intensity $I(x)$ along the length of a spectral line? (See § 6.14.)

4. What must be inferred, in connection with the spatial inhomogeneity in the arc, for the interpretation of the results of temperature measurements? (See § 6.16 and Chapter 8.)

Anticipating what will be discussed more fully in Chapter 7 (see particularly §§ 7.1 and 7.2), we state that the degree of ionization of an individual component of a gas mixture increases rather rapidly with temperature, namely according to a relationship of the type

$$\frac{\alpha_j}{1-\alpha_j} \, n_e = \text{constant} \times T^{\frac{5}{2}} \, 10^{-(5040/T)V_{ij}} \tag{6.48}$$

i.e. the Saha equation. Here α_j is the degree of ionization of the component labelled j, n_e the electron density, T the absolute temperature, and V_{ij} the ionization potential (eV).

If we consider the absolute population of an energy level of the neutral atom of a particular element as a function of the temperature, we see this level population pass through a maximum at a definite temperature T_M, provided that the total concentration of the substance, i.e. the sum of the concentrations of neutral atoms and (singly-charged) ions, is constant. This is a result of two opposing effects that act simultaneously: an increase of the relative level population according to Boltzmann's law and a decrease of the number of neutral atoms in favour of singly-charged ions after Saha's relationship. The numerical value of the optimum temperature T_M depends on the ionization potential of the element, on the excitation potential of the energy level, and on the electron concentration in the plasma and the behaviour of the latter as a function of temperature. We discuss the subject quantitatively in § 8.8. For the present, suffice it to realize that the population of a level, and consequently the intensity of a spectral line originating from that level, has a maximum value at a temperature T_M. This dependence of spectral-line intensity on temperature is displayed radially in any cross-sectional plane through the arc perpendicular to the axis. The complete picture is not a simple one, since the radial gradients of the temperature and the electron concentration, as well as the radial distribution of the concentration of the element under consideration, all participate in the function that expresses the resulting dependence of the intensity on the distance r to the axis.

If the temperature T_0 at the axis is below the optimum temperature T_M, the line intensity will be at a maximum at the axis. If T_0 exceeds T_M, the maximum intensity is found in a concentric, ring-shaped zone some distance away from the centre. So, spectral lines of elements of high ionization potential have their origin in the central core, whereas atom lines of elements of low ionization potential are emitted from the cooler fringe of the arc. Of course, there is a gradual transition as we pass from the one class of elements to the other.

We notice that in this general outline of radial intensity distributions we have implicitly restricted our considerations to the field of the normal low-current arc in air, where the second stage of ionization—even for the ion of the lowest ionization potential, Ba II—can be neglected. Therefore, ion lines of all elements are emitted from the centre of the arc. When the temperature in the arc is lowered, e.g. by introducing an alkali metal, the optimum-emission zones of atom lines of easily-ionized elements shift toward the axis (see further §§ 6.16 and 7.9, and Chapter 8).

Atom line emission from elements of low ionization potential in the flamy fringe is easily observed experimentally when a small quantity of an alkali metal salt is evaporated in a carbon arc. We then perceive the phenomenon that is known as a hollow flame. Lenard (1903, 1905) gives a minute description of hollow flames and his explanations are not so far from the truth, if allowance is made for the status of physical knowledge in his day. Also worth noting in this connection are the so-called spectrohelio-graphical investigations by Oldenberg (1913) and Berthold (1922). They used an arrangement of a monochromator and two rotating mirrors with which they took photographs of the arc. The photographs clearly reveal the spatial emission distribution of separate spectral lines and molecular bands. The published material including reproductions of their photographs are interesting historical documents.

An explanation of hollow arc-flames as brought out above has been given in a note-worthy paper by Ornstein and Brinkman (1931b). General discussions on the radial intensity distribution in the normal arc are found in various papers, e.g. by Brinkman (1937, pp. 48–52), by Kruithof (1943a, p. 25 ff., 1943b), and by Kruithof and Smit (1944); see also § 6.15. More recently, hollow flames have been studied by Eberhagen (1955a). According to him we should explain the phenomenon (exclusively) by ion migration. Curiously enough, he does not mention the arguments put forward above, though they are implicitly embodied in his equations. His experimental plots of radial distributions show maxima for the total concentration of his test element, strontium, and for the concentrations both of neutral atoms and of singly-charged ions, at an appreciable distance from the arc axis, thus supporting the view that the particle distribution also plays a role in the hollow-flame phenomenon.

We must be aware that computing radial concentration distributions of atoms and ions involves a number of intermediate steps that tend to amplify the systematic and random errors in the data from which we start. A fine example is furnished by Kruithof (1943a, p. 74). He measured the radial distribution of the temperature in a 7-amp d.c. carbon arc with sodium vapour added, according to the two-line method and employed a pair of Ba I lines and a pair of Ba II lines. The experimental points, indicated by o (for Ba I) and × (for Ba II), have been plotted in Fig. 6.17. To obtain a clear idea of the influence of errors in the temperature

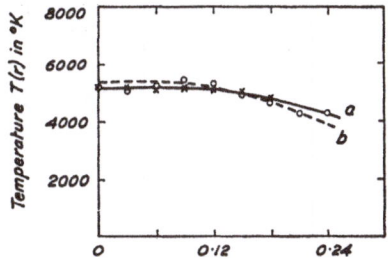

FIG. 6.17. The radial temperature distribution $T(r)$ in a 7-amp d.c. carbon arc with sodium vapour added, measured according to the two-line method using

    (a) The lines Ba I 4726 and Ba I 3910 : o
    (b) The lines Ba II 3892 and Ba II 4525 : ×

After Kruithof (1943a, p. 74).

determination, Kruithof considered two extreme curves fitted through the experimental points, the broken and the continuous lines in Fig. 6.17, for calculating the relative concentration distributions of Ba atoms and Ba ions along the arc radius. The results presented in Fig. 6.18 show clearly that a good deal of caution should be exercised when final conclusions are drawn.

Köstlin (1964), who employed the same type of arc as Eberhagen but burnt the sample in a nitrogen atmosphere, determined the radial distributions of neutral and singly-ionized calcium. The former distribution shows a marked maximum at a considerable distance from the axis, whereas the latter is spread out almost uniformly over the entire cross-section of the arc core (see Köstlin, 1964, Figs. 2 and 3).

We note that Eberhagen and Köstlin studied a horizontal arc discharge burning in a rotating glass tube, a so-called *Wälzbogen* after Schnautz (1939); they stabilized the discharge additionally by an axial blast of air or nitrogen. The diagrams of radial distributions show that this wall-and-air-stabilized arc discharge has a diameter which is surprisingly large if compared with that of the normal free-burning arc (see § 6.15 for further comments).

To conclude the brief survey given in this section we mention the remarkable experiments carried out by Vainshtein and co-workers in the U.S.S.R. They studied spatial distributions with the aid of a γ-ray camera and radioactive isotopes. In that way they obtained direct information on mass distribution in the arc. Further discussion of their work will occur when dealing with migration phenomena (see particularly § 9.3).

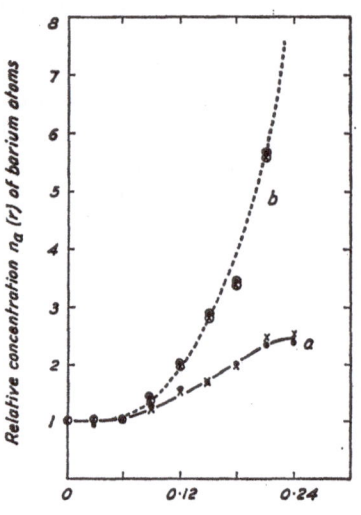

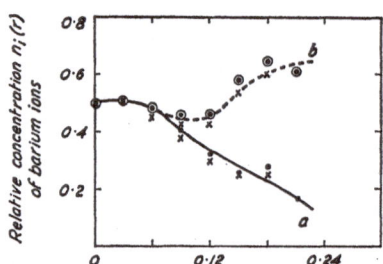

FIG. 6.18. Radial distributions of relative concentration of neutral barium atoms (left) and singly-charged barium ions (right) computed with each of the temperature distributions shown in Fig. 6.17 from the radial emission distributions of the same arc and spark lines that were employed for the temperature determination. Experimental points referring to different lines are indicated by different symbols. After Kruithof (1943a, p. 75).

§ 6.13   *The relation between radial emission distributions in the arc and observed relative intensity distributions in spectral lines*

We shall now see in what forms a radial emission profile in the arc can manifest itself in a longitudinal intensity distribution of a spectral line. Examining this interrelation in greater detail provides a useful introduction for considering the reverse course, that is deriving a radial emission profile in the arc from a one-dimensional intensity pattern along a spectral line (see § 6.14).

Whatever optical arrangement between light source and spectrograph is used, when dealing with a volume source like an arc, we always detect (line) integrated radiant energy coming from a range of depths on the line of sight, as has been explained in our general discussion of photometric and radiometric quantities in

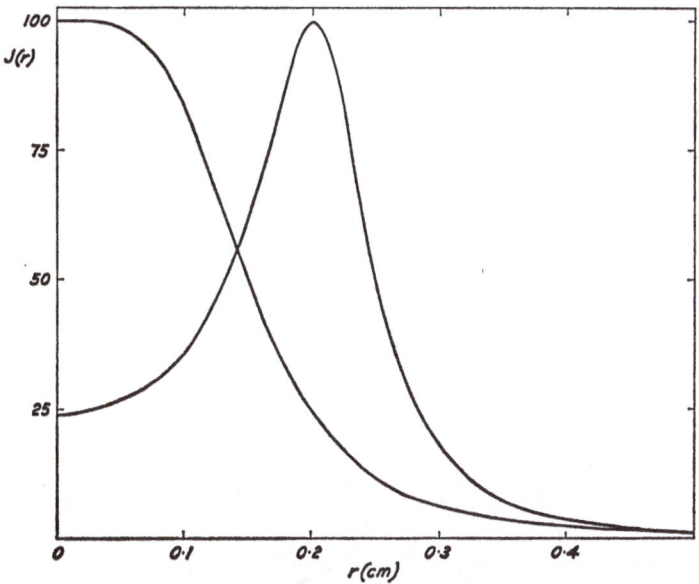

FIG. 6.19. Assumed radial distributions of the emission $J(r)$ per unit volume per unit solid angle, (a) for an ion line, and (b) for an atom line of an element of low ionization potential.

§ 1.2. The present exposition inherently builds on the definitions and their interpretation that have been discussed in § 1.2. We recall here, firstly, that intensity I has been defined as radiance, that is the radiant flux per unit projected area of the source per unit solid angle, for spectral lines conveniently taken as integrals over the spectral-line profile, and secondly, that a quantity, denoted as J, has been introduced, representing emissive power per unit volume of the source per unit solid angle. Radiance or intensity is emissive power integrated over the depth of the source along the line of sight. As we can only observe intensities, spatial emission distributions are concealed in projected intensity patterns. The problem is to get space-resolved information from properly chosen projections. If the radiating properties of the source are symmetrical, in particular cylindrically or spherically symmetrical, this retracing procedure is possible, provided that the source is optically thin so that self-absorption can be ignored (cf. § 1.2).

Before dealing with this matter in § 6.14, we shall discuss the transformation of an assumed radial distribution of the emission in the arc plasma into a longitudinal intensity pattern in a spectral line. This means, of course, that the optical system between source and spectrograph must be defined.

For our purpose here it will be sufficient to distinguish between two extreme ways of slit illumination, focusing the source on the slit or imaging it on the collimator lens. Evidently, as has already been discussed in § 1.2, the former method is used for analysing spatial distributions; the latter, which is the commonest method in spectrochemical analysis, is intended to record spectral lines of even intensity along their entire length. This intensity is an average taken over the part of the source that lies within the entrance aperture of the optical system.

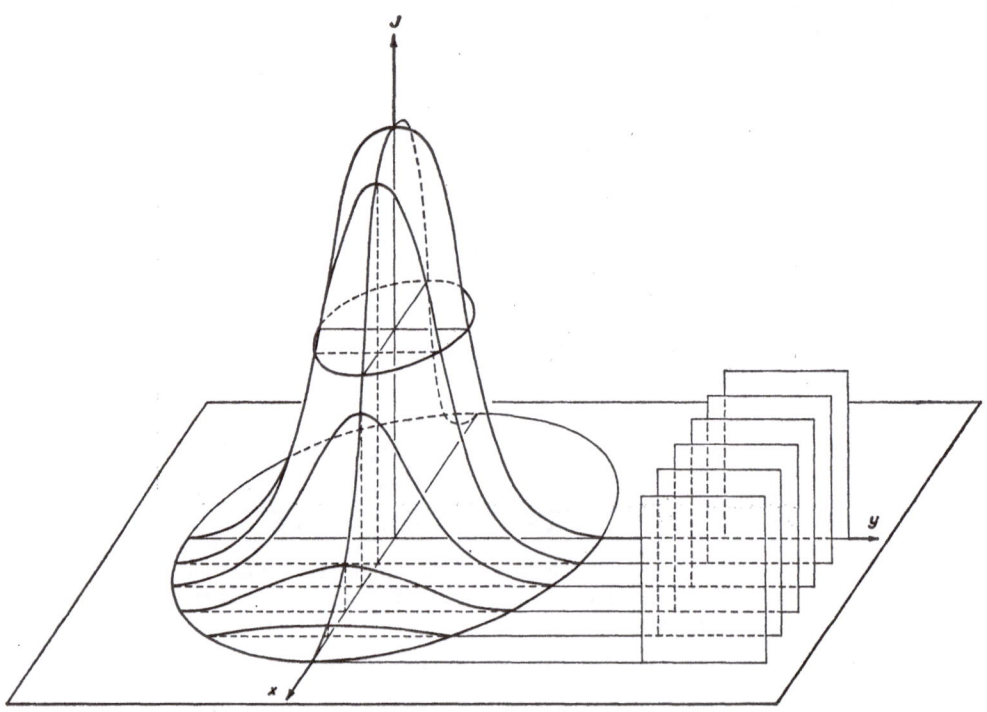

FIG. 6.20. Spatial representation of radial distribution of emission $J(r)$.
The solid body is obtained by the revolution of the curve a in Fig. 6.19 around the J axis. The figure shows curves of intersection of the solid of revolution with equidistant planes parallel to the Jy plane. These curves represent the emission profiles $J(x, y)$ for subsequent values of x. A similar set of profiles is presented in a plane diagram in Fig. 6.21.

If a vertically burning arc is imaged on the (vertical) slit, the spectra give information on the vertical intensity distribution. For studying the radiating properties in a horizontal cross-sectional plane the image of the arc must be rotated 90° by means of two correctly positioned plane mirrors or by an inverting prism in the light path. In the following we suppose the *vertical* arc to be focused *horizontally* on the slit. Just as in § 1.2, we shall discuss the matter in terms of orthogonal geometrical projections, which are easy to visualize and serve as a very good approximation of optical projections, since solid angles are small in common practice. As to the assumed orientation of the arc in the *xyz* coordinate system, we refer to Fig. 1.3. The slit thus coincides with the X axis in the plane of projection XZ. We symbolize the distribution along the slit as $I(x)$ and omit, for the sake of simplicity, labels that relate $I(x)$ to a particular cross-sectional plane at the height z in the arc.

The object is to investigate what longitudinal intensity pattern $I(x)$ is transferred on to the slit, and thus on to the spectral lines, for either of the radial emission distributions $J(r)$ shown in Fig. 6.19. These functions are representative

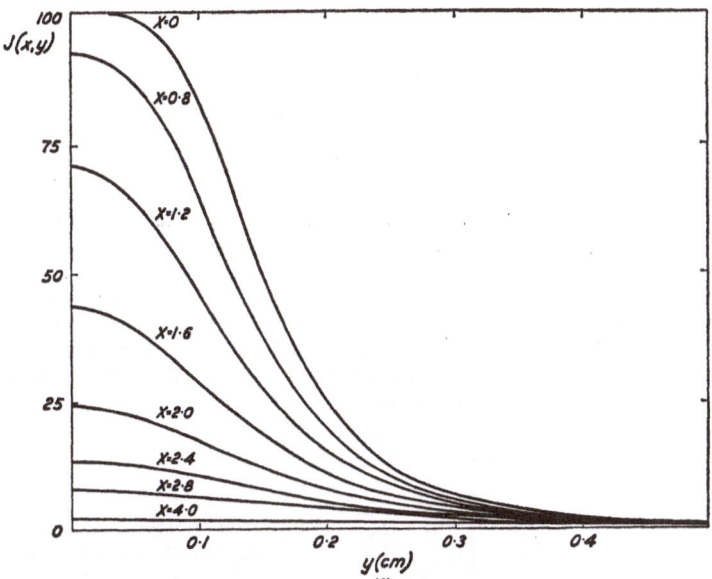

Fig. 6.21. Transverse emission distribution $J(x, y)$ on lines of sight parallel to the y axis and corresponding to the radial distribution a in Fig. 6.19. The figure shows the profiles $J(x, y)$ for $x = 0$, 0·8, 1·2, 1·6, 2·0, 2·4, 2·8, and 4·0 mm. Fig. 6.20 demonstrates how the curves of the type presented here fit a spatial diagram.

of the intensity distributions of ion lines (and of atom lines of substances of high ionization potential) and of atom lines of easily-ionizable elements. In Figs. 6.20 and 6.23 spatial representations of the radial distribution functions, viz. solids of revolution with parallel intersecting planes, have been sketched; needless to say the axis of rotation is the J axis. Similarly as in § 1.2, the y axis determines the direction of observation. If the axis is the line of sight, the emission profile $J(0, y)$, found by intersecting the Jy plane with the solid of revolution, is identical with $J(r)$. Displacing the line of sight parallel to the y axis gives rise to emission profiles $J(x, y)$ that are visualized by taking sections of the solid bodies as presented in Figs. 6.20 and 6.23. These sectional profiles are shown in simple, plane diagrams in Figs. 6.21 and 6.24.

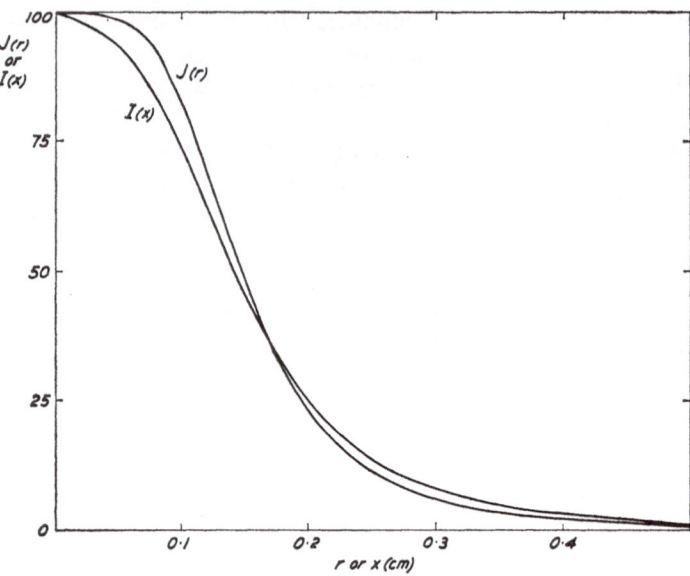

FIG. 6.22. Comparison of the radial distribution of emission $J(r)$ shown in Fig. 6.19 as curve a to its Abel transform $I(x)$. The latter is obtained by plotting integrals of emission profiles such as presented in Fig. 6.21 versus x. The functions $I(x)$ and $J(r)$ have been interrelated by putting the numerical value of $I(o)$ arbitrarily equal to that of $J(o)$.

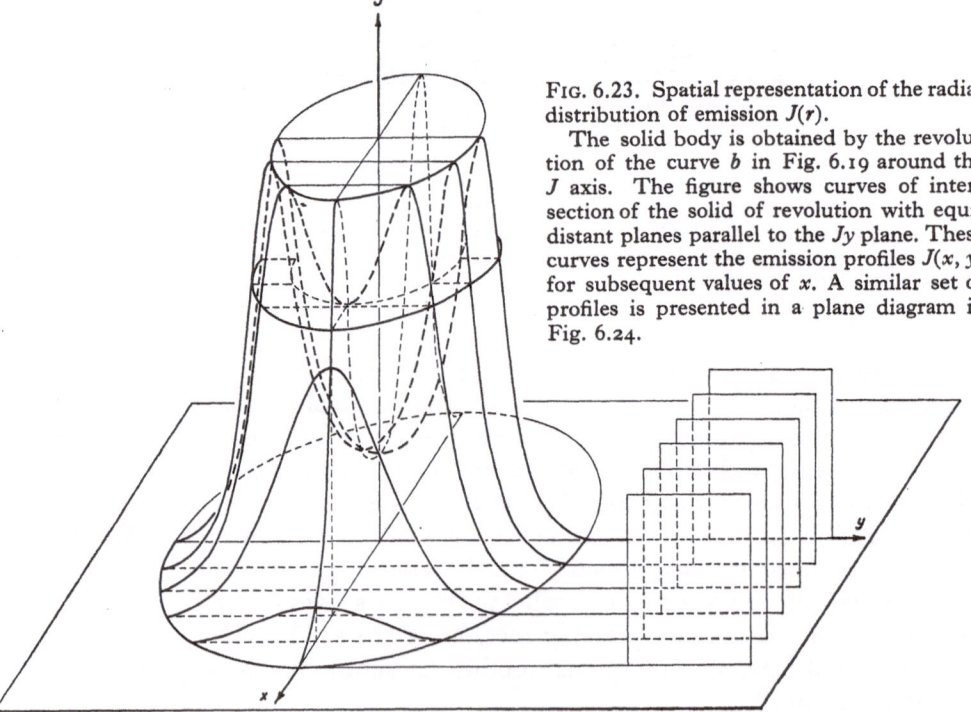

FIG. 6.23. Spatial representation of the radial distribution of emission $J(r)$.

The solid body is obtained by the revolution of the curve b in Fig. 6.19 around the J axis. The figure shows curves of intersection of the solid of revolution with equidistant planes parallel to the Jy plane. These curves represent the emission profiles $J(x, y)$ for subsequent values of x. A similar set of profiles is presented in a plane diagram in Fig. 6.24.

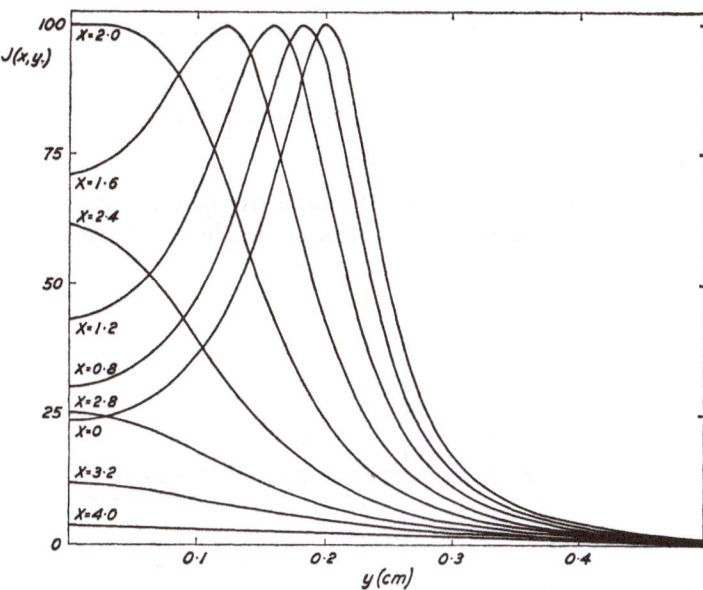

FIG. 6.24. Transverse emission distribution $J(x, y)$ on lines of sight parallel to the y axis corresponding to the radial distribution b in Fig. 6.18. The figure shows the profiles $J(x, y)$ for $x = 0, 0.8, 1.2, 1.6, 2.0, 2.4, 2.8, 3.2$, and 4.0 mm. Fig. 6.23 demonstrates how the curves of the type presented here fit a spatial diagram.

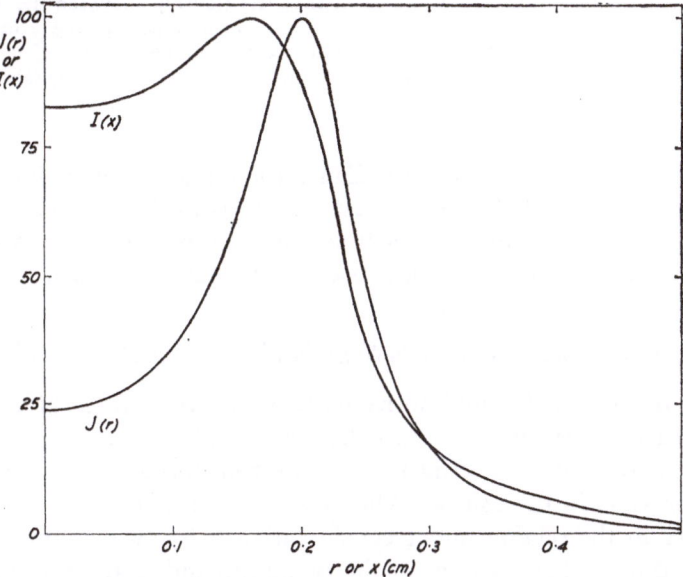

FIG. 6.25. Comparison of the radial distribution of emission $J(r)$ shown in Fig. 6.18 as curve b to its Abel transform $I(x)$. The latter is obtained by plotting integrals of emission profiles such as presented in Fig. 6.24 versus x. The functions $I(x)$ and $J(r)$ have been interrelated by making the numerical values of $I(x)$ and $J(r)$ equal in the maxima.

Actually we do not observe the $J(x, y)$ profiles but their integrals, viz.

$$I(x) = \int\limits_{-\infty}^{+\infty} J(x, y)\, dy = 2 \int\limits_{0}^{\infty} J(x, y)\, dy \qquad (6.49)$$

Here again, as in § 1.2, we have expressed the geometrical relationship between the emission per unit volume per unit solid angle in a horizontal section through

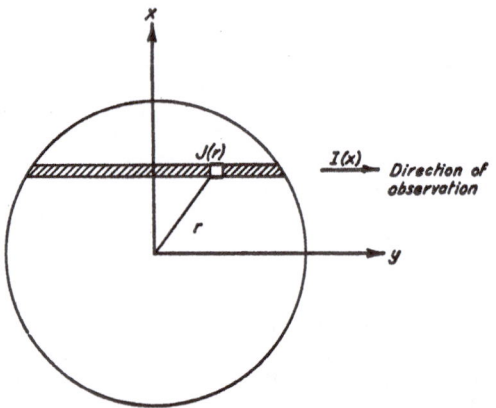

FIG. 6.26. Illustration of the geometrical relationships between the variables. A slice of plasma, circularly symmetric with respect to the z axis, that is normal to the paper, is observed in the y direction. On any line of sight parallel to the y axis, the intensity $I(x)$ is recorded as an integral of $J(r)$ taken over the depth of the slice.

the source, i.e. $J(x, y)$, and the radiance (or intensity) on a line parallel to the x axis at which the radiation from that section is projected, i.e. $I(x)$.

Figs. 6.22 and 6.25 present the functions $I(x)$ in conjunction with $J(r) = J(0, y)$ for the cases considered in Figs. 6.20, 6.21, and 6.23, 6.24.

§ 6.14 Deriving radial emission distributions in the arc from 'edge-on' spectra

If a vertical arc is focused horizontally on the vertical slit of a spectrograph, we obtain 'edge-on' spectra of a circular disk of plasma (see §§ 6.13 and 1.2). The intensity pattern in a spectral line can be transformed into a radial distribution of the emission in the source. When discussing this matter we postulate circular symmetry in the section under consideration. Also, we assume the source to be optically thin so that relations of the type (1.9) and (1.10) hold rigorously, that is self-absorption does not interfere. The problem is, given $I(x)$, to find $J(r)$, where $I(x)$ denotes radiance (or intensity) in the y direction at a distance x from the yz plane, and $J(r)$ is emissive power per unit volume per unit solid angle at a distance r from the z axis (see Fig. 6.26, cf. Figs. 1.3, 6.20 and 6.23).

In the preceding section we expressed $I(x)$ explicitly in $J(x, y)$, viz.

$$I(x) = 2 \int_0^\infty J(x, y) \, dy \qquad (6.50)$$

Inserting $J(r)$ instead of $J(x, y)$ gives

$$I(x) = 2 \int_x^\infty \frac{J(r) r \, dr}{\sqrt{(r^2 - x^2)}} \qquad (6.51)$$

where use has been made of the geometrical relation $x^2 + y^2 = r^2$.

Equation (6.51) is one form of Abel's integral equation and can be inverted into

$$J(r) = -\frac{1}{\pi} \int_r^\infty \frac{I'(x) \, dx}{\sqrt{(x^2 - r^2)}} \qquad (6.52)$$

Here $I'(x)$ is the first derivative of $I(x)$ with respect to x.

Fundamentals of integral equations in general and the Abel inversion in particular are discussed in specialized mathematical texts to which the reader is referred.*

Useful relationships involving Abel transforms have been summarized by Bracewell (1956). His paper also includes a table of functions $I(x)$ along with the corresponding inverse functions $J(r)$; the mathematical formulae are clarified by graphical representations. So, for example, a disk-shaped $J(r)$ function yields a semi-elliptic $I(x)$ pattern; a rectangular $I(x)$ transforms into a pocket-like $J(r)$, a parabolic $I(x)$ into a hemispheric $J(r)$, etc. (see Fig. 6.27).

Analytical expressions such as those of Fig. 6.27 can be enlightening for illustrative purposes; we must realize, however, that in practice $I(x)$ is obtained as a set of numerical data rather than as a mathematical function. For that reason various procedures have been developed for solving the integral equation graphically or numerically. The oldest method, employed by Hörmann (1935), Brinkman (1937, pp. 85–86), Kruithof (1943a, pp. 26–28; 1943b), Kruithof and Smit (1944), and others, is a lengthy graphical procedure that, according to its adherents, should yield reasonable results. Neither the original graphical procedure, nor a noteworthy improved alternative brought out by Huldt (1948b, pp. 72–74) have gone entirely out of use,† although numerical methods are more popular now (see below). Also, a combined graphical-numerical method has been produced (Friedrich, 1959).

Huldt's method avoids the transformation of the originally measured $I(x)$, e.g. into $I'(x)/x$ as in Brinkman's version, prior to the graphical integration; at the

---

* For example, Frank and Mises (1961), Hamel (1949), Magnus and Oberhettinger (1949), Whittaker and Watson (1948) and Margenau and Murphy (1943).

† See Terpstra (1956), Terpstra and Smit (1958), Eberhagen (1955a) and Eicke (1962).

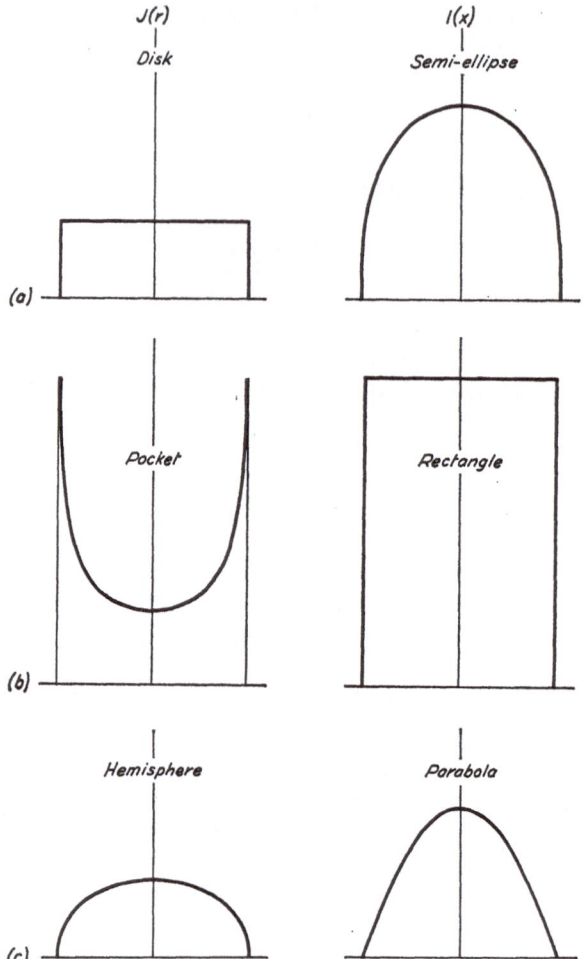

`FIG. 6.27. Graphical representation of the Abel transforms $J(r)$ and $I(x)$ for a few cases where both functions can be expressed in analytical form. The units of J and I have been arbitrarily chosen.

(a) $J(r) = 1$ $I(x) = 2\sqrt{(R^2 - x^2)}$

(b) $J(r) = \dfrac{R}{\sqrt{(R^2 - r^2)}}$ $I(x) = \pi R$

(c) $J(r) = \dfrac{\sqrt{(R^2 - r^2)}}{R}$ $I(x) = \dfrac{\pi(R^2 - x^2)}{2R}$

same time it allows a fair estimate of the accuracy. In general, the integration is not to be considered as the greatest source of error, which is illustrated by a self-consistency check reported by Kruithof (1943a, p. 28). In Fig. 6.28 the broken line is an assumed $J(r)$ curve; it has been integrated to give an $I(x)$ function which,

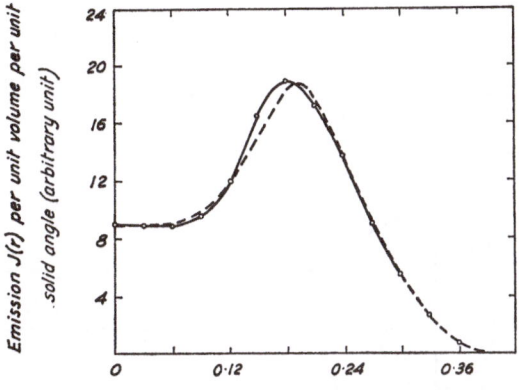

FIG. 6.28. Illustration of the inherent inaccuracy in the graphical Abel inversion procedure.

An assumed $J(r)$ curve (broken line) is integrated to give an $I(x)$ function, which, in turn, is inverted again graphically. The result is the continuous curve. After Kruithof (1943a, p. 28).

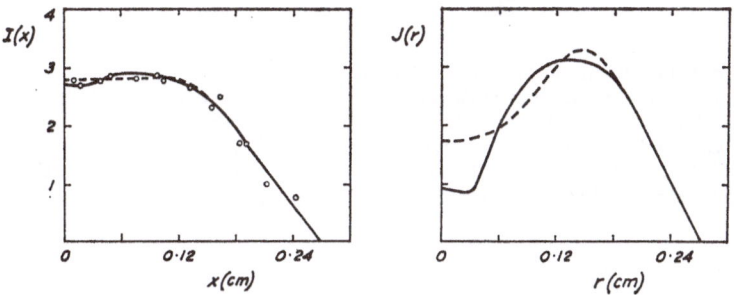

FIG. 6.29. Fitting either of the two $I(x)$ curves, the broken or the continuous one, through the experimental points (left half of figure) shows that the same set of data may give markedly different $J(r)$ distributions (right half of figure). Some caution, therefore, must be exercised when considering experimentally determined radial distributions. After Kruithof (1943a, p. 30).

in turn, was inverted again graphically; the result is the continuous $J(r)$ curve. Evidently, the original curve is very satisfactorily recovered.

More serious than inaccuracy of integration are random and systematic errors in the measurements. A striking example of the problems that can arise has been discussed in §6.12. It is once more demonstrated by the curves presented in Fig. 6.29. The experimental $I(x)$ points in the left half of the figure allow either the continuous or the broken curve to represent lines of best fit. Obviously, application of the Abel inversion to either curve yields remarkably differing radial distribution functions $J(r)$.

Let us now turn to numerical methods for solving the Abel integral equation. As an introduction to rigorous treatments of the subject, referred to below, we

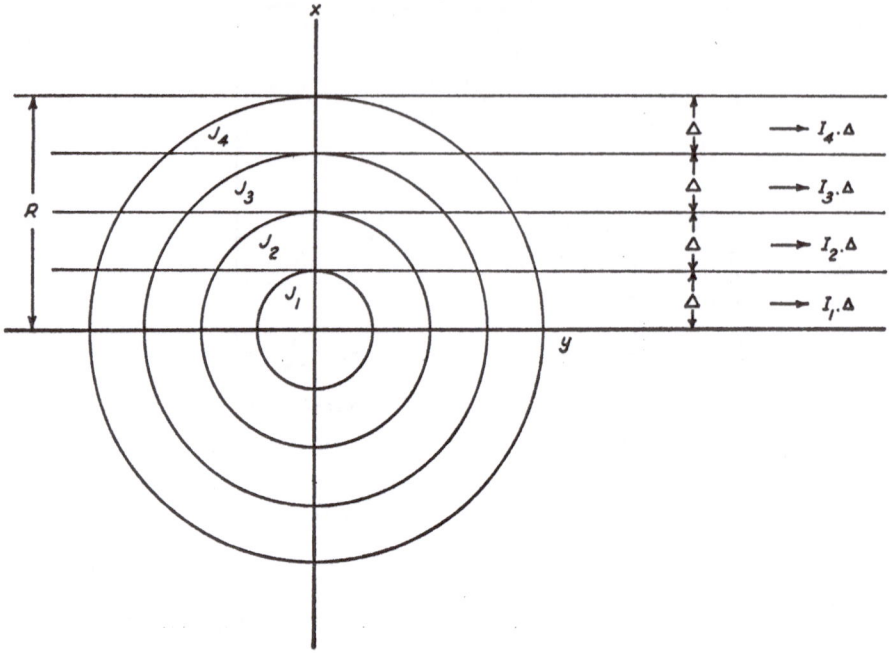

FIG. 6.30. Illustration of geometrical relationships.

The arc cross-section is divided into four circular zones which succeed at equal intervals of Δ ($= R/4$). The emission per unit volume per unit solid angle is taken to be constant within a zone; it assumes the values J_1, J_2, J_3, and J_4 for the four zones. The total emission in the y direction from the four sections parallel to the y axis is $I_1.\Delta$, $I_2.\Delta$, $I_3.\Delta$, and $I_4.\Delta$.

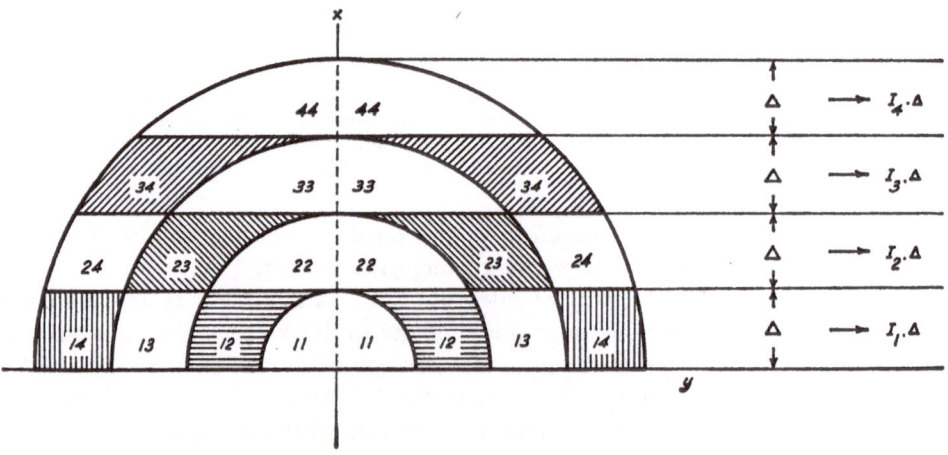

FIG. 6.31. A half-section of the arc divided into segments for computing interrelations between $J(r)$ and $I(x)$. The segments are denoted by the serial number of the transverse light path (first index) and by the radial zone number (second index). The *total* area of duplicate segments on either side of the x axis is considered.

shall consider the following simplified though instructive example. We divide the cross-sectional area of the arc into four concentric zones such as indicated in Fig. 6.30. The emissive power J is assumed to be constant within a zone and negligibly small outside the fourth zone. Thus the radiating properties of the arc are characterized by the four values of the emission per unit volume per unit solid angle, J_1, J_2, J_3, and J_4, and by the geometrical relations defined in Fig. 6.30. We record the intensity of radiation emitted in the y direction from each of the four sections parallel to the y axis marked in the figure. Let the total emission* from these transverse light paths be $I_1 . \Delta$, $I_2 . \Delta$, $I_3 . \Delta$, and $I_4 . \Delta$, successively; here I denotes radiance (or intensity) and $\Delta = R/4$. We now seek the relationships between I_n and J_k. Clearly, they are wholly determined by elementary geometrical relations. The different areas whose numerical values must be computed for expressing I_n explicitly into J_k are given in Fig. 6.31. Calculation gives the following results:

$$\left.\begin{aligned}
I_4 . \Delta &= 3 \cdot 6264\, J_4 . \Delta^2 \\
I_3 . \Delta &= 3 \cdot 0974\, J_3 . \Delta^2 + 3 \cdot 1030\, J_4 . \Delta^2 \\
I_2 . \Delta &= 2 \cdot 4567\, J_2 . \Delta^2 + 2 \cdot 6961\, J_3 . \Delta^2 + 2 \cdot 2373\, J_4 . \Delta^2 \\
I_1 . \Delta &= 1 \cdot 5708\, J_1 . \Delta^2 + 2 \cdot 2557\, J_2 . \Delta^2 + 2 \cdot 0605\, J_3 . \Delta^2 + 2 \cdot 0289\, J_4 . \Delta^2
\end{aligned}\right\} \quad (6.53)$$

Inversely we have

$$\left.\begin{aligned}
J_4 &= 0 \cdot 276\, I_4/\Delta \\
J_3 &= 0 \cdot 323\, I_3/\Delta - 0 \cdot 276\, I_2/\Delta \\
J_2 &= 0 \cdot 407\, I_2/\Delta - 0 \cdot 354\, I_3/\Delta + 0 \cdot 052\, I_4/\Delta \\
J_1 &= 0 \cdot 637\, I_1/\Delta - 0 \cdot 585\, I_2/\Delta + 0 \cdot 085\, I_3/\Delta - 0 \cdot 069\, I_4/\Delta
\end{aligned}\right\} \quad (6.54)$$

Equations (6.53) and (6.54) are conveniently summarized in tabular form (see Tables 6.5 and 6.6).

The data contained in equation (6.54) and in Table 6.6 exemplify a set of coefficients needed for the reduction of a measured $I(x)$ distribution to a radial $J(r)$ distribution. In the example under consideration, rather poor information on the radial function is collected, but it shows clearly how the desired transformation is effected; moreover, as will be discussed at the end of this section, it is of practical interest yet.

To establish more exact information on radial distributions the number of zones must be extended. If the assumption of constant J within a zone is maintained, alternative methods differ in the way of dividing the source region into zones.† When the number of them is sufficiently large, the complex shapes of the areas to be considered for computing coefficients reduce practically to simple rectangles

* The height of the section through the arc is entirely left out of consideration here. If final results have to be put on an absolute scale, this omission should be recognized.

† See, for example, Maecker (1953), Frie (1963b), Pearce (1958, 1960), and Hefferlin and Gearhart (1964).

whose lengths are found as the distances between successive zone circles measured on lines parallel to the y axis.

Further improvements in the practical elaboration of the Abel inversion are attained by using suitable interpolations of the $J(r)$ curve and by the application of techniques for numerical smoothing of the readings.* In general, the method

TABLE 6.5

n	k			
	1	2	3	4
1	1·5708	2·2557	2·0605	2·0289
2		2·4567	2·6961	2·2373
3			3·0974	3·1030
4				3·6264

Areas of segments marked in Fig. 6.31. The segments are denoted by the radial zone number (k) and by the serial number of the transverse light path (n). See also equation (6.53).

TABLE 6.6

n	k			
	1	2	3	4
1	0·63661			
2	−0·58453	0·40705		
3	0·08530	−0·35431	0·32285	
4	−0·06852	0·05203	−0·27624	0·27575

Coefficients for transforming measured radiances (I_n) into radial distributions of the emission (J_k). Labelling of coefficients is similar to that of segments in Fig. 6.31 and in Table 6.5. See also equation (6.54).

to be decided upon for use in a particular case will be a compromise between the desired accuracy and the cumbersomeness of the elaboration. In recent years the problem of transforming observed radiances into radial distributions of the emission has gained increasing attention, especially in the field of plasma spectroscopy. A logical step in this development is the application of interpolation schemes for electronic computers (see Wiese and Shumaker, 1961; Yokley and Shumaker, 1963; and Shumaker and Yokley, 1964).

Returning now to the four-zone model (Fig. 6.30) we shall consider the relative geometrical contribution of each of the four radial zones in different transverse light paths, namely the relative contribution of these zones to each of the intensities I_1, I_2, I_3, and I_4 and to I_1+I_2, $I_1+I_2+I_3$, and $I_1+I_2+I_3+I_4$, if $J_1 = J_2 = J_3 = J_4$. The results are entered diagramatically into Fig. 6.32. The data show how different zones will contribute if we select the radiation from a definite part

* See Nestor and Olsen (1960), Bockasten (1961), Edels, Hearne and Young (1962), Barr (1962), and Frie (1963b).

of the source with the aid of a diaphragm, for example, by focusing an intermediate image of the source on a diaphragm and imaging this in turn on the collimator lens. From Fig. 6.32 we see, for instance, that when choosing a narrow aperture so that only intensity I_1 is transmitted, we have about equal contributions of the four zones. (If the source is focused on the spectrograph slit, the contributions are exactly equal.) Increasing the aperture reduces particularly the relative contribution of the innermost zone, that is zone 1. In conclusion we must recognize that different ways of imaging the light source give different contributions of successive radial zones and so may influence the intensity distribution in the spectrum.

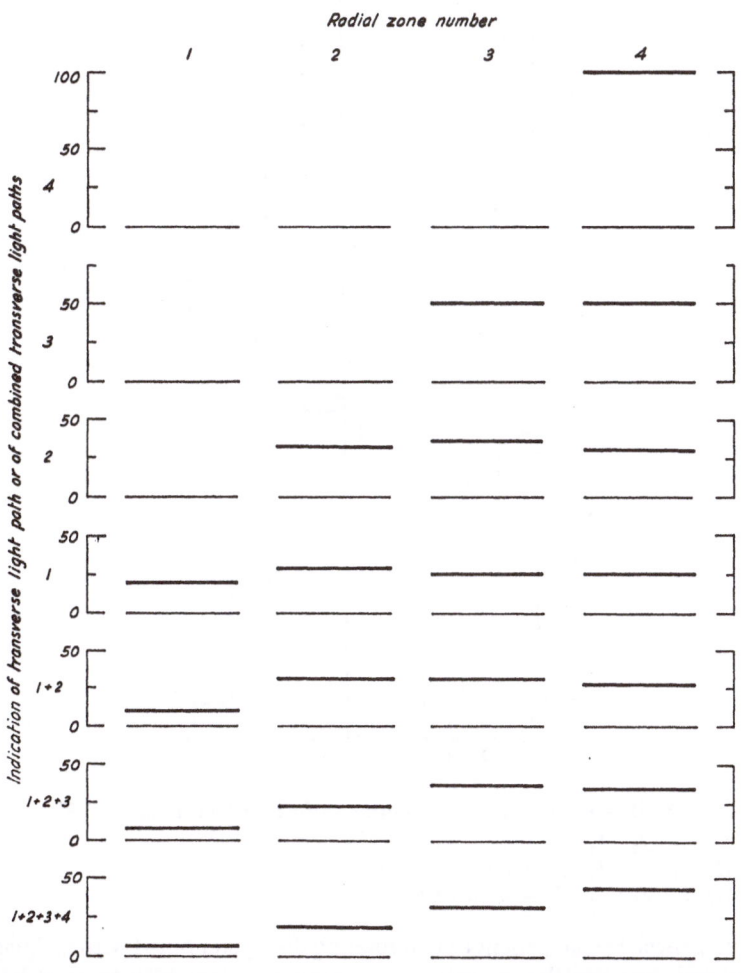

FIG. 6.32. Using the four-zone model (Fig. 6.30) to demonstrate the effect of different modes for spectrograph slit illumination. The diagram shows the relative geometrical contribution of each of the four radial zones in the transverse light paths 1 to 4, and in the combined transverse light paths 1+2, 1+2+3, and 1+2+3+4. The relative geometrical contribution equals the partial intensity supplied by the relevant zone if $J_1 = J_2 = J_3 = J_4$.

Fig. 6.33 gives another result deduced from the four-zone model and shows in a self-explanatory way the relationship between three distinct sets of J_k values and the corresponding inverted I_n values.

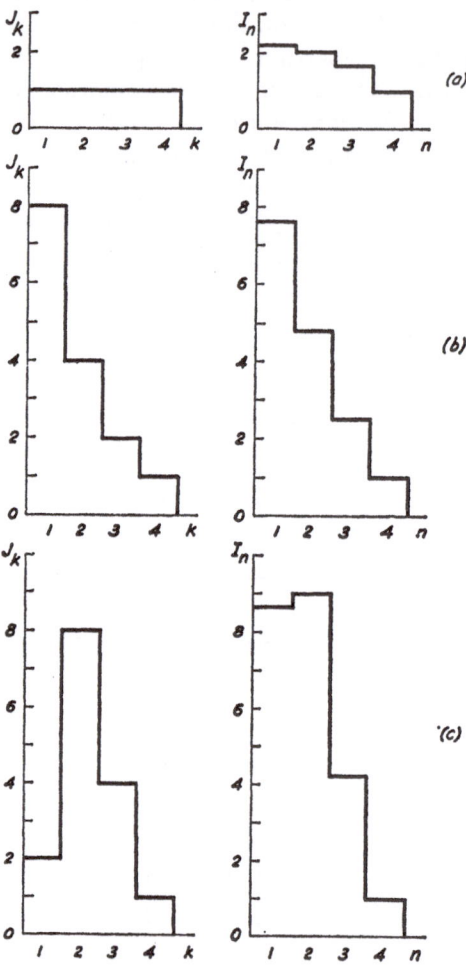

FIG. 6.33. Illustration of Abel transforms in the four-zone model (cf. Fig. 6.30).

(a) $J_1 : J_2 : J_3 : J_4 = 1 : 1 : 1 : 1$
(b) $J_1 : J_2 : J_3 : J_4 = 8 : 4 : 2 : 1$
(c) $J_1 : J_2 : J_3 : J_4 = 2 : 8 : 4 : 1$

By placing a special arrangement of two rotating disks, combined with an image inverter, in the light path between the arc and the spectrograph, four-step spectral lines are produced on the photographic plate which match exactly the four-zone model we are considering. The arrangement of rotating disks I have described previously (Boumans, 1959). It was originally intended for investigating axial distributions only; however, by inserting an image inverter (two accurately-positioned plane mirrors or an inverting

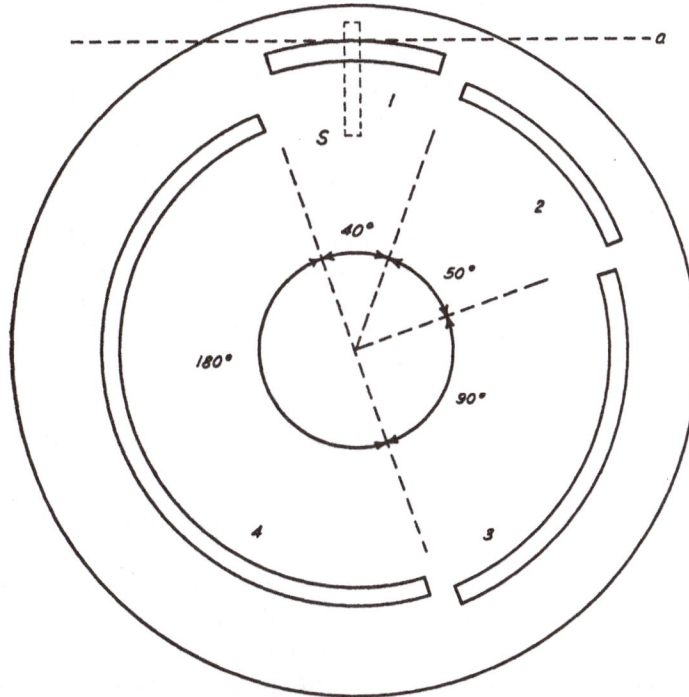

FIG. 6.34. Schematic representation of a disk as used in the rotating-disks arrangement for obtaining spectral lines divided into four evenly illuminated steps that correspond with four separate portions of the light source.

S = spectrograph slit (disk I) or a diaphragm several millimetres wide (disk II)
1, 2, 3, 4 = apertures
a = orientation of the arc axis in the image focused on disk II

The apertures have different arc lengths to compensate for the rapid decline of intensity at the arc surface from the centre (step 1) towards the edge (step 4). Compare with Fig. 6.30.

prism), it can be used for examining radial distributions as well. De Galan (1965a, 1966a) used it for studying the axial and radial distributions of particle concentrations in the arc (see § 9.4).

In principle, the device consists of two rotating disks (I and II), each provided with four apertures (1, 2, 3, and 4), as shown in Fig. 6.34. The disks are mounted on one shaft; disk I is located at the spectrograph slit, disk II in a plane that is imaged on the collimator lens by a lens at the slit. Disk II acts as a diaphragm for an intermediate image of the arc focused on it. Clearly, if disk I is removed, the slit is evenly illuminated along its entire length by the radiation that passes through that aperture in disk II which happens to be in the light path. By rotating disk II the portion of the arc from which radiation is transmitted is varied in steps. Now, if disk I is mounted on the shaft, in the same phase as disk II, the slit is divided into four steps, each of which is uniformly illuminated along its length; however, different steps receive light from different portions of the arc. Accordingly, we obtain four-step spectral lines. When the arc is imaged horizontally on disk II, such that just half its cross-section covers the four apertures, the densities of the separate steps of spectral lines correspond to the mean intensities I_1 to I_4 that are averaged over each of the four transverse light paths marked in Fig. 6.30.

An advantage of this method is that it yields evenly blackened spectral lines, whose densities correspond with *true* average intensities. In addition, the following must be considered. When a vertical arc is imaged horizontally in the usual way on the slit, the density of a spectral line falls off very rapidly along the length from the centre towards the ends of the line; consequently, the accuracy of measuring the longitudinal intensity pattern is rather limited. This difficulty is overcome to a considerable extent by using the arrangement we have just considered: the densities in the four steps can be made equal, or at least not so different, by adopting the arc lengths of the apertures, for example, as shown in the figure. A disadvantage, not to be overlooked either, is the low luminosity of the arrangement.

★ § 6.15 *Results of experimental determination of radial distribution of temperature in the arc*

Results of the determination of the radial temperature distribution in the normal low-current arc have been reported by several authors[*] who all used emission spectroscopic methods and so were unable to extend their measurements to the outskirts of the arc, as those methods fail at temperatures below 3500–4000°K. Information on the temperature distribution in the cooler fringe has been obtained by using interferometric and Schlieren techniques, developed by Schmitz (1949) and Sperling (1950). Radial temperature profiles calculated by solving the Elenbaas–Heller equation (see § 4.2) were established by Mannkopff (1943) (see also Maecker, 1951; and Roes, 1962, p. 113). Results of determinations in low-current arcs stabilized by a tangential gas stream are given by Aarts (1952, p. 163), Eberhagen (1955a), Köstlin (1964), and Vukanović (1964).

Typical results for the normal arc, that is the free-burning arc in air at atmospheric pressure, are presented in Figs. 6.35 to 6.38 and in Fig. 6.17. From the reports mentioned above we conclude on the whole that the temperature declines only slightly in the central zone of the arc, i.e. the region where the distance r to the axis is in a range between 0 and 1·5 to 3·0 mm. Here the temperature decreases from the centre outwards by only a few hundred degrees Kelvin. In the next zone, viz. when r is in a range between 1·5 to 3·0 mm and 2·5 to 4·0 mm, the radial gradient of the temperature is large. The margins in the boundaries of the zones allow for variations in the current strength and in the temperature at the axis. If the current is raised, the boundaries of the zones shift outward; how this lateral displacement of isothermal zone circles proceeds can be visualized from the temperature profiles calculated by Roes (see Fig. 6.37).

Similar changes take place in the outer, low-temperature layers, as has been demonstrated by Sperling (1950) (see Fig. 6.38). Brinkman studied the behaviour

[*] Hörmann (1935), Brinkman (1937, p. 48), Kersten (1941), Kersten and Ornstein (1941). Kruithof (1943a and b), Kruithof and Smit (1944), van Stekelenburg (1943), van Stekelenburg and Smit (1948), Huldt (1948b), Terpstra (1956), Terpstra and Smit (1958), and de Galan (1965a, 1966a).

of AlO bands and observed the lateral expansion of the 4000°K zone. We remark that AlO bands are emitted optimally when the gas temperature is between 3500 and 4000°K because AlO radicals dissociate into atoms at high temperatures and form into polyatomic aggregates at low temperatures. Thus temperature determination from AlO bands always yields numerical values ranging from 3500 to 4000°K. For that reason the bands are ideal for investigating spatial displacements of the 4000°K region in the arc (cf. § 6.16).

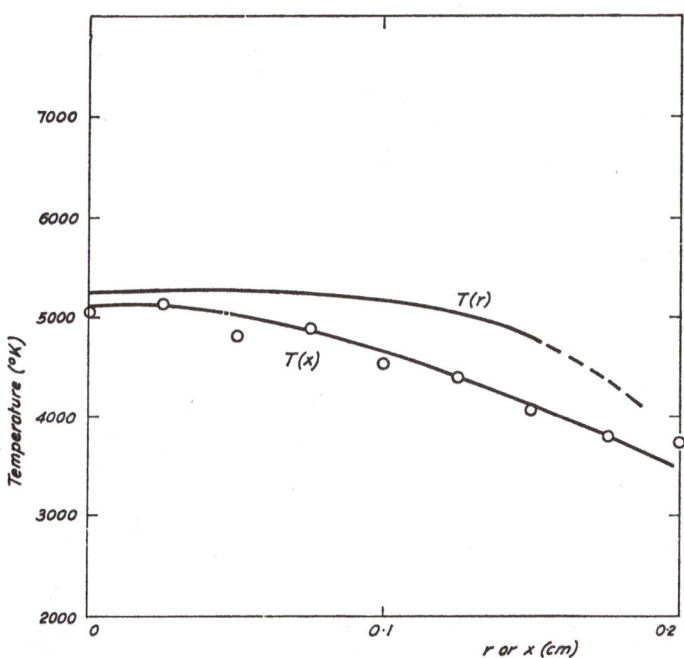

FIG. 6.35. Measured radial temperature distribution $T(r)$ in a 4-amp d.c. carbon arc in air at atmospheric pressure with potassium vapour added (according to van Stekelenburg, 1943, p. 22; cf. van Stekelenburg and Smit, 1948).

The effective temperature $T(x)$ averaged over successive transverse light paths is also given. Since the temperature was measured with the aid of CN bands, the functions $T(x)$ and $T(r)$ do not differ widely (see also § 6.16).

Customarily, we divide the arc into two zones with sharply defined boundaries, namely a more or less cylindrical region in the centre, denoted as the core, and the enveloping fringe, the flame (cf. § 2.2). Obviously, in view of the gradual fall in the temperature and in the other vital parameters, such as the electron concentration, the concept of core can be never defined satisfactorily. Identifying the core with the region from which the violet CN bands are emitted can sometimes serve as a useful approximation. We note, however, that the diameter of the violet part is smaller than that of the innermost zone of almost uniform temperature (see Brinkman, 1937, p. 45; and Smit, 1950, p. 65). Another objection against this definition is the fading of the CN band emission when the arc temperature is lowered by introducing, for example, an alkali metal. From a theoretical point of view the core radius should be defined as the distance from the axis at which the electron concentration has sunk to zero, in compliance with the conception of the core as the region through which the current flows. Since determining the radial

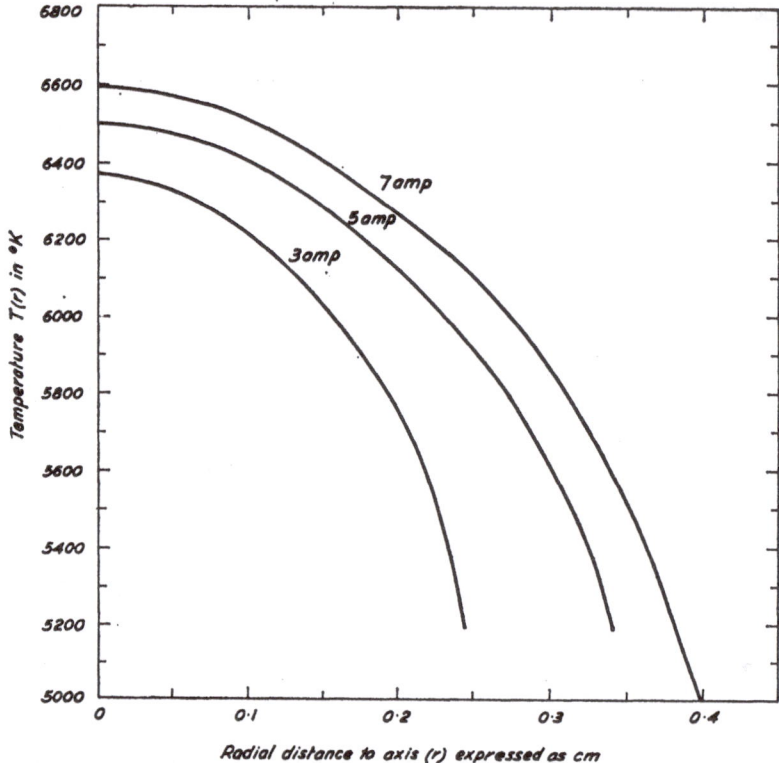

FIG. 6.36. Radial temperature profiles for a d.c. carbon arc in air at atmospheric pressure for current strengths of 3, 5, and 7 amp (after Roes, 1962, p. 113).

The temperature distributions have been calculated by using the Elenbaas–Heller equation (see § 4.2). The composition of the arc plasma was computed on the basis of the atomic ratio $N : C = 8·5 : 1$ for the 3-amp arc, $N : C = 11 : 1$ for the 5-amp arc, and $N : C = 14 : 1$ for the 7-amp arc, for a temperature of 5000°K (cf. § 11.1). The figure also shows implicitly to what extent isothermal zones are displaced outwards when the current is raised.

distribution of the electron concentration is tedious, one usually has recourse to simpler, though less exact, criteria for defining the core boundary, such as CN band emission (see § 3.8 and Chapter 8, however).

The first approximation of an arc is that of a gaseous body of *uniform* temperature and electron density. That uniform temperature is assumed to be identical with the effective temperature as derived from the intensities of spectral lines that originate in the central part of the arc (cf. § 6.16). The region in the arc to which,

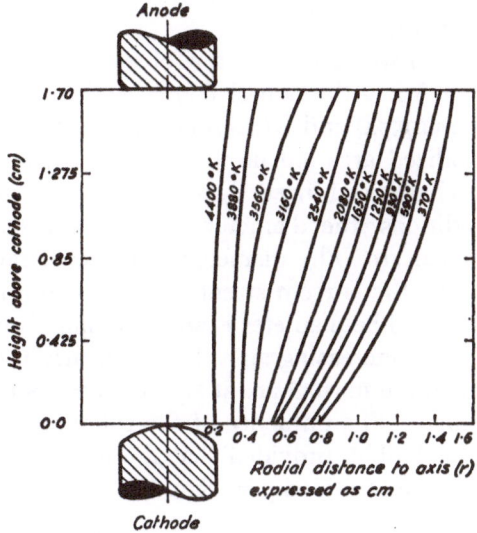

FIG. 6.37. Isotherms for the flame of a 10-amp d.c. carbon arc in air at atmospheric pressure. The curves have been established with the aid of a Schlieren procedure. From Sperling, *Zeitschrift für Physik*, 1950, **128**, 269. (Springer-Verlag, Berlin, Göttingen, Heidelberg).

in terms of this model, an even temperature distribution is assigned, is denoted as the core. Whether the model is adequate or not depends on the numerical value of the temperature and on the nature of the spectral line and the kind of element under consideration (cf. §§ 6.12 and 6.16, and particularly Chapter 8, where we deal

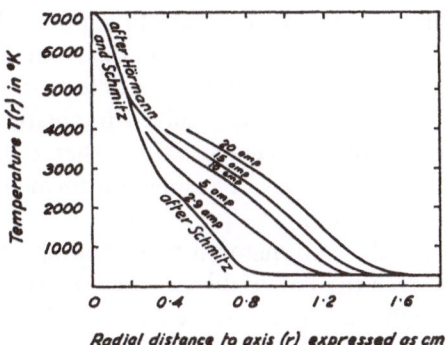

FIG. 6.38. Radial temperature distribution in the d.c. carbon arc in air at atmospheric pressure. In the composite diagram, measurements of different authors have been combined, viz. those of Hörmann (1935), Schmitz (1949), and Sperling (1950). From Sperling, *Zeitschrift für Physik*, 1950, **128**, 269. (Springer-Verlag, Berlin, Göttingen, Heidelberg).

with the appropriateness of effective temperatures, effective electron concentrations, and effective arc radii).

To conclude this section we mention a few observations on air-stabilized arcs. We have noted already in § 6.12 the remarkably large diameter of the arc discharge exploited by Eberhagen (1955a) and Köstlin (1964). A search for some systematic information on air-stabilized arcs led us to Aarts (1952, pp. 141–158 and 206), who studied an arc discharge stabilized by a spiral stream of air. He states that, under comparable conditions, the diameter of the stabilized arc appears to be larger than for the normal arc. The diameter of the core ranges from about 2·5 to 10·5 mm, depending on the current strength (4 to 10 amp) and on the axial air velocity (375 to 20 cm/sec). The diameter is an increasing function of the current; it decreases, however, with increasing air velocity. Nearly the same values of the diameter were found as in the normal arc at equal current strength when the axial air velocity was 250 cm/sec. The influence of the rotational component of the air velocity proved to be negligible, provided that the numerical value of that component was greater than five rotations per second.

★ § 6.16   *The interpretation of results of spectroscopic temperature measurements*

After our discussion of radial distributions in the preceding sections it will be clear that the interpretation of the results of spectroscopic temperature measurements requires special attention. In considering the subject we have in mind those procedures for temperature determination that use intensities of spectral lines [equation (6.19) or (6.20)]. This restriction is merely made for the sake of convenience.

Unless we employ space-resolved spectral-line intensities for computing temperatures, results do not give unambiguous and general information about the excitation properties of the source. We obtain some kind of average temperature, the meaning of which differs from the one case to another. This is mainly because different thermometric species are differently distributed through the source volume. The consequences of non-uniform particle distribution would be less serious if we were able to observe space-resolved intensities directly. Instead, we are able only to record involved average values that either must be analysed mathematically to give the desired space-resolved information or must be treated cautiously when being interpreted.

If an arc is imaged on the collimator lens of the spectrograph, spectral-line intensities are averaged over the entire depth and over the effective width of the light source. By focusing the arc on the slit we reduce the relative contribution of radiation from the outer layers to the mean intensities (cf. Fig. 6.32 and § 8.3). We cannot avoid, by an optical arrangement, averaging over the depth of the arc. The only remedy is spatial resolution of intensities by taking edge-on spectra and applying the Abel inversion (see § 6.14). As this elaborate procedure is practised but occasionally, it is worth considering the nature of the temperature values obtained by the commoner methods. The mean temperatures these procedures yield are not simply volume-averaged (cf. Feldman and Wittels, 1957), that is

$$T_{vol} = \frac{\int_{x_1}^{x_2} \int_{y_1}^{y_2} \int_{z_1}^{z_2} T(x, y, z)\, dx\, dy\, dz}{V} \tag{6.55}$$

where x_1, y_1, etc., are the boundaries of the region from which radiation contributes to the observed intensities, and V is the volume of that region. For, a volume-averaged value requires equal numbers of temperature signals to be emitted from each element of volume in the portion of the source under investigation. This situation is precluded inherently by the very procedure of spectroscopic temperature measurement, since the use of spectral-line intensities implies that temperature signals, i.e. photons, are not generated in equal numbers per unit volume in regions of different temperature. Accordingly, even if the particles that emit the signals were evenly distributed throughout the source, the measured temperature would not be of the type defined by (6.55); only if the particle distribution were such that the density were inversely proportional to the Boltzmann exponential factor of the level population would we expect a uniform generation of signals over the entire volume of the source.

To conclude, the observed temperatures are not volume-averaged, but *population-averaged*, i.e.

$$T_{pop} = \frac{\int_{x_1}^{x_2} \int_{y_1}^{y_2} \int_{z_1}^{z_2} T(x, y, z)\; n(T, x, y, z)\, dx\, dy\, dz}{N} \tag{6.56}$$

where n is the density or concentration (number per unit volume) of the particles that populate the upper levels of the spectral lines involved in the temperature determination, and N is the total number of those (excited) particles contained in the source.

The quantity defined by (6.56) is thus a weighted (or effective) temperature. There seems to be little objection against the use of this parameter, as it certainly describes excitation conditions in the source more effectively than, for example, the volume-averaged temperature. Yet, an effective temperature *alone* does not give sufficient information as to the real nature of the source. Since different species behave differently, an effective temperature determined for the one type of particle is not necessarily appropriate for the other. This complication arises from several causes. Here, we restrict discussion to those that originate in the creation and destruction of species by ionization, dissociation, and recombination. Material transport phenomena, such as diffusion, migration of ions in an electrical field, and convection, which can also play a role, are not considered.

6

The local equilibria involving ionization and dissociation are governed by the local temperature in the element of volume under consideration. For each separate species the degree of ionization or degree of dissociation is a function of the temperature; consequently, the radial decline of the temperature in the arc produces a radial variation of the relative concentrations of ions and atoms of each separate element; similarly, the variation of the degree of dissociation of any molecular species with temperature is displayed radially in the arc. The interest of dissociation is mainly confined to the arc flame, that of ionization to the core.

In § 6.12 it was explained that the simultaneous proceeding of excitation and ionization with temperature gives rise to *optimum temperatures* for the emission of distinct spectral lines. In the normal low-current arc, where 6500–7000°K can be regarded as an upper limit of the attainable temperature (see § 4.3), optima are reached only for atom lines and molecular bands; in extremely high temperature plasmas, optimum temperatures can be also assigned to the lines emitted by ions in subsequent stages of ionization.*

Restricting our discussion to the low-current arc we must recognize that in a cross-sectional plane normal to the axis, given the radial distribution of the temperature (and that of the electron concentration), each spectral line can be attributed to a concentric zone where it is emitted with maximum intensity. For ion lines this zone is always the central portion; for atom lines the location of the zone depends on the temperature at the axis, on the ionization potential of the element, and on the excitation potential of the line. As we shall see in § 8.8, it is the ionization potential V_i that predominantly determines the numerical value of the optimum temperature T_M; the dependence of T_M on the excitation potential V_q is less pronounced, though not negligible.

We emphasize that the concept of optimum temperature refers essentially to lines; relating numerical values of optimum temperatures to species is possible only if we do not want an exact answer. This abandoning of accuracy might be sometimes convenient for practical purposes. Hence we shall do so occasionally in the discussion below when speaking of the optimum temperature range of a species. Implicitly we assume, then, that only slight differences occur among excitation potentials of lines.

To summarize, in a given arc each species can be assigned a relatively small temperature range at which emission of atom lines is at an optimum. A particular thermometric species provides mainly information about the excitation conditions in that zone of the arc whose temperature range equals the optimum range for the species. Outside this zone the relevant species radiates but weakly, so that the corresponding temperature information carries little weight on the average value.

---

* The question is important in astrophysics (see e.g. Unsöld, 1955), in high-temperature plasma physics, and in the field of electric sparks (see among others Huldt, 1955; van Calker and Braunisch, 1956; Krempl, 1962; and Laqua and Hagenah, 1963). For determining radial temperature distributions in extremely-high temperature plasmas, Larenz (1951) initiated an absolute method that is founded on the concept of optimum temperatures (*Normtemperaturen*) (cf. Krempl, 1962; and Laqua and Hagenah, 1963).

All lines having an optimum temperature that exceeds the actual temperature T_0 at the axis, are emitted mainly from the core. If T_0 is high, say 6500°K, this holds true for ion lines as well as for atom lines of elements of high ionization potential; lowering T_0 widens the group of elements with atom line emission from the central portion of the arc. Conversely, if we intend to measure the maximum temperature of the arc, we should employ ion lines, or atom lines of elements of high ionization potential.*

Band emission is, as a rule, favoured at relatively low temperature; thus we mentioned in § 6.15 that the AlO radical radiates optimally from the arc flame. The dissociation of AlO at high temperature and the formation of polyatomic aggregates at low temperature lead to an optimum for AlO emission in the temperature range between 3500 and 4000°K (see also § 11.2).

An interesting example of widely differing effective temperatures measured in one and the same arc is furnished by Margoshes and Scribner (1963). They determined temperatures in a gas-stabilized arc in an argon atmosphere. The values found were: 7570°K from Ti II lines, 7680°K from Cr II lines, 5540°K from Ti I lines, and 5510°K from Fe I lines. Clearly, the atom lines indicate the temperature in a zone at some distance away from the axis; the results taken from the ion lines most probably refer to the centre of the arc.

Perhaps we should expect a significant difference between the Ti I and Fe I temperatures, in view of differing ionization potentials of the elements. Presumably, this detail is concealed as a result of systematic errors in the sets of gf-values employed. As those values for Ti I and Fe I were taken from Corliss and Bozman (1962), the high uniformity of their scale, discussed at the end of § 1.5, seems to be an obvious reason for the lack of disparity between the effective temperatures obtained for Ti I and Fe I.

Let us postpone further discussion of the interpretation of effective excitation temperatures until Chapter 8. There we shall systematically examine the appropriateness of effective values of the temperature, the electron concentration and the arc radius.

Finally, we note that the interpretation of temperature measurements can be complicated by changes in the excitation conditions with time. In the d.c. arc no special problems will be encountered in this respect; changes take place slowly, so that moving camera techniques and similar procedures are adequate means for recording variations with time. For the study of sparks and other transient discharges time-resolution techniques have been developed. Critical reviews of methods and results of time-resolved spectroscopy have been given by Laqua and Hagenah (1963), and by Harrington (1963).

* Remarkably, in the low-current arc, CN band emission is at an optimum in the vicinity of the axis; the same applies for C_2 bands. Hence these bands always give information about the temperature prevailing in the central portion (cf. Smit, 1950, p. 65). The average temperature derived from CN bands (arc axis focused on the spectrograph slit) has been established by Kersten (1941, pp. 33–35) and Kruithof (1943a, p. 42) to be only a 100°K below the temperature at the axis, obtained from space-resolved intensities. This order of magnitude is also demonstrated in Fig. 6.35. Compare $T(r)$ and $T(x)$ at $r = 0$ and $x = 0$, respectively.

★ § 6.17 *Dependence of arc temperature on current strength, electrode shape and grade, polarity, and gap width*

Fig. 6.39 shows a plot of temperature against current strength, as reported by Brinkman (1937, p. 38), for the violet core of the carbon arc in air at atmospheric pressure (gap width ≤ 1 cm; diameter of electrodes, 12 mm). The temperature was measured with the aid of CN bands; precision of a single determination is stated to be not better than ±250°K. Brinkman concludes that the temperature of the plasma in the carbon arc increases regularly from 6300 to 7000°K when the current is raised from 1 to 13 amp.

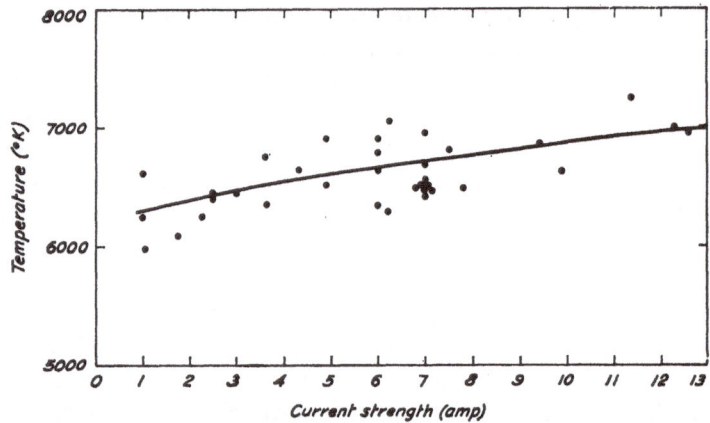

FIG. 6.39. Plot of the temperature in the violet core of a carbon arc in air versus the current strength (after Brinkman, 1937, p. 38).

Boeschoten (1953, p. 83) determined the temperature-current relationship for a carbon arc with zinc and copper vapour added. He used photoelectric equipment for measuring the ratio of the copper lines Cu 5153 and Cu 5782. His plot shows a rise of temperature from about 5300 to 6500°K for an increase of the current strength from 1·5 to 8·5 amp. The absolute values of the temperature—that have been adapted here as to agree with Schurer's transition probabilities for copper lines (see § 6.9)—appear to be somewhat high, in view of the optimum temperature for copper lines (see § 8.8) and the presence of metal vapour in the arc column.

For a graphite arc with a little zinc added as a thermometric species, I found a dependence of temperature on the current strength similar to that of Brinkman's results, namely an increase from 6130°K at 8 amp to 6330°K at 12 amp, thus not more than 50°K per amp. Operating conditions were similar to those described in § 6.7.

It is not easy to predict theoretically what happens to the temperature of an arc with alkali metal vapour if we vary the current strength externally by means of the ballast resistance. One has to consider a cycle of interacting factors. We remember firstly that the presence of an alkali metal in the arc lowers the temperature by an appreciable extent (see §§ 3.9, 7.7, and 7.8). Now, roughly, we have the following picture: the rate of volatilization of alkali affects the instantaneous concentration of the metal vapour in the arc; this concentration influences the electrical conductivity and the temperature of the plasma; these factors act upon the temperature of the electrode, which finally has repercussions on the rate of volatilization again. An increase in the current strength tends to enhance the temperature of the plasma; the effect may be quenched, however, by an increase in the rate of volatilization of alkali metal produced simultaneously. Therefore, within a certain range of current strength, the cycle could turn out to be more or less self-regulating.

Indeed we have experimental evidence that various factors and processes balance each other to some extent; an increase of the current strength in steps from 6 to 10 amp during the evaporation of NaCl from a furnace electrode (§ 6.7) brought about a rise of the temperature equal to that in the arc in air alone, that is about 50°K per amp.

The question has been raised whether rigorous stabilization of the current strength is likely to produce crucial improvements of spectrochemical precision in d.c. carbon arc procedures. Practice has proved that strict external control of the current strength does not repay the difficulty and expense of its practical realization (see particularly the bibliography by Ahrens and Taylor, 1961, pp. 38–40; also § 3.1). The rather poor outcome is in line with theoretical expectation. The temperature that is established in the arc column depends on a number of factors among which the composition and the chemical, physical, and mechanical state of the specimen arced are preponderant. Only an extremely large change of the current strength, brought about externally by varying the series resistance, might outweigh the effect of the specimen composition upon the temperature; small fluctuations in the current, even if of the order of 1 amp, effect the temperature but slightly. Therefore we cannot cancel, by external current control, temperature variations caused by a dominant internal effect like a compositional change of the plasma. In my opinion we must aim at producing a more or less steady state by volatilizing along with the sample a suitable buffer (see § 7.12) that carries the discharge. Once the arc is struck and the series resistance is set, the buffer should determine the current strength and keep it at a virtually constant level.

Experiments I carried out with steady evaporation of various substances showed the recorded temperature fluctuations to be within the limits of measuring precision, i.e. 20–85°K expressed as standard deviation (see §§ 6.6 and 7.7). During arcing the current strength remained constant at 10 amp, in most instances to within ± 0·3 amp. On passing from the one test substance to the other the setting of the ballast resistance had to be altered, of course, to secure the value of 10 amp in all cases.

To conclude, for obtaining steady excitations conditions it is more important that the samples should be diluted properly and that the electrode should be filled correctly, than that the current strength should be completely stabilized. When deciding upon an appropriate admixture we must consider the ionization potential of the diluting element and the factors that determine the smoothness of its evaporation into the arc; we mention, among others, the stability and the volatility of the compound of the element, and the influence of an additional agent, like graphite or carbon powder, upon the evaporation conditions (see particularly §§ 7.12, 10.2, and 10.3). Obviously, the electrode shape, especially that of the filled lower electrode, must also enter into the discussion now.* In our laboratory, strip chart recording of the gap voltage has become a common technique for making preliminary studies of the burning properties of test fillings (see § 7.14).

The shape of the electrode affects the rate of evaporation of the specimen and can thus have a secondary effect on the excitation conditions and on the steadiness of the arc burn. The grade of the electrode material has a similar influence; notably the difference in thermal conductivity between carbon and graphite should be borne in mind (see § 10.3). When the discharge is carried by an element of relatively low ionization potential, as in the buffered arc proposed above, the electrode material does not immediately affect the temperature in the arc. Further remarks on the importance of the grade of electrode material for spectrochemical analysis have been included in § 10.3.

Polarity can exert a marked influence on the arc temperature, when a substance is volatilized from the electrode cavity. As the anode assumes a higher temperature than the cathode, the rate at which the specimen is released may depend on the polarity of the supporting electrode.

In the carbon arc in air alone the gap width l has little influence on the temperature. Brinkman (1937, p. 38) determined the temperature to be independent of the arc length if the latter was between 2 and 10 mm. At greater electrode separation he found the temperature decreased as the arc lengthened (by about $300°$K when l was raised from 10 to 18 mm).

In a short arc, that is when l is less than 2 mm, the temperature tends to be higher than in the uniform column of a long arc. The reason for this is that the homogeneous column disappears when the gap width is less than 2 mm; the remaining electric field is the resultant of the anode and cathode falls (see § 3.1 and the end of § 3.3); fast electrons now also take part in the excitation, which no longer can be regarded as truly thermal (cf. § 5.4).

Interesting information on the axial variation of the temperature in arcs with metal vapour has been reported by de Galan (1965a). His measurements pertain to a 10-amp d.c. arc with a 10-mm gap width. The temperature of the arc was varied by evaporating KF, LiF, or Al_2O_3 from the anode into the discharge zone, resulting in mean temperatures of 5100, 5600, and $6000°$K respectively in the central part (cf. § 9.4). The axial distributions of the temperature, as well as those of the electron concentration (see § 7.8), are shown in Fig. 6.40. Obviously, a homogeneous column extends from 3 to 8 mm above the anode, whereas in the neighbourhood of the electrodes, particularly near the anode, the temperature and the electron concentration rise sharply. Such an increase of the temperature was also found by Righini (1935).

Keeping the gap width constant when arcing samples for spectrochemical analysis

---

* A great many interesting notes on sample preparation and electrode shapes, suggested by practical experience, have been accumulated by Ahrens and Taylor (1961, p. 41 ff.).

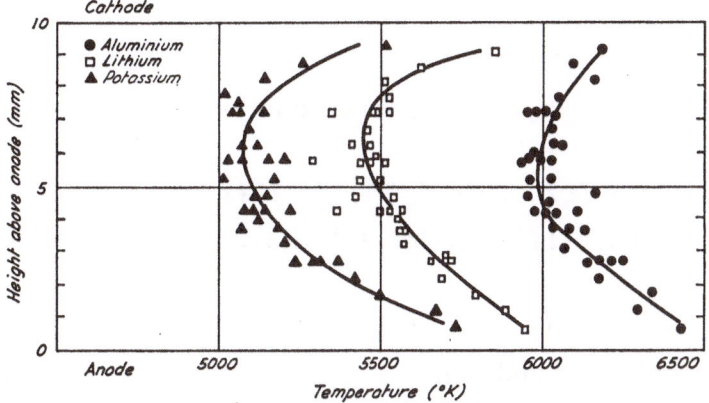

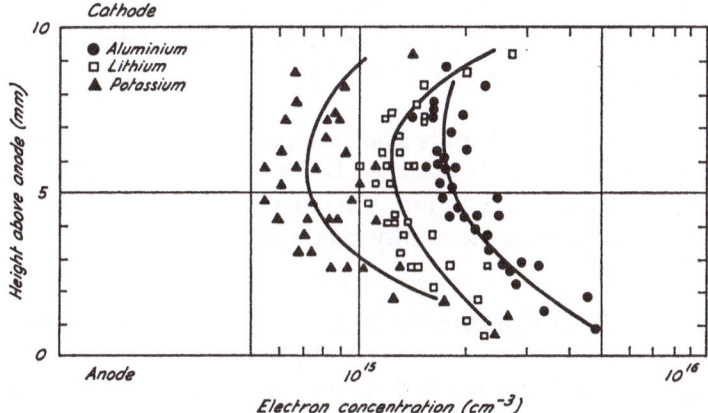

FIG. 6.40. Axial distributions of the temperature (upper diagram) and of the electron concentration (lower diagram) found in 10-amp d.c. carbon arcs with different metal vapours (after de Galan, 1965a, 1966a).

is, in contrast with the rigorous control of the current strength, rather stringent in view of the axial distribution of the temperature, the electron concentration, and the vapour concentrations. Moreover, it is advisable to ensure that the portion of the arc from which measurements are taken is always at the same height above the lower electrode.

CHAPTER SEVEN

IONIZATION PHENOMENA AND
THE DETERMINATION OF ELECTRON PRESSURE

★ § 7.1   *Introduction* (cf. §§ 3.8 and 3.9)

A necessary condition for the existence of the arc discharge is the partial ionization of the gas mixture in the discharge column. In the range of core temperatures at present under consideration, that is $T = 4000$–$7000°K$, about 0·01 per
cent to about 0·1 per cent of all particles exist in the ionic form, so that the partial
pressure of the electrons ranges from 10^{-4} to some 10^{-3} atm if the arc burns at
atmospheric pressure.

In the column of the arc there is *quasi-neutrality*, which means that the charge
carried by the free electrons is balanced by the total charge of the positive ions
(see § 3.8). In the normal carbon arc, where singly-charged ions only have to be
taken into account, the quasi-neutrality condition

$$n_e = \sum n_{ij} \tag{7.1}$$

applies. The label j distinguishes the particle densities or concentrations (n_i) of
different ionic species.

The concentrations of electrons and ions are uniquely related to the gas temperature and to the gas composition, since all charged particles originate from thermal
ionization of the gas; the electrical character of the arc is immaterial in this respect.
The current flow is indispensable for heating the gas and for maintaining the high
temperature; it has no direct bearing upon the ionization, however. If the gas,
enclosed in a furnace, were heated externally to the same temperature as in the
arc, it would be in a similar state of ionization. The only difference is the self-
regulation of the temperature and the electron concentration in the arc. For a
given gas composition, therefore, the temperature and the electron concentration
assume values such that the power dissipated per cm in the column equals the
heat lost per cm arc length in unit time by thermal conduction to the surroundings.
The governing fact is the very marked dependence of the electron concentration
on the temperature of the gas, its composition being assumed constant. Therefore, as has been explained more extensively in § 3.9, we expect that very temperature to be established at which the gas is ionized to such an extent that the resulting
electrical conductivity and power dissipation are adequate to keep up that temperature. This concept is essential for understanding the influence exerted upon the
arc temperature by substances of low ionization potential. This question has also

been raised in § 3.9; it will get further treatment in this chapter. The main object, however, is to deal with the bearing of ionization on the intensities of spectral lines. In this respect ionization is at least equally important as excitation, though this is not always realized by spectrochemists.

The elements whose concentrations in the samples are to be determined exist in the arc for some part as neutral atoms, for another as singly-charged ions. Spectral lines originate either from atoms or from ions and thus their intensities do not entirely depend on the total number of particles of the relevant element; inherently, the proportions of atoms and ions are likewise involved. For that reason we must thoroughly investigate how these proportions can be grasped in terms of ionization equations.

The basic concept of thermal ionization, introduced by Saha (1920, 1921a and b), is the application of the mass action law to ionization equilibria. For each component (labelled j) of a composite gas we consider an equilibrium of the form

$$\text{neutral atom} \rightleftharpoons \text{ion} + \text{electron} \qquad (7.2)$$

By applying the Nernst heat theorem, the equilibrium constant

$$K_{nj} = \frac{n_{ij}n_e}{n_{aj}} \qquad (7.3)$$

can be calculated as a function of the temperature. The resulting equation

$$K_{nj} = S_{nj}(T) \qquad (7.4)$$

is known as Saha's relationship.

In fact, thermal ionization is a special case of thermal dissociation (see Chapter 11). The mathematical treatment of the constant (7.4) is thus essentially implied in texts on statistical mechanics. Indeed, the function $S_{nj}(T)$ can be derived by elementary means. We refer the interested reader particularly to Unsöld (1955, p. 79 ff.), who treats the subject lucidly. He discusses also, in a concise historical survey, the different procedures developed in the literature for setting up the Saha equation. The relationship as it was originally derived by Saha is not complete, since partition functions are wanting. Moreover, in its initial form, Saha's formula applies exclusively to single gases. For composite ones a modified version of the original Saha equation must be used, as will become evident in the next sections (see also § 7.15).

In discussing ionization equilibria and their interrelation with spectral-line intensities it is convenient to introduce the concept of degree of ionization, viz.

$$\alpha_j = \frac{n_{ij}}{n_j} \equiv \frac{n_{ij}}{n_{aj}+n_{ij}} \qquad (7.5)$$

where n_{ij} and n_{aj} are the density or concentration (expressed as number per cm^3) of the singly-charged ions and the neutral atoms of the component labelled j;

n_j denotes the total concentration, i.e.

$$n_j = n_{aj} + n_{ij} \tag{7.6}$$

Using the degree of ionization α_j, the concentration of the atoms or of the ions can be expressed in terms of the total concentration n_j of the relevant species j, viz.

$$n_{aj} = (1 - \alpha_j)n_j \tag{7.7}$$

and

$$n_{ij} = \alpha_j n_j \tag{7.8}$$

Relating intensities of atom and ion lines to the total concentration of an element is formally accomplished by inserting (7.7) and (7.8) into our equations. Thus, in virtue of (1.26) and (6.5)—compare (6.17)—we now write for the intensity of an atom line

$$I_{qp} = K_{qp}(1 - \alpha_j)n_j \frac{\exp(-\epsilon_q/kT)}{Z_{aj}} \tag{7.9}$$

and for the intensity of an ion line

$$I_{qp}^+ = K_{qp}^+ \alpha_j n_j \frac{\exp(-\epsilon_q^+/kT)}{Z_{ij}} \tag{7.10}$$

where K and K^+ are constants that depend on the nature of the line q $\to$ p and on the geometry of the source. For instance,

$$K_{qp} = \frac{d}{4\pi} g_q A_{qp} h\nu_{qp} \tag{7.11}$$

In equations (7.9) to (7.11) ϵ is the excitation energy, Z the partition function (§ 6.2), T the absolute temperature, k the Boltzmann constant, g the statistical weight (§§ 1.4 and 5.5), A the transition probability (§ 1.3), h the Planck constant, and ν the frequency. The use of distinctive labels (a, i, and $+$) is self-explanatory. Note that the excitation energy of an ion line is taken from the ground level zero of the ion; it is not defined as the sum of the ionization energy of the atom and the excitation energy of the ion line from the ground level zero of the ion.

The expression for intensity is composed of three functions of temperature, namely the degree of ionization α_j, the Boltzmann exponential factor, and the partition function Z. In the temperature range considered here, Z will either be regarded as constant or its dependence on the temperature will be taken into account, to a first approximation, by applying a correction to the excitation energy in the Boltzmann factor (see § 6.3).

More serious difficulties occur for the function α_j. To calculate it as a function of the temperature, data about the electron concentration (n_e) or about the electron pressure (p_e) of the gas are also required. From equations (7.3), (7.4), and (7.5) it follows that*

$$\frac{\alpha_j}{1-\alpha_j} n_e = S_{nj}(T) \tag{7.12}$$

which is equivalent to

$$\alpha_j = \frac{S_{nj}(T)}{n_e + S_{nj}(T)} \tag{7.13}$$

or to

$$1 - \alpha_j = \frac{n_e}{n_e + S_{nj}(T)} \tag{7.14}$$

So far, the expressions given in conjunction with the Saha equation are entirely *elementary*. Other elementary relationships will be used in subsequent sections. They do not touch the field of statistical mechanics, which concerns particularly the functions $S_{nj}(T)$ and $S_{pj}(T)$ (see § 7.2). Often, the Saha equation has caused confusion in the literature, chiefly, in my opinion, because authors have not recognized the elementary nature of much of the subject. Normal physical chemistry should not be ignored when considering a 'highly physical' problem such as the ionization of the plasma in an arc. Once more I recall that ionization in the arc bears a close resemblance to the dissociation of weak acids in aqueous solution (§ 5.1). The arc gas, i.e. the mixture of N_2, O_2, CO, C_2, N, O, and C, is the 'solvent' in which the metal vapours are embedded. The concentration of dissolved material in the arc is low; it amounts to about 1 per cent at most (see § 7.9).

A well-known problem in physical chemistry is the calculation of *pH*-values and the evaluation of the dissociation equilibria in an aqueous solution that contains a number of weakly dissociated acids, given their analytical concentrations and the dissociation constants. In principle, this problem is easily solved; there is only the practical difficulty of solving by elementary means the set of equations involved. It is much simpler to evaluate the dissociation equilibria, given the dissociation constants and a measured *pH*-value. Degrees of dissociation (β_j) can be computed then for an arbitrary large number of electrolytes. Each numerical value β_j follows from an independent equation

$$\frac{\beta_j}{1-\beta_j} [H^+] = K_j \tag{7.15}$$

* Similarly, with the partial pressure p_e we have

$$\frac{\alpha_j}{1-\alpha_j} p_e = S_{pj}(T) \tag{7.12n}$$

$$\alpha_j = \frac{S_{pj}(T)}{p_e + S_{pj}(T)} \tag{7.13n}$$

and

$$1 - \alpha_j = \frac{p_e}{p_e + S_{pj}(T)} \tag{7.14n}$$

Numerical values of the function S_{pj} have been listed in Appendix 6 (see also § 7.2).

if equilibria of the type

$$HA_j \rightleftharpoons H^+ + A_j^-$$ (7.16)

are considered.

The equation that we shall use for evaluating the degrees of ionization in the arc, i.e. (7.12), is similar to (7.15). The only difference is that here we are particularly interested in changes of the equilibrium constants with the temperature; in an aqueous solution we are commonly concerned with only one temperature, say 20°C.

As well as computing the pH-value of the solution referred to above, we can tackle the problem of evaluating the electron concentration in the arc. The problem is more involved here, especially when the carbon arc in air is considered. Even if no material is added, there is a fairly large set of equations to be dealt with, since various dissociation equilibria must also be evaluated (see § 11.1). A serious difficulty is specifying the elementary composition of the gas, particularly the amount of carbon present. When a substance is vaporized into the arc plasma, specifying the composition becomes even more puzzling. Therefore, it is preferable in our calculations to use measured values of the electron concentration or of the electron pressure.

★ §7.2   *Dependence of ionization constants on the temperature: the Saha formula*

The dependence of the ionization constant S_{pj} on the temperature is expressed by Saha's relationship:

$$\frac{p_{ij} p_e}{p_{aj}} = S_{pj} \equiv \frac{(2\pi m)^{\frac{3}{2}}(kT)^{\frac{5}{2}}}{h^3} \frac{2Z_{ij}}{Z_{aj}} \exp(-\epsilon_{ij}/kT)$$ (7.17)

where m is the electronic mass, k the Boltzmann constant, h the Planck constant, Z the partition function, ϵ the ionization energy, and p the partial pressure. The indices a, i, and e refer to the neutral atom, the ion, and the electron; the label j specifies the element or the compound. The factor 2 that immediately precedes the ratio of partition functions is the statistical weight of a free electron ('spin orientation').

The practical version of the Saha formula is

$$\frac{p_{ij} p_e}{p_{aj}} = S_{pj} \equiv 6 \cdot 58 \cdot 10^{-7} \; T^{\frac{5}{2}} \frac{Z_{ij}}{Z_{aj}} 10^{-(5040/T)V_{ij}}$$ (7.18)

or in logarithmic form

$$\log S_{pj} = \frac{5}{2} \log T - \frac{5040}{T} V_{ij} + \log \frac{Z_{ij}}{Z_{aj}} - 6 \cdot 18$$ (7.19)

where the ionization potential V_{ij} is expressed as electron volts, the electron pressure p_e as atmospheres.*

For facilitating calculations we have tabulated $5040/T$, $\frac{5}{2} \log T$, and $\frac{3}{2} \log T$ as functions of T in the temperature range 4000–8000°K (see Appendices 3, 4, and 5). Also included is a table of S_{pj} for different values of T and $\bar{V}_{ij}$ (see Appendix 6).

★ § 7.3 *Partition functions in Saha's ionization relationship*
Apparent ionization potentials

The original logarithmic Saha equation does not contain the term $\log 2Z_{ij}/Z_{aj}$, which was introduced later as a result of thermodynamical considerations. We shall not concern ourselves with this rigorous treatment here; suffice it to say that the partition functions account for the discrepancy between the true ionization energy, i.e. the energy difference between the ground states of the atom and the ion, on the one hand, and the difference in mean energy of the ionic and atomic populations, on the other, each having a Boltzmann distribution of energy levels.

Often, when managing the Saha formula (7.19) mathematically, it is inconvenient to retain the term $\log Z_{ij}/Z_{aj}$. Dropping it entirely is incorrect, of course, as this is equivalent to putting the ratio of partition functions arbitrarily equal to unity. There is an alternative, however, which can be used for a moderate range of temperature (Boumans, 1961, 1962).

In (7.19) we write $\log Z_{ij}/Z_{aj}$ as a correction of the ionization potential V_{ij}, inserting

$$\log \frac{Z_{ij}}{Z_{aj}} = -\frac{5040}{T} \delta_j \qquad (7.20)$$

* We recall the relationships (3.18) and (3.19) for converting density into partial pressure, namely

$$n = \frac{p}{kT} \qquad (1)$$

which in practical form reads

$$n = 7\cdot340 \cdot 10^{21} \frac{p}{T} \qquad (2)$$

where n is expressed in cm^{-3} and p in atm. Examples of numerical values of n for a few values of p and T have been listed in Table 3.1. An abridged conversion table has been appended.

If the total pressure $P = 1$ atm, the numerical values of partial pressures (atm) express, at the same time, proportions of the total. For this reason, partial pressures are frequently used in preference to concentrations.

Using equations (1) and (2) we convert (7.17) to (7.19) into

$$\frac{n_{ij}n_e}{n_{aj}} = S_{nj} \equiv \frac{(2\pi mkT)^{\frac{3}{2}}}{h^3} \frac{2Z_{ij}}{Z_{aj}} \exp(-\epsilon_{ij}/kT) \qquad (7.17n)$$

$$\frac{n_{ij}n_e}{n_{aj}} = S_{nj} \equiv 4\cdot83 \cdot 10^{15} T^{\frac{3}{2}} \frac{Z_j}{Z_{aj}} 10^{-(5040/T)V_i} \qquad (7.18n)$$

$$\log S_{nj} = \frac{3}{2} \log T - \frac{5040}{T} V_{ij} + \log \frac{Z_{ij}}{Z_{aj}} + 15\cdot684 \qquad (7.19n)$$

so that the term $\log Z_{ij}/Z_{aj}$ can be omitted, if V_{ij} is replaced by $\tilde{V}_{ij}$

$$\tilde{V}_{ij} = V_{ij} + \delta_j \tag{7.21}$$

The new quantity $\tilde{V}_{ij}$ can be called then the *apparent ionization potential.*[*]

The correction δ cannot be truly constant, that is independent of T. This is evident for those elements where Z_a and Z_i are virtually constant at the temperatures considered here ($T < 6500°K$). In a more general way, it is seen by writing (6.14) in the form

$$\log \frac{Z_i}{Z_a} = \log \frac{Z_i{}'}{Z_a{}'} - \frac{5040}{T}(\zeta_i - \zeta_a) \tag{7.22}$$

So, only if $Z_i{}'/Z_a{}' = 1$, we have $\delta = \zeta_i - \zeta_a =$ constant. Mostly, however, we have $\zeta_i - \zeta_a \approx 0$ (see Table 6.2) and $Z_i{}'/Z_a{}' \neq 1$.

In spite of these objections, the use of constant values for apparent ionization potentials is reasonably well justified if we consider a fairly small range of temperature, say 5000–6000°K. The data of Table 7.1 show that the difference of δ_j at 5000 and 6000°K exceeds 0·1 eV only in a few cases. For this reason the table lists mean values of the apparent ionization potential ($\tilde{V}_{ij}$) for the whole temperature range. Whenever necessary, individual δ-values, instead of mean values, may be used at either end of the temperature range (cf. Table 7.6 p. 191).

In summary, for many purposes it is preferable to use apparent ionization potential values, such as those in columns 5 and 10 of Table 7.1. Their use enables us to omit the term $\log Z_{ij}/Z_{aj}$ from the Saha equation, which is advantageous when manipulating the equation mathematically for general considerations (see §§ 8.5 to 8.9, for instance). We must realize that the δ-correction is an approximation. However, there is more justification for using the δ-correction than for omitting, as so often is done, the partition functions. In the temperature range 5000–6000°K, the apparent ionization potentials ($\tilde{V}_{ij}$) in Table 7.1 give a clearer idea how easily the elements ionize than do the true potentials (V_{ij}), and so it is a useful concept, especially if we *compare* the ionization properties of elements, since we are inclined to exclude the partition functions from our picture; only in a rigorous calculation is their influence shown. The apparent potentials show immediately and explicitly how the different elements actually behave in thermal ionization. It is immaterial then that, for example, the true situation for aluminium at 5000°K is best approximated by an apparent ionization potential of 6·74 eV and that at 6000°K by 6·90 eV; the vital point is that the value of 6·8 eV is preferable to 5·98 eV in the entire range of $T = $ 5000–6000°K.

The δ-correction is substantial for a number of elements. For Al, B, Bi, Cs, Cu, Ga, In, K, Ni, and Tl it ranges from 0·4 to 0·9 eV. Negative values also occur, e.g. for the alkaline earths and for Zn, Cd, and Hg.

---

[*] The correction of the ionization potential introduced here is only a convenient though approximate way to allow for the partition functions. It is redundant in a rigorous treatment, in contrast with the true corrections due to the lowering of the ionization potential within a plasma (see Ecker and Weizel, 1956; Theimer, 1957; Griem, 1962, 1964b; Lochte-Holtgreven, 1963; and Drawin and Felenbok, 1965). The latter correction is negligible under the conditions considered here.

TABLE 7.1

Element	δ_j 5000°K	δ_j 6000°K	V_{ij}	$\bar{V}_{ij}$	Element	δ_j 5000°K	δ_j 6000°K	V_{ij}	$\bar{V}_{ij}$
Ag	0·30	0·36	7·57	7·96	Na	0·31	0·38	5·14	5·49
Al	0·76	0·92	5·98	6·8	Nb	0·09	0·11	6·88	6·98
As	−0·12	−0·15	9·81	9·67	Ni	0·46	0·50	7·63	8·11
Au	0·33	0·39	9·22	9·58	Os	0·05	0·08	8·5	8·6
B	0·77	0·93	8·30	9·15	P	−0·29	−0·33	10·48	10·17
Ba	−0·22	−0·15	5·21	5·03	Pb	−0·13	−0·06	7·41	7·32
Be	−0·29	−0·33	9·32	9·01	Pd	−0·40	−0·37	8·33	7·95
Bi	0·58	0·68	7·29	7·92	Pt	0·31	0·32	9·0	9·3
C	0·19	0·24	11·26	11·48	Rb	0·34	0·50	4·18	4·6
Ca	−0·28	−0·29	6·11	5·82	Re	—	0·05	7·87	7·90
Cd	−0·30	−0·36	8·99	8·66	Rh	0·24	0·30	7·46	7·73
Co	0·05	0·07	7·86	7·92	Ru	0·16	0·20	7·36	7·54
Cr	0·16	0·21	6·76	6·95	Sb	0·15	0·13	8·64	8·78
Cs	0·39	0·58	3·89	4·4	Sc	−0·28	−0·30	6·54	6·25
Cu	0·36	0·46	7·72	8·13	Se	0·25	0·31	9·75	10·03
Fe	−0·19	−0·21	7·87	7·67	Si	0·22	0·29	8·15	8·41
Ga	0·71	0·86	6·00	6·8	Sn	0·21	0·27	7·34	7·58
Ge	0·23	0·30	7·90	8·17	Sr	−0·25	−0·20	5·69	5·47
Hf	0·04	0·04	6·8	6·84	Ta	−0·12	−0·12	7·88	7·76
Hg	−0·30	−0·36	10·43	10·10	Te	0·16	0·19	9·01	9·19
In	0·61	0·76	5·78	6·5	Ti	−0·27	−0·27	6·82	6·55
K	0·34	0·46	4·34	4·74	Tl	0·38	0·50	6·11	6·55
La	−0·06	−0·02	5·61	5·54	V	0·04	0·06	6·74	6·79
Li	0·32	0·40	5·39	5·75	W	−0·04	−0·05	7·98	7·93
Mg	−0·29	−0·33	7·64	7·33	Y	−0·13	−0·12	6·38	6·26
Mn	−0·08	−0·11	7·43	7·33	Zn	−0·30	−0·36	9·39	9·06
Mo	0·06	0·08	7·10	7·17	Zr	−0·12	−0·11	6·84	6·73

The δ-correction of ionization potentials for fifty-four elements in the temperature range 5000–6000°K. The δ_j-values at 5000 and 6000°K have been computed with equation (7.20) from the partition functions listed in Table 6.1. The mean values of δ_j were used for calculating the apparent ionization potentials ($\bar{V}_{ij}$) from the true values (V_{ij}). The latter were taken from Moore (1958, p. XXXIV) and Corliss (1962b and c).

§7.4 *Experimental determination of the degree of ionization of the components of a plasma*

The degree of ionization of a plasma component can be determined from the intensity ratio of an ion–atom line pair, if the temperature of the plasma is simultaneously measured. From equations (7.9) and (7.10) for the intensities of atom and ion lines we derive the following expression

$$\log\frac{\alpha_j}{1-\alpha_j} = \log\frac{I_{qp}^+}{I_{qp}} - \log\frac{g_q^+ A_{qp}^+ \nu_{qp}^+}{g_q A_{qp}\nu_{qp}} - \frac{5040}{T}(V_q^+ - V_q) + \log\frac{Z_{ij}}{Z_{aj}} \tag{7.23}$$

where the practical form of the exponential factor [cf. (6.6)] has been inserted.

Clearly, the degree of ionization α_j can be computed from the measured values of I^+/I and T, provided that the ratio of partition functions and the ratio of gA-values are known. The scarcity of accurate data on transition probabilities of proper ion–atom line pairs limits the application of equation (7.23). Corliss (1962b and c) found, while investigating the N.B.S. copper arc, discussed at length in § 1.5, that only eleven elements are suited for the purpose (see § 7.6). Therefore, to obtain information on the degree of ionization of a large number of elements over a wide range of temperature we profitably return to the Saha equation, as will be explained in the next section.

★ §7.5   *Calculation of the degrees of ionization with the aid of the Saha equation*
The role of electron pressure

Calculating the degrees of ionization of individual vapour components can be accomplished by using the Saha equation, if the temperature T and the electron pressure p_e (or the electron concentration n_e) of the gas are known. According to equations (7.12n) and (7.19) we have*

$$\log \frac{\alpha_j}{1-\alpha_j} = \log \frac{S_{pj}}{p_e} \equiv -\log p_e + \frac{5}{2}\log T - \frac{5040}{T}V_{ij} + \log \frac{Z_{ij}}{Z_{aj}} - 6 \cdot 18 \qquad (7.24)$$

To ease calculation when using this equation it is to our advantage that we recognize a few elementary relationships of $\log[\alpha/(1-\alpha)]$. If we put

$$\log \frac{\alpha}{1-\alpha} = A \qquad (7.25)$$

for the sake of convenience, it follows that

$$\log \alpha = A - \log(1 + 10^A) \qquad (7.26)$$

and that

$$\log(1-\alpha) = -\log(1 + 10^A) \qquad (7.27)$$

These equations are represented graphically in Fig. 7.1. Obviously, if $A > 2$,

$$\log \alpha \approx 0$$

and

$$\log(1-\alpha) \approx -A$$

whereas if $A < -2$,

$$\log \alpha \approx A$$

and

$$\log(1-\alpha) \approx 0$$

* Analogously, with the electron concentration:

$$\log \frac{\alpha_j}{1-\alpha_j} = \log \frac{S_{nj}}{n_e} \equiv -\log n_e + \frac{3}{2}\log T - \frac{5040}{T}V_{ij} + \log \frac{Z_{ij}}{Z_{aj}} + 15 \cdot 684 \qquad (7.24n)$$

A most convenient procedure is to compute log [α/(1−α)] and to read log α and log (1−α) immediately from a table. Such a table, listing log(1−α) and log α as functions of A for −2 < A < 2, has been added as Appendix 7.

Let us consider the dependence of the degree of ionization on the temperature, the electron pressure, and the ionization potential. For facilitating our discussion we take the term log Z_{ij}/Z_{aj} in equation (7.24) up into the ionization potential, in virtue of our considerations in § 7.3. Thus we write

$$\log \frac{\alpha_j}{1-\alpha_j} = -\log p_e + \frac{5}{2}\log T - \frac{5040}{T}\bar{V}_{ij} - 6\cdot18 \qquad (7.28)$$

Accordingly, log [α_j/(1 − α_j)] is seen to be a linear function of the apparent ionization potential $\bar{V}_{ij}$ with p_e and T as parameters. Straight lines depicting this

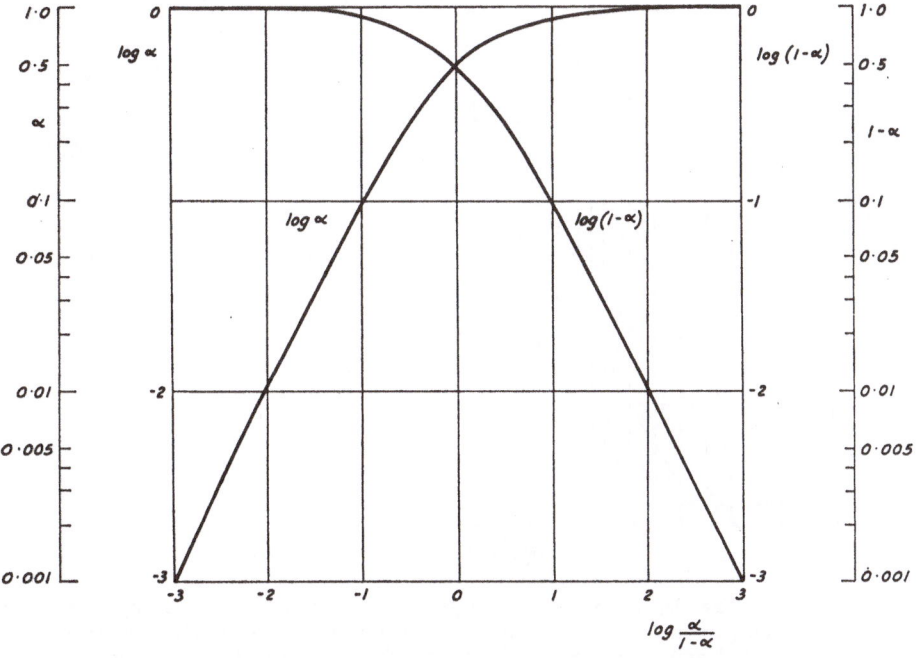

FIG. 7.1. An illustration of the dependence of log α and log(1−α) on log[α/(1 − α)]. When calculating, it is preferable to obtain values for log α and log(1 − α) from a table (see, for example, Appendix 7).

function are shown in the lower left-hand part of Fig. 7.2 for T = 4500, 5500, and 6500°K, each with p_e = 4 . 10⁻⁴ and 4 . 10⁻³ atm. To find α_j as a function of $\bar{V}_{ij}$ we have 'reflected' the straight lines in a self-explanatory way against the curve α_j = f{log[α_j/(1 − α_j)]} given in the upper left-hand diagram and obtained the curves shown in the right-hand diagram.

The results show the marked influence of the electron pressure on the degree

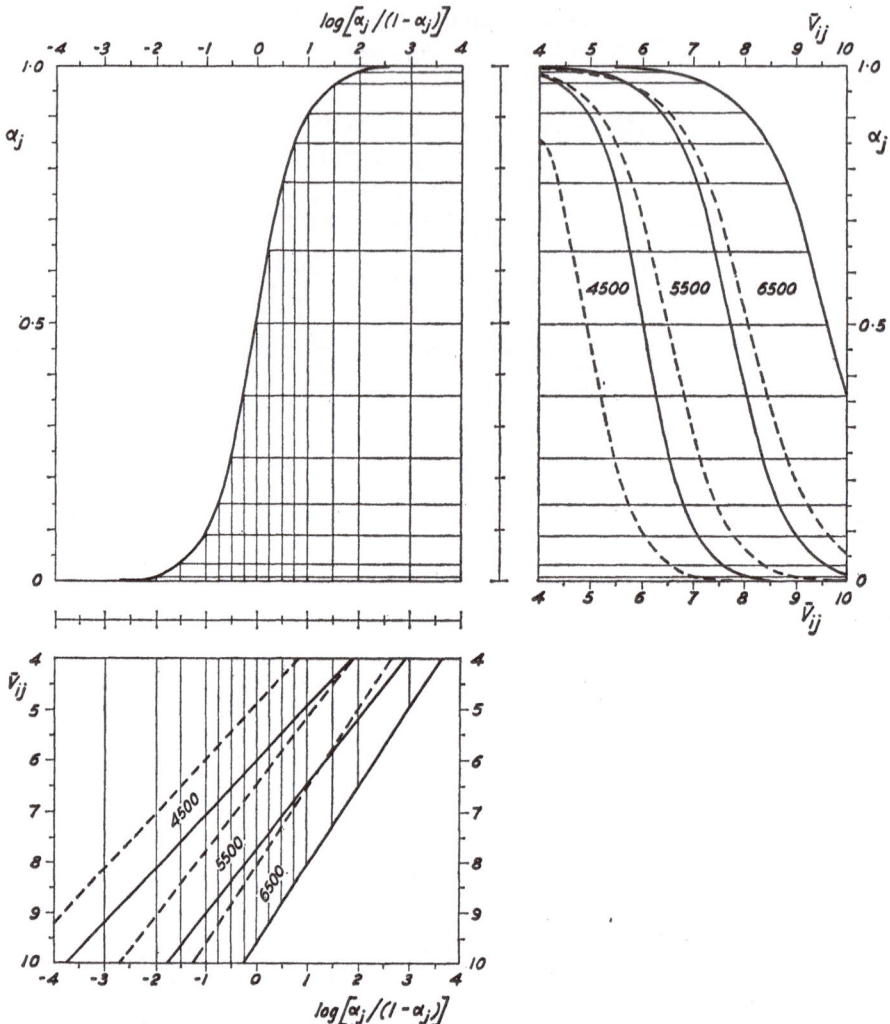

FIG. 7.2. The dependence of the degree of ionization α_j on the apparent ionization potential $\bar{V}_{ij}$ (§ 7.3), for $T = 4500$, 5500, and $6500°$K, each with $p_e = 4.10^{-4}$ and 4.10^{-3} atm. The relationships are depicted in the upper right-hand figure. The diagrams on the left demonstrate the procedure by which the curves $\alpha_j = F(\bar{V}_{ij})$ on the right have been derived from straight lines $\log[\alpha_j/(1 - \alpha_j)] = \varphi(\bar{V}_{ij})$ by 'reflection' against the curve $\alpha_j = f\{\log[\alpha_j/(1 - \alpha_j)]\}$.

----- $p_e = 4.10^{-3}$ atm ——— $p_e = 4.10^{-4}$ atm

of ionization. *Increasing p_e by a factor of 10 has roughly the same depressing effect on α_j as reducing the temperature by 1000°K.*

For detailed orientation, Table 7.2 shows the degree of ionization as a function of the temperature for a set of ionization potentials and for two values of the electron pressure, $p_e = 4.10^{-4}$ and 4.10^{-3} atm.

TABLE 7.2

Percentage of ionization of elements ($= 100\,\alpha_j$) as a function of the apparent ionization potential V_{tj} for the temperature range 4500–6500°K. The upper values are for an electron pressure $p_e = 4 \cdot 10^{-3}$ atm, the lower ones for $p_e = 4 \cdot 10^{-4}$ atm. The values have been computed from equation (7.28). Use was made of the abridged conversion table $\log [\alpha_j/(1-\alpha_j)] \to \log \alpha_j$ (Appendix 7).

Each cell gives the upper value / lower value.

T(°K)	V_{tj} (eV) 4·0	4·5	5·0	5·5	6·0	6·5	7·0	7·5	8·0	8·5	9·0	9·5	10·0
4500	99/88	95/67	85/36	61/13	30/4	10/1	3/0	0/0	0/0	—/—	—/—	—/—	—/—
4600	99/91	97/73	88/44	69/18	39/6	15/1	5/1	1/0	1/0	—/—	—/—	—/—	—/—
4700	100/92	98/78	92/52	76/24	48/8	21/3	7/1	2/0	1/0	0/0	—/—	—/—	—/—
4800	100/94	98/83	93/60	81/31	57/12	29/4	10/1	3/0	1/0	0/0	—/—	—/—	—/—
4900	100/96	99/86	95/67	86/38	65/16	37/5	15/2	5/1	2/0	1/0	0/0	—/—	—/—
5000	100/97	99/89	97/73	89/45	72/21	45/8	20/3	7/1	2/0	1/0	0/0	—/—	—/—
5100	100/97	99/92	98/78	92/53	80/27	54/10	27/4	13/1	4/0	1/0	0/0	—/—	—/—
5200	100/98	99/93	98/82	93/60	83/33	62/14	35/5	15/2	5/1	2/0	1/0	0/0	0/0
5300	100/98	100/94	99/85	96/67	87/40	69/18	43/7	20/2	8/1	3/0	1/0	0/0	0/0
5400	100/99	100/96	99/88	97/72	90/47	75/23	51/9	26/3	11/1	4/0	1/0	1/0	0/0
5500	100/99	100/97	99/91	97/78	92/54	80/29	59/12	33/5	15/2	6/1	2/0	1/0	0/0
5600	100/99	100/98	99/92	98/81	94/61	85/36	66/16	41/7	20/2	8/1	3/0	1/0	0/0
5700	100/99	100/98	100/94	98/85	96/67	90/47	72/21	49/9	25/3	11/1	4/0	2/0	1/0
5800	100/100	100/98	100/95	99/88	97/72	91/49	78/26	56/11	32/4	15/2	6/1	2/0	1/0
5900	100/100	100/99	100/96	99/90	97/77	92/55	82/31	64/15	39/6	19/2	8/1	3/0	1/0
6000	100/100	100/99	100/97	99/92	98/80	94/62	86/38	70/19	47/8	25/3	11/1	4/1	2/0
6100	100/100	100/99	100/98	100/93	98/84	96/68	89/44	75/23	54/11	31/4	15/2	6/1	3/0
6200	100/100	100/99	100/98	100/94	99/87	97/72	91/51	80/29	62/14	39/6	19/2	9/1	4/0
6300	100/100	100/100	100/98	100/96	99/89	97/77	93/57	84/34	68/17	45/8	25/3	12/1	5/1
6400	100/100	100/100	100/99	100/97	99/91	98/80	94/62	87/40	73/21	52/10	31/4	15/2	8/1
6500	100/100	100/100	100/99	100/97	100/92	98/83	96/68	89/46	78/26	59/13	37/6	19/2	9/1

TABLE 7.3

$T(°K)$	$p_e(10^{-3}$ atm$)$	$\log p_e$	$\Delta \log p_e$
5100	3·78	−2·42	+0·73
5400	1·45	−2·84	+0·31
5700	0·71	−3·15	0·00
6000	0·57	−3·24	−0·09₅
6300	0·47	−3·33	−0·18

Numerical values of T and p_e based on an experimentally established correlation between these quantities for a given type of graphite arc with metal vapour (§§ 7.7 to 7.10). The data have been used as a basis for Fig. 7.3.

An elaborate example provided by my investigation of the temperature and the electron pressure in graphite arcs with metal vapours (see §§ 7.7 to 7.10) discloses the part played by the numerical value of the electron pressure in calculating atom and ion line intensities [equations (7.9) and (7.10)]. Under the conditions prevailing in the experiment, a correlation between T and p_e reproduced in Fig. 7.6 was found. Some basic data are given in Table 7.3. For each temperature listed in the table we have considered two values of the electron pressure, viz. the p_e value given in the table and a constant p_e value, i.e. $p_e = 0·71.10^{-3}$ atm, which is the p_e value for $T = 5700°$K. Fig. 7.3 shows the ratios $(1 - \alpha_j)_2 : (1 - \alpha_j)_1$ and $(\alpha_j)_2 : (\alpha_j)_1$ as functions of V_{ij}, where the labels 1 and 2 refer to constant and variable electron pressure respectively. The figure makes clear what deviations result in calculated intensities of atom and ion lines if changes in the electron pressure such as observed in an actual experiment are not considered. Evidently, the discrepancies depend largely on the apparent ionization potential V_{ij} of the relevant element.

Although we have only considered one example, we now realize that, for computing line-intensity variations with changing excitation conditions, knowledge of the behaviour of the electron pressure is doubtless as important as that of the temperature. Therefore, we must keep in mind that the electron pressure is a fundamental parameter for the quantitative evaluation of matrix affects.

The form of the ionization relationships is such that not only the *relative* magnitude of changes, but also the *absolute* value of the electron pressure, acts upon the results. To see this we shall consider the equations that express the relation between a slight change of $\log p_e$ and the corresponding variations of $\log(1 - \alpha)$ or $\log \alpha$ (at constant temperature). From the Saha equation (7.12n) we can derive by expanding $\log(1 - \alpha)$ in a Taylor series and neglecting the higher terms

$$d \log(1 - \alpha) \approx \alpha \, d \log p_e \qquad (7.29)$$

For most purposes it is sufficiently accurate for estimating the magnitude of the effect of a change of the electron pressure on $\log(1 - \alpha)$ and thus on logarithmic

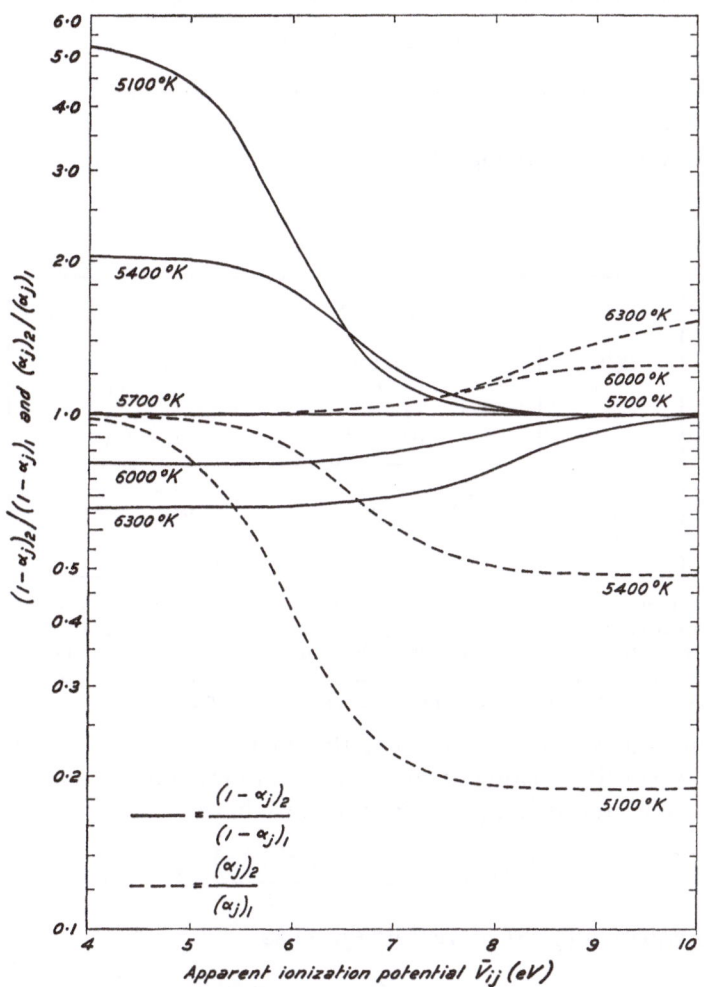

FIG. 7.3. An illustration of the influence of the numerical value of the electron pressure upon the logarithmic intensities of atom and ion lines.

The effect is demonstrated by plotting at five temperatures ($T = 5100, 5400, 5700, 6000, 6300°K$) the ratios $(1 - \alpha_j)_2 : (1 - \alpha_j)_1$ and $(\alpha_j)_2 : (\alpha_j)_1$ as functions of the apparent ionization potential ($\bar{V}_{ij}$) of the relevant element. The values of $(1 - \alpha_j)_2$ and $(\alpha_j)_2$ have been calculated with p_e values observed in an actual experiment (see Table 7.3), whereas the values of $(1 - \alpha_j)_1$ and $(\alpha_j)_1$ are based on a constant value of p_e (= $0.7 . 10^{-3}$ atm).

The figure exemplifies the magnitude of the correction that must be applied to calculated atom and ion line intensities [equations (7.9) and (7.10)], if we, instead of assuming a plausible mean value of the electron pressure, take into consideration variations of this pressure such as those reported for an actual experiment.

line intensities.* A corresponding expression pertaining to ion lines reads

$$d \log \alpha \approx -(1-\alpha)\, d \log p_e \tag{7.30}$$

The influence of the absolute value of p_e on the relation between $d \log(1-\alpha)$ or $d \log \alpha$ and $d \log p_e$ is included in the factor of proportionality, α or $(1-\alpha)$, whose value was shown in Figs. 7.2 and 7.3 to depend markedly on the absolute value of p_e.

TABLE 7.4

$T(°K)$	$V_{ij} = 6$ eV		$V_{ij} = 9$ eV	
	α	relative error in $1-\alpha$ (per cent)	α	relative error in $1-\alpha$ (per cent)
4500	0·30	5·4	≈ 0	≈ 0
5500	0·90	13·6	0·05	1·1
6500	0·99	13·1	0·50	9·3

Numerical examples of equation (7.32) expressing the error in $(1-\alpha)$ as a function of the error in the temperature. The values listed above are based on a relative error of 1 per cent in the temperature. The error in $(1-\alpha)$ expressed as a percentage has been obtained by multiplying $d \log(1-\alpha)$ by 230. In the examples the electron pressure is assumed to be 10^{-3} atm.

Concluding this section we consider the changes of $\log(1-\alpha)$ and $\log \alpha$ that result from a (small) variation of the temperature (at constant electron pressure). Neglecting higher terms in the Taylor series we have

$$d \log(1-\alpha) \approx \frac{\partial \log(1-\alpha)}{\partial T}\, dT \tag{7.31}$$

In virtue of equation (7.28) we rewrite (7.31) as

$$d \log(1-\alpha) \approx -\alpha \left(0·434 \times \tfrac{5}{2} + \frac{5040}{T} V_i \right) \frac{dT}{T} \tag{7.32}$$

When dealing with line-intensity variations we must remember that the Boltzmann exponential factor also depends on the temperature; consequently, equation (7.32) is combined with a further equation in order to find the effect of a change of the temperature upon line intensity (see § 8.8). If, on the other hand, we are immediately interested in degree of ionization, equation (7.32) is adequate. Numerical examples in Table 7.4 illustrate the fairly large sensitivity of $(1-\alpha)$ for errors in T, when α is large.

* The absolute correction of $d \log(1-\alpha)$ due to the higher terms in the Taylor series is largest for $\alpha \approx 0·5$. For example, it amounts to 0·026 when $d \log p_e = 0·300$. Hence the approximation (7.29) gives the change of $(1-\alpha)$ correct to within 6 per cent, even if the considered variation of p_e is as large as a factor of 2.

★ §7.6 *A spectroscopic method using the Saha equation for determining the electron pressure in the arc* (cf. § 3.8)

In §7.4 we explained how the degree of ionization of a vapour component is related to the log intensity ratio of an ion–atom line pair and to the temperature [equation (7.23)]. We also expressed the degree of ionization as a function of the temperature and the electron pressure, i.e. Saha's relationship (7.24). The equations readily suggest a procedure for determining spectroscopically the electron pressure; for, by eliminating the left-hand side of (7.23) and (7.24), that is $\log[\alpha_j/(1-\alpha_j)]$, we obtain a relation between

1 The electron pressure.

2 The intensity ratio of the ion–atom line pair.

3 The temperature.

After rearranging we have

$$\log p_e =$$

$$-\log \frac{I_{qp}^+}{I_{qp}} + \log \frac{g_q^+ A_{qp}^+ \nu_{qp}^+}{g_q A_{qp} \nu_{qp}} - \frac{5040}{T}(V_{ij} + V_q^+ - V_q) + \frac{5}{2}\log T - 6\cdot18 \qquad (7.33)$$

So, the electron pressure can be determined when the temperature and the intensity ratio I^+/I are measured. Note that this procedure in fact involves the adjustment of a parameter value for $\log p_e$ so that both equation (7.23) and the Saha equation yield one and the same value for the degree of ionization of the component whose spectral lines are used. We assume, once this adjustment has been made, that the parameter value and the Saha equation are appropriate for computing the degrees of ionization for substances other than the original test element.

When deriving p_e from (7.33) we must be aware of the possibility of large errors of a systematic and a random nature. Errors of the former kind are caused by an inaccurate value for the ratio of transition probabilities, A^+/A, and by an inaccurate temperature value, the latter also due to incorrect fundamental data, for instance, a ratio of transition probabilities, needed for the method discussed in § 6.5. Unfortunately, neither the magnitude nor the sign of those constant errors can be indicated. We can only calculate the error in p_e by estimating the magnitude of the error in A^+/A and T. So, from equation (7.33) we see immediately that p_e is in error by the same factor as A^+/A; for example, an inaccuracy of a factor of 2 in A^+/A produces a systematic deviation of the true value of p_e by a factor of 2.

To average out the influence of systematic errors in A^+/A, Corliss (1962b and c) 'adjusted' for the N.B.S. copper arc the Saha equation using ion–atom line pairs of eleven elements. He computed eleven values of $\log n_e$, the mean of which was taken to be an unbiased estimate of this quantity. At the same time he evaluated separate values of $\log(n_{ij}/n_{aj}) = \log[\alpha_j/(1-\alpha_j)]$ for the eleven elements, using equation (7.23). Fig. 7.4

shows his plot of log $(n_{ij}Z_{aj}/n_{aj}Z_{ij})$ against the ionization potential V_{ij}.* Note that the points plotted in this figure were obtained empirically. The plot supposes solely the appropriateness of the Boltzmann distribution of atom and ion levels, i.e. the validity of equation (7.23). Therefore, there was no need to use Saha's relationship for determining the degrees of ionization, since an empirical line through the points suits the purpose just as well. Instead, as was mentioned in § 1.5, Corliss preferred to use the Saha equation

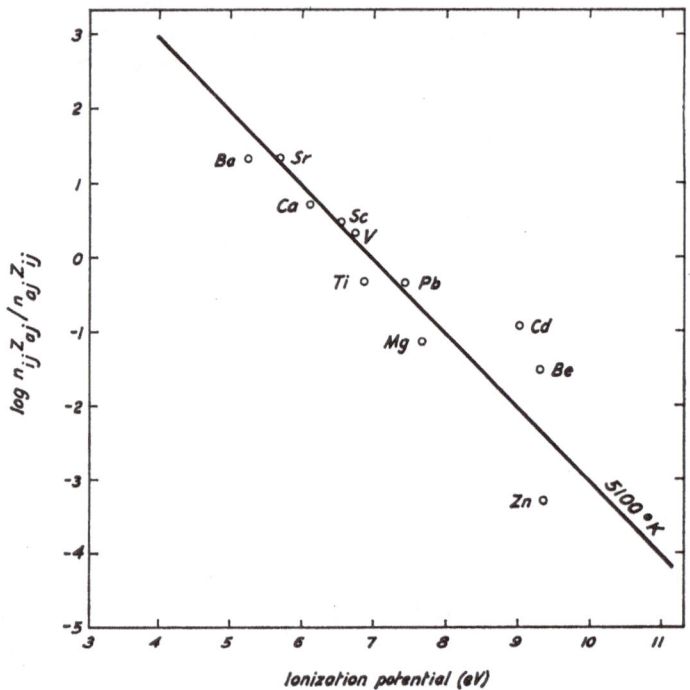

FIG. 7.4. Ionization in the N.B.S. copper arc. The points represent the experimentally determined values of log $n_{ij}Z_{aj}/n_{aj}Z_{ij}$ for eleven elements. The straight line represents Saha's ionization equation for a temperature of 5100°K and an electron concentration of $2\cdot4.10^{14}$ cm^{-3} (after Corliss, 1962b).

after he had inserted the mean value of log n_e. It is interesting how well the straight line plot of the Saha equation with this mean value of log n_e and an effective temperature of 5100°K fits that of the experimental points, thus supporting the appropriateness of Saha's relationship with effective values. On the other hand, we must not overlook the magnitude of the deviation of several points from the line, which is of the order of unity on the log scale. Most probably, the deviation is due to systematic errors in the ratios A^+/A. Anyhow, it demonstrates the risk we run when choosing a line pair and adopting literature values of transition probabilities for determining the electron pressure.

The error in log p_e caused by a systematic error dT in the temperature is derived

* The plot is similar to the bottom left-hand diagram of Fig. 7.2. Turning Fig. 7.4 clockwise through 90° gives it the same orientation. The only difference is the way in which the ratio of partition functions has been considered. In Fig. 7.2 it appears with the ionization potential in the form of the δ-correction (§ 7.3), in Fig. 7.4 it is with log (n_{ij}/n_{aj}).

from (7.33) to be

$$d \log p_e = \left[0 \cdot 434 \times \tfrac{5}{2} + \frac{5040}{T} (V_{ij} + V_q^+ - V_q) \right] \frac{dT}{T} \qquad (7.34)$$

So, for example, with the ion–atom line pair Mg 2796/2780 an error of 100°K at $T = 5500°$K gives rise to a systematic deviation of 0·10 in $\log p_e$, which is equivalent to a factor of 1·26 in p_e. For the ion–atom line pair Mg 2796/2852 these numbers become 0·15 and 1·41 respectively. A relatively small error in T is thus seen to contribute largely to the inaccuracy of p_e.

Next we have to consider the *random* errors in p_e that originate from errors in measuring the intensity ratio I^+/I and the temperature. If they are fully independent, the variance of $\log p_e$ is

$$\mathrm{var}(\log p_e) = \mathrm{var}\left(\log \frac{I^+}{I} \right) + [f(T)]^2 \frac{\mathrm{var}(T)}{T^2} \qquad (7.35)$$

where variance is the square of the standard deviation (σ), and $f(T)$ is equal to the expression in the square brackets of equation (7.34). Usually, the second term in (7.35) will predominate. Suppose, for instance, that $\log I^+/I$ is measured with a precision of 0·04, expressed as standard deviation, which corresponds to a coefficient of variation (= relative standard deviation expressed as a percentage) of nearly 10 per cent for I^+/I. If the standard deviation of the temperature measurement amounts to 100°K at $T = 5500°$K, the precision of $\log p_e$ is 0·11 and 0·15$_5$ respectively for the line pairs of magnesium mentioned above. The error in T alone would give rise to standard deviations of 0·10 and 0·15. The example underlines the high sensitivity of p_e for errors in the temperature values that are inserted in the fundamental equation (7.33).

Finally we must mention the complications associated with the interpretation of the results of measuring n_e or p_e when the source exhibits spatial inhomogeneity. These complications are essentially of the same type as those discussed in § 6.16 with respect to the appreciation of spectroscopic temperature values. Thus, if we omit spatial resolution of spectral-line intensities, we acquire a population-averaged or effective value of n_e or p_e, whose meaning varies with the source characteristics and with the line pair employed. We shall elucidate this numerically in Chapter 8.

Applications of formula (7.33) for measuring n_e or p_e in the arc have been described by several authors,[*] who all used space-resolved intensities to obtain n_e or p_e as a function of the distance to the arc axis. In other cases, authors[†] confined themselves to the measurement of effective values of n_e or p_e.

[*] Kruithof (1943a), Kruithof and Smit (1944), Huldt (1948b and c), Eberhagen (1955a) (cf. Eberhagen, 1955b; Eicke, 1962; and Köstlin, 1964), Vukanović (1964), and de Galan (1965a, 1966a).

[†] See, for example, Vukanović (1960), Boumans (1961, 1962), Corliss (1962b and c), and Boumans and Rouws (1965).

★ § 7.7 *Results of the measurement of the temperature and the electron pressure in graphite arcs with metal vapour*

Using the procedure described in the preceding section and that discussed in §§ 6.5 to 6.8, I made a systematic investigation of the influence exerted upon the effective temperature and the effective electron pressure of a d.c. graphite arc by the presence of different metal vapours (see Boumans, 1961, 1962; and Boumans and Rouws, 1965).

To summarize, we volatilized KCl, NaCl, LiCl, LiF, In_2O_3, Tl_2O_3, Al_2O_3, PbO, Ni CuO, Sb_2O_4, and graphite, from a furnace-shape graphite anode into a 10-amp d.c arc (see § 6.7). Supplementary results were obtained recently with 'normal' electrodes (Fig. 7.14) filled with various mixtures of LiF, NaCl, and graphite, or of LiF, KCl, and graphite. By volatilizing zinc and magnesium along with the matrices we were able to measure simultaneously the temperature and the electron pressure, using zinc as the thermometric element and magnesium as the manometric element.

Spectra were taken with a Hilger large quartz spectrograph. The optics in front of the slit selected radiation from a portion of the arc as small as 2×2 mm (see Fig. 6.10); yet optimum luminosity was maintained. Once the arc had been struck, a more or less steady volatilization of the matrix and the elements zinc and magnesium began after about half a minute in most instances. Therefore, no special difficulties were encountered in keeping the current strength constant at 10 amp; once the ballast resistance had been set for a particular matrix, only slight readjustments were necessary when arcing (cf. § 6.17). The technique, a camera with its racking plate, provided up to twelve micro-spectra of each arc; their exposure time varied according to the matrix and was either 10, 15, or 25 sec.

We decided, in principle, upon the use of the ion–atom line pair Mg 2796/2780 as the most suitable for the purpose. The advantage of the small wavelength difference is offset by a high or moderately high intensity ratio. For enhancing the precision and the accuracy of its determination from densities on a photographic plate, the two-step spectrogram with a step ratio of 10, as produced in order to measure the intensity ratio of the Zn I lines accurately (§ 6.6), was a welcome expedient, though not entirely sufficient owing to the fairly large range of excitation conditions covered in the experiment. For that reason we resorted to another ion–atom line pair also, Mg 2796/2852, which was used for temperatures beyond 5800°K, since the intensity ratio of the other pair becomes too high then. The use of the resonance line Mg 2852 necessitated special precautions to be taken for the magnesium vapour concentration in the arc in order to avoid interfering self-absorption.

The next problem was to select the best values of transition probabilities. Initially we adopted the absolute values cited by Allen (1955, p. 63 ff.) for the line Mg II 2796 after Biermann and Lübeck (1948) and for the line Mg I 2852 after Treffzt (1949, 1950); both values, $2 \cdot 6 \cdot 10^8$ and $4 \cdot 6 \cdot 10^8$ sec^{-1} respectively, were obtained by the self-consistent field approximation. In a later stage of the investigation we changed to a more recent value for the atom line, viz. $3 \cdot 04 \cdot 10^8$ sec^{-1}, determined by Demtröder (1962) through lifetime

measurements.* For the other atom line, Mg I 2780, we determined the value of ΣgA†
experimentally with respect to the gA-value of Mg I 2852.

The numerical values of the spectral-line constants of the magnesium lines
are listed in Table 7.5. Substitution of the constants for the line pair

TABLE 7.5

λ(Å)	A (sec^{-1})	$(\Sigma)g$	$(\Sigma) gA_{rel}$	V_q (eV)	designation
Mg II 2795·5	2·6.10^8	4	10·4	4·41	+
Mg I 2779·8		8	33·1	7·14	1
Mg I 2852·1	3·04.10^8	3	9·1	4·33	2

The numerical values of the line constants [wavelength (λ), transition probability (A), statistical weight (g), and excitation energy (V_q)] inserted in equation
(7.33) for measuring the electron pressure with the aid of ion–atom line pairs
of magnesium.

Mg 2796/2780 into equation (7.33) gives

$$\log p_e = -\Delta Y_1 - 4\cdot91\,\frac{5040}{T} + \frac{5}{2}\log T - 6\cdot68 \qquad (7.36)$$

For the pair Mg 2796/2852 we have

$$\log p_e = -\Delta Y_2 - 7\cdot72\,\frac{5040}{T} + \frac{5}{2}\log T - 6\cdot11 \qquad (7.37)$$

In these equations

$$\Delta Y = \log \frac{I^+}{I} \qquad (7.38)$$

with index 1 and 2 referring to Mg 2780 and to Mg 2852 respectively. For the
ionization energy of magnesium we inserted the value $V_{ij} = 7\cdot64$ eV after
Moore (1958).

Results of the measurements were worked out as follows. Each micro-spectrum
yielded a value of T and a value of ΔY, either ΔY_1 or ΔY_2. These results were, in
principle, divided into groups according to the matrix; indeed this could be done
in most cases where the results clustered at random about matrix mean values.
In a few instances, however, subdivision was imperative; in making such sub-
divisions the gap voltage (that was recorded during the entire arc burn with an
electronic strip chart recorder: see § 7.14) proved to be a valuable and objective
help.

* Considering various literature values Garstang (1962) decided upon $A = 3\cdot5.10^8$ sec^{-1} for
further work.
Recently, Lurio (1964) rejected Demtröder's value. His value, which was also derived from lifetime measurements, again is in the neighbourhood of the one calculated by Treffzt.
† According to Moore (1950, 1949) the line Mg I 2779·832 is composed of two transitions,
$J = 2 \to J = 2$ and $J = 1 \to J = 1$, of almost equal energy.

For each group we calculated the mean of T and ΔY. We converted mean values of ΔY_2 into $\overline{\Delta Y_1}$ using the mean temperature and the equation

$$\Delta Y_1 = \Delta Y_2 + 2 \cdot 81 \frac{5040}{T} - 0 \cdot 57 \tag{7.39}$$

which is essentially the Boltzmann formula for the two atom lines. Before computing the numerical values of the electron pressure, a preliminary plot of $\overline{\Delta Y_1}$ versus $5040/T$ was made so that the random errors could be smoothed out more satis-

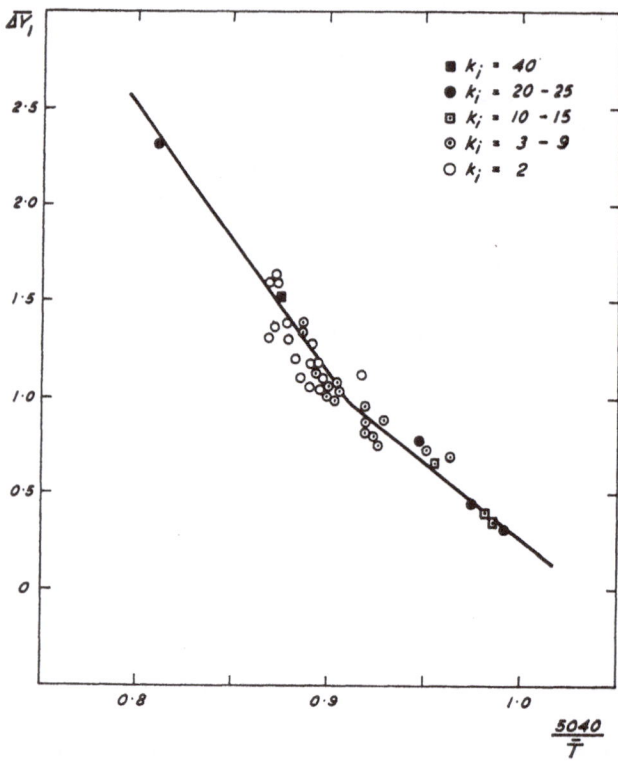

Fig. 7.5. Plot of experimentally determined values of $\overline{\Delta Y_1}$ versus $5040/\overline{T}$ for graphite arcs with different metal vapours.
   $\Delta Y_1 = \log I^+/I_1$, where I^+ is the intensity of Mg II 2796 and I_1 is the intensity of Mg I 2780. The points represent mean values that are based on k_i single measurements of ΔY_1 and T. The correlation between $\overline{\Delta Y_1}$ and $5040/\overline{T}$ has been approximated by two intersecting straight lines.

factorily. The graph is presented in Fig. 7.5. It brings out the original results as well as supplementary results recently obtained by Miss C. J. J. Rouws for arcs containing mixtures of Li and K, and of Li and Na, in varying proportions. The correlation between $\overline{\Delta Y_1}$ and $5040/T$ that appears from the figure can be approximated by two straight lines,

$$\overline{\Delta Y_1} = -a \frac{5040}{T} + b \tag{7.40}$$

with $a = 13\cdot58$ and $b = 13\cdot58$ for $T < 5600°K$ and $a = 8\cdot33$ and $b = 8\cdot88$ for $T > 5600°K$.

By substituting (7.40) into (7.36) we get two relations that together depict a correlation between the electron pressure and the temperature. The correlation is

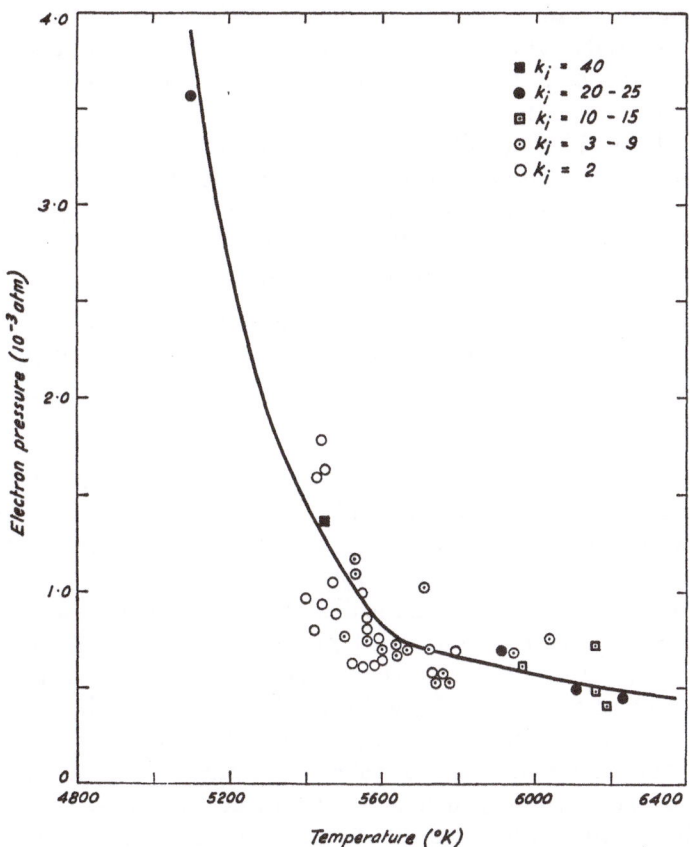

Fig. 7.6. A correlation established experimentally between the electron pressure p_e and the temperature T in graphite arcs with metal vapour. The p_e values for separate points have been computed with equation (7.36) from mean values of T and $\overline{\Delta Y_1}$. The curve has been drawn in accordance with the straight line plots of $\overline{\Delta Y_1}$ versus $5040/T$ produced in Fig. 7.5.

Quantitatively, the correlation between p_e and T holds true for narrowly defined operating conditions only. Qualitatively, it is symptomatic for the influence exerted on the excitation conditions by metal vapours. When the partial pressure of an element of low ionization potential in the plasma of a carbon arc whose current strength is kept constant is increased beyond a critical level of about 10^{-4} atm, a relatively large fall in the temperature is produced initially; a further increase of the vapour concentration reduces the temperature only slightly more and makes it approach a value that is more or less characteristic for the ionization potential of the relevant substance; the chief effect of further increasing the vapour concentration is raising the electron pressure. See also § 7.9.

shown graphically in Fig. 7.6; the separate p_e values, computed with (7.36) from the various mean values of T and ΔY_1, have also been inserted (cf. Table 7.6, p. 191). The points are seen to scatter to a fairly large extent along the curve. This dispersion does *not* mainly result from measuring errors, but must be considered to be an essential feature. This, inherently, prohibits us from identifying the correlation between T and p_e with a functional relationship. Moreover, the correlation is not generally valid; yet, this does not prevent us from discussing it in detail, since it serves as a useful example for elucidating several aspects of the mechanism of the arc that have a more general scope.

Functional relationships are likely to exist between the temperature and the gas composition, on the one hand, and between the electron pressure and the gas composition, on the other. In our experiment a correlation between T and p_e also emerged from the interrelations. Before we attempt to unravel these involved interrelations, we shall summarize the results of the experiment.

1 Introducing metal vapour into the arc causes the well-known drop in the temperature; it is very pronounced when the vapour is of an alkali metal.

2 The lowering of the temperature tends to be associated with an increase in the electron pressure.

The increase in the electron pressure will be shown in subsequent sections to depend to a large extent on the concentration of the metal vapour in the arc. In general, for interpreting the influence of the metal vapour on the excitation conditions, two quantities must be observed; the ionization potential of the metal and the vapour concentration. So, we must realize that, even when an arc discharge is said to be carried by a particular metal, there is a *range* of excitation conditions involved. A first indication in support of this statement is shown in Fig. 7.7, where the single observations of ΔY_1 and $5040/T$ for alkali arcs have been plotted instead of the mean values used in Fig. 7.5. The plot illustrates the occurrence of a range of excitation conditions for each of the three alkali metals. It will be seen in § 7.9 that the variations, such as become evident from the scattering of the points along the line in Fig. 7.7, are due to changes in the metal vapour concentration.*

In conclusion, we point out that the correlation between the temperature and the electron pressure shown in Fig. 7.6 is symptomatic, but incomplete. The merits and imperfections of the facts established hitherto will be dealt with more extensively in the next sections. Anyhow, the experiment shows quantitatively that the entry of a metal vapour into the arc can exert a powerful influence on both the temperature and the electron pressure, particularly when the metal has a low ionization potential.

---

* These changes were not particularly expected at this stage of the investigation, but occurred more or less as a secondary effect, owing to slow variations of the rate of evaporation over the arcing period. For lithium, an additional variation was produced by arcing different salts, LiF and LiCl. The difference in volatility of the fluoride and the chloride gives rise to markedly different evaporation rates and vapour concentrations. This accounts for the wide range of excitation conditions for lithium.

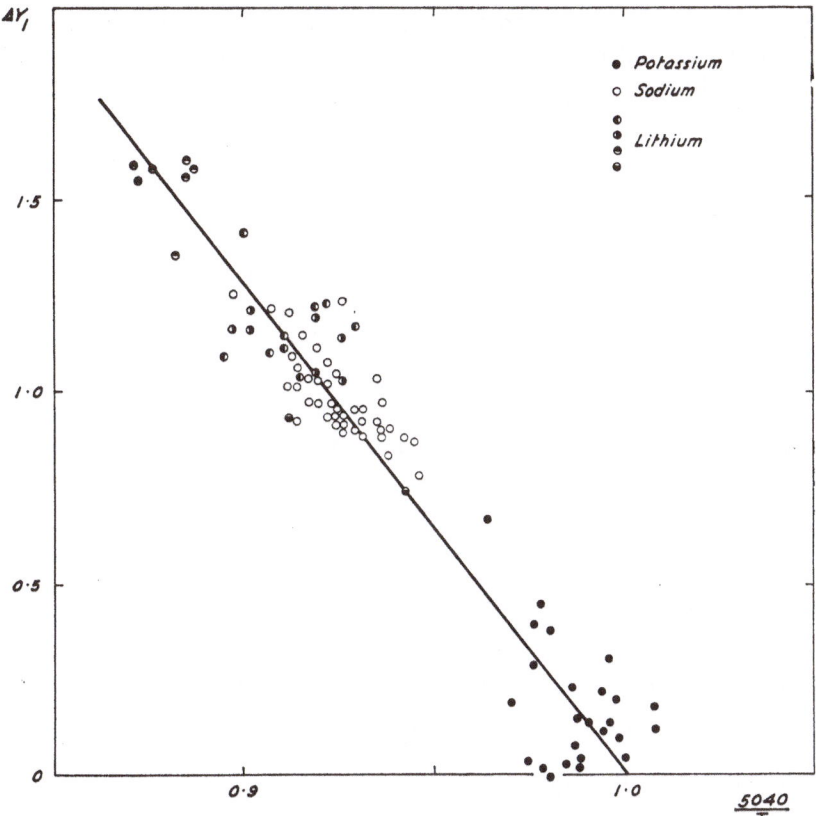

FIG. 7.7. Individual measurements of ΔY_1 for alkali metal graphite arcs plotted against $5040/T$ (compare Fig. 7.5). The scattering of the points along the line gives a first indication that for each of the alkali metal arcs there is a range of excitation conditions involved. The occurrence of such ranges is shown in § 7.9 to be associated with variations of the metal vapour concentration in the arc.

★ § 7.8 *The relation between the composition of the arc plasma, its temperature, and its electron pressure*

As a starting-point of our discussion let us consider that:

1 The electron pressure of a composite gas in thermal equilibrium is an unambiguous function of the temperature, given the elemental composition of the gas. We can compute it if we know the numerical values of the atomic and molecular constants such as the energy of ionization and the energy of dissociation.

2 Conversely, the temperature is completely fixed by the elemental composition, if the electron pressure is assigned a definite and invariant value.

We illustrate the tenor of these statements with the help of models that correspond to the conditions in the core of the carbon arc in air (cf. § 11.1).

Consider a mixture of two monoatomic gases (labelled 1 and 2) having ionization potentials of 14 and 6·5 eV respectively. To evaluate the electron pressure p_e of the composite gas as a function of the temperature the following set of equations applies:

1 Two Saha equations [cf. (7.3) and (7.17)]:

$$\frac{p_{i1}p_e}{p_{a1}} = S_{p1}(T) \tag{7.41}$$

and

$$\frac{p_{i2}p_e}{p_{a2}} = S_{p2}(T) \tag{7.42}$$

2 Two equations expressing the partial pressures p_1 and p_2 of the components as the sum of the partial pressures of neutral atoms and singly-charged ions [cf. (7.6)]:

$$p_1 = p_{a1} + p_{i1} \tag{7.43}$$

and

$$p_2 = p_{a2} + p_{i2} \tag{7.44}$$

3 An equation that relates the partial pressures to the total pressure P:

$$P = p_1 + p_2 + p_e \tag{7.45}$$

4 The condition for quasi-neutrality

$$p_e = p_{i1} + p_{i2} \tag{7.46}$$

So, we have six equations and nine variables, T, P, p_1, p_2, p_{a1}, p_{a2}, p_{i1}, p_{i2}, and p_e. Given T, P, and p_1, we can compute the unknowns, among which is p_e. In all cases to be considered here, p_e turns out to be about 10^{-3} at maximum when $P = 1$ atm and $T < 8000°K$. It is practical then to ignore p_e in the equation for the total pressure (7.45) and to fix the elemental composition of the gas by stating p_1 and p_2 such that their sum is 1 atm. The evaluation now proceeds as follows.

The partial pressures p_{i1} and p_{i2} are eliminated from the Saha equations (7.41) and (7.42) and from the quasi-neutrality condition (7.46). Multiplying both sides of the resulting equation by p_e yields

$$p_e^2 = p_{a1}S_{p1} + p_{a2}S_{p2} \tag{7.47}$$

The partial pressure of the neutral atoms p_{aj} can be expressed in terms of p_j through the degree of ionization [see equation (7.7)], such that

$$p_e^2 = (1 - \alpha_1)p_1 S_{p1} + (1 - \alpha_2)p_2 S_{p2} \tag{7.48}$$

Remembering (7.14n) we convert (7.48) into

$$p_e^2 = \frac{p_e}{S_{p1} + p_e}p_1 S_{p1} + \frac{p_e}{S_{p2} + p_e}p_2 S_{p2} \tag{7.49}$$

The use of the S_{pj} table (Appendix 6) facilitates numerical calculations. The table lists S_{pj} as a function of T in the temperature range 4000–7000°K for $\bar{V}_{ij} = 4 \cdot 5$, $5 \cdot 0, 5 \cdot 5, \ldots, 10, 11, \ldots, 15$ eV.*

In general, equation (7.49) is of the third degree in p_e. Here, however, for the apparent ionization potentials $\bar{V}_{i1} = 14$ and $\bar{V}_{i2} = 6 \cdot 5$ eV, the first term at the right in equation (7.49) is negligible in the temperature range below 7000°K when $p_2 = 10^{-3}$ atm, thus making $p_1 \approx P = 1$ atm. Component two is said now to carry the discharge; it supplies virtually all the electrons.

The dependence of p_e on T calculated with the second term at the right of equation (7.49) is depicted in Fig. 7.8. The electron pressure approaches the value 10^{-3} atm, i.e. the partial pressure p_2 of the constituent of low ionization potential. In fact, p_e does not reach a constant level, but it continues to rise, since the first term of equation (7.49) becomes of interest at higher temperatures, while the second term remains constant ($\approx p_e p_2$). Species two is fully ionized then.

The behaviour of p_e as a function of T in a wider range of temperatures is illustrated in Fig. 7.9 for 3 values of the total pressure and for $\bar{V}_{i1} = 15$, $\bar{V}_{i2} = 7$ eV, and $p_1 : p_2 = 10^4 : 1$. Here also, the lower portions of the curves are almost exclusively determined by the ionization of the component of low ionization potential, thus in spite of its very low concentration.

The examples illustrate by what mathematical method the electron pressure of a gas of given composition and temperature is computed in principle. As a rule, more difficulties of a practical nature are encountered when solving the equation for p_e. Its general form,

$$p_e^2 = \sum_{j=1}^{j=k} \frac{p_e}{S_{pj}+p_e} p_j S_{pj} \tag{7.50}$$

is an equation of degree $k+1$, where k is the number of components involved.

If in the temperature range under consideration constituents that exist partly in molecular form must be taken into account, the additional problem of dealing with a set of dissociation equilibria must be faced. This actually arises with the carbon arc in air (see §§ 11.1 and 3.9). The ionization potentials of the major constituents of the pure carbon arc plasma (N_2, N, O, and CO) are very high, that is of the order of 14 eV; those of the minor constituents, viz. C and NO, are moderately high, namely $11 \cdot 26$ and $9 \cdot 25$ eV respectively. When no material is added, or when metals of high ionization potential only are present, C and NO play a substantial part in the ionization of the gas, though the role of the major constituents is not entirely negligible either. Here these complications can be kept in the background, however, as we are considering primarily the carbon arc with metals of low or moderately low ionization potential, so that the metals chiefly furnish the electrons, whereas the contributions of C and NO can be practically ignored. Numerical examples below illustrate this.

* Note that we used apparent ionization potentials (§ 7.3) in this connection, thus avoiding the assumption that Z_{ij}/Z_{aj} in the Saha equation is unity. We must remember, however, that numerical values of $\bar{V}_{ij}$ for temperatures below 5000 and above 6000°K will depart, in general, from the values tabulated in § 7.3.

7

So far we have been concerned with calculating the electron pressure of a gas of given composition and temperature. Now we must turn our attention to the reverse problem, which is the more important one, namely the evaluation of the temperature of a gas of given composition and electron pressure. In § 3.9 we considered the constancy of the electron pressure in the arc. For reasons of self-consistency the electron pressure in the core of the normal carbon arc covers a

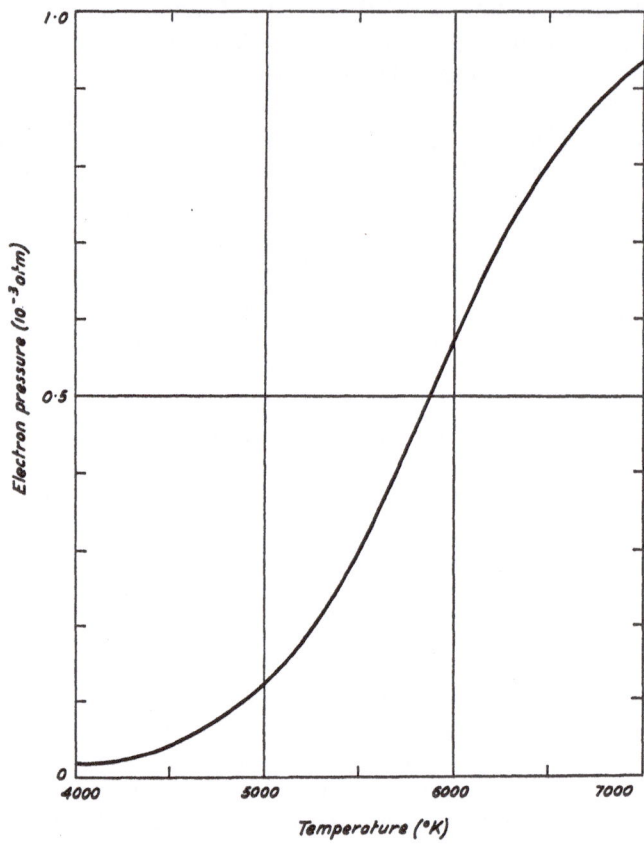

FIG. 7.8. Calculated dependence of the electron pressure p_e on the temperature T for a gas composed of two constituents, labelled 1 and 2 and having apparent ionization potentials $V_{i1} = 14$ and $V_{i2} = 6\cdot5$ eV. The total pressure $P = 1$ atm and the partial pressure $p_2 = 10^{-3}$ atm. The second constituent supplies virtually all the electrons, in spite of its low concentration.

restricted range of values, between some 10^{-4} to some 10^{-3} atm. On the strength of this delimitation we can understand the influence exerted upon the arc temperature by a substance of low ionization potential. Let us consider this in terms of a rigorous constant p_e prior to investigating the consequences of slight changes of p_e within the range just mentioned.

Evidently, a similar set of equations as in the preceding problem applies here. As before, we reduce them to a single expression, (7.49), or more generally,

(7.50). From the latter we see immediately that the temperature of the gas is determined if the electron pressure and the composition of the gas (i.e. the complete set of partial pressures p_j) are given; for the Saha constants S_{pj} contain the temperature as the only variable.

If we compute for a composite gas the temperature as a function of the partial pressure of one of its components, given the electron pressure and the partial pressures of all other constituents, we encounter the difficulty of solving equation (7.50) for T. Therefore, we reverse the procedure and compute the partial

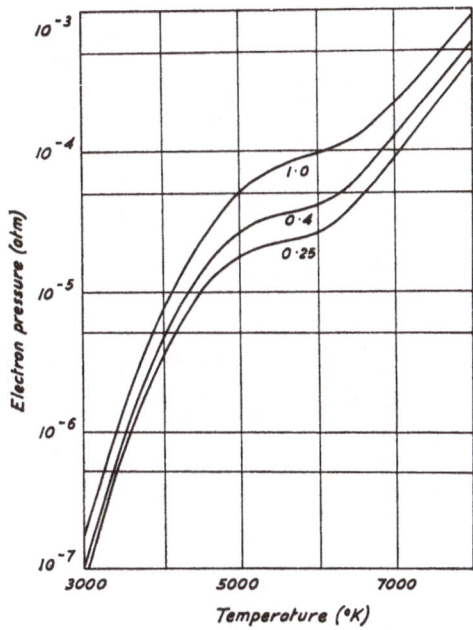

FIG. 7.9. Calculated dependence of the electron pressure p_e on the temperature T for a gas composed of two constituents, labelled 1 and 2 and having apparent ionization potentials $V_{i1} = 15$ and $V_{i2} = 7$ eV. The ratio of the partial pressures $p_1 : p_2 = 10^4 : 1$. The figure gives curves for three values of the total pressure, namely P = 1, 0·4, and 0·25 atm. (after Brinkman, 1937 p. 68; cf. Ornstein, Brinkman, and Beunes, 1932).

pressure of the relevant component as a function of T, again given the electron pressure and the partial pressures of all other constituents. In reality, the computed partial pressure is the independent variable, the temperature the dependent one.

First we shall consider a gas composed of two species having apparent ionization potentials of 14 and 11 eV; if $p_e = 4.10^{-4}$ atm, $P = 1$ atm, and $T = 6600°K$, equation (7.50) reads

$$16 \cdot 10^{-8} = 4 \cdot 77 . 10^{-8}(1 - p_2) + 9 \cdot 11 . 10^{-6} p_2$$

where numerical values for S_{pj} have been taken from Appendix 6. The partial pressure p_2 is thus evaluated to be $1 \cdot 24 . 10^{-2}$ atm.

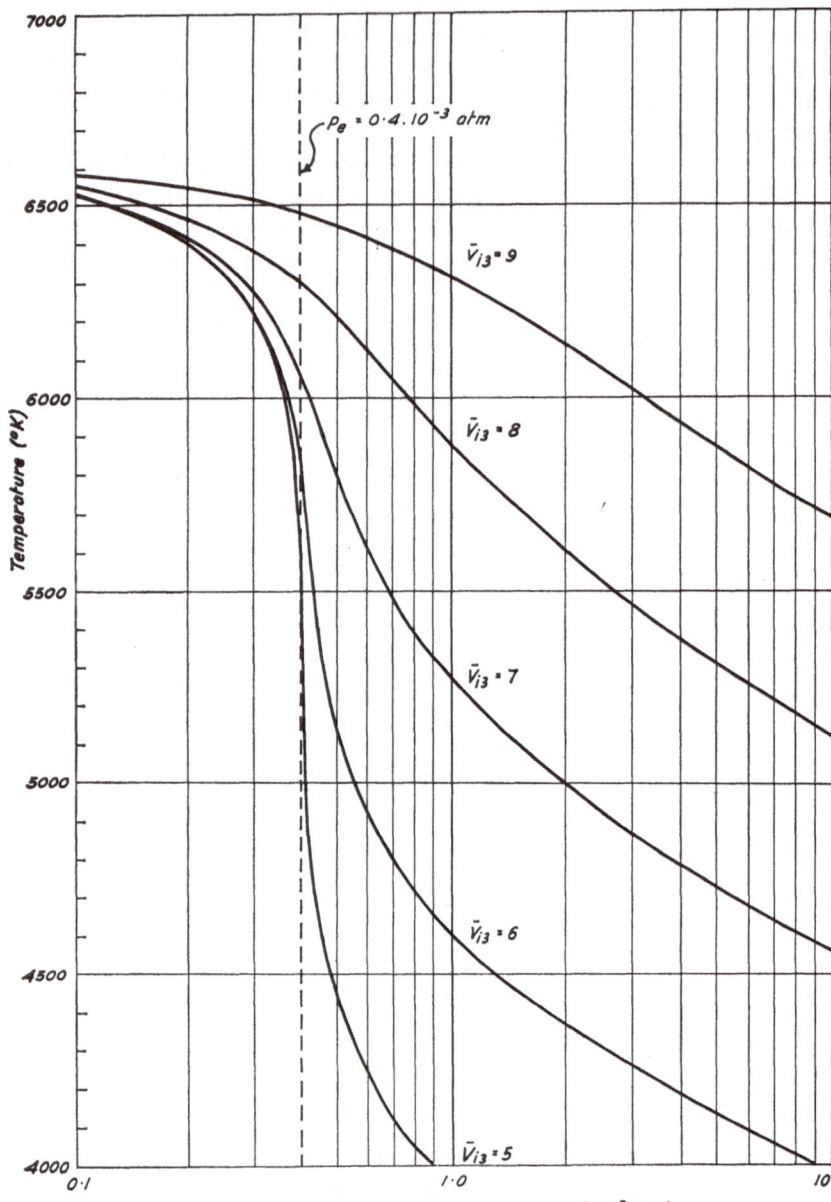

FIG. 7.10. Calculated dependence of the arc temperature on the partial pressure of added constituents of different ionization potential (partial pressure on log scale). The plasma of the arc in air alone is depicted schematically by a major constituent of apparent ionization potential $\bar{V}_{i1} = 14$ eV and a minor one of apparent ionization potential $\bar{V}_{i2} = 11$ eV, representing the bulk consisting of N_2, N, O, and CO, on the one hand, and free carbon vapour, on the other. The content of the latter is taken to be of the order of 1 per cent (cf. § 11.1). The electron pressure is supposed to be rigorously constant ($= 4.10^{-4}$ atm).

In view of what has been said above about the ionization potentials of N_2, N, O, CO, and C, we are justified in making an approximation for the complex gas in the carbon arc in air based on the fictitious, two-component gas just considered. The content of free carbon, that we evaluated to be approximately 1 per cent at $T = 6600°K$ and $p_e = 4 . 10^{-4}$ atm, also matches the actual conditions very well, as will be shown in § 11.1. In further calculations we shall consider the bulk of the arc gas (N_2, N, O, and CO) to be a single constituent, i.e. component one, having an ionization potential $V_{ij} = 14$ eV. Free carbon is represented by constituent two, having an ionization potential $V_{i2} = 11$ eV and a partial pressure $p_2 = 1 \cdot 24 . 10^{-4}$ atm. This partial pressure is regarded as constant, that is independent of the temperature. Again we assume the total pressure $P = 1$ atm and the electron pressure $p_e = 4 . 10^{-4}$ atm. We now add a third constituent and compute its partial pressure p_3 as a function of the temperature, using equation (7.50). Let constituent three be a substance having an apparent ionization potential of 9, 8, 7, 6, and 5 eV successively. The results of the calculations are presented in Figs. 7.10 and 7.12. The diagrams show how substances of different ionization potential would affect the arc temperature if the electron pressure were to remain rigorously equal to $4 . 10^{-4}$ atm. Note how steeply the curves run—indeed they have an inflection point there—when p_3 becomes equal to p_e. Particularly if the third constituent is of low ionization potential, the magnitude of the temperature drop is seen to depend to a great extent on the value of p_3.

When p_3 is below 10^{-4} atm, the curves in Fig. 7.10 tend to become horizontal, which means that for partial pressures of added substances below 10^{-4} atm the arc temperature is virtually unaffected by the addition of the extraneous element. The scale of the partial pressure is related to the content of the metal in a sample by considering that a partial pressure of about 10^{-2} atm for a metal vapour is the maximum attainable in a normal carbon arc (cf. §§ 7.9 and 9.4). So, the working conditions for major constituents of samples loaded into the electrode cavity corresponds to a range of partial pressure between 10^{-3} and 10^{-2} atm; hence, a content of about 1 per cent of an element, say an alkali metal, in a sample marks the point at which the relevant element begins to influence the excitation conditions. For the aggregate of vapours leaving the electrode, including carbon vapours, the critical percentage is about $0 \cdot 1$ per cent (see § 7.9). Note that such a critical impurity level is not disclosed if we plot the curves of Fig. 7.10 on a normal linear scale (Fig. 7.12) instead of a log scale.

How does the schematic, theoretical picture outlined in Figs. 7.10 and 7.12 compare with experiment? For this comparison we shall use primarily the results given in the previous section. Although they are not likely to apply generally, they serve as a convenient tool for determining the agreement and the discrepancy between theory and experiment, thus giving access to a more concrete formulation of the observed phenomena and the questions that still await complete elucidation.

A first glance at the empirical data reveals that actually the introduction of various substances into the arc can lower the temperature appreciably. Dropping

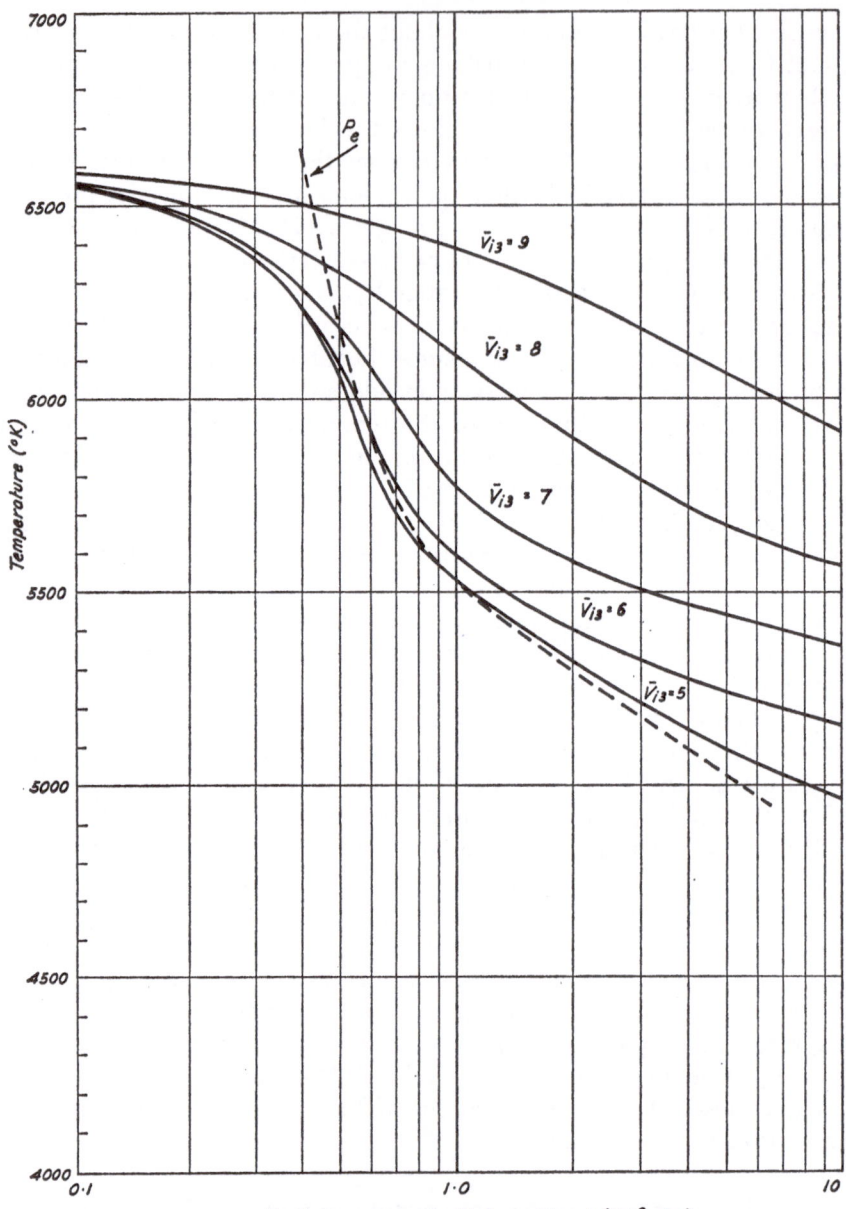

FIG. 7.11. Calculated dependence of the arc temperature on the partial pressure of added constituents of different ionization potential (partial pressure on log scale). In contrast to Fig. 7.10 where p_e was assumed constant, the electron pressure varies systematically with the temperature according to Fig. 7.6.

For orientation, the scale of partial pressure can be related to the metal content in a sample by considering that a partial pressure of about 10^{-2} atm for a metal vapour is the maximum attainable in a normal carbon arc, that is when the electrode charge consists entirely of a (volatile) compound of the relevant metal (cf. § 7.9).

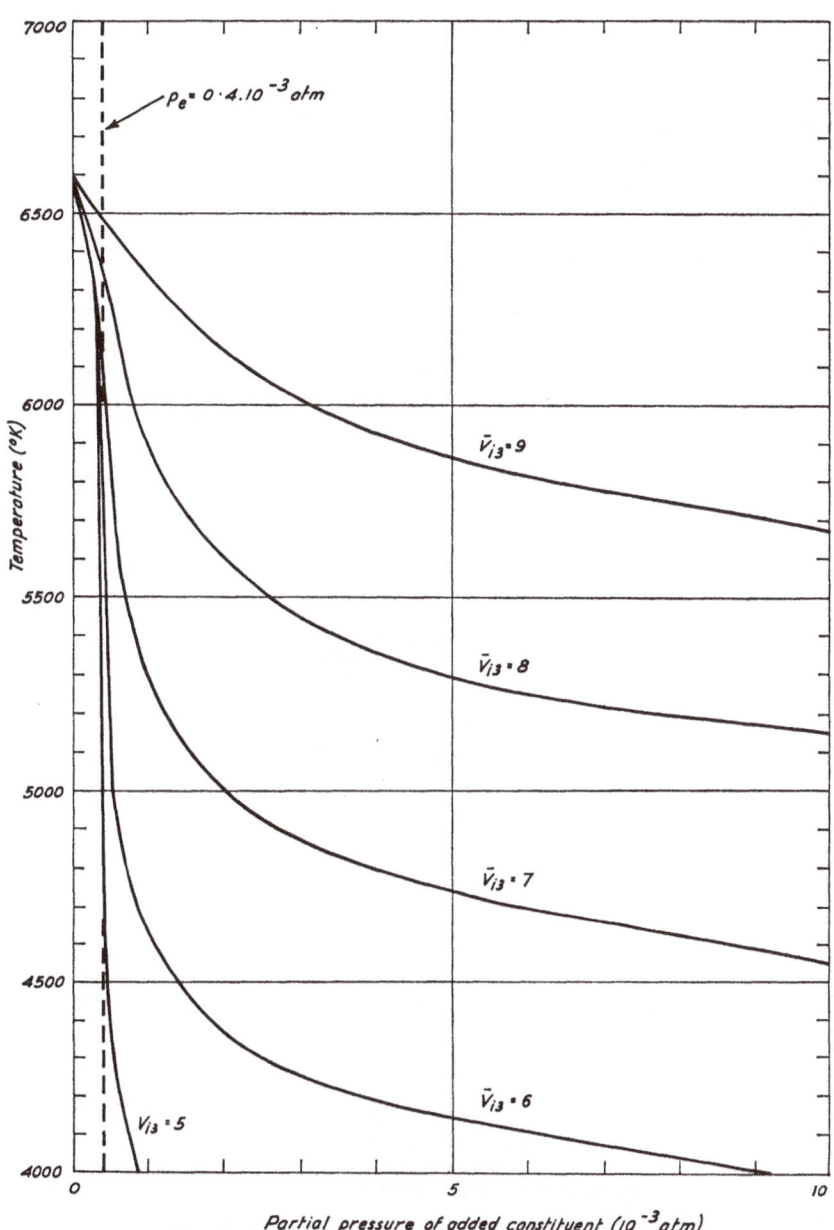

FIG. 7.12. Calculated dependence of the arc temperature on the partial pressure of added constituents of different ionization potential (partial pressure on normal, linear scale). See also Fig. 7.10.

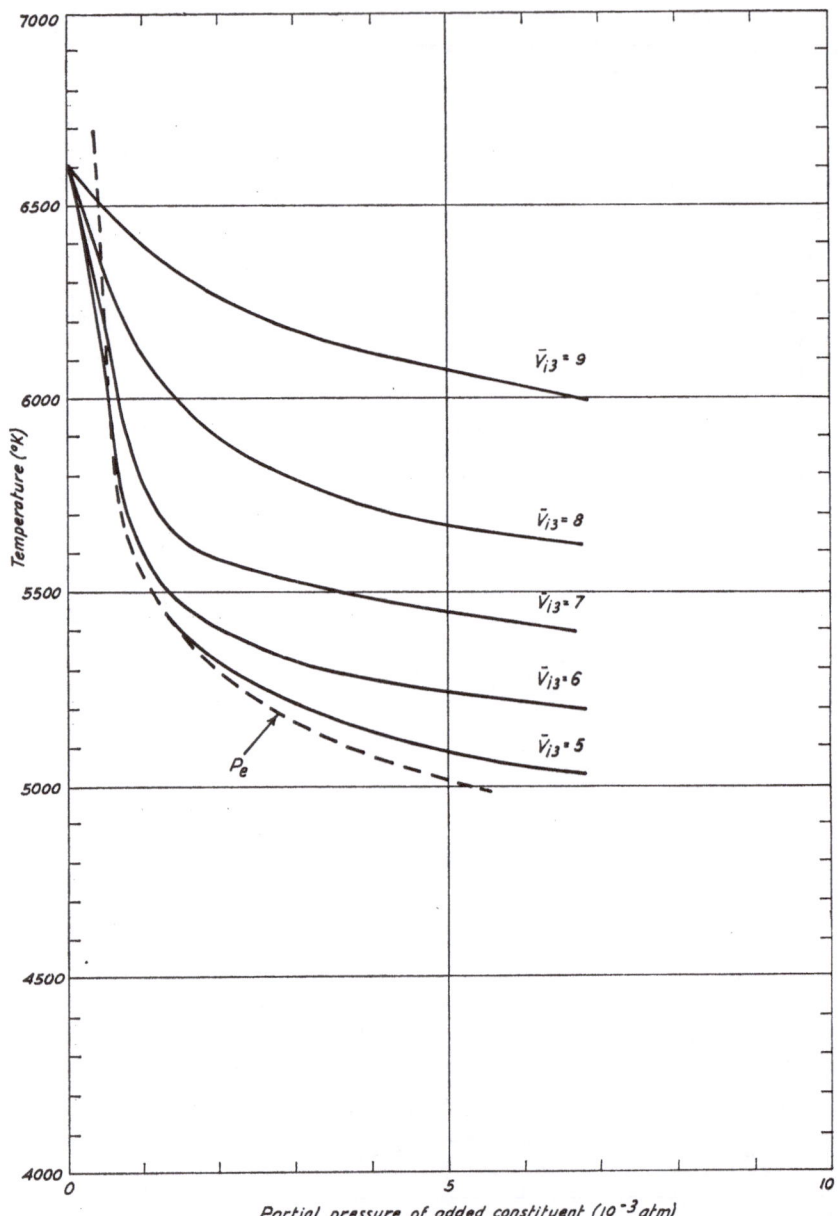

FIG. 7.13. Calculated dependence of the arc temperature on the partial pressure of added constituents of different ionization potential (partial pressure on normal, linear scale). In contrast to Fig. 7.12 where p_e was assumed constant, the electron pressure varies systematically with the temperature according to Fig. 7.6. See also Fig. 7.11.

seems to be less extreme than predicted by theoretical arguments. However, experiment also demonstrates that the electron pressure is not rigorously constant. Both these discrepancies are interrelated: if the entry of a substance of low or moderately low ionization potential into the arc plasma causes an increase of the electron pressure, its influence on the temperature is inherently smoothed down. So, for instance, when the correlation between p_e and T discussed in § 7.7 is taken to be a truly functional relationship, we obtain the curves produced in Figs. 7.11 and 7.13. They show that an increase of the electron pressure has a stabilizing effect on the temperature.

At this point it is important that we know what has been explained and what still remains to be argued. Assuming the appropriateness of an empirical correlation between p_e and T, we were able to show that the temperature runs less steeply with the partial pressure of an added component than in the absence of such a correlation, i.e. when p_e was unaffected by the addition of the substance. The nature of the correlation is uncertain, in other words, we did not account for the mechanism that determines the values of p_e and T separately. Why did we find, for example, $T = 5100°K$ and $\bar{p}_e = 3.6 . 10^{-3}$ atm in an arc with potassium vapour, and $T = 5720°K$ and $\bar{p}_e = 0.7 . 10^{-3}$ atm in one lithium vapour arc and $T = 5440°K$ and $\bar{p}_e = 1.8 . 10^{-3}$ atm in the other?

A full theoretical account would require an elaborate arc theory, which will not be attempted here. Anyhow, the experimental data needed for treating the problem quantitatively in terms of arc theory (in the sense defined in § 4.1) are wanting. Although we are thus forced to leave the matter at the qualitative considerations of § 3.9, a further discussion, mainly from a phenomenological point of view, presented in the next section, will be enlightening.

★ § 7.9 *The influence of the ionization potential and the vapour density of an added element upon the temperature and the electron pressure in the arc*
Estimating partial pressures and densities of metal vapours

Bearing in mind the arguments of the previous section, we shall now argue in favour of the following hypothesis.

When the partial pressure of an element of low ionization potential in the plasma of a carbon arc whose current strength is kept constant is increased beyond a critical level of about 10^{-4} atm, a relatively large fall in the temperature is produced initially; a further increase of the vapour concentration reduces the temperature only slightly more and makes it approach a value that is more or less characteristic for the ionization potential of the relevant substance; the chief effect of further increasing the vapour concentration is to raise the electron pressure.

It will be profitable to consider first the principle of estimating the vapour concentration of an element in the arc core from the electron pressure (and the temperature) and to use this procedure to obtain those estimates for the arcs discussed in § 7.7.

If a single substance (l) can be assumed to carry the discharge, that is when this species furnishes practically all the charge carriers, the general condition for quasi-neutrality

$$p_e = \Sigma p_{ij} \qquad (7.51)$$

reduces to

$$p_e \approx p_{il} \qquad (7.52)$$

It can be shown that below 6000°K at least 90 per cent of the electrons come from the added substance. So, we do not commit a great error if we suppose equation (7.52) to hold rigorously. For computing the partial pressure p_l of the substance in question we now use the equalities

$$\alpha_l p_l = p_{il} \approx p_e \qquad (7.53)$$

Numerical values of the degree of ionization α_l to be inserted in (7.53) follow from the Saha equation, (7.24) or (7.28), by substituting the measured values of p_e and T.*

This procedure for estimating the concentration of metal vapour in the arc has been long known. It was applied, for instance, by van Stekelenburg (1943), Huldt (1948b), and Prilezhaeva and Goryachev (1950). Other methods will be considered in § 9.4. All methods lead to the conclusion that the total contribution of the metal vapours to the plasma composition is, at the most, about 1 per cent; often it reaches the level of a few tenths of a per cent only.

The relative amount of carbon vapour in a carbon arc in air is usually of the order of 10 per cent. The carbon is present dominantly as CO; the content of atomic carbon is of the order of 1 per cent of the total plasma composition (cf. § 11.1).†

Table 7.6 summarizes the results for the arcs explored in § 7.7. In addition to the partial pressure p_l of the relevant matrix element, we have listed besides the concentration n_l, the electron pressure p_e, the electron concentration n_e, the mean temperature T, and the apparent ionization potential V_{il}.

Inspection of the results for arcs with lithium shows that the correlation between the vapour concentration, on the one hand, the temperature and the electron pressure, on the other, accords closely with the hypothesis stated at the beginning of this

---

* Alternatively, we reduce (7.47) to

$$p_e{}^2 = p_{al} S_{pl} \qquad (1)$$

which in logarithmic form reads

$$2 \log p_e = \log p_{al} + \frac{5}{2} \log T - \frac{5040}{T} V_{il} - 6 \cdot 18 \qquad (2)$$

Here V_{il} is the apparent ionization potential (§ 7.3). Given p_e and T, the pressures p_{al} and p_{il} are readily evaluated from (2) and (7.52) respectively; p_l follows subsequently as $p_{al} + p_{il}$.

† On the magnitude of the proportion of sample vapours in the arc plasma misconceptions have been existing and they appear to have not yet entirely vanished (see § 7.16).

section. The evidence is not convincing, because the lithium vapour pressure has *not* been independently measured. However, qualitative examination of the intensity of a lithium line showed it to fit in the picture.

The physical reasons for the difference in lithium vapour concentration in the cases considered are, firstly, that from identical electrodes, LiCl evaporates more rapidly than LiF because they have different boiling points (1353 and 1676°C respectively) and, secondly, that the material is released more easily from the normal open electrode (Fig. 7.14) than from the furnace electrode (Fig. 6.9).

TABLE 7.6

matrix*	$\bar{V}_{ij}\ddagger$ (eV)	$\bar{T}$ (°K)	$\bar{p}_e \cdot 10^3$ (atm)	$\bar{n}_e \cdot 10^{-15}$ (cm^{-3})	$\bar{p}_l \cdot 10^3$ (atm)	$\bar{n}_l \cdot 10^{-15}$ (cm^{-3})
KCl	4·68	5100	3·6	5·1	4·0	5·8
NaCl	5·49	5450	1·4	1·8	1·5	2·0
LiC†	5·75	5440	1·8	2·4	2·3	3·1
LiCl	5·75	5530	1·1	1·4	1·2	1·6
LiCl	5·75	5530	1·2	1·5	1·3	1·8
LiF	5·75	5720	0·7	0·9	0·7	0·9
Tl$_2$O$_3$	6·55	5730	0·6	0·7	0·7	0·9
In$_2$O$_3$	6·54	5910	0·7	0·9	0·8	1·0
PbO	7·35	5940	0·7	0·8	1·1	1·4
Tl$_2$O$_3$	6·55	5970	0·6	0·8	0·7	0·8
Sb$_2$O$_4$	8·78	6040	0·8	0·9	7·3	8·8
Al$_2$O$_3$	6·90	6110	0·5	0·6	0·5	0·7
Sb$_2$O$_4$	8·78	6160	0·7	0·9	4·8	5·7
Ni	8·13	6160	0·6	0·6	1·0	1·2
CuO	8·18	6190	0·4	0·5	0·8	0·9
graphite	11·5	6230	0·4	0·5	—§	—§

* The matrices were evaporated from a furnace electrode (Fig. 6.9) except for LiCl†. The matrices Al$_2$O$_3$, Ni, and CuO evaporate at a low rate from the furnace electrode.

† LiCl was volatilized from an electrode of the type shown in Fig. 7.14.

‡ Apparent ionization potentials according to § 7.3 are given here. The numerical values have been adapted closest to the mean temperature for the relevant matrices so as to bring them into best agreement with the ratios of partition functions, Z_{tj}/Z_{aj}.

§ In view of the presence of the test elements, zinc and magnesium, in the plasma, the discharge cannot be assumed to be carried by carbon alone.

Partial pressures and concentrations of metal vapours in the arc column. The table lists for each matrix the apparent ionization potential ($\bar{V}_{tl}$) of the matrix element, the mean temperature ($\bar{T}$) observed in the relevant arc, and the mean values of the electron pressure ($\bar{p}_e$), the electron concentration ($\bar{n}_e$), the partial pressure ($\bar{p}_l$) of the matrix element, and its vapour concentration ($\bar{n}_l$).

More extensive evidence regarding the appropriateness of the hypothesis has been furnished by recent experiments with LiF–graphite arcs (Boumans and Rouws, 1965). Mixtures of LiF and graphite, in varying proportions, were arced from electrodes of the type shown in Fig. 7.14. Racking-plate studies provided information on the changes with time of the log intensity (Y_{Li}) of the line Li 2475, of the log intensity ratio (ΔY_1) of the ion–atom line pair Mg 2796/2780, and of the log intensity of the background (Y_u), the last quantity being measured in the vicinity

of 2780 Å. In Fig. 7.15, typical plots of Y_{Li}, ΔY_1, and Y_u versus time are reproduced. Fig. 7.16 shows clearly the close correlation between ΔY_1 and Y_{Li}.*

The critical reader will have noticed that in our argument we have regarded
the intensity of a spectral line as an adequate measure of the metal vapour concentration in the arc core. Obviously, it is true that the observed intensities of
Li 2475 have been modified by self-absorption; the main objection, however, is

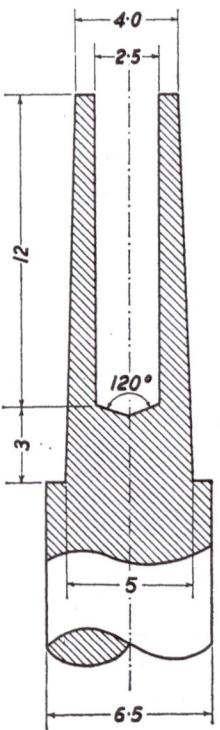

FIG. 7.14. The electrode shape normally used by the author for spectrochemical analysis with a LiF+graphite (1+4) buffer admixture of samples
(dimensions are in millimetres).
　　The electrode charge consists of 1 to 10 mg of sample, mixed with 70 to
60 mg of buffer admixture. Burning to completion takes about 5 min.

to our relating *atom* line intensity for elements of low ionization potential to total
metal concentration in the arc core. Under the conditions of the experiments,
lithium is ionized to a large extent in the arc core; hence our inferences have

* The scales of p_e and T at the right have been constructed on the basis of the interrelations
between p_e, T and ΔY_1 (see § 7.7). The appropriateness of these interrelations for this type of
electrode was verified separately; indeed a number of the points included in Fig. 7.6 were obtained
from the experiment that has just been described.
　Further diagrams for lithium vapour arcs are given in §§ 7.10 and 7.14 (Figs. 7.30 and 7.31)
along with reproductions of the plates. I also give in § 7.14 diagrams pertaining to a pure graphite
arc (Fig. 7.32) and an arc with potassium vapour fed from a furnace electrode (Fig. 7.29). The plots
show that the temperature and the electron pressure have a similar trend to the intensity of a
spectral line of the relevant metal.

been made from lithium lines (and potassium lines: see § 7.14) originating mainly in the flame-like fringe of the arc. A recent study of the radial vapour distribution in the arc, undertaken by de Galan (1965a, 1966a) (cf. § 9.4), indicates that, in the region of interest for spectroscopy, metal vapours are diffused uniformly in the arc cross-section. Therefore, we have reason to believe that the intensity variations

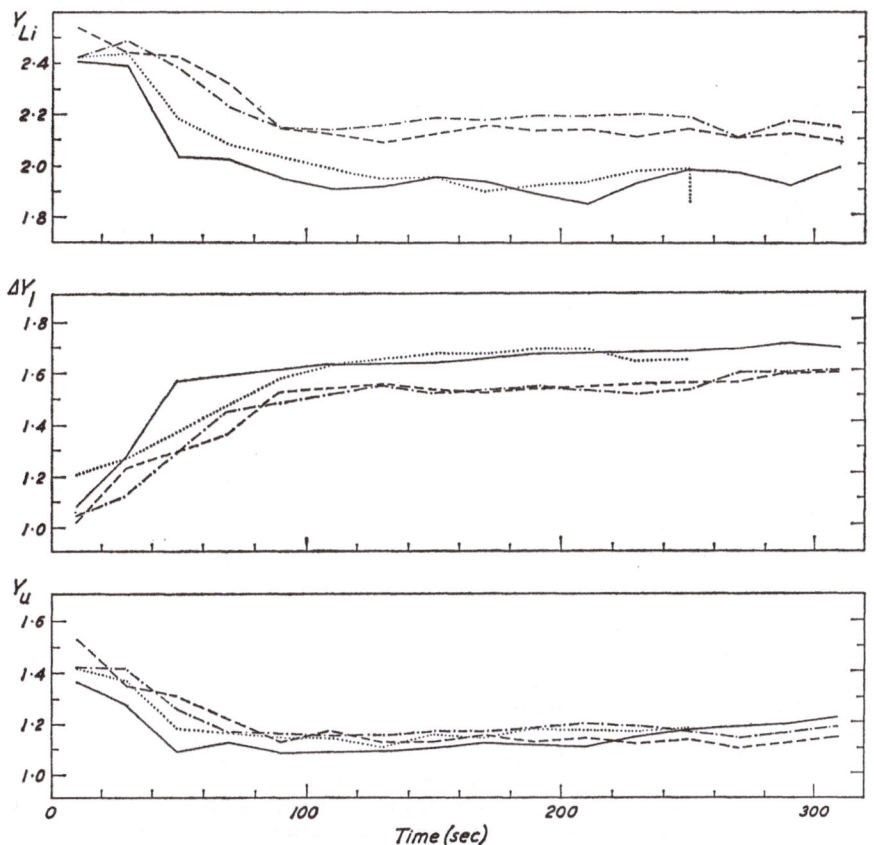

FIG. 7.15. Racking-plate plots of
1 The log intensity (Y_{Li}) of the line Li 2475.
2 The log intensity ratio (ΔY_1) of the lines Mg II 2796 and Mg I 2780.
3 The log background intensity (Y_u) in the vicinity of 2780 Å.
    The diagrams depict the change of excitation conditions during the volatilization of LiF + graphite mixtures in varying proportions. Electrode type as in Fig. 7.14.

of potassium and lithium lines indicate changes in the alkali metal vapour density in the *core*.

To substantiate the hypothesis, de Galan and Boumans (1965) demonstrated that the relationship between the temperature, the electron pressure, and the total metal vapour concentration applies also to aluminium. As this element emits atom lines chiefly from the core, the total aluminium concentration (ions and atoms in the core can be calculated with reasonable accuracy from the intensity of an atom line by applying the Boltzmann and Saha formulae [see equation (9.13)]. The arguments are shown in a self-explanatory way by the diagrams of Fig. 7.17.

Now it is important that we appreciate thoroughly that the electron pressure depends to a considerable extent on the metal vapour concentration; for, if we do not, we are faced with serious contradictions in the observations. De Galan (1965a,

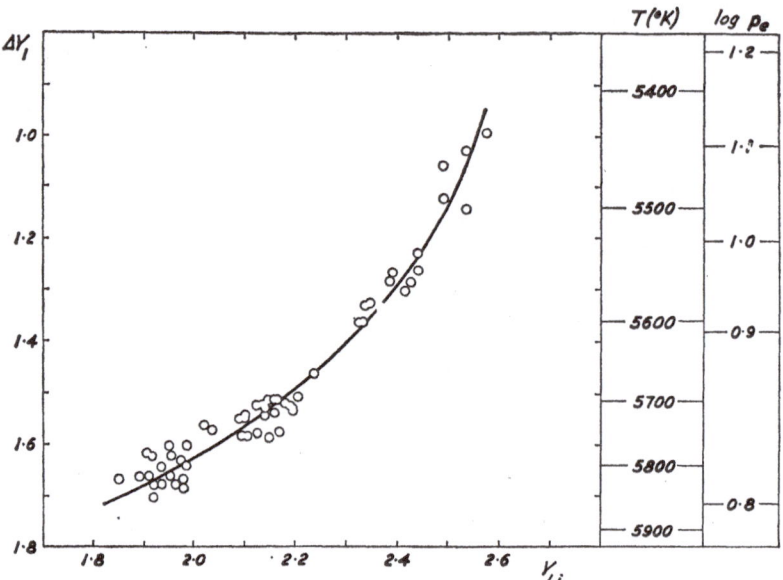

FIG. 7.16. Plot of the intensity ratio (ΔY_1) of the ion–atom line pair Mg 2796/2780 versus the log intensity (Y_{Li}) of the line Li 2475.

The data pertain to micro-spectra of 25-second exposure time, taken during the volatilization of LiF+graphite mixtures from electrodes of the type sketched in Fig. 7.14 (cf. Fig. 7.15 for the relevant racking-plate graphs).

The temperature and electron pressure (in 10^{-4} atm) scales on the right are based on the empirical interrelations between p_e, T, and ΔY_1, demonstrated in § 7.7.

1966a), for instance, explored arcs with aluminium, lithium, and potassium vapour; with his operating conditions, which were different from mine, he found a *positive* correlation between the temperature and the electron pressure (see Fig. 6.40). This curious contrast with my findings turned out to be due to differences in metal vapour concentration. De Galan, using normal electrodes (Fig. 7.14), achieved a fairly high aluminium vapour concentration when arcing a mixture of aluminium oxide and graphite, which led to a relatively high electron pressure at an elevated temperature (circa 6000°K); the potassium vapour content in his potassium arc, on the contrary, proved to be rather low when the element was introduced as the fluoride with graphite admixture; so, at the reduced temperature

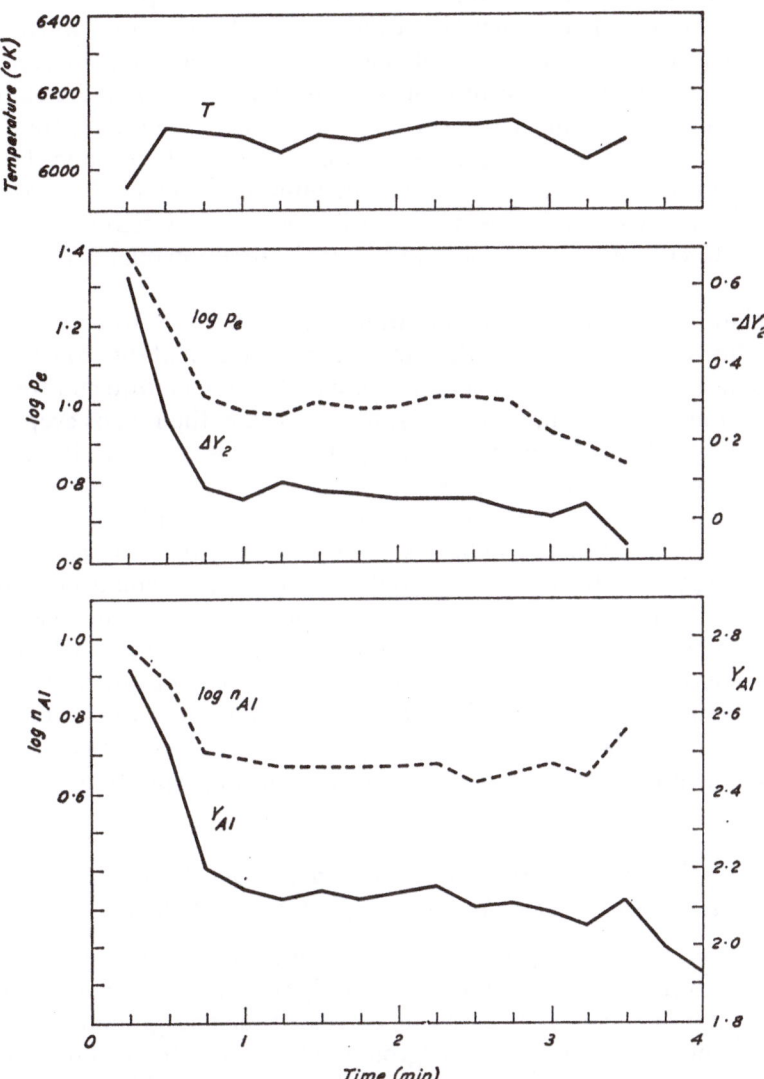

FIG. 7.17. Racking-plate graphs pertaining to an arc with an Al_2O_3 + graphite matrix. The diagrams show the variation with time of

1 The temperature (T).
2 The log intensity ratio (ΔY_2) of the ion–atom line pair Mg 2796/2852.
3 The log electron pressure (log p_e), where p_e is in 10^{-4} atm.
4 The log relative intensity (Y_{Al}) of the line Al 2652.
5 The log relative concentration (log n_{Al}) of aluminium in the arc (ions + atoms) as derived from Y_{Al}.

The plots demonstrate that the high concentration of aluminium in the early arcing periods gives rise to a substantial enhancement of the electron pressure.

of the potassium arc (circa 5100°K) a relatively low electron pressure was found.*

The results of de Galan suggest that caution must be exercised when we make generalizations about empirical correlations between the temperature and the electron pressure. Thus the correlation described in § 7.7 should not be made absolute. Particularly we must bear in mind that, in the high temperature range of the interval explored, an elevated electron pressure can be produced when the vapour concentration of a metal of medium ionization potential is made high; by contrast, a low electron pressure is feasible with a low temperature when a relatively small portion of an element of low ionization potential is present in the arc.

To conclude, in addition to the ionization potential of the major constituent of a specimen introduced into an arc, the vapour concentration of that constituent is of crucial significance for excitation conditions, viz. the temperature and the electron pressure, that are imposed upon the plasma. Therefore, the rate of evaporation of a sample must be considered to be a possible cause of variation in quantitative spectrochemistry. The rate of entry can affect line intensities indirectly via the vapour concentrations of constituents that tend to exert a marked influence on the excitation conditions. If those conditions are stabilized by the addition of a spectro-scopic buffer (§ 7.12), the volatilization rate of the sample should be immaterial, except for its interference via the degree of self-absorption of lines (see §§ 5.7 and 10.2). Unfortunately, entirely efficaceous buffering is not so easily achieved, since the vapour concentration of the buffer element must be kept under close control. Before we go further into this matter we shall consider a few aspects of the back-ground intensity. As its behaviour is closely related to the variations of the excita-tion conditions at present under discussion, it seems appropriate to consider this question now.

★ § 7.10   *Empirical relationships between the spectral-background intensity, the electron pressure, the temperature, and the metal vapour concentration*

When measuring the temperature and the electron pressure (§ 7.7), we were obliged to measure the background intensity adjacent to the zinc and magnesium lines so that the line intensities could be corrected for background. Since even a superficial glance at the overall background in the spectra manifested several pecularities, it appeared worth while to examine more thoroughly the set of data for the intensity of the background. This was done, with particular reference to the wavelengths in the vicinity of 3072 and 2780 Å; the conclusions pertain to the background as a whole.

The main features are summarized in Figs. 7.18 and 7.19. The former plots temperature against the mean value of the background intensity (I_u) at 3072 Å for graphite arcs with different metal vapours; the latter figure gives a similar representation for individual measurements (I_u) for potassium vapour and lithium

* I also note that de Galan's values of temperature and electron pressure are effective mean values averaged over the entire burning period; in my experiment, the observations of separate racking-plate exposures were averaged.

vapour arcs. Plotting the individual observations for the other arcs served no useful purpose, as they differ little from the mean value, unlike the wider dispersion of the individual measurements for the alkali metal arcs. In these cases the variations are essentially systematic.

Evidently, the background intensity increases regularly with the temperature when T is beyond $5700°K$, whereas below $5700°K$, that is in the alkali metal vapour

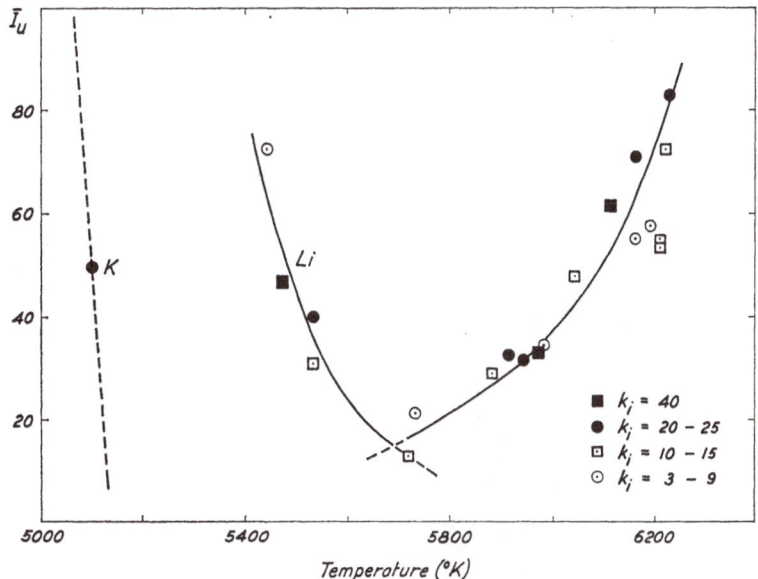

FIG. 7.18. Mean values of the background intensity $(\bar{I}_u)$ at 3072 Å plotted against the mean temperatures as measured in graphite arcs with different metal vapours. Each point represents an average of k_i single measurements in racking-plate micro-spectra. The continuous curve with negative slope at the left pertains to measurements in arcs with lithium vapour. The broken line refers to potassium vapour arcs; it has been drawn on the basis of individual measurements (see Fig. 7.19).

arcs, a steep rise occurs when the temperature decreases.[*] Besides, visual inspection of the spectra showed generally a continuous background for alkali arcs and a band structure for the other arcs.

A plausible explanation of the observed changes in the background intensity is the following. In the temperature range beyond $5700°K$ the background in our type of arc is chiefly one of band emission. Owing to the enhanced population of excited levels at higher temperatures, the intensity of the bands tends to increase rather rapidly, in spite of the counteracting effect of the increasing degree of dissociation of the molecular species. The sharp rise of the background intensity, when alkali metals are introduced in high concentration, seems to be attributable to the

[*] Although in the presence of much alkali metal vapour enhanced background intensity must have been observed frequently, it has, to the best of my knowledge, been noted only by Schuttevaer (1943, p. 22) and by Rusanov, Khitrov, and Batova (1959).

appearance of continuous emission caused by free-free transitions of electrons in the electric fields of the ions (*Bremskontinuum*) and by free-bound transitions (radiative recombinations) (see § 7.11).

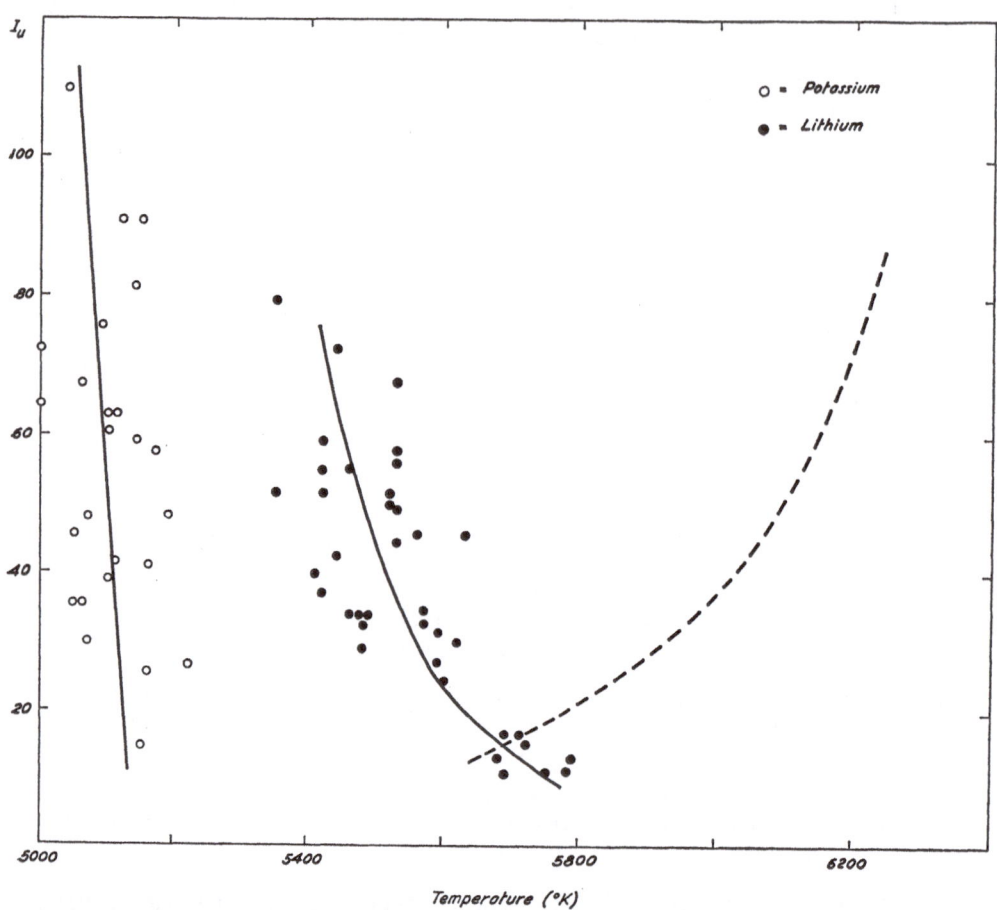

FIG. 7.19. Individual measurements of the background intensity (I_u) at 3072 Å in racking-plate micro-spectra plotted against the relevant temperatures. The broken line represents the variation of the mean intensity of the background with temperature for the other arcs investigated (see Fig. 7.18).

This picture is substantiated by analysing the fluctuations of the temperature (T), of the log background intensity (Y_u), and of the log intensity ratio (ΔY) of the magnesium ion–atom line pairs (see Boumans and Rouws, 1965). Briefly the line of argument is as follows.

1 The standard deviations ($\hat{\sigma}_T$) of the temperature computed from individual measurements of the micro-spectra proved to be independent of T and thus of the metal vapour.

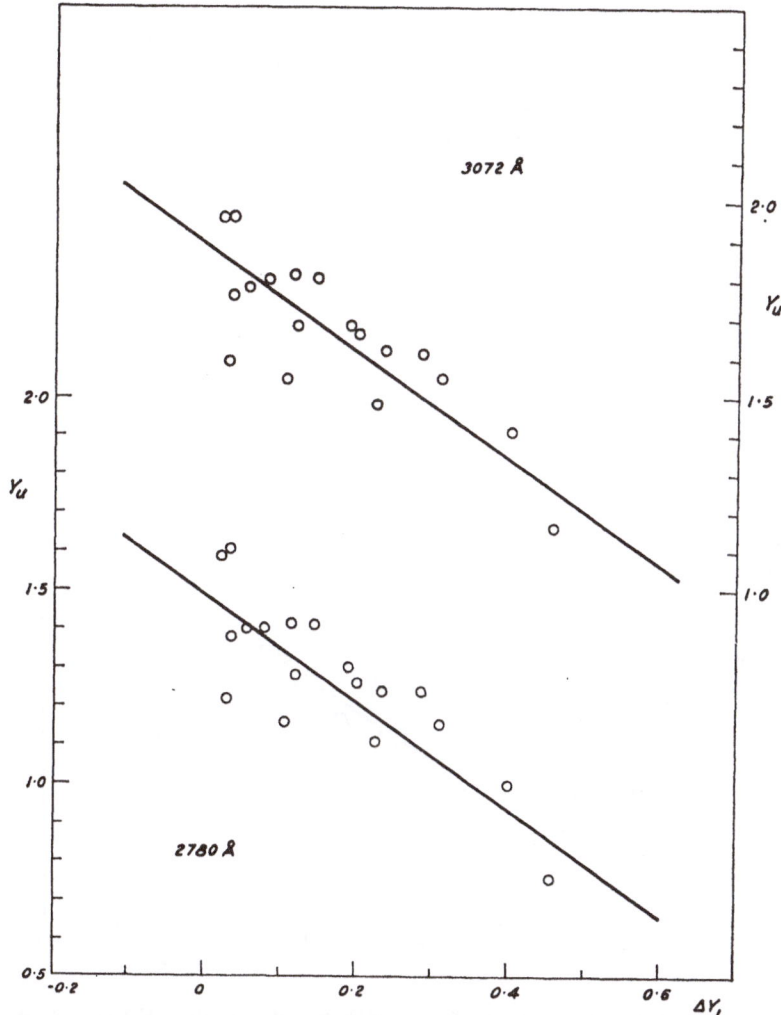

FIG. 7.20. Correlation between the log background intensity (Y_u) and the log intensity ratio (ΔY_1) of the ion–atom line pair Mg 2796/2780 in graphite arcs with potassium vapour. The points represent individual observations in racking-plate micro-spectra. The background intensity was measured at 2780 and 3072 Å.

2 The standard deviations $\hat{\sigma}_{\Delta Y}$ and $\hat{\sigma}_{Y_u}$ for the alkali arcs turned out to be significantly greater than those for the other arcs.

3 The high value of $\hat{\sigma}_{\Delta Y}$ for alkali arcs cannot be explained on the basis of $\hat{\sigma}_T$ alone.

4 Therefore, the scattering of ΔY in alkali arcs is due to variations of the electron pressure.

5 If a substantial fraction of the total background intensity is produced by

continuous emission, such as indicated above, the high variability of Y_u in the alkali arcs is caused by the electron pressure variations.

If this reasoning is correct, we must expect Y_u and ΔY to be closely correlated. Figs. 7.20 and 7.22 show this to be so. Various interrelations were explicitly verified for LiF–graphite arcs. One of the correlation diagrams (Fig. 7.16) depicts the dependence of the log intensity ratio ΔY_1 of the ion–atom line pair Mg

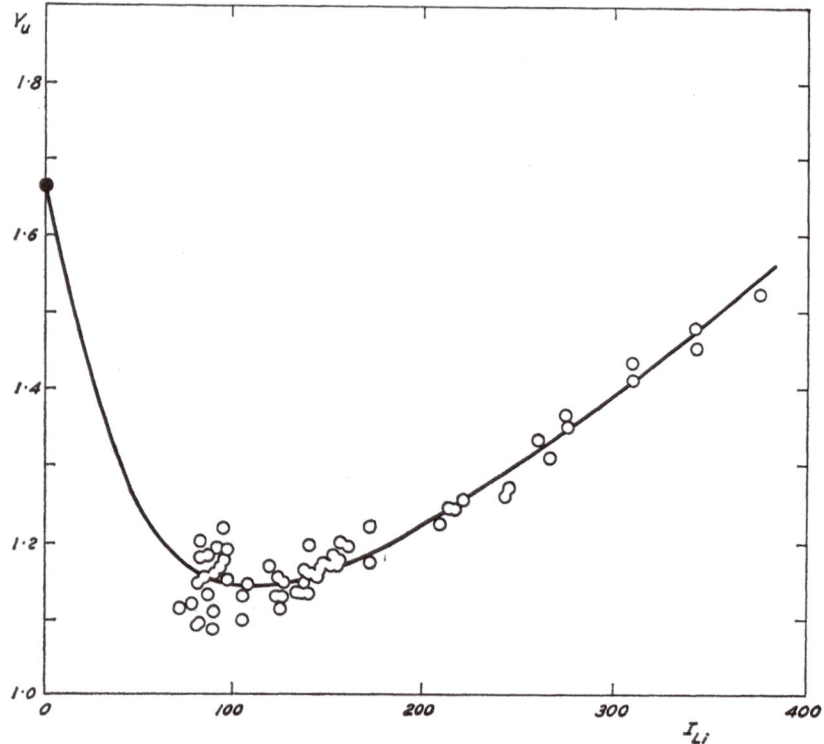

FIG. 7.21. Correlation between the log background intensity (Y_u) at 2780 Å and the intensity (I_{Li}) of the line Li 2475. The points represent individual observations in racking-plate micro-spectra of 25-second exposure time, taken during the volatilization of LiF+graphite mixtures from electrodes of the type sketched in Fig. 7.14 (cf. Fig. 7.15).

2796/2780 on the log relative intensity (Y_{Li}) of the line Li 2475. We also report the correlation between log background intensity Y_u and the intensity of the line Li 2475 (Fig. 7.21) and that between Y_u and ΔY_1 (Fig. 7.22).

To summarize, Figs. 7.16, 7.21, and 7.22 demonstrate plainly what changes in the excitation conditions take place in a lithium–graphite arc when the lithium vapour concentration varies. *Note that when T is between 5700 and 5800°K, the background intensity is at a minimum. The excitation conditions prevailing in the vicinity of this minimum are favourable for detecting and determining most elements in spectrochemical analysis. They present a good compromise between detectability and stability* (see §§ 7.12 and 8.8).

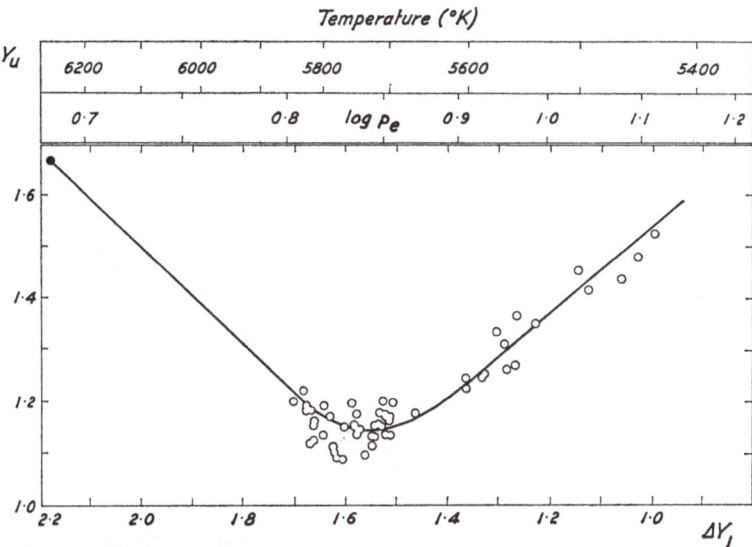

FIG. 7.22. Correlation between the log background intensity (Y_u) and the log intensity ratio (ΔY_1) of the ion–atom line pair Mg 2796/2780. The points represent individual observations in racking-plate micro-spectra of 25-sec exposure time, taken during the volatilization of LiF + graphite mixtures from electrodes of the type sketched in Fig. 7.14 (cf. Figs. 7.15 and 7.21).

The temperature and electron pressure (in 10^{-4} atm) scales at the top are based on the empirical interrelations between p_e, T, and ΔY_1, demonstrated in § 7.7.

§ 7.11 Spectral background and continuous emission

Broadly speaking, spectral background is the spectrum on which the spectral lines of atoms and ions are superimposed. We distinguish between continuous background and that consisting chiefly of molecular bands. The nature of the bands and the dispersion and resolving power of the spectroscopic equipment decide whether band spectra give us the impression of continua or not. Intrinsically, clear distinctions can be drawn, of course. Below we shall consider a few topics related to continuous emission; a thorough treatment of the contribution of band spectra to the background is postponed until § 11.2.

Continuous background is due to scattered radiation within the spectrograph or to continuous radiation that enters the slit and is dispersed in the spectrum. Continuous spectra are emitted either as blackbody radiation by the glowing electrode tips and by incandescent particles in the arc column, or as true continua by the arc plasma.

The intensity distribution of blackbody radiation follows Planck's law. Accordingly, the intensity is at a maximum at a distinct wavelength, λ_{max} (Wien's law). For instance, λ_{max} is computed to be about 7000 Å at 4000°K, which is the sublimation temperature of carbon (see §§ 2.2 and 10.3). Towards the ultra-violet the intensity falls off rapidly with decreasing wavelength. When prisms are used, the apparent effect is further enhanced because of the increasing dispersion at shorter wavelengths. If adequate means are taken to prevent stray light from the electrode tips obscuring the spectrum, the portion of the background to be identified as blackbody radiation is small compared with other contributions, predominantly band emission.

Adding metal vapours to the plasma may cause an electron continuum, for example, when a relatively high alkali metal vapour concentration is reached in the arc (§ 7.10). Golling (1957) noticed a tendency for electron continua to develop when elements that have a dense energy-level diagram were present, such as Fe, Cr, V, Ti, Mo, Ni, and Co. He discussed some aspects of the underlying theory.*

The origins of the continua are dominantly free-free transitions of electrons in the fields of the ions (*Elektronenbremskontinuum*) and free-bound transitions (series limit continuum). In the latter case the continuous emission is effected by radiative recombination of electrons and ions. According to theory, superposed continua display the following features. Up to a definite frequency limit v_l the emitted intensity is independent of the radiation frequency; it is proportional to the product of electron and ion concentrations in the plasma:

$$I_{v \leq v_l} = K \frac{n_e n_i}{\sqrt{(kT)}} \tag{7.54}$$

where K is a constant of proportionality, k the Boltzmann constant, n_e the electron concentration, n_i the concentration of the ions, and T the absolute temperature.

From the frequency limit the intensity drops exponentially toward shorter wavelengths:

$$I_{v > v_l} = K \frac{n_e n_i}{\sqrt{(kT)}} \exp \frac{h(v_l - v)}{kT} \tag{7.55}$$

The numerical value of the frequency limit depends on the energy-level diagram of the ion. Maecker and Peters (1954) have divided the elements into three groups according to the distribution of their terms. If the terms follow closely up to the series limit of the main series, v_l is high and the continuum extends over the visible and ultra-violet spectral regions. This applies for elements with the highest multiplicity, among which are many transition elements. When the terms are widely separated, emission independent of v is only to be expected in the infra-red. In between these two groups is a group of elements whose discrete terms are located close to each other except in the far ultra-violet.

Lochte-Holtgreven (1958) notes that observations do not always agree very well with theory. Also, experiments must be conducted to prove the classification of Maecker and Peters. Golling (1957) investigated the influence of the matrix element on the intensity of continuous background and on the limits of detection

---

* More detailed studies and references will be found in astrophysics literature (e.g. Unsöld, 1955, p. 163 ff.), in studies on the high-current arc (Maecker and Peters, 1954; Finkelnburg and Maecker, 1956, pp. 354–356; and Lochte-Holtgreven, 1957, 1960) and in plasma physics (Lochte-Holtgreven, 1958, p. 328; and Griem, 1962, 1964b).

in spectrochemical analysis and was able to disclose a notable deterioration of the detection limits with matrix elements that have a dense energy-level diagram, thus giving rise to high background intensity. Therefore, proper dilution of samples that contain elements such as iron, titanium, vanadium, etc. as major constituents will often result in more favourable line-to-background ratios for trace elements sought (cf. § 7.12). Carrier techniques and preliminary chemical separation procedures form valuable alternatives, particularly if very low detection limits must be reached (cf. § 10.4).

It will be of interest to investigate more thoroughly how the facts observed by Golling are interrelated with those I obtained (§ 7.10).

§ 7.12 *Spectroscopic buffers*

At the end of § 1.1 we discussed the concept of spectroscopic buffers. We defined the buffer as a substance with which the sample is diluted to make the excitation conditions virtually independent of the properties of the substance to be analysed. Excitation was seen to encompass a number of processes, broadly divided into two groups, those governing the entry of material into the discharge zone and those defining the plasma conditions.

Particularly in routine methods of spectrochemical analysis, the use of an appropriate buffer is indispensable. Certainly, fluctuating excitation conditions do not influence analytical results if reference elements are used; however, if the ratio between the intensity of an internal standard line and the analysis line is to be constant for large variations in the excitation conditions, the internal standard must be so nearly identical with the analysis element (see § 8.9) that theoretically for each element sought there ought to be a separate internal standard. This is undesirable, of course, in a routine method. An essential advantage of the spectroscopic buffer, even if it does not fully annul all variations in excitation conditions, is that it limits the need for different internal standards drastically, so that one to three reference elements are adequate. Also, by the appropriate choice of the buffer a good compromise with regard to the detection limit of the different elements can be obtained.

In choosing a buffer, we must remember that the greater the dilution of the samples the less their influence on the excitation conditions. At the same time we must not forget that dilution ultimately lowers detection limits. We strive, therefore, to obtain a compromise where an optimum quantity of sample enters the plasma, while minimally disturbing the conditions of the plasma.

The plasma conditions are principally defined by the temperature and the electron pressure. In § 7.8 (Figs. 7.10 to 7.13) we concluded that the temperature of a pure carbon arc is strongly affected by the sample if it contains elements of low ionization potential, particularly alkali metals, in a concentration over 1 per cent. So, if this interference is to be annulled, either the sample must be introduced very slowly into the discharge column or the temperature of the plasma must be maintained at a relatively low level by the presence of a constant amount of alkali metal vapour. In the latter case the sample can be diluted far less than in the former. Moreover, a moderate lowering of the temperature favours the emission of atom lines of a great many elements (see especially §§ 8.8 and 8.9) and reduces the background intensity (see § 7.10).

Let us consider what happens when we add a buffer element with an apparent ionization potential $V_{i3} = 6$ eV to the plasma of a carbon arc, i.e. a gas composed of a major constituent of ionization potential $V_{i1} = 14$ eV and a minor one (free carbon) of partial pressure $1 \cdot 24 . 10^{-2}$ atm with $V_{i2} = 11$ eV (see § 7.8). Lithium represents such a buffer rather well as it has an apparent ionization potential of 5.75 eV between 5000 and 6000°K (Table 7.1). Computing the temperature as a function of the partial pressure of the buffer element, according to the procedure outlined in § 7.8, yields the continuous curve of Fig. 7.23 if we assume the validity of the interrelation between T and p_e shown in Fig. 7.6.*

Down to about 5600°K the temperature falls rapidly as the lithium vapour pressure ($p_{Li} < 10^{-3}$ atm) increases. Then, a typical buffer effect, shown by the decreasing slope of the curve, begins to appear. This self-buffering is closely associated with the rise of the electron pressure (see the broken curves in Fig. 7.23). The curve labelled p_e depicts the experimental correlation between the electron pressure and the temperature; that denoted by b shows the dependence of the temperature on the lithium vapour pressure if the electron pressure were to be constant ($= 4 . 10^{-4}$ atm).

Now we come to the buffer effect exerted by lithium when another substance (fourth component) is added. Let us assume that the fourth component is an element with an apparent ionization potential $V_{i4} = 4 \cdot 5$ eV which identifies it very nearly with the elements of lowest ionization potential, caesium, rubidium, and potassium, that have $V_{ij} = 4 \cdot 4$, $V_{ij} = 4 \cdot 6$, and $V_{ij} = 4 \cdot 74$ eV respectively (Table 7.1). We first suppose that the arc burns at 6200°K with $0 \cdot 43 . 10^{-3}$ atm lithium vapour pressure (point 1 on the continuous curve a in Fig. 7.23); then we add gradually the fourth component, say caesium, and compute the temperature as a function of the caesium vapour pressure. If we plot the temperature as a function of the sum of the vapour pressures of lithium and caesium, we obtain, of course, a curve that initially falls very steeply and then stabilizes at lower temperatures. The curve has not been included in the figure; it lies in between the continuous curve for lithium (a) and the broken curve for the electron pressure (p_e). Starting from $T = 5600$°K and $p_{Li} = 0 \cdot 92 . 10^{-3}$ atm and with caesium added, we see that the action of the extraneous element differs little from that of lithium (compare the continuous curve a with the broken curve a_2). Similar curves are obtained if the partial pressure of the buffer is increased further before adding caesium (branch a_3 and a_4). Although buffering becomes more effective the larger the lithium concentration, it is evident also that below about 5500°K, that is when p_{Li} is over 10^{-3} atm, little will be gained by increasing further the lithium vapour density.

---

* This assumption is likely to hold well, particulary for lithium, between 5800 and 5400°K. Outside this temperature range, whatever deviations from the curve in Fig. 7.23 occur, they do not appear to have any influence on the conclusions we shall arrive at.

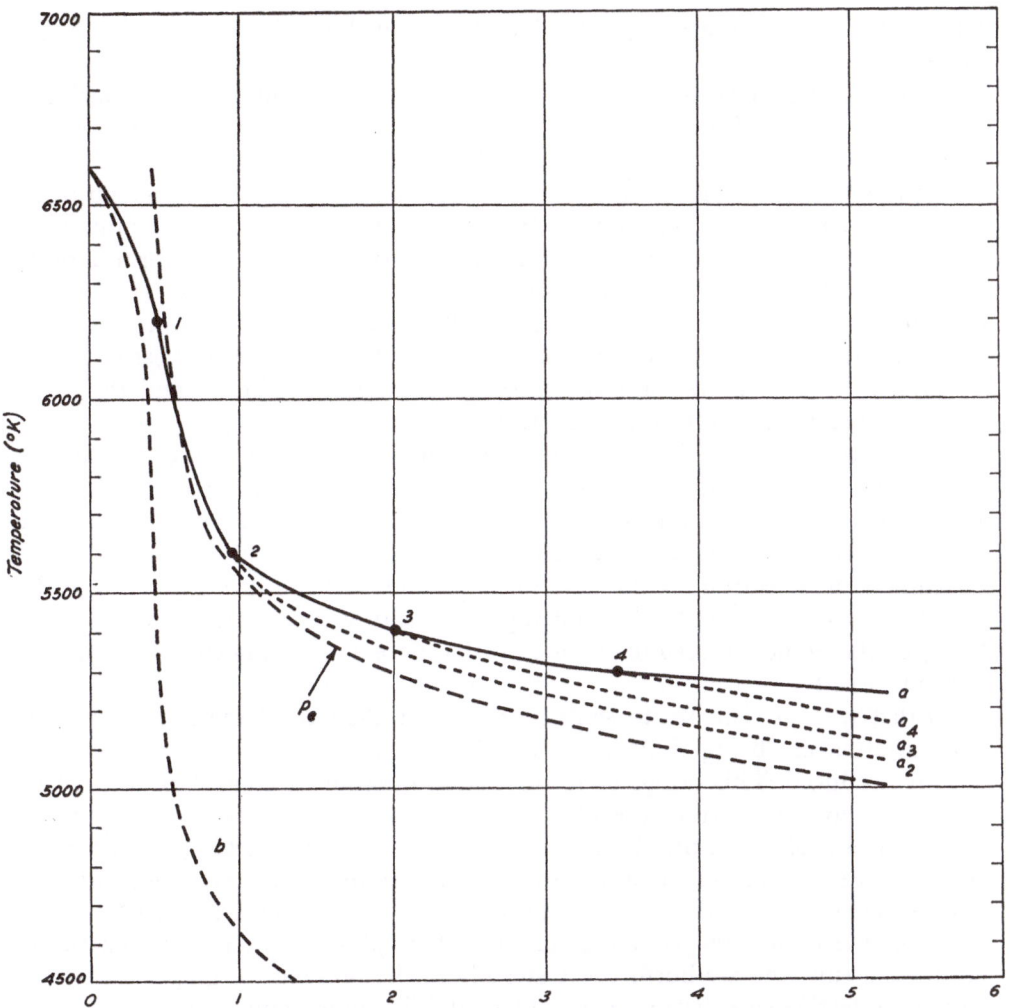

Fig. 7.23. An illustration of self-buffering and the buffering effect in an arc with lithium vapour. The broken curve b depicts the dependence of the temperature on the lithium vapour pressure if the electron pressure p_e were rigorously constant ($= 4.10^{-4}$ atm). Actually, a negative correlation between temperature and electron pressure is found in the lithium vapour arc (broken curve p_e). The rise of p_e with increasing lithium vapour pressure causes the temperature to drop less steeply (curve a). This self-buffering effect becomes very marked after an initial steep decline of the temperature to about $5600°$K.

The influence of an added component of low ionization potential ($V_{ij} = 4·5$ eV) upon the temperature differs according to the lithium vapour pressure. The buffer effect of lithium begins to appear at $p_{Li} \approx 10^{-3}$ atm, that is when $T = 5600°$K. Having this situation (point 2 on the continuous curve) and adding the easily-ionizable constituent in increasing concentration, we obtain the curve a_2 where the partial pressure now refers to the sum of the lithium vapour pressure and that of the extraneous element. Analogously, curves a_3 and a_4 are achieved if we commence adding at the points 3 and 4 respectively.

Obviously, those circumstances tend to be optimal for lithium as the buffer element.

There are other factors to be considered, however, in addition to the stability of the temperature. From our observations of the behaviour of electron pressure and background intensity in arcs with lithium vapour (see Figs. 7.21 and 7.22), we inferred that excitation conditions are favourable when the temperature is between 5700 and 5800°K, for then the background intensity is at a minimum. Thus the optimum temperatures for effective buffering (5500°K) and for minimum background intensity (5700–5800°K) do not coincide. The most satisfactory compromise should be examined empirically. Naturally, the compromise will also depend on the requirements of the analysis. To make the task easier it will be profitable to examine some more facets of the lithium buffer and the more practical side of the spectroscopic buffers, in general.

The preceding exposition was made in terms of partial pressure, which is a convenient concept for theoretical considerations, but is not satisfactory in practice. Besides the selection of the most suitable operating conditions purely from my own experience narrows the field of discussion too much. The essential changes of temperature, electron pressure, and background intensity brought about in the arc plasma by increasing the lithium vapour density will be generally observable; however, the numerical results obtained with different operating conditions do not coincide exactly.

So that we can visualize to some degree what changes of the lithium vapour concentration are being dealt with, we shall consider how these variations are usually produced. This brings us to the crucial question of controlling the volatilization of specimens from the electrode cavity (see especially §§ 10.2 to 10.4). The grade and shape of the electrodes, admixtures, and the packing of the sample all influence the rate of volatilization of the test materials. Here we consider only the effects achieved by diluting the specimen with graphite powder.*

Often in routine analysis an admixture of graphite powder is automatically used. As a rule, the arc burns more steadily in the presence of graphite powder, and also selective volatilization is less marked. The improvement of the excitation conditions is attributed to the enhanced electrical and thermal conductivity of the substance in the crater, to the assisted chemical reduction of oxides, and to the prevention of globule formation and fusion (Ricard, 1953a and b). If there is a fairly large excess of graphite, further advantages can be expected. Beintema and Kroonen (1955) point out that the graphite or the graphite–buffer mixture governs not only the flow of the vapours but also the rate of volatilization of the test sample, and so specific properties of the sample do not influence the entry of the constituent elements into the test volume.

A graphite admixture can be profitably combined with a spectroscopic buffer.

* Graphite powder is normally used in preference to carbon. Extensive comparison between graphite and carbon admixtures have not been made, as they have for the behaviour of graphite and carbon as material for the crater- and the counter-electrode (see § 10.3). Observations by Spindler (1961) suggest that carbon powder is not better anyway. Although it packs well and lengthens the volatilization time of samples, it sticks to all the tools used.

If the proportion by weight of graphite to buffer compound is chosen carefully, the rate of entry of the buffer element can be controlled, and so we can obtain the whole sequence of excitation conditions depicted by the continuous curve in Fig. 7.23. Instead of the partial pressure of the buffer element we have the ratio by weight of graphite to buffer compound.* In practice, it is not easy to produce a well-defined relation between the partial pressure and this ratio by weight. It may happen that in the initial stage of arcing much of the buffer vapour is released (see Fig. 7.15), so that, in the following period of more or less steady volatilization, the rate of entry of the buffer element is not closely correlated with the original ratio by weight of graphite to buffer substance. The conditions of the initial burning period are often decisive for the entire characteristic of the evaporation. For example, covering the electrode load with a small layer of graphite powder changes the picture considerably (see e.g. Beintema and Kroonen, 1955; and Beintema, 1957). It is related to the establishment of a definite temperature distribution in the electrode. The initial liberation of vapour can be suppressed to some extent by using pellets that fit tightly in the electrode cavity (Ordelman, Smit, and Tolk, 1965).

In summary, samples are effectively diluted by a mixture of a buffer substance (in the strictest sense) and graphite powder. Thus the rate of volatilization of both the sample and the buffer can be adjusted, while the excitation conditions can be made widely independent of the properties of the sample itself. Once the proper type of electrode has been chosen, we must attempt by varying the proportions by weight of sample, buffer compound, and graphite, to create the most favourable circumstances for the analysis. In the ideal case, the concentration of the buffer element in the discharge zone is constant throughout the arcing period so that all the constituents of the specimen are uniformly excited, irrespective of fractional distillation effects, which can be never entirely avoided.

Although numerous types of admixtures are employed in spectrochemical analysis, the use of a lithium compound has gained much popularity for routine methods, and justly so. The reduction of the temperature by the addition of lithium vapour to the carbon arc plasma is moderate, so that, considering the whole aggregate of spectral lines and elements, a fine compromise for the detection limits of the different elements is established. The proportion of sample to buffer should be kept fairly low, however, if a high alkali content is expected, and stability of excitation is crucial. Usually, the greater the stress on detection limits the less stringent the precision and accuracy, and vice versa.

Another advantage of lithium is the medium volatility of the element and its compounds; thus by adding the correct amount of graphite to the lithium salt, a sufficient amount of the buffer element can be kept in the arc until the sample has completely volatilized. Normally lithium is added as the carbonate. In the literature I found one case in which the fluoride was employed (Vorob'ev, 1961).

---

* Evidently, the chemical form of the buffer element is most important. For example, we report in § 7.14 the different excitation conditions for a LiF and a LiCl arc, which are due to the difference in the volatility of the fluoride and the chloride.

I observed a marked difference between the behaviour of lithium carbonate, on the one hand, and lithium chloride or fluoride, on the other (Boumans, 1961, 1962, 1962a). Buffered with the carbonate the arc tends to wander continuously on the outer walls of the supporting lower electrode (anode), whereas chloride or fluoride dilution of the sample results in a smoother burning in most instances.* In my experience, an admixture of lithium fluoride and graphite, for example, in the proportion of 1 to 4 if used with the electrode type of Fig. 7.14, has proved to be a very useful buffer.

I note that my object of using lithium fluoride in preference to the carbonate is primarily to secure a smooth entry of the buffer element; though *fluoridation* of the sample occurs, it is not especially intended here. Several authors aiming at the fluoridation of samples for promoting the volatilization of elements have added metal fluorides for the purpose; for instance, Brandenstein, Janda, and Schroll (1960) (see also Schroll, Brandenstein, Janda, and Rockenbauer, 1960) employed a 1 : 1 mixture of AlF_3 and NaF for determining traces of boron in graphite. By this means they were able to attain a detection limit for boron as low as 10^{-8} per cent, when applying a double-arc procedure. Kosheleva and Kuznetsova (1959) used CaF_2 and NaF for determining gallium, germanium, indium, and thallium in pyrites, cinders, and dust. Also of interest is the use of teflon powder (polymer of $CF_2 = CF_2$) proposed by Ginzburg, Glukhovetskaya, and Lerner (1963). In general, the possibility of using thermochemical reactions should be recognized. This highly important question, which is gaining interest, is studied systematically by Schroll (1963) and co-workers (see § 10.4).

Concluding this section, we mention Holdt's experimental study of the influence of buffers on the accuracy and sensitivity of spectrochemical results.† Holdt examined the buffer efficiency of lithium carbonate and copper oxide when added in increasing proportion to a test sample containing eighteen elements in matrices of either B_4C or NaCl. As a measure of buffer efficiency he used an adequate quantity defined as

$$\frac{1}{18} \sum_1^{18} |\log I_{Na} - \log I_B| = |\delta_M|$$

where I_{Na} and I_B are the intensities of spectral lines of the eighteen elements in NaCl and B_4C matrix respectively. Thus by the use of δ_M he obtained a good average for a large variety of elements. His results are remarkable and are, in general, in close accordance with the theoretical exposition given at the beginning of this section. Holdt's interpretation of the results remains vague and is not quite to the point, which does not reduce, however, the usefulness of his experimental results.

Finally, we point out the benefits of using an intensity ratio of an ion–atom line pair to indicate variations in excitation conditions. Examples are found in Holdt's

---

* This different behaviour is most probably related with the decomposition of the carbonate in the electrode cavity; halides are known to volatilize as the compound (cf. § 10.3).

† See Holdt (1962) and Holdt, Maritz, and Mark (1962). This study is continued by Maritz and Strasheim (1964a and b).

paper and e.g. in the paper by Beintema and Kroonen (1955). After the extensive discussion in the preceding sections it is needless to say that the variations of such an intensity ratio reflect fluctuations of both the temperature and the electron pressure. It seems worth while to consider this in greater detail (see § 7.13).

§ 7.13 *Sensitivity of atom–atom and ion–atom pair intensity ratios to variations in the temperature and the electron pressure*

The usefulness of an intensity ratio (I^+/I) of an ion–atom line pair as an indicator for changes in the excitation conditions has long been recognized. For example, Gerlach and Schweitzer (1930) enunciated the concept of fixation pairs for ascertaining the constancy of excitation conditions. Nowadays it is customary for spectrochemists to check excitation conditions by means of an ion–atom line pair. The preceding discussion (see particularly § 7.6) has made it clear that the intensity ratio of an ion–atom line pair is not solely determined by the temperature, but that it, in contrast with the intensity ratio of an atom–atom or an ion–ion pair (§ 6.5), depends also on the electron pressure. This must be born in mind when rigorous conclusions are drawn; for general considerations it is not very harmful to associate changes of I^+/I with variations of the temperature alone, as indeed such a ratio is very sensitive to temperature fluctuations.

As an example let us consider a plot of the temperature versus the log intensity ratio (ΔY_1) for the lines Mg II 2796 and Mg I 2780 (Fig. 7.24). The continuous curves depict equation (7.36), viz.

$$\Delta Y_1 = -\log p_e - 4 \cdot 91 \frac{5040}{T} + \frac{5}{2} \log T - 6 \cdot 68 \qquad (7.56)$$

for different parameter values of $\log p_e$. Thus we see what temperature changes can be inferred from variations of the log intensity ratio ΔY_1, provided that the electron pressure is constant. For instance, if $p_e = 10^{-3}$ atm, an increase of I^+/I by a factor of 3, say from 10 to 30, corresponds to a rise of the temperature from about 5300 to 5800°K. Actually, a variation of the temperature is often accompanied by some change of the electron pressure, so that the relationship between T and ΔY is depicted by a curve that intersects those of the ideal set of Fig. 7.24. For example, when the negative correlation between T and p_e, which I observed, (Fig. 7.6) is appropriate, the T–ΔY curve is represented by the dashed line in the figure. If, on the contrary, p_e rises with increasing temperature, a curve such as depicted by the dotted line expresses the relation between T and ΔY.

Figs. 7.25 and 7.26 show in detail the sub-effects of which the variations of ΔY for the ion–atom line pairs, Mg 2796/2780 and Mg 2796/2852, are composed. For the log intensity Y^+ of the ion line we write in virtue of (7.10):

$$Y^+ = \log K_{qp}^+ + \log n_j - \log Z_{ij} + \log \alpha_j - \frac{5040}{T} V_q^+ \qquad (7.57)$$

Similarly, for the atom line:

$$Y = \log K_{qp} + \log n_j - \log Z_{aj} + \log(1-\alpha_j) - \frac{5040}{T}V_q \qquad (7.58)$$

In these equations K_{qp} is the line constant defined by (7.11), n_j the total concentration of magnesium particles (atoms and ions) in the plasma, Z the partition function, α_j the degree of ionization of magnesium, and V_q the excitation potential.

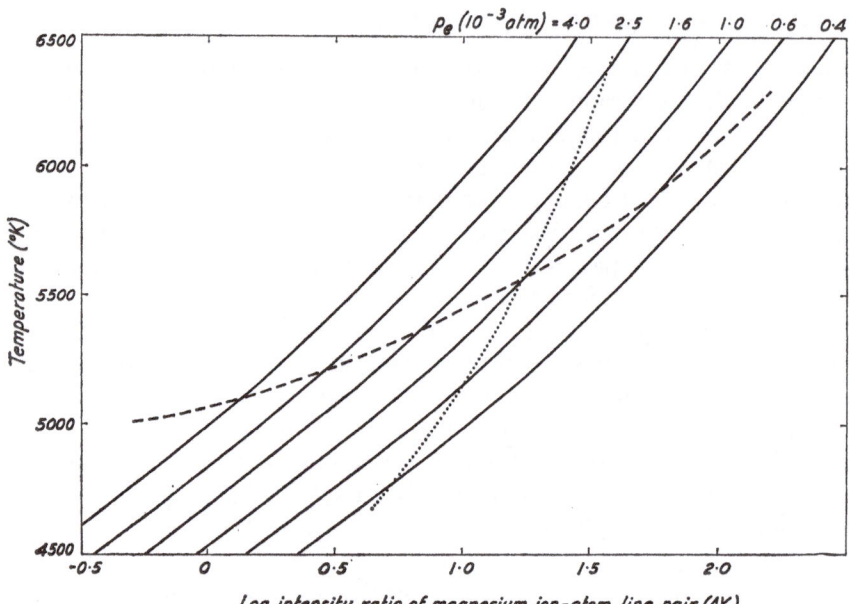

FIG. 7.24. Relationships between the temperature and the log intensity ratio (ΔY_1) of the ion–atom line pair Mg 2796/2780 for different values of the electron pressure (p_e). As a rule, a variation of the temperature is accompanied by some change in the electron pressure. When the negative correlation between T and p_e illustrated in Fig. 7.6 is appropriate, the relationship between T and ΔY_1 is represented by the broken curve. If a positive correlation between T and p_e holds true, a curve such as depicted by the dotted line applies.

Diagram a in Figs. 7.25 and 7.26 depicts how the log intensities of the ion and the atom line would behave as functions of the temperature if the Boltzmann factors solely were considered, in other words if the degree of ionization was assumed to be constant. In reality, the variation of the degree of ionization α_j with the temperature is a vital feature that cannot be ignored. To evaluate its influence, using equation (7.24), we must know how p_e behaves as a function of T. In Figs. 7.25 and 7.26 two cases are considered:

1 $p_e = $ constant $= 0.7 . 10^{-3}$ atm (diagrams c to e).

2 p_e follows the correlation shown in Fig. 7.6 (diagrams f to h).

The figures show in a self-explanatory way the behaviour of α_j, $(1-\alpha_j)$, Y^+, Y, and ΔY as functions of T. Clearly, the influence of the ionization is essential.

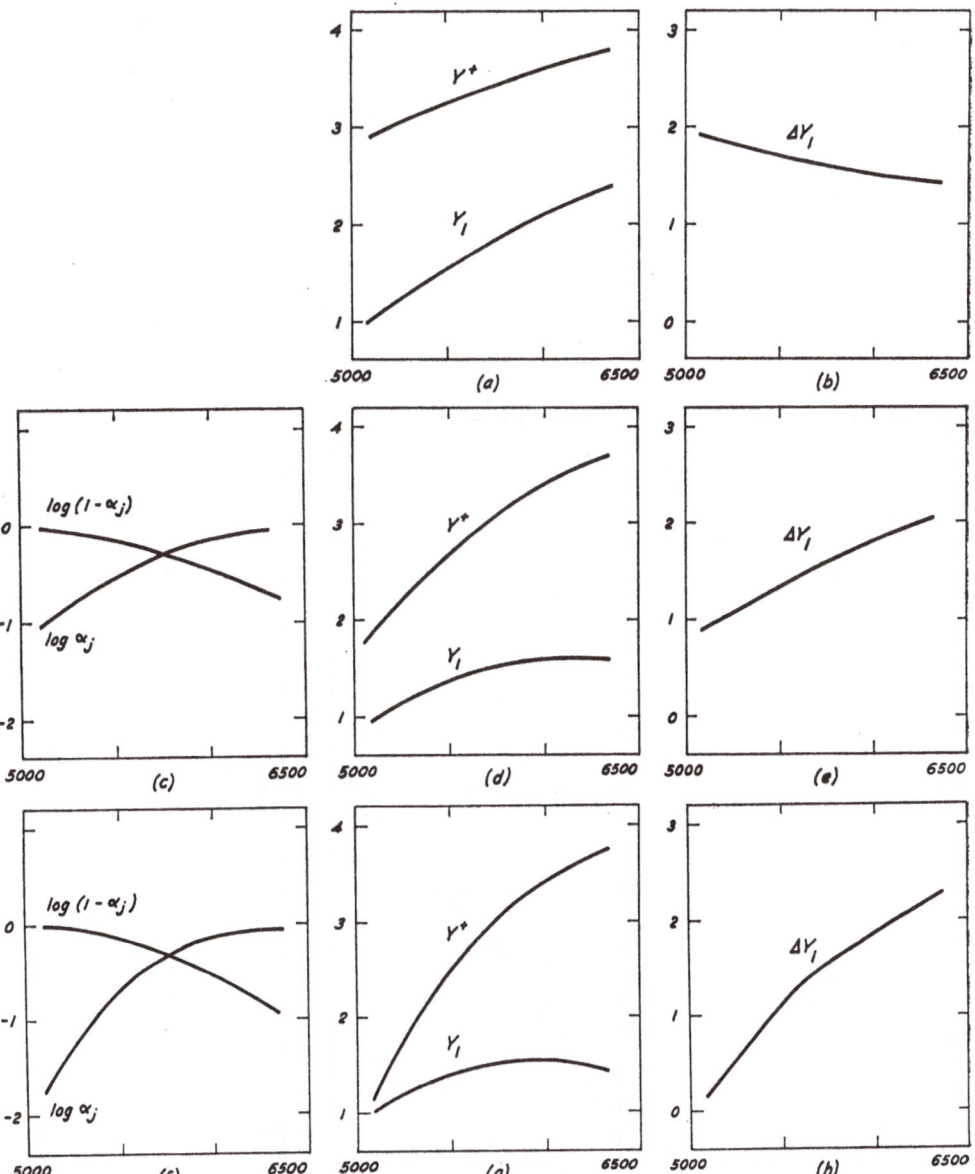

FIG. 7.25. Dependence of the log intensities Y^+ and Y_1 and of the log intensity ratio ΔY_1 of the lines Mg 2796 and Mg 2780 on the temperature.

(a) and (b) Degree of ionization (α_j) assumed to be constant.

(c) to (e) Degree of ionization varying only with the temperature, assuming the electron pressure to be constant ($= 0.7 . 10^{-3}$ atm).

(f) to (h) Degree of ionization varying with both the temperature and the electron pressure. The negative correlation between T and p_e according to Fig. 7.6 is assumed to be appropriate.

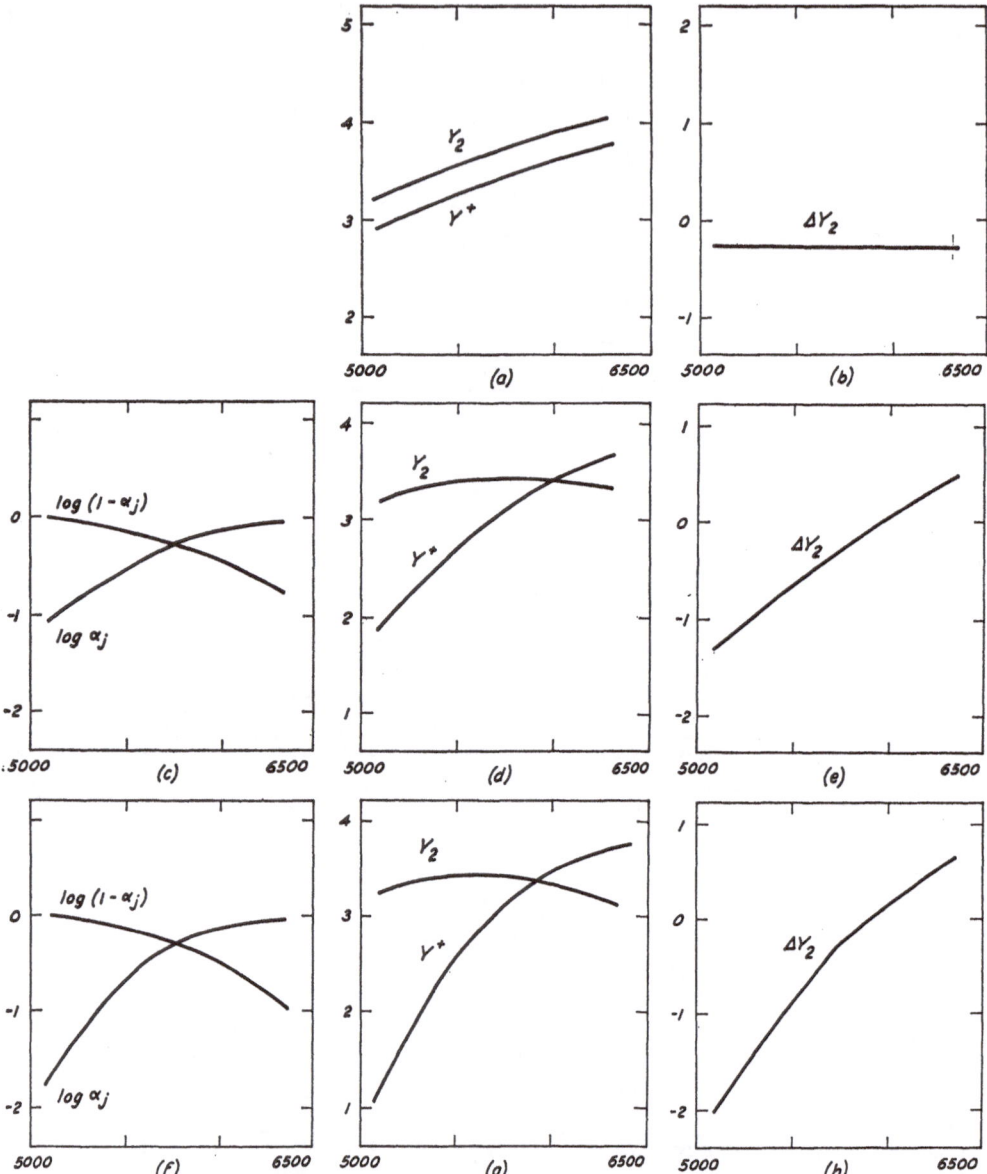

FIG. 7.26. Dependence of the log intensities Y^+ and Y_2 and of the log intensity ratio ΔY_2 of the lines Mg 2796 and Mg 2852 on the temperature.

    (a) and (b)   Degree of ionization (α_j) assumed to be constant.

    (c) to (e)   Degree of ionization varying only with the temperature, assuming the electron pressure to be constant (= $0 \cdot 7 . 10^{-3}$ atm).

    (f) to (h)   Degree of ionization varying with both the temperature and the electron pressure. The negative correlation between T and p_e according to Fig. 7.6 is assumed to be appropriate.

Let us now consider an atom–atom line pair and compare the sensitivity of its intensity ratio to temperature fluctuations with that of an ion–atom pair. By rearranging equation (6.22) we find for an atom–atom (or ion–ion) line pair

$$d\Delta Y = +\frac{5040}{T}(V_a - V_b)\frac{dT}{T} \tag{7.59}$$

where V_a and V_b are the excitation potentials of the lines labelled a and b.

TABLE 7.7

Spectrum	Wavelength (Å)	V_{ij}	V_q^+ or V_q	$V_{ij}+V_q^+-V_q+1\cdot08$
V II	3102·3	6·74	4·34	8·26
V I	3184*		3·9	
Ti II	3372·8	6·82	3·67	7·35
Ti I	2956·1		4·22	
Zr II	3496·2	6·84	3·57	7·98
Zr I	3519·6		3·51	
Mg II	2795·5	7·64	4·41	8·80
Mg I	2852·1		4·33	
Mg II	2795·5	7·64	4·41	5·99
Mg I	2779·8		7·14	
Ca II	3933·7	6·11	3·14	7·41
Ca I	4226·8		2·92	
Sr II	4077·7	5·69	3·03	7·12
Sr I	4607·3		2·68	
Ba II	4554·0	5·21	2·71	6·77
Ba I	5535·5		2·23	

* Any of the multiplet components 3183.41, 3183.96+3183.98, or 3185.40 Å can be used.

Characteristic values for judging the effect of temperature variations upon the intensity ratios of ion–atom line pairs. The table lists wavelengths and excitation potentials (V_q^+ or V_q) of lines, the ionization potential of the element (V_{ij}) and the quantity ($V_{ij}+V_q^+-V_q+1\cdot08$) occurring in equation (7.60).

From (7.33) when putting $\log I^+/I = \Delta Y^+$ and differentiating,* we derive for an ion–atom line pair

$$d\Delta Y^+ = -d\log p_e - \frac{5040}{T}(V_{ij}+V_q^+ - V_q+1\cdot08)\frac{dT}{T} \tag{7.60}$$

where V_{ij} is the ionization potential of the element, V^+ the excitation potential of the ion line, and V_q that of the atom line.

* Strictly speaking, this result ought to be derived from the intermediary of expansion in a Taylor series (cf. § 7.5); since we confine ourselves to small variations, higher terms of the Taylor series can be omitted so that the rigorous treatment leads to the same formula as the simplified derivation.

8

For comparing the sensitivity to temperature changes of ΔY with that of ΔY^+ we thus have to compare the numerical values of the factors in brackets, $(V_a - V_b)$ and $(V_{ij} + V^+ - V_q + 1 \cdot 08)$ respectively. When choosing examples to make this comparison we must consider also that for atom–atom or ion–atom line pairs to be useful for determining the excitation conditions, their intensities should not be too different, neither should be their wavelengths. Numerical data for ion–atom line pairs that fulfil these requirements reasonably in the range of temperature considered here have been summarized in Table 7.7. Accordingly, the factor $(V_{ij} + V^+ - V_q + 1 \cdot 08)$ is seen to vary from 6 to 9 eV. Useful atom–atom line pairs having a difference in excitation potential of this magnitude are not known; for the particularly useful pairs Zn 3076/3072 and Zn 3076/3282, the difference $(V_a - V_b)$ is only about 4 eV (see §§ 6.5–6.6); for suitable copper lines, $(V_a - V_b)$ amounts to about 2 eV only (see § 6.9).

So, we conclude that for useful ion–atom line pairs the sensitivity of their intensity ratio to temperature changes is substantially higher than for useful atom–atom line pairs.

★ § 7.14   *Use of the total voltage drop of an arc for ascertaining the excitation conditions*

In connection with our discussion on the electric field strength and the total potential drop in a d.c. carbon arc (§§ 3.2 and 3.3) I give here some results of voltage measurements that I obtained when investigating the excitation conditions in metal vapour arcs described in §§ 7.7 to 7.10. When taking racking-plate spectra for determining the temperature and the electron pressure during the evaporation of different metallic compounds from the furnace electrode, we recorded the gap voltage with an electronic strip chart recorder.

To arrive at some correlation between the voltage and an excitation parameter, it is required that 'trivial actions' upon the voltage, such as variations of the current strength and the arc length, are suppressed. This condition was fulfilled sufficiently well in the experiment at present under discussion, although the recorded voltage-time curves were not entirely free from a slight influence of current fluctuations.

Typical examples of voltage-time records are shown in Fig. 7.27. According to the matrix we obtained curves with large or small oscillations (types a and b and c to f respectively) that succeeded each other rapidly (a to d) or slowly (e and f). Moreover, the mean voltage exhibited either a good constancy with time (a, c and e) or varied gradually (b, d and f). Often, for b, d and f, it was not possible to show that the temperature in the arc varied systematically with the voltage, although the former was measured with reasonable precision so that actual changes were not likely to have been caused by random errors (cf. § 3.3). In spite of the fairly large intervals that were frequently covered by the voltage at one mean temperature, there existed a well-defined correlation between the median values of the voltage and the temperature, as shown in Fig. 7.28. We have indicated in the diagram the range of the voltage for each point; for comparison, the confidence limits of the mean temperatures have been also given (5 per cent level of significance).

From the correlation between the arc voltage and the arc temperature given in Fig. 7.28, it might appear that the gap voltage presents itself as a fine parameter for measuring the temperature. Yet, we should not be too optimistic; particularly we must avoid correlating fairly small changes of the voltage, say ± 5 V, with temperature variations. Only the overall level of the potential drop is indicative, in my experience, of the level of the temperature. Besides, in the low temperature range, represented by the alkali metal arcs, the voltage is rather insensitive to temperature changes (see Fig. 7.28), while it fails almost completely to disclose even large variations in the electron pressure (see below).

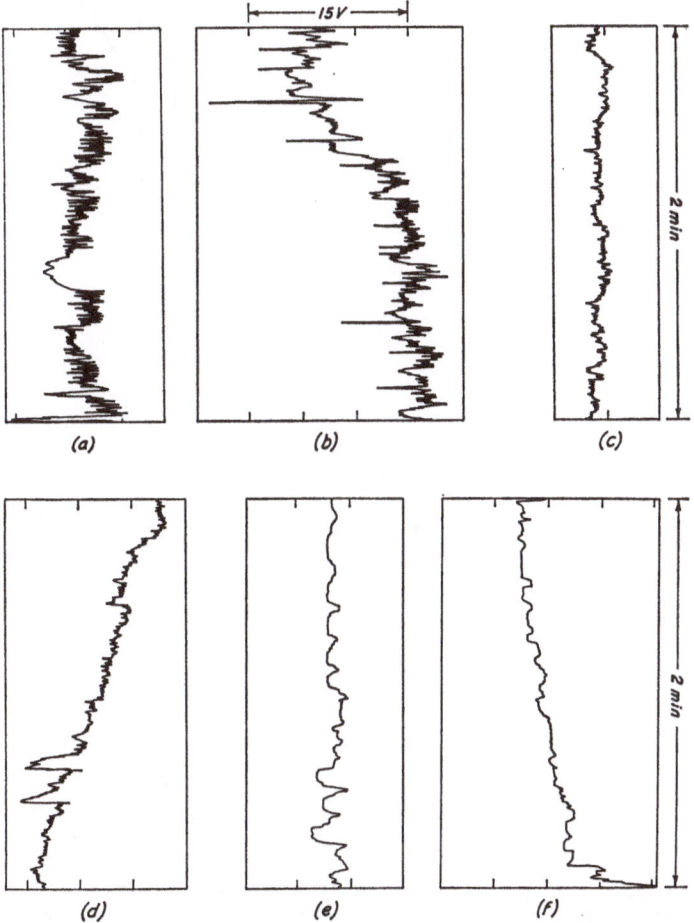

FIG. 7.27. Typical examples of voltage-time records, of which there are several types depending on the matrix used. Types *a* and *b* have large and rapid oscillations, types *c* and *d* small and rapid oscillations, and types *e* and *f* small and slow oscillations (alkali metal vapour arcs). In addition, for *a*, *c*, and *e*, the mean value of the voltage exhibits a good constancy with time, whereas for *b*, *d*, and *f*, the average voltage changes regularly with time.

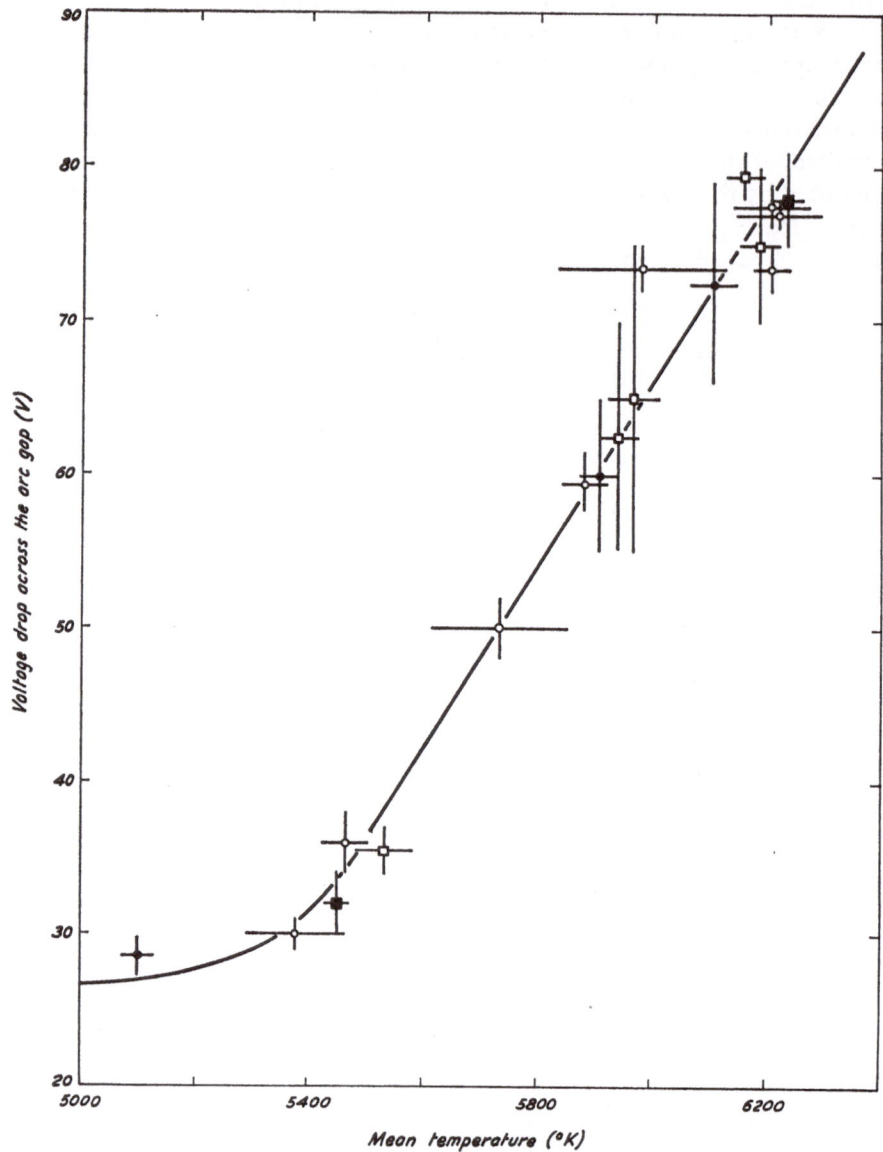

FIG. 7.28. Plot of median values of the gap voltage ($\overline{V}$) against the mean temperature ($\overline{T}$) observed in graphite arcs with different metal vapours. The vertical lines indicate the range covered by the voltage during the time that racking-plate spectra were taken for measuring the temperature. The individual values of the temperature were scattered at random about the mean values produced in the diagram; those pertaining to the same mean value did not show a correlation with the voltage. The precision of the mean temperatures is seen from the horizontal lines in the figure; they represent the confidence limits of the means at the 5 per cent level of significance.

For the reader to obtain some idea about the changes of arc voltage, temperature, electron pressure, and related quantities during the evaporation of different matrices from furnace electrodes, I give in Figs. 7.29 to 7.34 and Plates 1 to 4 detailed information on the variations of the excitation conditions in a graphite arc with either KCl, LiCl, LiF, or graphite as the matrix. In taking the spectra, the operating conditions were the same as in §§ 7.7 and 6.7. Accordingly, the matrices were volatilized from furnace electrodes, at a constant current of 10 amp and a gap width of 8 mm. The arc was stabilized by a graphite ring at the cathode. Zinc and magnesium were evaporated along with the matrices so as to introduce a thermometric and a manometric element.

Figs. 7.29 to 7.34 show in a self-explanatory way the changes of the excitation conditions that were observed. The nature of these changes and their interrelations have been discussed at length in §§ 7.7 to 7.10, and, therefore, a few general remarks suffice here.

The levels of voltage and temperature are seen to correlate well. In the temperature plots fairly large fluctuations seem to occur. This scattering is chiefly due to measuring errors; for, the plots of the logarithmic intensity ratio (ΔY) of the magnesium ion–atom line pair, those of the logarithmic intensities of the lithium (Y_{Li}) and potassium (Y_{K}) lines, and those of the log background intensity (Y_u) do not reflect such fluctuations as the temperature graph. On the other hand, the fluctuations in the temperature diagrams show clearly in the plots of the electron pressure; this is to be expected, as the values computed for the electron pressure (p_e) depend largely on the temperature values, inserted into the pertinent equation (see § 7.6). Despite error amplification in the resulting p_e values, the diagrams plainly demonstrate a substantial rise of p_e in the KCl and the LiCl arc, which is not associated with pronounced temperature variations. This increase of p_e is not indicated by the arc voltage.

There is no need to point out the obvious similarities between plots of ΔY, Y_{Li} (or Y_{K}), and Y_u. The relations between these quantities have been thoroughly discussed in §§ 7.9 and 7.10. Note also that results for the first stage of arcing tend to deviate, clearly because it takes some time for the distillation of the matrix from the furnace electrode to set in, so that in the initial stage the excitation conditions change gradually from those prevailing in the pure graphite arc to those of the arc with the relevant metallic constituent.

The excitation conditions for the four arcs are compared in Figs. 7.33 and 7.34, where the plots of T, $\log p_e$, ΔY, Y_{Li} (or Y_{K}), and Y_u have been assembled to provide an overall picture. Plates 1 to 4 give an idea of the appearance of the spectra when inspected visually. Particularly, the overall level of the background intensity, as well as the intensities of the lithium and potassium lines are clearly revealed. The difference in background character—continuous or discontinuous—is not easily illustrated in reproductions.

It is, with reservations, that I recommend the use of voltage-time records for ascertaining the excitation conditions. Therefore, too much emphasis must not be put on the detailed quantitative interpretation of voltage variations. The voltage

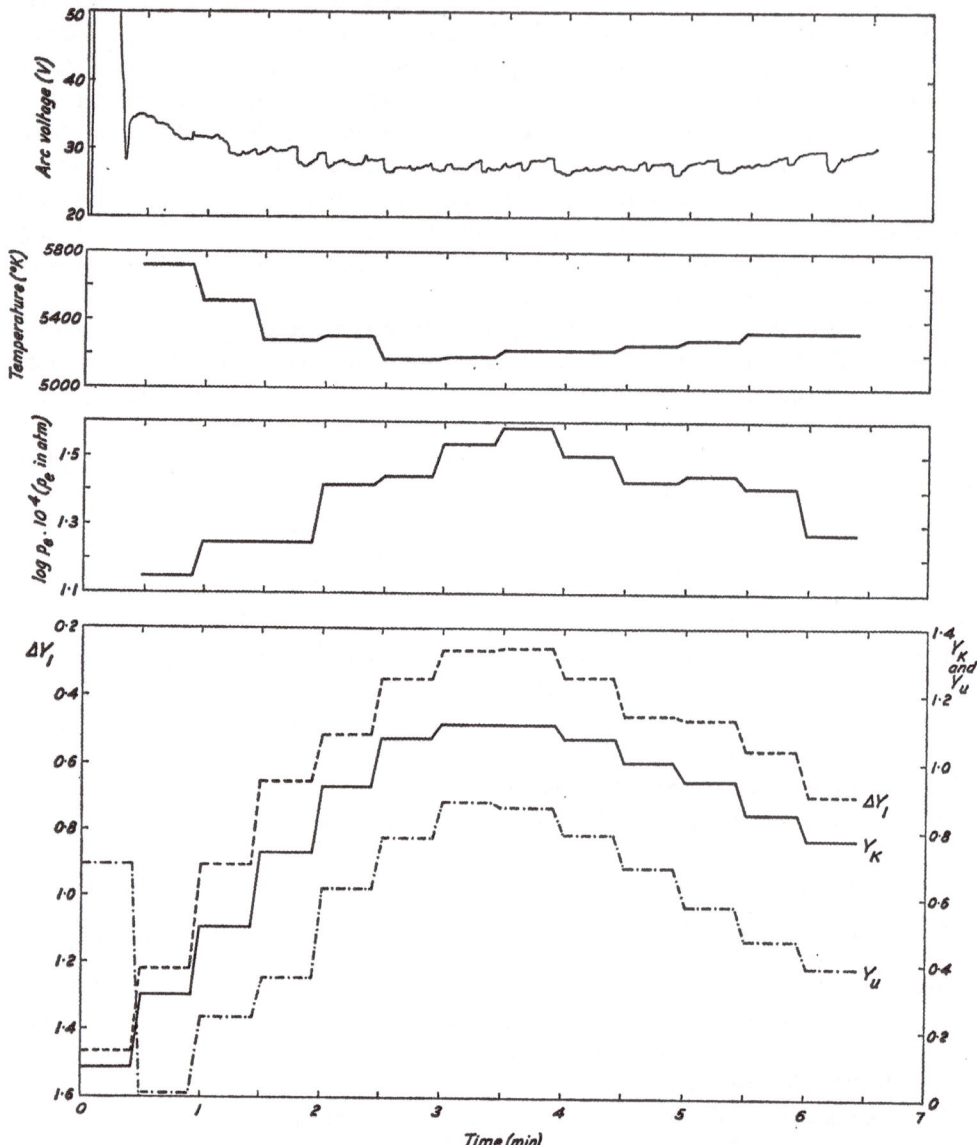

FIG. 7.29. Observed changes of the excitation conditions during the volatilization of KCl from a furnace electrode (Fig. 6.9).

The diagrams show successively the variations of the arc voltage, the temperature, the log electron pressure, and the log intensity ratio (ΔY_1) of the ion–atom
line pair Mg 2796/2780, the log intensity (Y_K) of the line K 3217, and the
log background intensity (Y_u) in the vicinity of 3280 Å. Intensity scales of the
potassium line and of the background are not comparable.

The exposure time of the racking-plate spectra was 25 sec. The spectrograph
slit was shut for 5 sec during camera positioning between consecutive exposures.
See Plate 1 for reproduction of spectra.

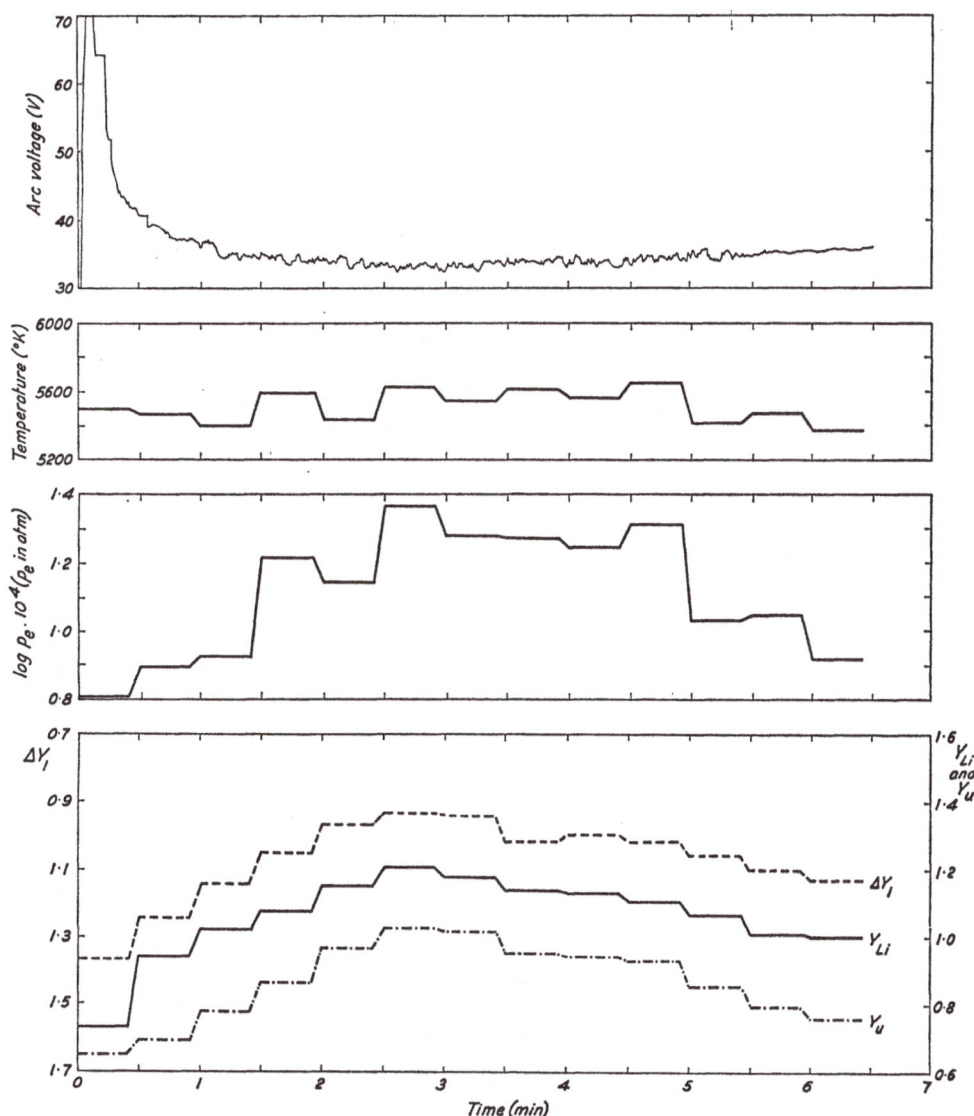

FIG. 7.30. Observed changes of the excitation conditions during the volatilization of LiCl from a furnace electrode (Fig. 6.9).

The diagrams show successively the variations of the arc voltage, the temperature, the log electron pressure, and the log intensity ratio (ΔY_1) of the
ion–atom line pair Mg 2796/2780, the log intensity (Y_{Li}) of the line Li 2741,
and the log background intensity (Y_u) in the vicinity of 3280 Å. Intensity
scales of the lithium line and of the background are not comparable; the scales
of Y_{Li} and Y_u compare with those of Fig. 7.31, however.

The exposure time of the racking-plate spectra was 25 sec. The spectrograph
slit was shut for 5 sec during camera positioning between consecutive exposures.
See Plate 2 for reproduction of spectra.

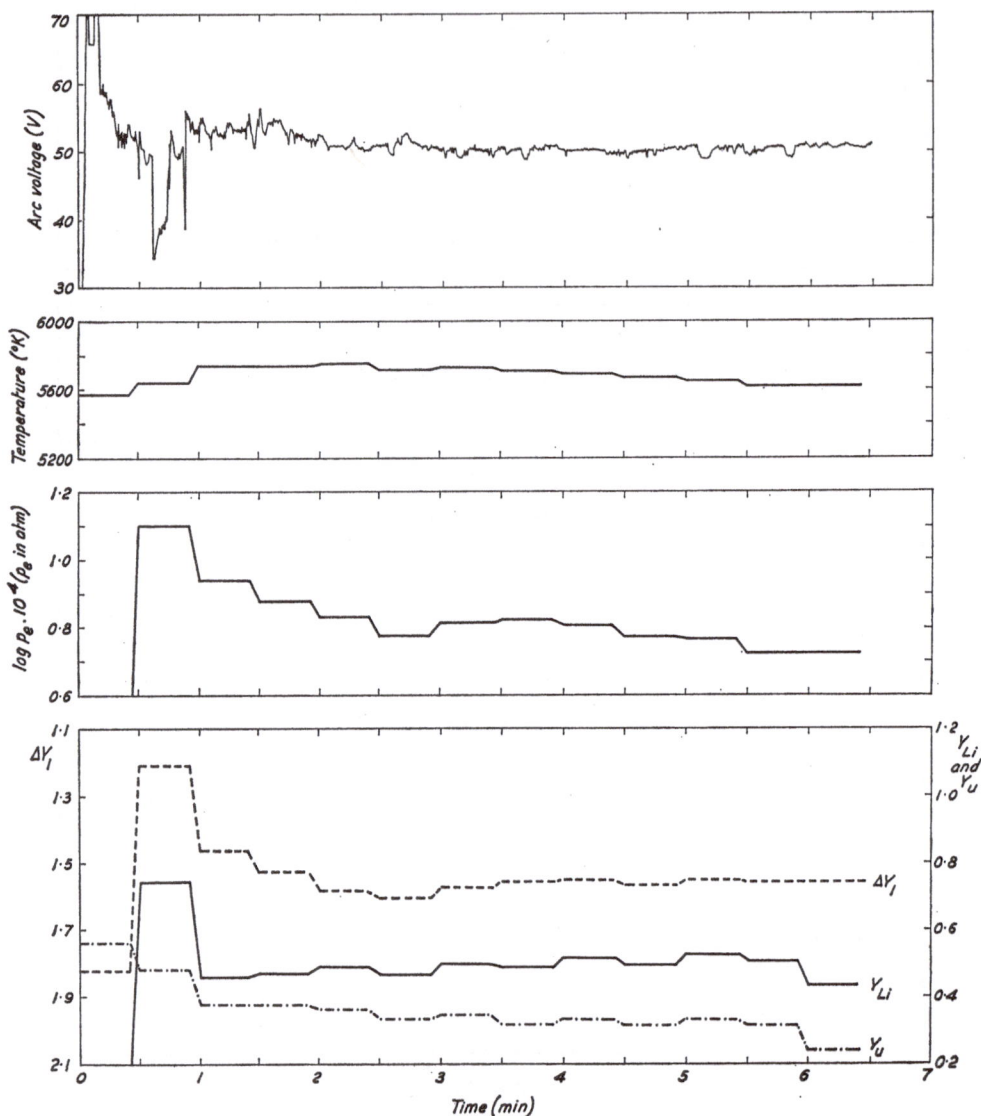

FIG. 7.31. Observed changes of the excitation conditions during the volatilization of LiF from a furnace electrode (Fig. 6.9).

The diagrams show successively the variations of the arc voltage, the temperature, the log electron pressure, and the log intensity ratio (ΔY_1) of the ion–atom line pair Mg 2796/2780, the log intensity (Y_{Li}) of the line Li 2741, and the log background intensity (Y_u) in the vicinity of 3280 Å. Intensity scales of the lithium line and of the background are not comparable; the scales of Y_{Li} and Y_u compare with those of Fig. 7.30, however.

The exposure time of the racking-plate spectra was 25 sec. The spectrograph slit was shut for 5 sec during camera positioning between consecutive exposures. See Plate 3 for reproduction of spectra.

may be a fine criterion for the preliminary examination of the behaviour of different specimens, matrices, buffers, and diluents. Thus, highly valuable information can be gained without even taking spectra. The safest guide in interpreting the voltage-time records is the worker's experience, based upon comparisons of racking-plate spectra with voltage-time records taken simultaneously. This will teach him to rate the latter at their true worth. In selecting a proper buffer admixture (see § 7.12) we may use as a criterion the constancy of the voltage level during the entire burning period as well as its invariance from the one sample to another.

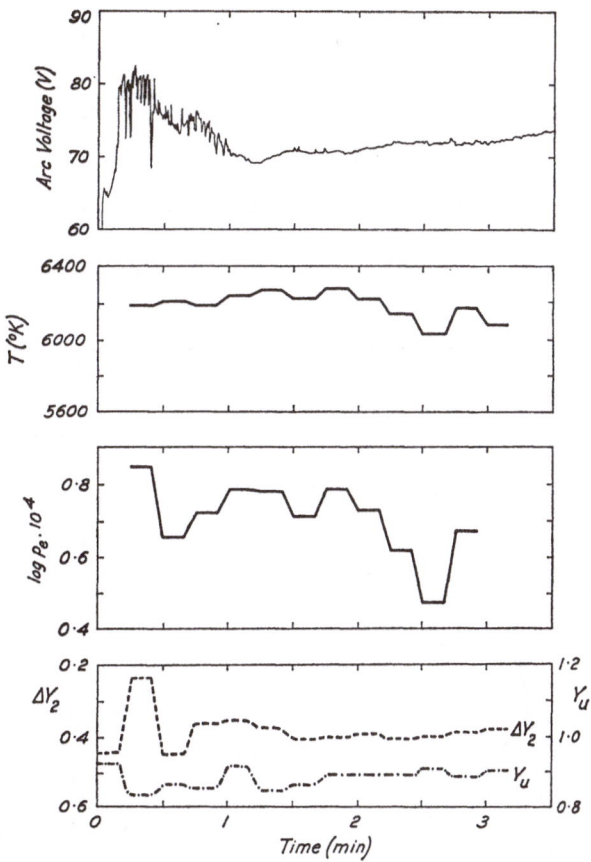

FIG. 7.32. Observed changes of the excitation conditions in an arc between graphite electrodes with only the thermometric and manometric species (zinc and magnesium) present in small quantities.

The diagrams show successively the variations of the arc voltage, the temperature, the log electron pressure, and the log intensity ratio (ΔY_2) of the ion-atom line pair Mg 2796/2852 and the log background intensity (Y_u) in the vicinity of 3280 Å.

The exposure time of the racking-plate spectra was 10 sec. The spectrograph slit was shut for 5 sec during camera positioning between consecutive exposures. See Plate 4 for reproduction of spectra.

Note, however, that this constancy does not guarantee that variations of the excitation conditions are fully excluded. Conversely, a gradual change of the voltage, say over 10 V, does not necessarily mean that the excitation conditions will vary markedly. Such variations of the voltage are primarily due to changes of the anode and cathode falls, not of the field strength (see § 3.2), which explains why variations of the temperature in the arc column often cannot be demonstrated to

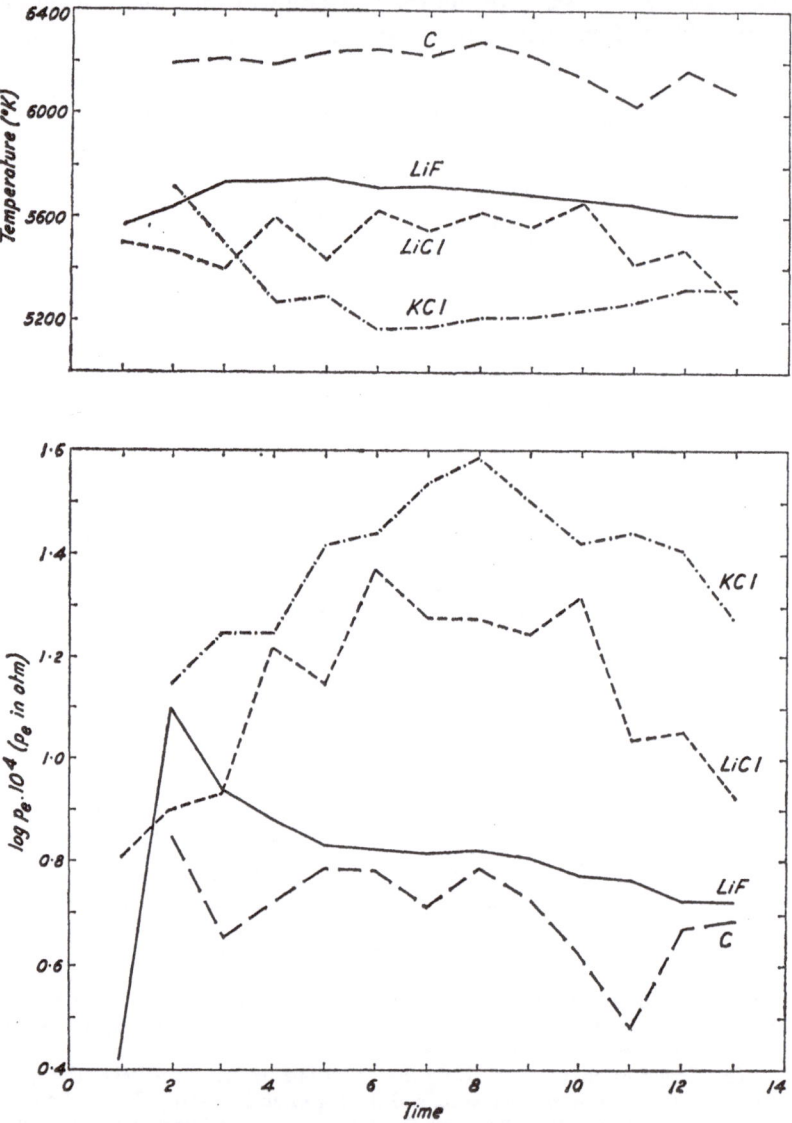

Fig. 7.33. Comparisons of the temperature and the electron pressure in the four arcs considered in Figs. 7.29 to 7.32, viz. arcs of KCl, LiCl, LiF, and pure graphite. The unit of time on the abscissa is 30 sec for all the arcs except graphite for which it is 15 sec.

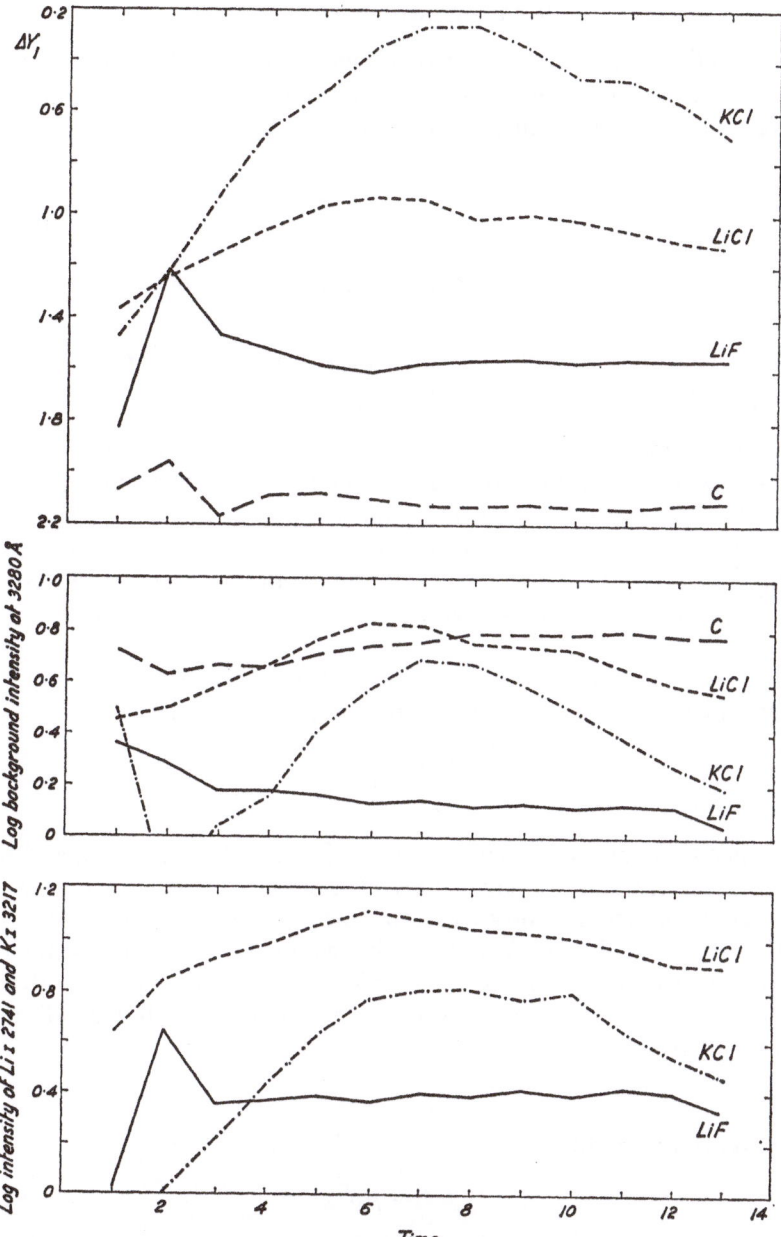

FIG. 7.34. Comparisons for the four arcs considered in Figs. 7.29 to 7.32, viz. arcs of KCl, LiCl, LiF, and pure graphite. The graphs represent (1) ΔY_1, the log intensity ratio of Mg II 2796 and Mg I 2780, (2) Y_u, the log background intensity of 3280 Å, (3) Y_K and Y_{Li}, the log intensities of the lines K 3217 and Li 2741.

For the graphite arc, ΔY_1 values have been calculated from measured values of ΔY_2 and T [equation (7.39)]. The unit of time on the abscissa is 30 sec for all the arcs except graphite for which it is 15 sec. Background intensity refers to equal exposure times in all cases.

accompany well-defined changes of the voltage. Anyhow, when the voltage covers a large range during the evaporation of a sample or when it varies erratically, we should have no illusions about the smoothness of the excitation.

Remarks on the use of the voltage when deciding upon a buffer are given by Doerffel and Geyer (1964). Their study continues the work of Hegemann and Schöntag (1953); it is of empirical character and states a few observations, which can be readily interpreted when we consider the theory brought out in the preceding sections.

§ 7.15 *Ionization of a single gas compared with that of a composite one*
Total degree of ionization of a gas mixture

After our treatment of the ionization of the composite arc gas in the preceding sections it appears superfluous to deal with the ionization of a single gas. However, for two reasons we must consider this subject. Firstly, the concept of the effective ionization energy of a composite gas fits in with the representation of a gas mixture as a single gas (see § 7.16); secondly, a misconception, which considers components of the arc plasma to be identical with single gases, has been introduced into the literature.

Recalling our discussion on the fundamental concepts of ionization in §§ 7.1 and 7.2, we consider the Saha equation for partial pressures, viz.

$$\frac{p_{ij}p_e}{p_{aj}} = S_{pj} \qquad (7.61)$$

where S_{pj} is a function of the temperature defined by (7.17), p_a, p_i, and p_e are the partial pressures of neutral atoms, singly-charged ions, and electrons respectively and j is a label that denotes the species to which the quantities pertain.

Equation (7.61) applies *both* to a single gas and to separate components of a composite gas; clearly, so far no distinctions need to be made. Differentiation begins, however, as soon as we bring up the quasi-neutrality condition. For a single gas we have

$$p_e = p_i \qquad (7.62)$$

Hence (7.61) can be reduced to

$$p_e^2 = p_a S_p \qquad (7.63)$$

where the index j is omitted to show that the equations refer to a single gas only.

For a composite gas we have

$$p_e = \Sigma p_{ij} \qquad (7.64)$$

and

$$p_e^2 = \Sigma p_{aj} S_{pj} \qquad (7.65)$$

[cf. (7.47)].

Even if only one component of a gas mixture ionizes perceptibly, there is a substantial difference between the ionization of that substance when it is present alone at pressure P or when it is a constituent of a composite gas of total pressure P. For a single gas we compute the electron pressure from (7.63) to be

$$p_e = -S_p + \sqrt{(S_p{}^2 + PS_p)} \tag{7.66}$$

if we remember that $p_a = P - 2p_e$.

Fig. 7.35 presents p_e for a single gas of apparent ionization potential $\bar{V}_i = 6.5$ eV and total pressure $P = 1$ atm as a function of temperature. For comparison the figure also shows the p_e curve calculated in § 7.8 for the two-component gas

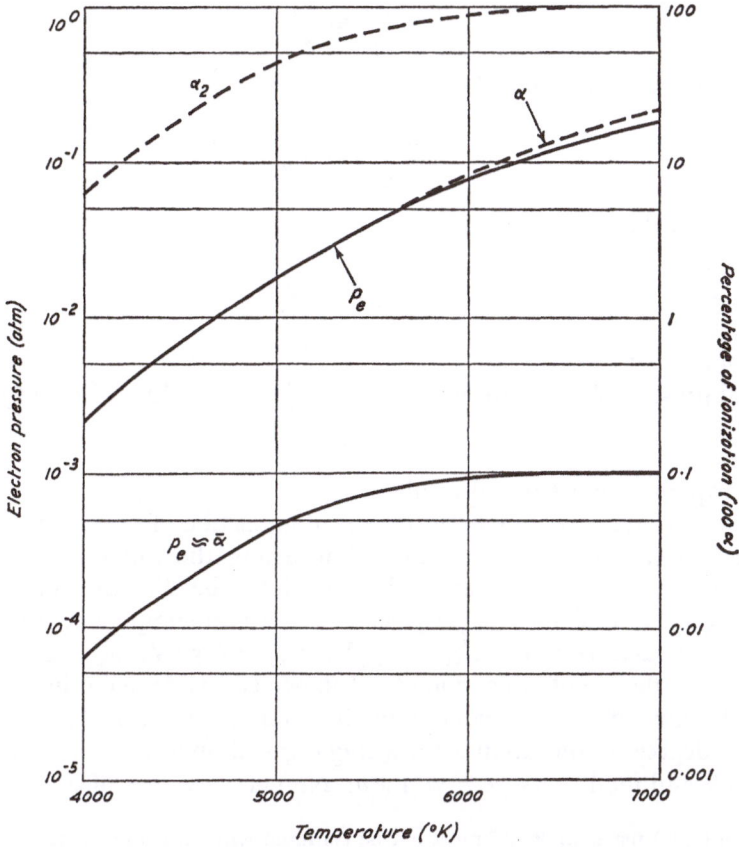

FIG. 7.35. Plot of the electron pressure (p_e) and the degree of ionization (α) for a single gas with apparent ionization potential $\bar{V}_i = 6.5$ eV and pressure $P = 1$ atm. For comparison the figure shows the electron pressure (p_e) or the total degree of ionization ($\bar{\alpha}$) for a two-component gas of total pressure $P = 1$ atm consisting of a major constituent with apparent ionization potential $\bar{V}_{i1} = 14$ eV and a minor one with $\bar{V}_{i2} = 6.5$ eV, the latter having a partial pressure $p_2 = 10^{-3}$ atm. The degree of ionization (α_2) of component two is also given.

characterized by the apparent ionization potentials $V_{i1} = 14$ and $V_{i2} = 6 \cdot 5$ eV, by the partial pressure $p_2 = 10^{-3}$ atm of the second component, and by the total pressure $P = 1$ atm.

A more important aspect of the subject is the degree of ionization. Remembering the Saha equation containing the degree of ionization α_j [equation (7.12n)], we write for a single gas

$$\frac{\alpha}{1-\alpha} p_e = S_p \tag{7.67}$$

where

$$\alpha = \frac{p_i}{p} \equiv \frac{p_e}{p} \tag{7.68}$$

Equation (7.67) can be reduced to

$$\frac{\alpha^2}{1-\alpha} p = S_p \tag{7.69}$$

which is put in the form

$$\frac{\alpha^2}{1-\alpha^2} P = S_p \tag{7.70}$$

if we consider that $P = p + p_e$.*

For total pressure $P = 1$ atm we have for numerical values of α and p_e:

$$\alpha \approx p_e \tag{7.71}$$

a well-known and almost trivial result.

A misconception is to compute the degree of ionization of a component of the arc gas using (7.70), putting $P = 1$ atm and inserting the numerical value of the ionization energy of the relevant substance into the S_p function. Obviously, (7.12n) should be used for the purpose. For the two-component gas considered above and characterized by $V_{i1} = 14$ eV, $V_{i2} = 6 \cdot 5$ eV, $p_2 = 10^{-3}$ atm, and $P = 1$ atm, the degree of ionization of substance two is depicted in Fig. 7.35 as a function of temperature. This curve must be compared with the broken one representing the degree of ionization α for a single gas of apparent ionization potential $V_i = 6 \cdot 5$ eV and total pressure $P = 1$ atm. Numerical values of α_2 and α are seen

---

* Equation (7.70) for a single gas must not be confused with an approximate expression for a composite gas that applies if one constituent, say the one labelled l, can be regarded to carry the discharge. The quasi-neutrality condition then reads

$$p_e \approx p_{il} = \alpha_l p_l \tag{1}$$

so that equation (7.12n) for constituent l can be reduced to

$$\frac{\alpha_l^2}{1-\alpha_l} p_l \approx S_{pl} \tag{2}$$

which contains the partial pressure p_l instead of the total pressure P.

to differ by more than an order of magnitude at low temperature, whereas both approach unity at a high temperature. The serious discrepancy is also evident in Fig. 7.36, where the corresponding plots of $(1 - \alpha_2)$ and $(1 - \alpha)$ are shown.

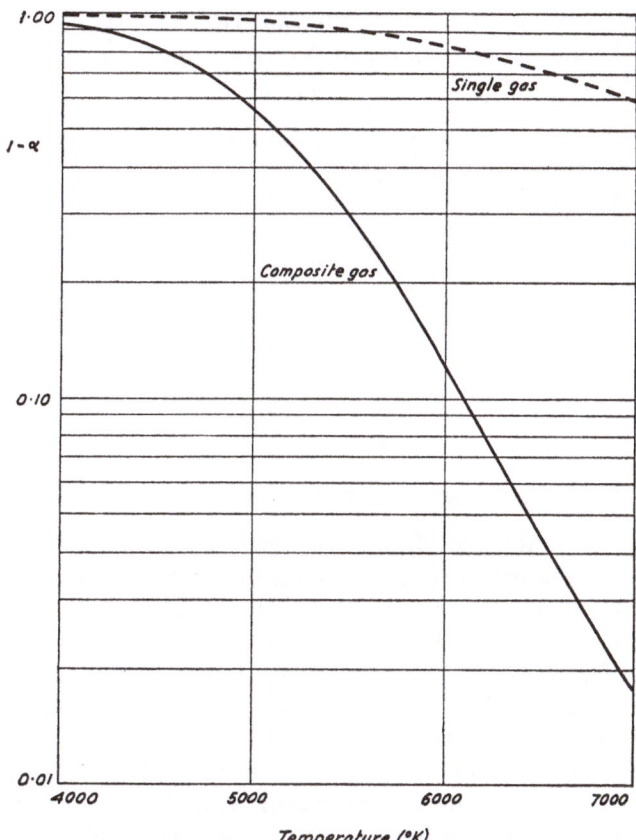

FIG. 7.36. Comparison between $(1 - \alpha)$ for a single gas and $(1 - \alpha_2)$ for a minor constituent of a composite gas, viz. the two-component gas defined in Fig. 7.35.

The quantity $(1 - \alpha_2)$ represents the proportion of neutral atoms of the minor constituent with respect to the total amount (sum of atoms and ions) of this component. The quantity $(1 - \alpha)$ represents this proportion for when the gas denoted as component two is present *alone* at a pressure of 1 atm.

Finally we mention a few formulae involving the total degree of ionization of a composite gas. It is defined as

$$\bar{\alpha} = \frac{\Sigma p_{ij}}{\Sigma p_j} = \frac{p_e}{\Sigma p_j} \tag{7.72}$$

Since $\Sigma p_j = P - p_e$, equation (7.72) can be written as

$$\bar{\alpha} = \frac{p_e}{P - p_e} \tag{7.73}$$

or as

$$p_e = \frac{\bar{\alpha} P}{1 + \bar{\alpha}} \qquad (7.74)$$

In the low-current carbon arc in air at atmospheric pressure, p_e was seen to be of the order of 10^{-3} atm. Hence we have in very good approximation

$$p_e \approx \bar{\alpha} P \qquad (7.75)$$

The use of total degree of ionization $\bar{\alpha}$ in ionization equations, which is preferred by some authors, is thus equivalent to that of electron pressure p_e. In terms of $\bar{\alpha}$, equation (7.12n) becomes

$$\frac{\alpha_j}{1 - \alpha_j} \frac{\bar{\alpha}}{1 + \bar{\alpha}} P = S_{pj} \qquad (7.76)$$

or approximately

$$\frac{\alpha_j}{1 - \alpha_j} \bar{\alpha} P = S_{pj} \qquad (7.77)$$

§ 7.16   *The concept of effective ionization energy*

For discussing the influence of easily-ionized substances upon the excitation conditions some authors prefer to use the concept of effective ionization energy of the arc plasma.* In physical arc theory (§ 4.1) the use of this concept is often convenient. For spectrochemical physics, however, I do not see any advantage in applying it; I prefer to discuss the excitation conditions in terms of temperature and electron pressure, as I have done in §§ 7.7 to 7.10. The concept of effective ionization energy has spread in the literature and therefore I shall show the formal identity of discussing excitation conditions in terms of effective ionization energy and electron pressure or in terms of temperature and electron pressure.

The effective ionization energy of a composite gas of given temperature is defined to be the ionization energy which a single gas ought to have for it to be ionized at this temperature to the same extent as the composite gas (cf. Huldt, 1948b and c).   According to this definition and in view of (7.70) and (7.19) we have

$$\log \bar{S}_p \equiv \log \frac{\bar{\alpha}^2}{1 - \bar{\alpha}^2} P = \frac{5}{2} \log T - \frac{5040}{T} V_{\text{ieff}} - 6.18 \qquad (7.78)$$

where $\bar{\alpha}$ is the total degree of ionization, P is the total pressure (in atm), T is the

* In the spectrochemical literature the concept of effective ionization energy has been introduced by Semenova (1945, 1946). It reappeared, among others, in a discussion between Ginzburg and Glukhovetskaya (1962a and b) and Semenova and Levchenko (1962).

I note that Semenova's original papers contain some erroneous conceptions, which she rejected later, as appears from the article by Semenova and Levchenko (1962); they cite a few papers which unfortunately I have not been able to obtain.

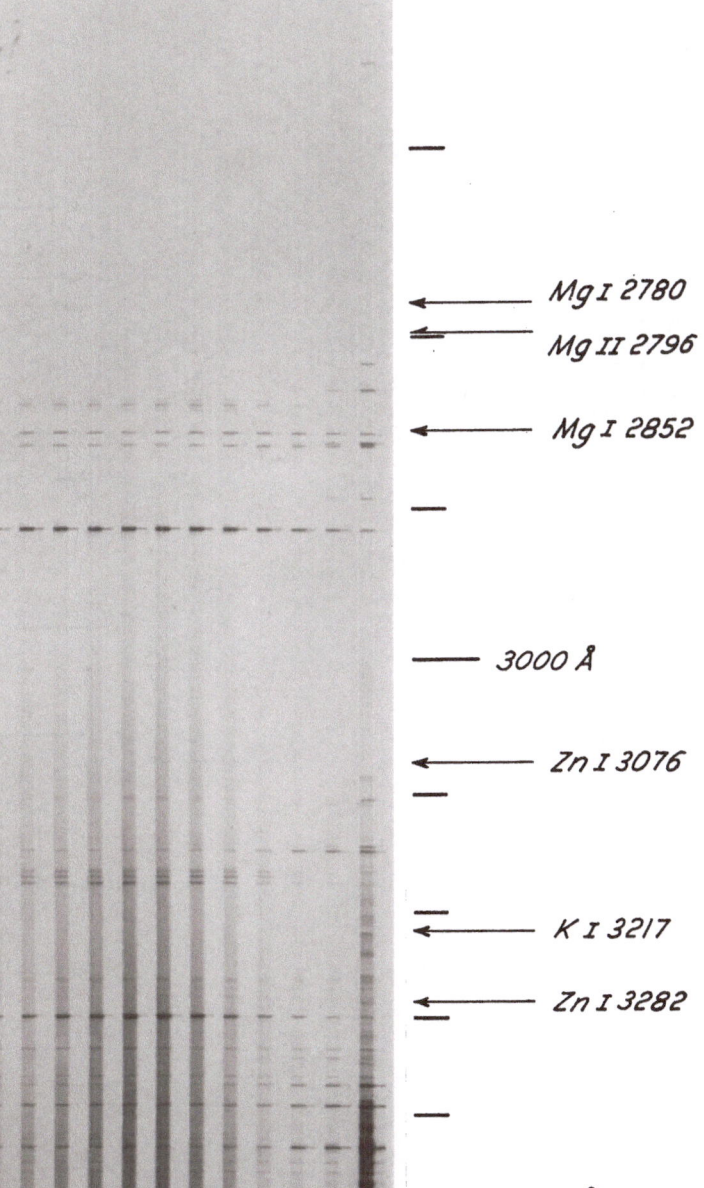

PLATE 1. Reproduction of plate representing the racking-plate spectra that pertain to the diagrams of Fig. 7.29 (KCl arc). The two-step spectra (step ratio = 10) were taken with a Hilger large quartz spectrograph adjusted for the wavelength range 2500–3600 Å. Exposure time of each spectrum was 25 sec. (It has not been possible to reproduce all the lines on the original plates.)

2500 Å

Mg I 2780
Mg II 2796
Mg I 2852

3000 Å

Zn I 3076

K I 3217

Zn I 3282

3500 Å

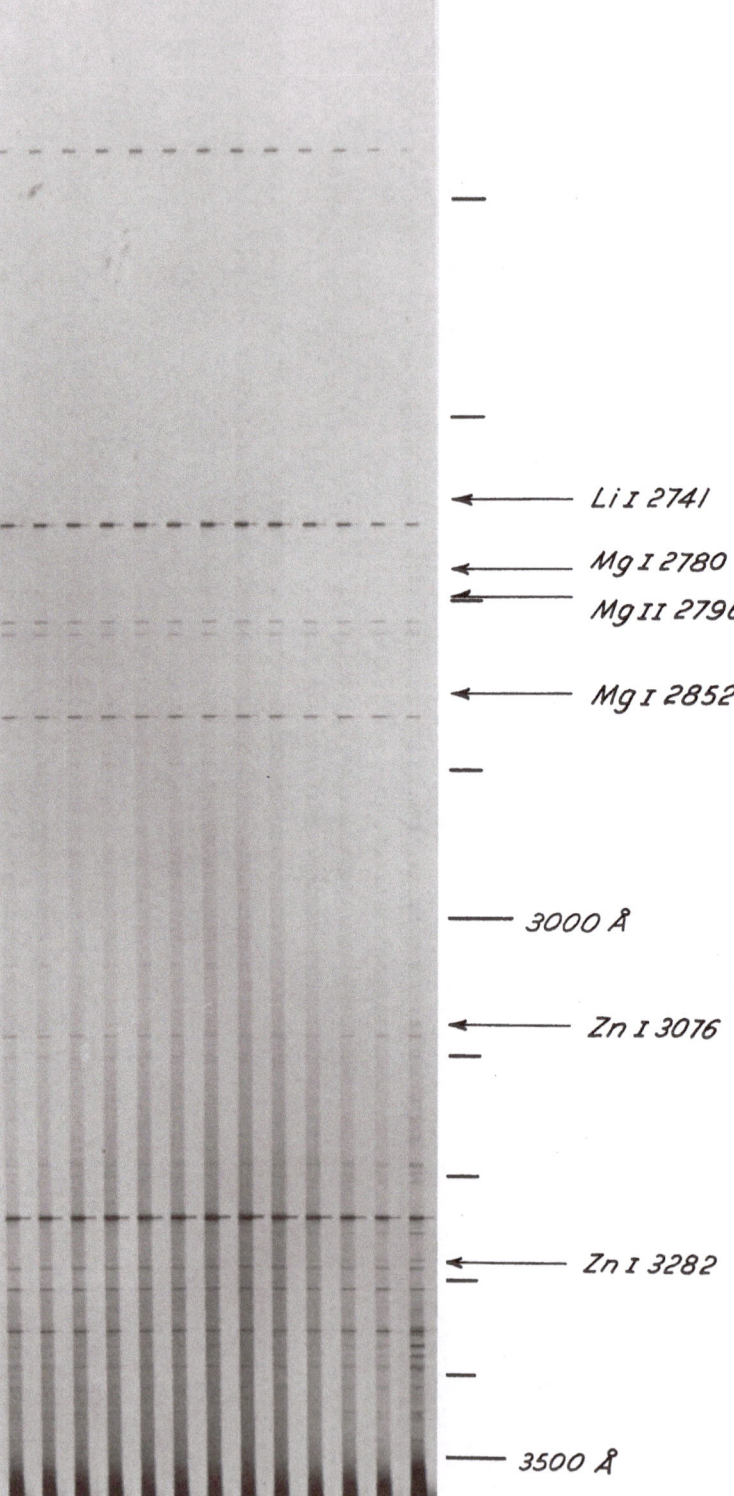

PLATE 2. Reproduction of plate representing the racking-plate spectra that pertain to the diagrams of Fig. 7.30 (LiCl arc). The two-step spectra (step ratio = 10) were taken with a Hilger large quartz spectrograph adjusted for the wavelength range 2500–3600 Å. Exposure time of each spectrum was 25 sec.

2500 Å

Li I 2741

Mg I 2780

Mg II 2796

Mg I 2852

3000 Å

Zn I 3076

Zn I 3282

3500 Å

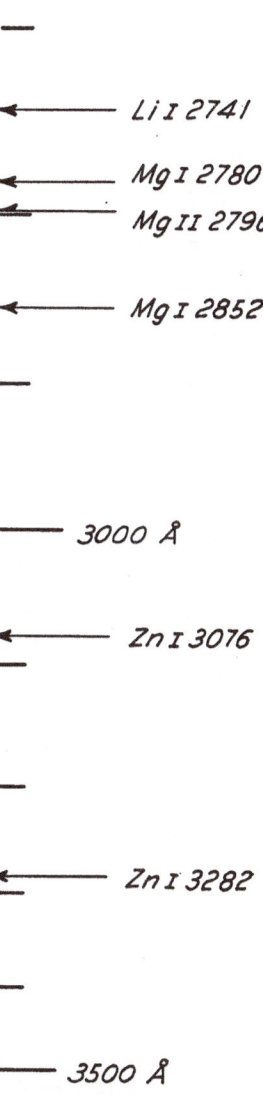

PLATE 3. Reproduction of plate representing the racking-plate spectra that pertain to the diagrams of Fig. 7.31 (LiF arc). The two-step spectra (step ratio = 10) were taken with a Hilger large quartz spectrograph adjusted for the wavelength range 2500–3600 A. Exposure time of each spectrum was 25 sec.

2500 Å

Li I 2741

Mg I 2780

Mg II 2796

Mg I 2852

3000 Å

Zn I 3076

Zn I 3282

3500 Å

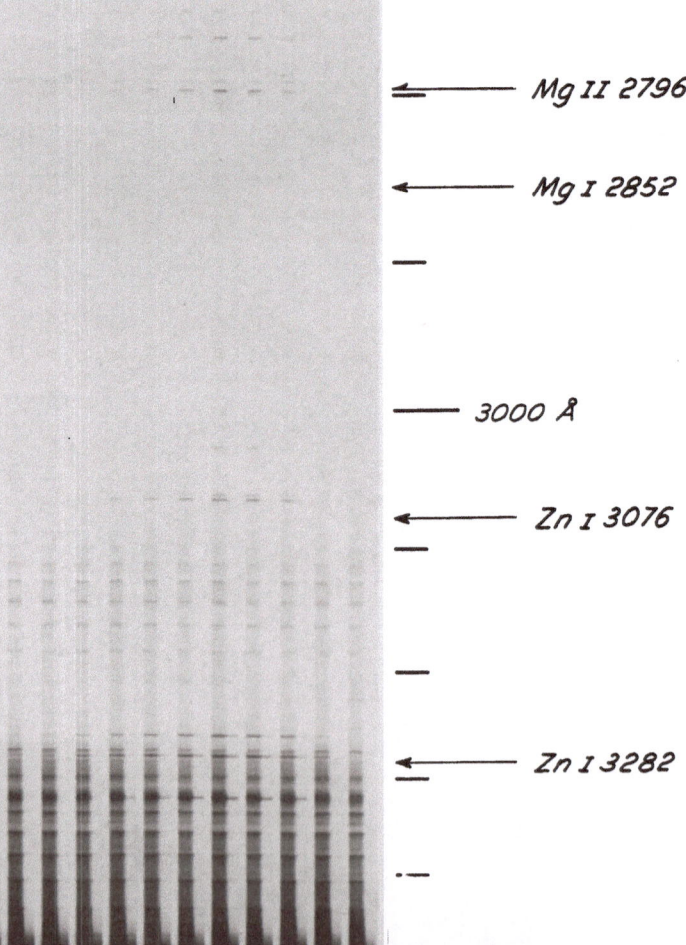

PLATE 4. Reproduction of plate representing the racking-plate spectra that pertain to the diagrams of Fig. 7.32 (pure graphite arc). The two-step spectra (step ratio = 10) were taken with a Hilger large quartz spectrograph adjusted for the wavelength range 2500–3600 A. Exposure time of each spectrum was 10 sec.

2500 Å

Mg II 2796

Mg I 2852

3000 Å

Zn I 3076

Zn I 3282

3500 Å

absolute temperature (in °K), and V_{ieff} is the effective ionization potential (in eV). The ratio of partition functions occurring in (7.19) is put equal to unity. So, by introducing the total degree of ionization and the effective ionization potential, we transfer the Saha equation for a single gas, i.e. equation (7.70), to the gas mixture in the arc.

Recalling that in the low-current carbon arc in air at atmospheric pressure the electron pressure p_e is of the order of 10^{-3} atm, we may neglect $\bar{\alpha}^2$ in comparison to unity in the denominator at the left-hand side of (7.78); inserting then (7.75) into (7.78) gives

$$\log \frac{p_e^2}{P} = \frac{5}{2} \log T - \frac{5040}{T} V_{\text{ieff}} - 6 \cdot 18 \qquad (7 \cdot 79)$$

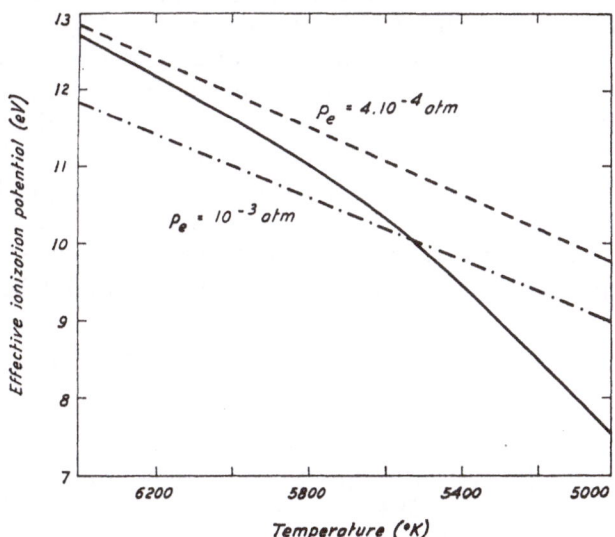

FIG. 7.37. Effective ionization potential (in electron volts) as a function of the temperature according to equation (7.80).

The dashed line and the dotted-dashed line pertain to constant electron pressure $p_e = 4.10^{-4}$ and $p_e = 10^{-3}$ atm respectively. The continuous line refers to the correlation between T and p_e shown in Fig. 7.6. It is assumed that changes of the temperature are brought about by changes of the plasma composition.

Solving for V_{ieff} finally yields

$$V_{\text{ieff}} = \frac{T}{5040} \left(\frac{5}{2} \log T - 2 \log p_e + \log P - 6 \cdot 18 \right) \qquad (7.80)$$

where the term $\log P$ has been retained solely for formal reasons; actually it vanishes when $P = 1$ atm. In that case the effective ionization potential is seen to be a function of the temperature and the electron pressure.

Characterizing the excitation conditions by V_{ieff} and p_e is thus formally the same as characterizing it by T and p_e. The interrelation between V_{ieff} and T is shown in Fig. 7.37 for three cases, namely when

   1 p_e is constant and equal to $4 \cdot 10^{-4}$ atm.

   2 p_e is constant and equal to 10^{-3} atm.

   3 p_e is correlated with the temperature according to the relationship shown in Fig. 7.6.

Obviously, in the temperature range considered here, V_{ieff} is a linear function of T when p_e is constant.

Let us now relate V_{ieff} with the plasma composition. To that end we recall the calculations involving the relationship between the temperature and the plasma composition produced in § 7.8. The results, shown in Figs. 7.10 to 7.13, represent the dependence of the arc temperature on the partial pressure of added constituents of different ionization potential. Using equation (7.80) or Fig. 7.37 we can transform the curves presented in Figs. 7.10 to 7.13 into graphs that represent the relationship between the partial pressure of an added constituent and V_{ieff}. Fig. 7.38 shows some examples for an element with an ionization potential of 6 eV.* Clearly, saying that the addition of a substance lowers the effective ionization potential of the plasma is identical with saying that it lowers the temperature. To indicate precisely how V_{ieff} or T changes with the amount of the substance added requires that we are informed about the behaviour of the electron pressure.

The effective ionization potential can also be expressed explicitly in terms of the plasma composition. Applying (7.65) we rewrite equation (7.80) as

$$V_{ieff} = \frac{T}{5040} \left(\frac{5}{2} \log T - \log \Sigma \frac{p_{aj}}{P} S_{pj} - 6 \cdot 18 \right) \qquad (7.81)$$

which in virtue of (7.18) reduces to

$$V_{ieff} = -\frac{T}{5040} \log \Sigma \frac{p_{aj}}{P} 10^{-(5040/T)V_{ij}} \qquad (7.82)$$

if again the ratio of partition functions is put equal to unity. Note particularly that p_{aj} is the partial pressure of the neutral atoms of the constituent labelled j and does not represent the total partial pressure p_j $(= p_{aj}+p_{ij})$ of that constituent.

Using equation (7.82) for computing the curves in Fig. 7.38 actually involves similar calculations as those produced in § 7.8, so that further discussion is redundant. We notice only the following. If all terms summed in (7.82) can be neglected in comparison with that for the component (*l*) of lowest ionization potential, we have

$$V_{ieff} = V_{il} - \log \frac{p_{al}}{P} \qquad (7.83)$$

* In compliance with the view expressed by Semenova and Levchenko (1962) the figure demonstrates that easily-ionized impurities in the plasma do not affect V_{ieff} if their partial pressure is below 10^{-4} atm. This is equivalent to saying that the excitation conditions are not influenced if the proportion of such impurities is below 0·01 per cent of the total plasma composition, corresponding to a proportion of 0.1 per cent with respect to the vapours (including carbon) released from the electrode (cf. §§ 7.8 and 7.9).

Recalling that a sample volatilizing from the electrode cavity contributes at the most about 1 per cent to the plasma composition (§§ 7.8, 7.9 and 9.4), we see that $V_{i\text{eff}}$ exceeds V_{if} at least by 2 eV.*

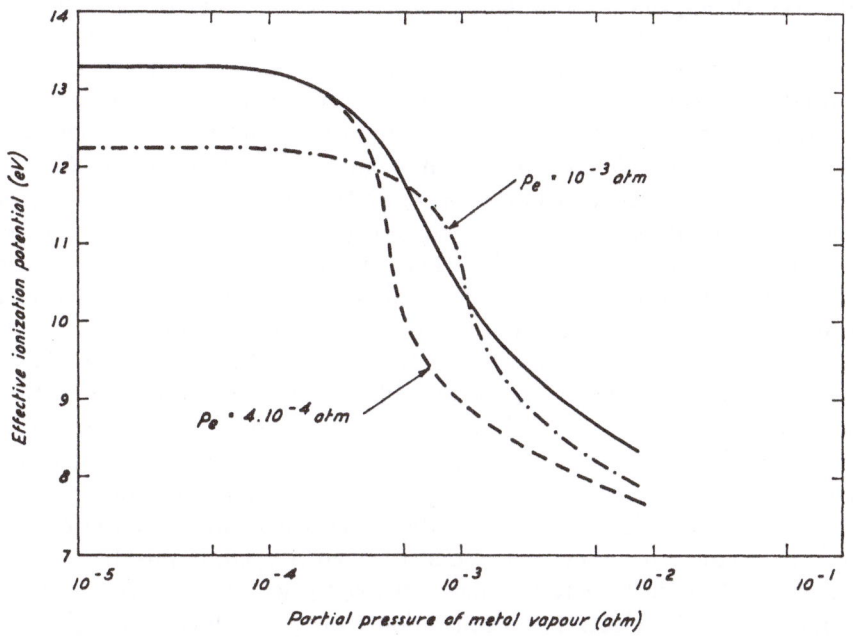

FIG. 7.38. Effective ionization potential as a function of the partial pressure of an added metallic constituent with apparent ionization potential $\bar{V}_{if} = 6$ eV. The curves correspond to those produced in Fig. 7.37.

It is seen that the first perceptible influence of the added material on the effective ionization potential begins to appear when the vapour pressure of the extraneous element reaches a value of 10^{-4} atm. This critical level is equivalent to a 1 per cent content in the sample; evidently, if we recall that the partial pressure of a major constituent of a sample loaded into the electrode crater amounts to about 10^{-2} atm at the most.

§ 7.17 *The determination of the radial distribution of the electron concentration*

Our discussion on electron concentration hitherto has either pertained to uniform gases or concerned effective values of the electron concentration in the arc such as determined spectroscopically without the use of space-resolved information. The simplified way of dealing with electron concentration in terms of effective values is of considerable practical interest, as will be shown in detail

* It appears from a paper by Mochalov and Raff (1956) that in the Russian literature a misconception with regard to the composition of the arc plasma has spread. Curiously enough, it was believed that the electrode vapour constitutes the principal component of the arc plasma, whereas the surrounding medium (air) is of minor importance. Ginzburg and Glukhovetskaya (1962a and b) seem to refer to the old misconception when they identify the effective ionization potential of the arc gas with the ionization potential of an added constituent.

in the next chapter, where we consider the use of effective values of tempera-ture and electron pressure for computing spectral-line intensities. It will also become evident then that numerical values of the effective electron concentration as derived by spectroscopic measurement do not coincide with volume-averaged values involved in approximate equations describing the current transport (cf. § 3.8). If the Saha method of § 7.6 is applied, spectroscopic values of $\bar{n}_e$ depend highly on the ionization potential of the element and on the excitation potential of the lines used. Therefore, we must be aware that the spectroscopic values of $\bar{n}_e$ are primarily parameter values which enable us to calculate the right magnitude for the effective degrees of ionization of components of the plasma (cf. § 7.6).

If we measure the radial distribution $n_e(r)$ of the electron concentration, our measurements essentially concern the *same* quantity as that involved in the current transport equation (3.31). Consequently, when dealing with the radial distribution $n_e(r)$, equation (3.31) can be adopted as an auxiliary relationship. It can either be used as a check of results furnished by spectroscopic techniques alone, or be applied for determining unknown spectroscopic quantities, e.g. transition probabilities. So, for instance, equation (3.31) is used as an additional relationship in the methods developed by Eberhagen (1955b), Eicke (1962), and Köstlin (1964) for determining transition probabilities. It was also applied by Kruithof (1943a) and by Kruithof and Smit (1944) in their study on the appropriateness of Saha's relationship.

For the spectroscopic determination of the radial distribution of the electron concentration $n_e(r)$ the procedure discussed in § 7.6 can be used. It must be modified, however, so that all intensity data are space-resolved prior to computing values of the temperature and the electron concentration from the relevant equations. If, for example, an atom–atom line pair of zinc is used for temperature measurement (§§ 6.6 and 6.7) and an ion–atom line pair of magnesium for the determination of n_e (§ 7.7), the intensity distribution $I(x)$ along the spectrograph slit (edge-on projection) must be converted for each of the four lines, with the aid of the Abel inversion, into a radial emission distribution $J(r)$ (see §§ 6.12 to 6.14). Then, the radial distributions of the pertinent ratios can be computed, from which the desired distributions $T(r)$ and $n_e(r)$ are derived. The main practical difficulty arises from error amplification. Precision of space-resolved emission profiles $J(r)$ tends to be rather limited and so is that of the distributions $T(r)$ and $n_e(r)$. Therefore, inter-pretation of results should be always based on experimental distributions that are averaged over a sufficiently large number of measurements.

A recent investigation on the radial distributions $T(r)$ and $n_e(r)$ in d.c. carbon arcs with metal vapours undertaken by de Galan (1965a, 1966a) will be considered in § 9.4.

CHAPTER EIGHT

SPECTRAL-LINE INTENSITY
AND SPATIAL SOURCE INHOMOGENEITY

★ §8.1 *Introduction*
A model of the arc

When discussing in the previous chapter the dependence of spectral-line intensity on temperature and electron pressure we tacitly assumed the source to be homogeneous. In fact, as we mentioned in § 3.8 and found in §§ 6.12 to 6.16, we must consider radial distributions in an arc. In this chapter we shall attempt to incorporate the radial decline of temperature (T) and electron pressure (p_e) into our picture of line intensities. We shall resort to a model that approximates the actual conditions well enough to permit conclusions which have general validity. Interestingly, our calculations, which were made for instruction purposes, yield some remarkable results which enable us to define the scope of the effective values of temperature and electron pressure. We shall demonstrate in subsequent sections that a slice of plasma through the arc can be approximated, with respect to the emission of spectral lines, by a circular disk of effective temperature $\hat{T}$, of effective electron pressure $\hat{p}_e$, and of effective radius $\hat{R}$. This holds for all ion lines and for most atom lines except the low-level lines of elements with a low ionization potential.*

We shall adopt the following model of the arc. The parabolic function

$$T = T_0 - a(b - T_0)r^2 \qquad (8.1)$$

depicts the radial temperature profile, where T_0 is the temperature at the axis, a and b are constants, and r is the radial coordinate. By the proper choice of the constants a and b the function can be made to describe the widening of the high temperature region as T_0 increases. Obviously, the function cannot hold true for large r, but this causes no difficulty; the emission is negligible when r is large, so that calculations never extend to the very outskirts of the arc, where the temperature falls less steeply with r than is predicted by (8.1).

To fix numerical values for the constants we considered the temperature profile calculated by Roes for a 7-amp d.c. arc in air (see Fig. 6.36). Accordingly we adopted the following condition: $T_0 = 6600°K$, while $T = 5000°K$ at $r = 4$ mm. As a second condition we introduced $T_0 = 5000°K$, while $T \approx 4000°K$ at

---

* Some arc model calculations similar to those mentioned here have been reported by Hefferlin and Gearhart (1964).

$r = 3$ mm, to have a slightly decreasing discharge diameter with falling temperature at the axis.

From these conditions we determined the constants a and b to be 10^{-2} mm^{-2} and $16\,600°$K respectively. Temperature profiles computed from these values are shown in Fig. 8.1 for different values of T_0. Note that the curves of the actual

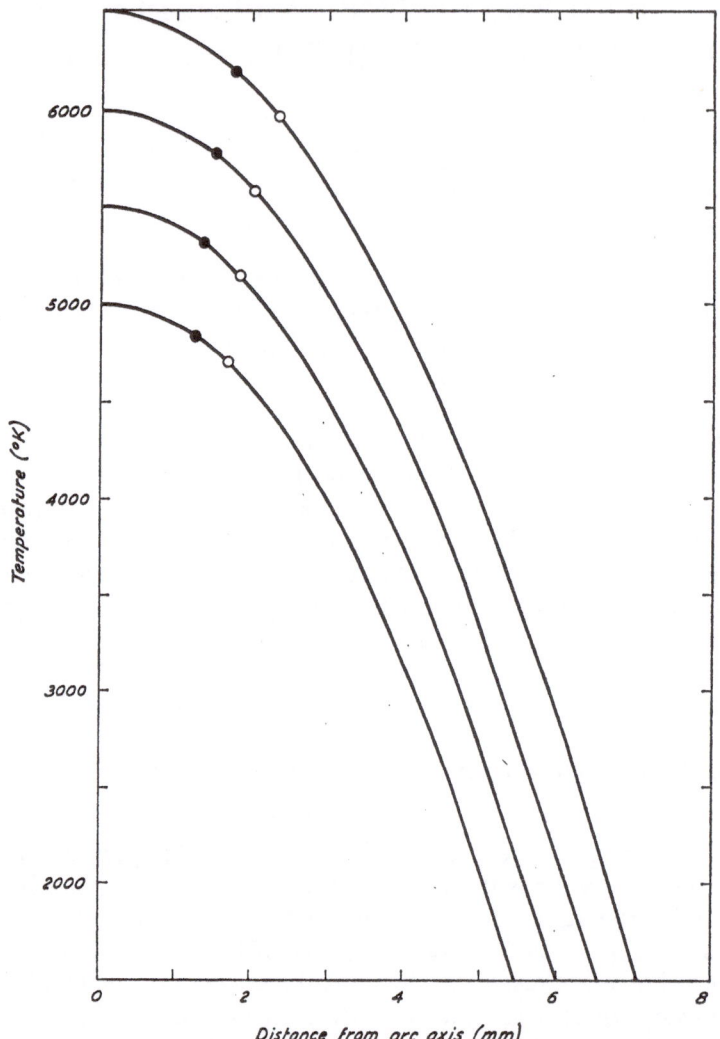

Fig. 8.1. Radial distributions of the temperature according to equation (8.1) as assumed for the model arc considered throughout this chapter.

The figure shows the radial temperature profiles with temperatures at the axis: $T_0 = 6500$, 6000, 5500, and $5000°$K. Spectroscopic, population-averaged temperatures determined from the intensity ratio of the atom–atom line pair Zn 3072/3076 have been inserted into the graphs, namely as

● for when the arc is focused on the spectrograph slit, and as

☉ for when the arc is focused on the collimator lens (see § 8.3).

temperature distributions tend to run less steeply in the central portion of the arc, while, at some distance from the arc axis, their gradient is steeper.

Next we approximate the radial distribution of the electron concentration n_e by a Gaussian function

$$n_e(r) = n_e(o) \, \exp[-c(d-T_0)r^2] \tag{8.2}$$

where $n_e(o)$ is the electron concentration at the axis and c and d are constants. c and d are determined by the condition that the 4000°K zone coincides with the circle at which electron concentration has dropped to one tenth of its value $n_e(o)$ at the axis. In conjunction with the adopted temperature profiles, c and d are computed to be 10^{-4} mm^{-2} degree^{-1} and 7500°K respectively. Relative distributions of electron concentration, based on these numerical values, are given in Fig. 8.2 for $T_0 = 6500$, 6000, 5500, and 5000°K.

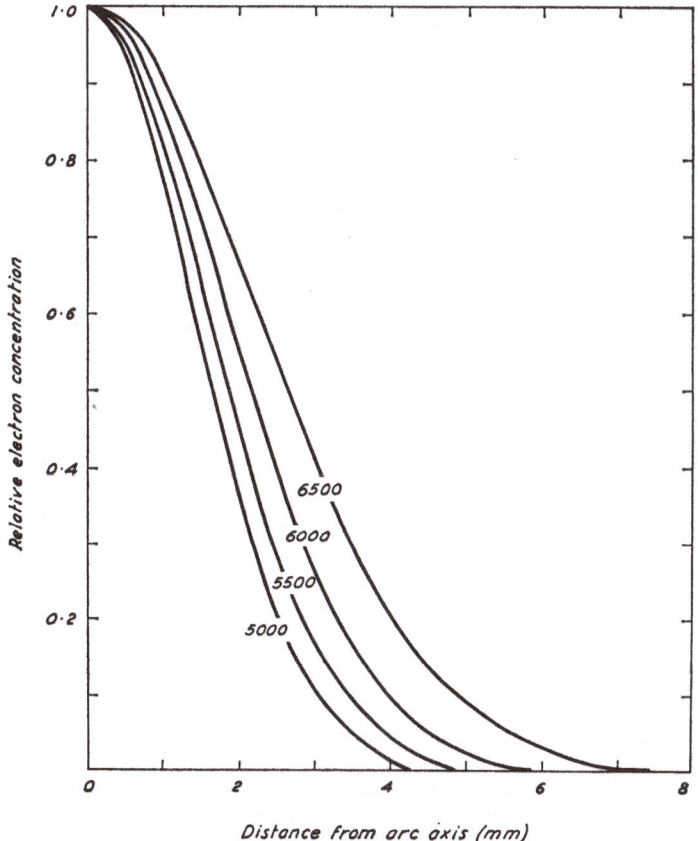

FIG. 8.2. Relative radial distributions of the electron concentration (n_e) according to equation (8.2) as assumed for the model arc considered throughout this chapter. The four $n_e(r)$ profiles correspond to the four temperature distributions characterized by $T_0 = 6500$, 6000, 5500, and 5000°K, and shown in Fig. 8.1.

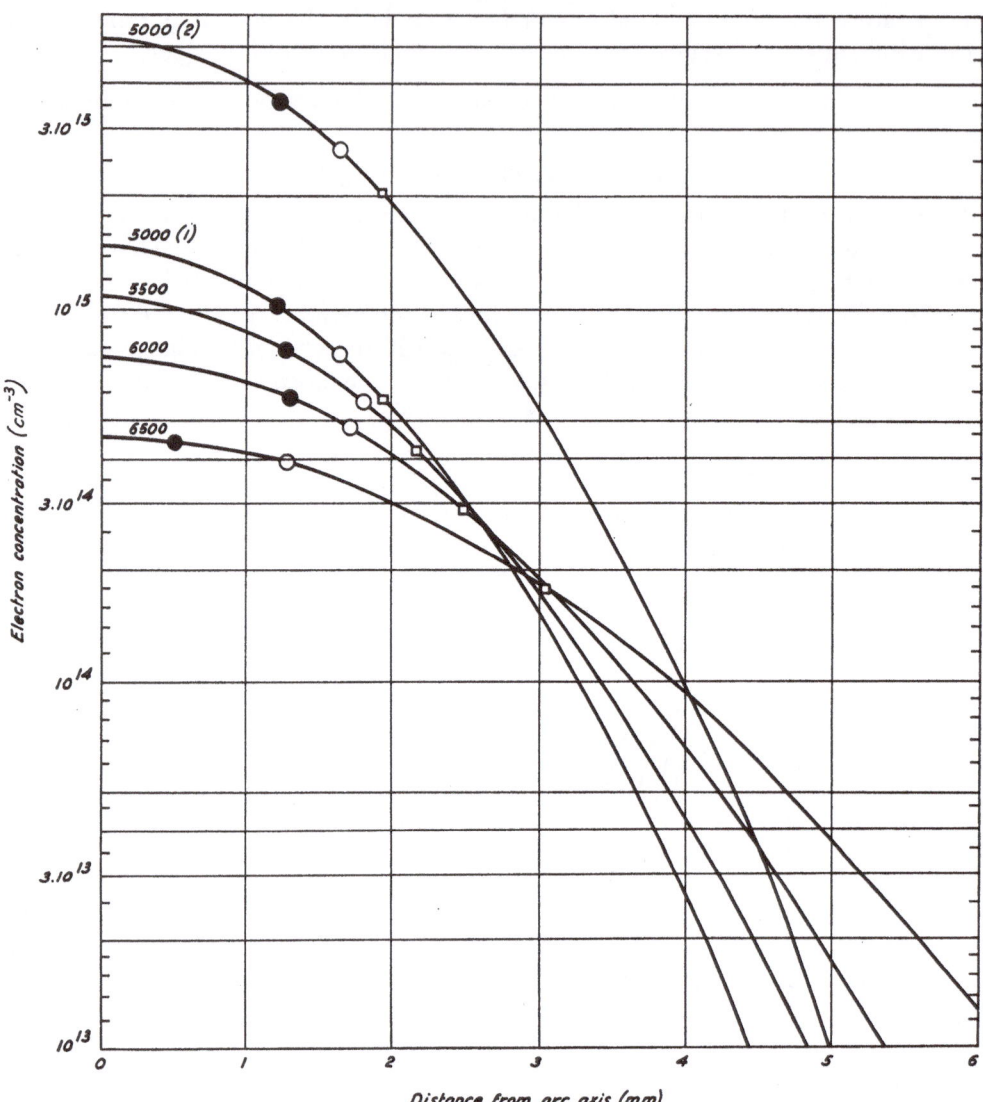

FIG. 8.3. Absolute radial distributions of the electron concentration (n_e) according to equation (8.2). The absolute $n_e(r)$ profiles have been computed from the relative ones in Fig. 8.2 by assigning to $n_e(o)$ the values listed in Table 8.1. These, in turn, are primarily founded on the assumption of a slow decrease of electron pressure with rising temperature, which fits in qualitatively with a trend detected experimentally under specified conditions (see §§ 7.7 to 7.10). Moreover, at $T_0 = 5000°K$, an additional $n_e(r)$ profile with markedly enhanced electron concentration is considered.

Mean values of n_e have been included in the graphs (cf. § 8.3), viz.

○ Spectroscopic values determined from the intensity ratio of a magnesium ion–atom line pair when the arc is focused on the spectrograph slit.

● Similar spectroscopic values determined when the arc is imaged on the collimator lens.

□ Volume-averaged values taken over a cylindrical region that contains 90 per cent of the electrons.

To assign plausible values to $n_e(0)$ we assume that changes in the temperature result from variations in the plasma composition, *not* from changes in the current strength, and that the electron concentration rises as the temperature falls [which was detected experimentally (§§ 7.7 to 7.10)]. For convenience, we take a simple, negative, linear correlation between the temperature and the electron pressure at the axis. In addition, we shall consider what happens if $p_e(0)$ at $T_0 = 5000°K$ is enhanced by a factor of 3·5. The adopted numerical values have been summarized in Table 8.1. The order of magnitude of p_e agrees closely with the effective values

TABLE 8.1

T_0 (°K)	$p_e(0) \cdot 10^3$ (atm)	$n_e(0) \cdot 10^{-15}$ (cm^{-3})	$\log n_e$
6500	0·4	0·45	14·655
6000	0·6	0·73	14·866
5500	0·8	1·07	15·029
5000 (1)	1·0	1·47	15·167
5000 (2)	3·5	5·14	15·711

Adopted values for the electron pressure $p_e(0)$ at the arc axis for different values of the temperature T_0 at the centre. The values of the electron concentration $n_e(0)$ listed have been computed from $p_e(0)$ and T_0 with the aid of (3.19).

determined experimentally. Fig. 8.3 shows the absolute distributions of electron concentration that will be now used for calculating radial intensity profiles of spectral lines.

★ § 8.2 *Outline of proposed intensity calculations using the arc model of radial distributions*

With the arc model described in the previous section as a basis, we intend to consider

1 The radial distribution of the emission per unit volume per unit solid angle for different atom and ion lines of various elements.

2 The intensities of these lines as obtained from the radial emission distribution by integrating either over the entire depth of the source or both over its depth and its width.

3 The appropriateness of the procedure to treat integrated intensities in terms of effective temperatures and electron pressures.

To make these calculations we assume the following.

1 Each element is distributed uniformly throughout a cross-section of the arc, that is, the sum of the concentrations of atoms and ions is constant throughout a cross-section: $n_j = n_{aj} + n_{ij}$. This postulate has been substantiated experimentally by a recent investigation undertaken by de Galan (1965a, 1966a; see § 9.4).

2 The formation of molecular species can be ignored. This assumption does not hold true in the arc flame, for example when the temperature is below 3500–4000°K; for, many elements are known to form stable radicals or molecules with oxygen in the arc envelope (see § 11.3). If we neglect the formation of molecules and radicals, the calculated radial emission distributions of atom lines tend to spread too widely in the arc. Therefore, our conclusions on the scope of effective temperatures and electron pressures tend to be more rigorous than is necessary when considering the appropriateness of effective values. So, there is good reason to believe that deviations are less pronounced in practice than theory suggests.

Now we are in a position to evaluate the radial emission distribution for each spectral line selected for the purpose. In view of equations (1.11) and (7.9) we have for an atom line of species j

$$J_{qp} = \frac{1}{4\pi} g_q A_{qp}(1 - \alpha_j) n_j \frac{\exp(-\epsilon_q/kT)}{Z_{aj}} h\nu_{qp} \qquad (8.3)$$

where J is the emission per unit volume per unit solid angle, g_q the statistical weight of the upper level q, A_{qp} the probability of the transition q → p, α_j the degree of ionization of the element labelled j, n_j the total concentration of that element (number of atoms and ions in unit volume), ϵ_q the excitation energy of the upper level q, Z_{aj} the partition function of the atom, ν_{qp} the frequency of the transition q → p, k the Boltzmann constant, and h the Planck constant.

Putting

$$C_{qp} = \frac{1}{4\pi} g_q A_{qp} h\nu_{qp} \qquad (8.4)$$

we rewrite (8.3) in practical and logarithmic form as

$$\log J_{qp} = \log C_{qp} + \log n_j + \log(1 - \alpha_j) - \frac{5040}{T} V_q - \log Z_{aj} \qquad (8.5)$$

As C_{qp} is constant and n_j is postulated to be independent of r, we must dispose of the functions $\alpha_j(r)$, $T(r)$, and $Z_{aj}(r)$ for computing the radial emission distribution $J(r)$.

The degree of ionization $\alpha_j(r)$ is calculated from (7.24n), viz.

$$\log \frac{\alpha_j}{1 - \alpha_j} = -\log n_e + \frac{3}{2}\log T - \frac{5040}{T} V_{ij} + \log \frac{Z_{ij}}{Z_{aj}} + 15 \cdot 684 \qquad (8.6)$$

and the radial distributions $n_e(r)$ and $T(r)$ according to the model given in § 8.1. The calculation is readily made with a desk calculator.* If the partition functions occurring in equations (8.5) and (8.6) cannot be regarded as constant in the temperature range under consideration, radial distribution functions will have to be

---

* For the benefit of readers willing to perform calculations on the basis of the model, we have tabulated the basic values $5040/T$ and $\frac{3}{2}\log T - \log[\exp -c(d - T_0)r^2]$ as functions of r for the axis temperatures of 6500, 6000, 5500, and 5000°K (see Appendix 8).

used. We shall show, however, that this inconvenience can be safely overcome by applying *apparent* ionization and excitation potentials.

The equivalent expressions (8.3) to (8.5) for ion lines read

$$J_{qp}^+ = \frac{1}{4\pi} g_q^+ A_{qp}^+ \alpha_j n_j \frac{\exp(-\epsilon_q^+/kT)}{Z_{ij}} h\nu_{qp}^+ \tag{8.7}$$

$$C_{qp}^+ = \frac{1}{4\pi} g_q^+ A_{qp}^+ h\nu_{qp}^+ \tag{8.8}$$

and

$$\log J_{qp}^+ = \log C_{qp}^+ + \log n_j + \log \alpha_j - \frac{5040}{T} V_q^+ - \log Z_{ij} \tag{8.9}$$

From the radial distribution of the emission (J) per unit volume per unit solid angle we compute the intensities of spectral lines by integration over a part of the source. Two types of integrals will be dealt with:

$$\text{1} \qquad I_s = 2 \int_0^\infty J(r)\,dy = 2 \int_0^\infty J(r)\,dr \tag{8.10}$$

where, in conformity with our definitions in §§ 1.2 and 6.13, the y axis is the direction of observation. We take the vertical arc to be focused vertically and symmetrically on the vertical slit of the spectrograph (compare Fig. 1.3 with the slit along the z axis); for the sake of convenience we suppose the observed portion of the arc to be uniform along its length.

$$\text{2} \qquad I_c = 2 \int_0^\infty \int_0^\infty J(r)\,dy\,dx = 2\pi \int_0^\infty r J(r)\,dr \tag{8.11}$$

which expresses spectral-line intensity as obtained by integrating the emission over the entire depth and width of the arc; it thus depicts the recorded intensity when the vertical arc is focused on the collimator lens of the spectrograph. Of course we assume the image to fall entirely within the effective aperture of this lens.

★ § 8.3 *Effective temperature and effective electron pressure of the model arc*

Let us proceed to find the effective temperature of the model arc as determined from the atom lines Zn 3072 and Zn 3076. The procedure for temperature measurement with these lines has been discussed in §§ 6.5 and 6.6.

Using equations (8.5) and (8.6) we compute the radial distributions of the emission $J(r)$ for both zinc lines. With the four radial temperature profiles, characterized by $T_0 = 6500$, 6000, 5500, and 5000K° (§ 8.1), we thus obtain for Zn 3076 the curves shown in the left half of Fig. 8.4. (The curves for different values of the electron pressure at $T_0 = 5000°K$ coincide, because the ionization

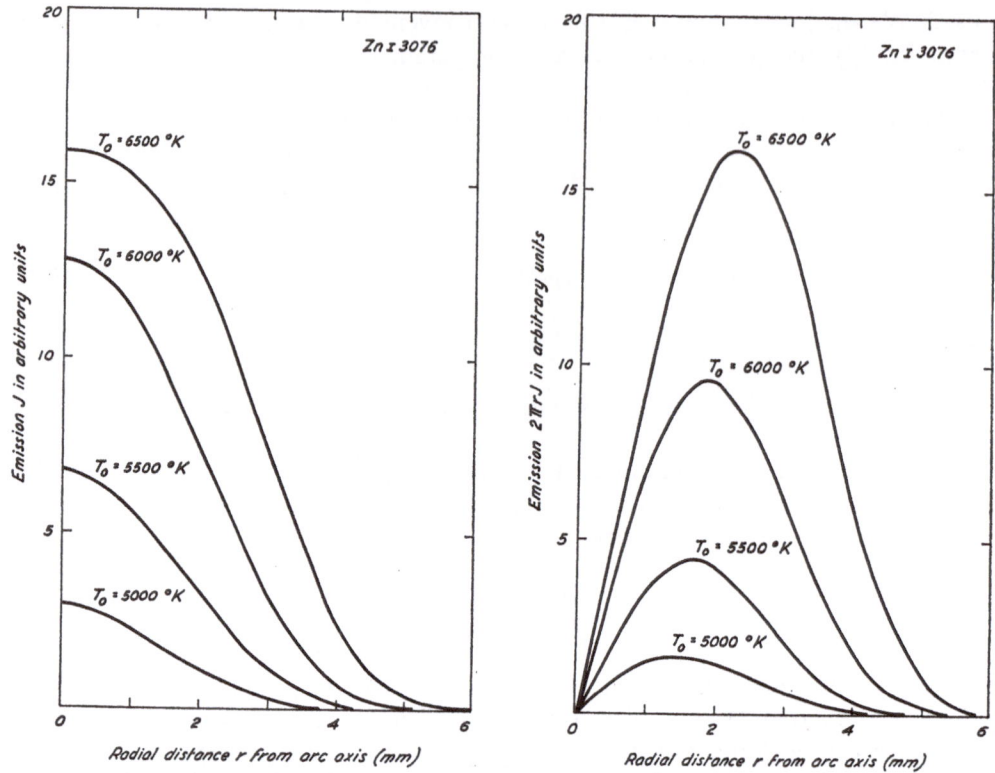

FIG. 8.4. Radial distribution of the emission of the atom line Zn 3076, pertaining to the five cases comprised in the model arc, viz.

$T_0 = 6500°K$, $p_e(o) = o \cdot 4 . 10^{-3}$ atm $T_0 = 5000°K$, $p_e(o) = 1 \cdot 0 . 10^{-3}$ atm
$T_0 = 6000°K$, $p_e(o) = o \cdot 6 . 10^{-3}$ atm $T_0 = 5000°K$, $p_e(o) = 3 \cdot 5 . 10^{-3}$ atm
$T_0 = 5500°K$, $p_e(o) = o \cdot 8 . 10^{-3}$ atm

defined in § 8.1. In view of the high ionization potential of zinc, the emission distributions for low and high electron pressure at $T_0 = 5000°K$ coincide.

In the left-hand graph, emission $J(r)$ per unit volume per unit solid angle is plotted; in the right-hand one, the plot of the function $2\pi r J(r)$, representing emission per unit solid angle from a ring of radius r having unit height and thickness, is given. The latter function occurs in calculations of intensities that are integrated over the entire cross-sectional area of the arc (arc imaged on collimator lens).

of zinc is negligible at this low temperature.) Integration of the curves yields intensities I_s similar to those we observe when imaging the model arc on the spectrograph slit. Effective temperatures ($\hat{T}$) are computed by inserting the ratio of integrated intensities of Zn 3072 and Zn 3076 into equation (6.21).* Results are given in Table 8.2.

* Mark that it is not necessary to know the relative transition probabilities of the lines; the only requirement is to use the same relative gA/λ values both for evaluating the intensities and for determining the effective temperature. Obviously, this applies solely to the model arc, where we compute $\hat{T}$ with respect to a given radial distribution of the temperature. Thus either the ratio of intensities at a given temperature or the ratio of transition probabilities can be arbitrarily fixed.

For computing $\hat{T}$ from intensities that are obtained by integrating the emission $J(r)$ over the entire cross-section of the arc, that is from intensities (I_c) such as recorded when the arc is focused on the collimator lens, we consider in view of (8.11) the function $2\pi r J(r)$. It depicts emission per unit solid angle from a circular ring of radius r having unit height and thickness. Complementing the $J(r)$ curves

TABLE 8.2

T_0 (°K)	6500	6000	5500	5000 (1)	5000 (2)
$\hat{T}$(°K)	6190	5760	5300	4830	4830
	5950	5570	5130	4690	4690
$\log I^+/I$	0·52	−0·13	−0·89	−1·78	−2·33
	0·29	−0·29	−1·02	−1·91	−2·46
$\log \hat{p}_e \cdot 10^4$	0·56	0·66	0·75	0·83	1·37
	0·49	0·56	0·60	0·68	1·23
$\hat{p}_e$ (10^{-3} atm)	0·37	0·46	0·56	0·67	2·36
	0·31	0·36	0·40	0·48	1·71
$\hat{n}_e \cdot 10^{-15}$ (cm^{-3})	0·44	0·59	0·78	1·03	3·58
	0·38	0·48	0·57	0·76	2·68
$\bar{n}_e \cdot 10^{-15}$ (cm^{-3})	0·18	0·29	0·42	0·57	2·01
$\hat{n}_e/\bar{n}_e$	2·46	2·04	1·86	1·79	1·78
	2·18	1·66	1·36	1·31	1·32

Effective temperatures, effective electron pressures and concentrations, and volume-averaged electron concentrations for model arcs. For each of the model arcs considered in § 8.1 and denoted by the temperature T_0 at the axis, the table lists

1 The effective temperature ($\hat{T}$) as derived from the intensity ratio of the atom–atom line pair Zn 3072/3076.
2 The log intensity ratio I^+/I of the ion–atom line pair Mg 2796/2852.
3 The effective electron pressure $\hat{p}_e$ and its logarithm as derived from the intensity ratio of the ion–atom line pair and the effective temperature.
4 The effective electron concentration $\hat{n}_e$ computed with (3.19) from $\hat{P}_e$ and $\hat{T}$.
5 The volume-averaged electron concentration $\bar{n}_e$ as defined in the text.
6 The ratio $\hat{n}_e/\bar{n}_e$.
Results are given for the relevant arcs focused on the spectrograph slit (upper values in each row) and for the arcs imaged on the collimator lens (lower values in each row).

for the line Zn 3076 on the left in Fig. 8.4, curves $2\pi r J(r)$ are presented at the right. Clearly, as a result of increasing volume of successive rings, the curves on the right have maxima at some distance from the arc axis. The temperature information carried by the integrated line intensities

$$I_c = 2\pi \int r J(r) \, dr$$

is weighted accordingly, so that the values of $\hat{T}$ derived from intensities I_c are lower than those found from intensities I_s (see Table 8.2).

Exactly in the same way as with the atom–atom line pair Zn 3072/3076 we now proceed with the ion–atom line pair Mg 2796/2852 to find values for the effective electron pressure $\hat{p}_e$ using equation (7.33).

The calculated relative radial distributions $J(r)$ and $2\pi r J(r)$ for the magnesium lines are produced in Figs. 8.5 and 8.6. By integration we find the intensities I_s and I_c. The logarithmic ratios $I_s{}^+/I_s$ and $I_c{}^+/I_c$ have been listed in Table 8.2*

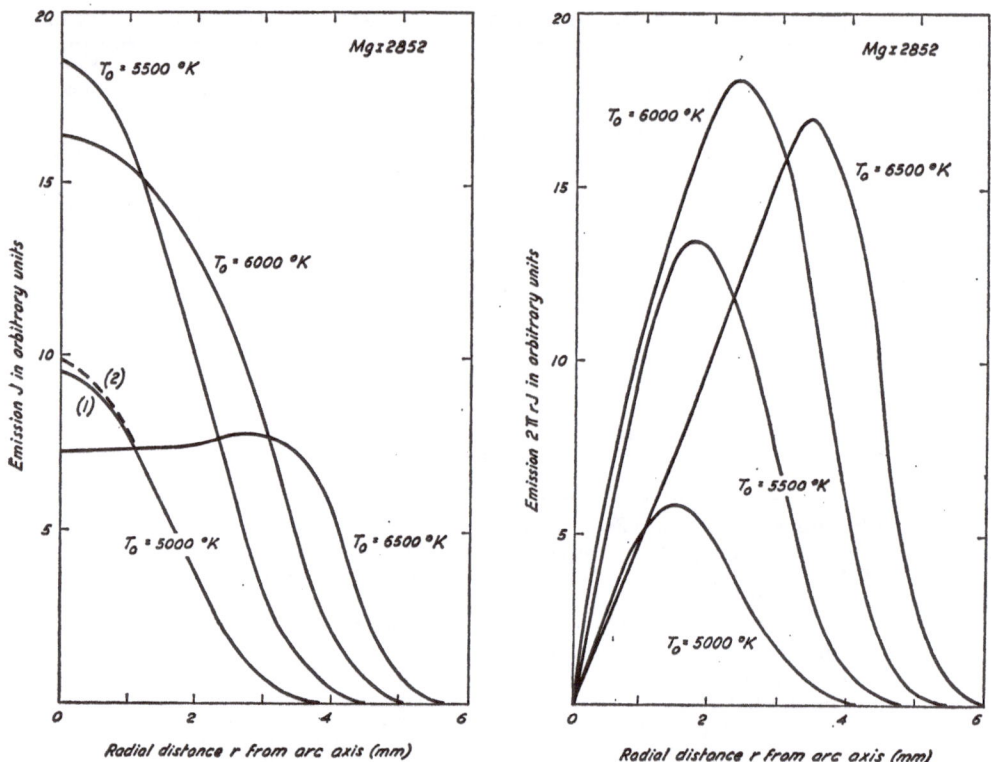

FIG. 8.5. Radial distribution of the emission of the atom line Mg 2852 for the five cases comprised in the model arc (cf. Fig. 8.4). Note that the intensity of the magnesium line passes through a maximum in the temperature range considered here (cf. §§ 8.7 and 8.8).

along with the values of the effective electron pressure $\hat{p}_e$ computed from (7.33) by inserting the relevant numerical values of $\hat{T}$ and I^+/I. In summary Table 8.2 demonstrates what differences among the effective values of T, p_e, and n_e result from two alternative ways of integrating emission per unit volume per unit solid angle.

In addition to effective spectroscopic values, we have listed for comparison

* To link up smoothly with the experimental values displayed in Chapter 7 we used true values for relative transition probabilities of magnesium lines. Similarly, however, as when determining $\hat{T}$ from a known radial distribution of $\hat{T}$, we could employ arbitrary gA-values when dealing with the evaluation of the effective electron pressure from a known radial distribution $p_e(r)$.

volume-averaged values of the electron concentration $\bar{n}_e$, defined as follows. The column radius R to be related to the current transport (see § 3.8) is taken to be $R_{0.1}$, that is the distance from the axis at which the electron concentration has become equal to one tenth of its maximum value $n_e(0)$ at the axis. This radius $R_{0.1}$ is used to define the volume over which n_e is averaged. For the model under discussion it is readily seen that 90 per cent of the current transport takes place

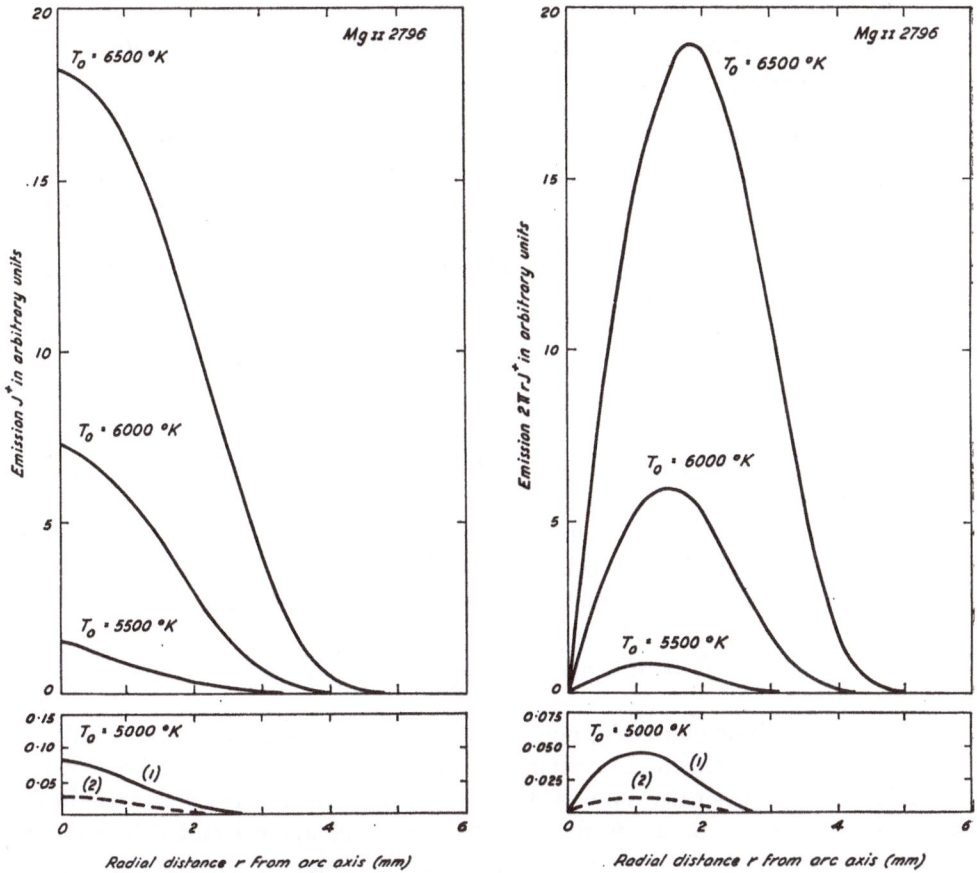

FIG. 8.6. Radial distribution of the emission of the ion line Mg 2796 for the five cases comprised in the model arc (cf. Fig. 8.4). Note that the intensity of the ion line changes very rapidly with temperature (cf. § 8.9).

within the circle of radius $R_{0.1}$. The volume average $\bar{n}_e$ is uniquely related to the value of n_e at the axis, viz. by

$$\bar{n}_e = 0.391\, n_e(0)$$

Naturally, $\bar{n}_e$ does not equal either type of effective value, nor does it show a rigorous correlation with either of them, as is seen from the ratio $\hat{n}_e/\bar{n}_e$ in Table 8.2. A few of the tabulated results have been also indicated in the graphs representing the arc model (Figs. 8.1 and 8.3) to disclose the difference among the numerical values. The existence of such discrepancies is a matter of course and should be consequently considered as a mere fact, that has been demonstrated in detailed form in terms of the model. The more important side of the question is to consider the appropriateness of calculated effective spectroscopic values of temperature and electron pressure for computing line intensities.

★ § 8.4 *The appropriateness of effective temperature and electron pressure values The effective arc radius*

The calculations in the preceding section provided integrated intensity values, I_s and I_c, for individual spectral lines in five arcs with different radial distributions of temperature and electron pressure. The intensity values were needed to evaluate the effective temperatures and electron pressures. We now ask whether the reverse procedure can be applied, that is computing intensities $\hat{I}_s$ and $\hat{I}_c$ for individual spectral lines from effective temperatures $\hat{T}$ and effective electron pressures $\hat{p}_e$. To answer this question let us first compare the nature of the quantities I_s and $\hat{I}_s$, e.g. for an atom line.

Remembering (8.3), (8.4), and (8.10), we have

$$I_s = 2 \int J(r)\, dr \equiv 2 C_{\mathrm{qp}} n_j \int \frac{(1-\alpha_j)\exp(-\epsilon_q/kT)}{Z_{aj}}\, dr \qquad (8.12)$$

where α_j, T, and Z_{aj} all are known functions of r, given $T(r)$ and $p_e(r)$. Note that the integral at the right has the dimension of length.

Computing the effective intensity $\hat{I}_s$ is by definition conducted as

$$\hat{I}_s = 2 C_{\mathrm{qp}} n_j \frac{(1-\hat{\alpha}_j)\exp(-\epsilon_q/k\hat{T})}{\hat{Z}_{aj}} \hat{R}_s \qquad (8.13)$$

Obviously, an additional quantity now enters into discussion, the *effective radius* $\hat{R}_s$. It is defined as the radius of a circular slice of plasma, uniform throughout, which produces the same intensity along a diameter as its non-uniform prototype in the arc. The conditions of the uniform gaseous layer are characterized by the effective temperature $\hat{T}$ and by the concentration n_j of particles of species j having an effective degree of ionization $\hat{\alpha}_j$.

For the five cases comprised in the arc model of § 8.1 we computed effective radii $\hat{R}_s$ for a number of spectral lines, using the condition

$$I_s = \hat{I}_s \qquad (8.14)$$

so that

$$\hat{R}_s = \frac{I_s}{2 C_{\mathrm{qp}} n_j (1-\hat{\alpha}_j) \dfrac{\exp(-\epsilon_q/kT)}{\hat{Z}_{aj}}} \qquad (8.15)$$

Results have been summarized in Table 8.3, along with the actual and the apparent ionization potentials of the elements and the actual and the apparent excitation potentials of the lines (cf. §§ 6.3 and 7.3). Inspection of the results in Table 8.3 shows primarily that the effective radius $\hat{R}_s$ is reasonably constant at each temperature, in other words, $\hat{R}_s$ is to a first approximation independent of the line and

TABLE 8.3

Temperature T_0 at the arc axis (°K)	6500	6000	5500	5000	5000				
Effective temperature $\hat{T}$(°K)	6190	5760	5300	4830	4830				
Effective electron pressure $\hat{p}_e$ (10^{-3} atm)	0.37	0.46	0.56	0.67	2.36				
Spectral line	$\hat{R}_s$ (mm)					V_{ij} (eV)	$\bar{V}_{ij}$ (eV)	V_q (eV)	$\bar{V}_q$ (eV)
Zn I 3076	3.23	3.04	2.72	2.46	2.46	9.39	9.06	4.01	4.01
Zn I 3072	3.22	3.06	2.73	2.50	2.50			8.08	8.08
Mg I 2852	3.44	3.01 (2.98)	2.69	2.45 (2.45)	2.45	7.64	7.33	4.33	4.28
Mg II 2796	3.49	3.08 (3.06)	2.77	2.47 (2.34)	2.51			4.41	4.41
Li I 2741	3.43	3.16 (3.36)	2.84	2.51 (2.53)	2.46	5.39	5.75	4.50	4.40
Li I 6708	12.5	6.9	4.7	3.4	3.04			1.84	1.74
Ti I 2956		3.44 (3.54)		2.49 (2.51)		6.82	6.55	4.22	3.72
Ba I 2785	4.41	3.36	3.30	2.88	2.74	5.21	5.03	4.43	3.48
Ba I 5535	9.4	5.0	4.2	3.0	2.9			2.23	1.28
Ba II 2771	3.55	3.02	2.70	2.44	2.46			7.16	6.81

Effective radii $\hat{R}_s$ computed with (8.15) for a number of spectral lines when emitted by each of the five model arcs considered in § 8.1. The arc is assumed to be imaged on the spectrograph slit.

On the right, actual (V_{ij}) and apparent ($\bar{V}_{ij}$) ionization potentials of elements (§ 7.3) and actual (V_q) and apparent ($\bar{V}_q$) excitation potentials of lines (§ 6.3) have been tabulated.

Results in parentheses have been calculated from the approximation involved in the use of apparent potentials; the other results are found in a rigorous treatment; for barium, however, an intermediate way was followed (see p. 247).

of the element, except for the low-level lines Li I 6708, Ba I 2785, and Ba I 5535. Therefore, we conclude that the concepts of effective temperature and effective electron pressure are appropriate. Only for low-level atom lines of elements of low ionization potential are deviations to be expected, particularly at elevated effective temperatures. This is not surprising, since, at high temperatures in the central core of the arc, low-level lines are predominantly emitted from zones at a considerable distance from the axis (see Fig. 8.7, which gives radial distributions for lithium and barium lines). Actually, the situation tends to be more favourable than the results suggest; for, in our model arc the region where the gradients of

9

temperature and electron concentration are small is less extended than in a true arc; in addition, formation of molecules in the arc envelope reduces emission of atom lines so that, in fact, contributions from the outer zones are smaller than our calculations indicate.

The results in Table 8.3 also show that the effective radius $\hat{R}_s$ increases regularly with the effective temperature $\hat{T}$, namely by a factor of 1·3 to 1·4 when $\hat{T}$ rises

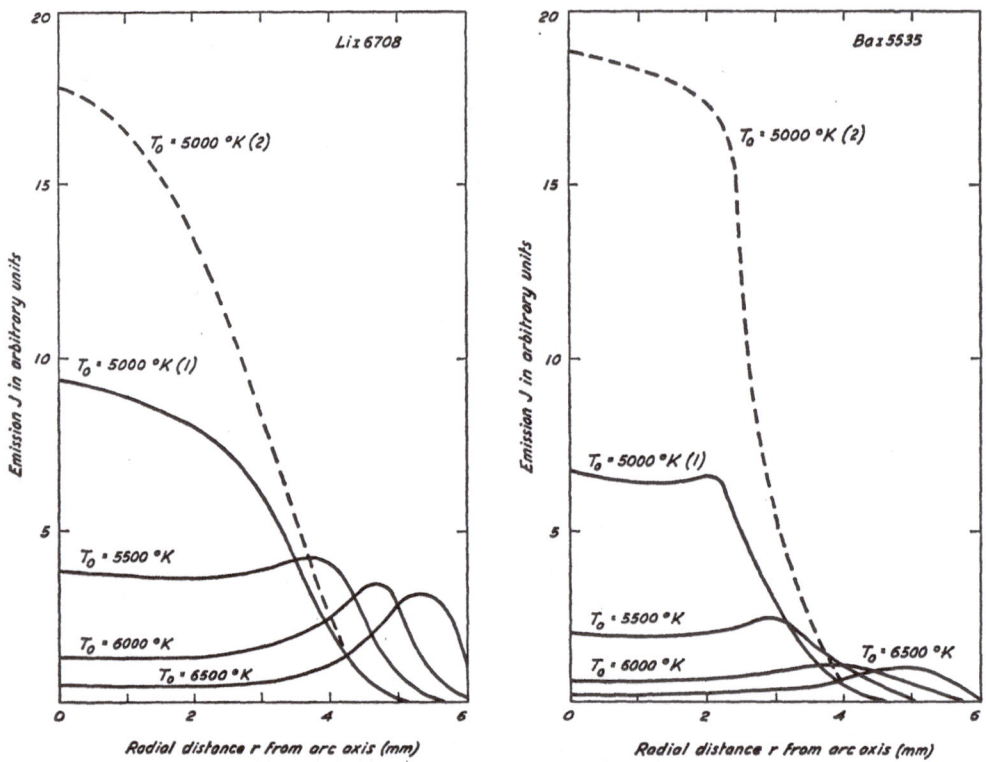

FIG. 8.7. Radial distribution of the emission of the atom lines Li 6708 and Ba 5535 for the five cases comprised in the model arc (cf. Fig. 8.4). Note that these low-level lines are predominantly emitted from the arc flame when the temperature in the core is high.

from 4800 to 6200°K. Electron pressure seems to have no effect (compare the results at $\hat{T} = 4830°$K for $\hat{p}_e = 0·67 . 10^{-3}$ and $\hat{p}_e = 2·36 . 10^{-3}$ atm).

Before we proceed to consider the bearing of the ionization and excitation potential for the appropriateness of the effective quantities $\hat{T}$, $\hat{p}_e$, and $\hat{R}_s$, a few comments on the method of calculating the results given in Table 8.3 must be made. Intensities I_s of the lines were inherently evaluated by integrating radial emission profiles, such as those shown for zinc, magnesium, lithium, and barium lines in Figs. 8.4 to 8.7. In computing radial distributions a practical difficulty is encountered, owing to the dependence of the partition functions Z_{aj} and Z_{ij} on temperature. For zinc, magnesium, and lithium the question is not puzzling, since in the temperature range under discussion the partition functions are well approximated by constants. For elements where this is not so, we feel inclined to resort to apparent ionization and excitation potentials (cf. §§ 6.3 and 7.3). This expedient would at first glance seem critical here, because we extend our calculations to much lower temperatures than provision was made for when establishing the concept of apparent potentials. However, the outer, cooler regions of the arc, where departures begin to become appreciable, contribute only little to the total intensities of most spectral lines, whence apparent potentials tend to be appropriate at least in all those cases where the concepts of effective temperature and electron pressure turn out to be appropriate. We verified this explicitly for the atom line Ti 2956 and applied both the rigorous calculation of radial emission profiles and the approximated one based upon apparent ionization and excitation potentials. Results of the latter calculation, given in parentheses in Table 8.3, are seen to agree very well with those obtained by the more accurate procedure.

Conversely, it appeared useful to check also the appropriateness of the apparent potentials for cases where the partition functions are virtually constant, e.g. magnesium and lithium; for, it is convenient if all elements can be treated similarly.* The figures in parentheses in Table 8.3 show clearly that apparent potentials can also be safely used for elements with partition functions that are practically constant. Obviously, the reason again is that in general the fringe of the arc contributes little to the total intensities of most lines, so that the departures do scarcely show themselves in the integrated values I_s.

For the atom line Ba 5535 emission is still appreciable in the arc flame; therefore we allowed for a regular rise of $\bar{V}_q$ from 1·28 to 2·23 eV in the range of temperature between 4500 and 3000°K, when we considered approximately the decline of the partition function Z_{aj} with temperature. Altogether a more rigorous treatment serves no useful purpose here, because $\hat{R}_s$ values deviate.

Hitherto we have dealt mainly with effective quantities derived from intensities I_s obtained by integrating the emission $J(r)$ over r. These intensities correspond to those measured experimentally when the arc is imaged on the spectrograph slit. For spectrochemical purposes, however, the arc is commonly focused on the collimator lens. Observed intensities are defined then by (8.11). Written in the form of equation (8.12) this becomes

$$I_c = 2\pi \int rJ(r)\,dr \equiv 2\pi C_{qp} n_j \int \frac{(1-\alpha_j)\exp(-\epsilon_q/kT)}{Z_{aj}} r\,dr \qquad (8.16)$$

The corresponding effective intensity $\hat{I}_c$ reads

$$\hat{I}_c = 2\pi C_{qp} n_j \frac{(1-\hat{\alpha}_j)\exp(-\epsilon_q/k\hat{T})}{\hat{Z}_{aj}} \hat{R}_c^{\,2} \qquad (8.17)$$

Using these equations we computed, in a similar manner as was done for $\hat{R}_s$, the effective radius $\hat{R}_c$ for the majority of the lines dealt with in Table 8.3. Interestingly,

---

* For elements such as magnesium and lithium, we are concerned with the partition functions—actually with the quotient Z_{ij}/Z_{aj}—mainly for computing the degree of ionization [equation (8.13)]. As the quotient is constant, the apparent ionization potential at low temperature may differ considerably from that at high temperature. So, for example, a δ-correction of 0·33 adapted for $T = 5500°K$ should read 0·18 at $T = 3000°K$.

we found that in all cases where effective quantities are appropriate the ratio $\hat{R}_s/\hat{R}_c$ was equal to $1\cdot30$ to within a few per cent, in other words, it turned out to be independent of the temperature, of the element, and of the line. Only for low-level atom lines, such as Li I 6708 and Ba I 5535, does the ratio $\hat{R}_s/\hat{R}_c$ depart substantially from $1\cdot30$, increasing, e.g., for the lithium line from $1\cdot3$ at $T_0 = 5000°K$ to $2\cdot0$ at $T_0 = 6500°K$.

★ § 8.5   *Systematic variation of the effective radius $\hat{R}_s$ with the excitation potential of a spectral line and with the ionization potential of an element*

In the preceding sections we have considered the usefulness of effective values of the temperature ($\hat{T}$), of the electron pressure ($\hat{p}_e$), and of the arc radius ($\hat{R}_s$ or $\hat{R}_c$) for evaluating spectral-line intensities. On the basis of effective temperatures determined with the aid of an atom–atom line pair of zinc and effective electron pressures measured by means of an ion–atom line pair of magnesium we were able to show the appropriateness of $\hat{T}$, $\hat{p}_e$, and $\hat{R}_s$ (or $\hat{R}_c$) for a number of representative spectral lines. Only for low-level atom lines of elements of low ionization potential were departures noted.

TABLE 8.4

V_{ij}	V_q			
	5·0	4·0	3·0	2·0
9·0	*2·96*	*3·02*	*3·16*	3·44
8·0	*2·90*	*3·01*	*3·20*	3·57
7·0	*2·93*	*3·16*	*3·58*	4·41
6·5	*3·04*	*3·38*	*4·03*	5·37
6·0	*3·12*	*3·54*	4·41	6·53
5·5	*3·13*	*3·55*	4·60	7·67
5·0	*2·98*	*3·36*	4·36	7·82
4·5	—	*3·18*	3·97	7·59

Effective radii $\hat{R}_s$ (in mm) for fictitious atom lines of different apparent excitation potential (V_q) pertaining to elements of different apparent ionization potential (V_{ij}). Results refer to the model arc of § 8.1 characterized by the temperature $T_0 = 6000°K$.

The italic figures indicate the region of V_{ij} and V_q where $\hat{R}_s$ can reasonably be regarded as constant.

In this section we shall approach the problem systematically and calculate effective radii $\hat{R}_s$ for fictitious atom lines having apparent excitation potentials of 5, 4, 3, and 2 eV successively and pertaining to elements having apparent ionization potentials of 9, 8, 7, 6, 5, and 4·5 eV. As before, we used in our calculations the four temperature distributions comprised in the model of § 8.1, to which correspond the effective temperatures $\hat{T} = 6190$, 5760, 5300, and 4830°K. In the last case we considered only one value of $\hat{p}_e$, namely, $\hat{p}_e = 0·67 . 10^{-3}$ atm (cf. Table 8.2). Results are given in Fig. 8.8 and in Table 8.4. For each effective temperature the figure clearly shows the trend of the effective radius $\hat{R}_s$ with the apparent excitation potential ($\bar{V}_q$) of the line and with the apparent ionization potential ($\bar{V}_{ij}$) of the element. Evidently, as the effective temperature $\hat{T}$ rises, the more $\hat{R}_s$ depends on $\bar{V}_q$ and $\bar{V}_{ij}$. This is readily understood; for, at the higher temperatures, the zone from which temperature information is furnished by the zinc lines, i.e. the core, does not coincide with the zones of maximum emission of low-level atom lines of easily-ionized elements, i.e. the flamy fringe. At the lower temperatures, emission from the outskirts becomes negligible and the emission centres of different lines all are located predominantly in the arc core (see e.g. Figs. 8.4 to 8.7).

Now let us draw a few general conclusions with respect to the appropriateness of the effective values of $\hat{T}$, $\hat{p}_e$, and $\hat{R}_s$. In view of our inability to determine precise $\hat{R}_s$ values experimentally, let us consider log $\hat{R}_s$ values within a range of about 0·1 to be constant. On this basis we conclude from Fig. 8.8 that at the highest temperature considered here, i.e. $\hat{T} \approx 6200°K$, effective values of $\hat{T}$, $\hat{p}_e$, and $\hat{R}_s$ are appropriate only for atom lines having an apparent excitation potential $\geqslant 4$ eV. At $\hat{T} \approx 5800°K$, the critical excitation level for elements having an ionization potential $\bar{V}_{ij} > 6·5$ eV has changed to 3 eV, and at $\hat{T} \approx 5300°K$ to 2 eV. For the easily-ionized elements ($\bar{V}_{ij} < 6·5$ eV) the critical level of 3 eV is reached at $\hat{T} \approx 5300°K$.

Summarizing, on the strength of the arc model of § 8.1 we have shown the appropriateness of effective temperatures ($\hat{T}$) and electron pressures ($\hat{p}_e$) measured with zinc and magnesium as the thermometric and manometric species respectively. The concepts can be applied for all ion lines and for atom lines having an apparent excitation potential $\geqslant 4$ eV. For low-level lines it depends on the effective temperature of the arc and on the ionization potential of the element whether the concepts are appropriate or not.

The appropriateness of effective temperatures and electron pressures means that we are allowed to replace actual radial distributions of T and p_e by individual, effective values of $\hat{T}$ and $\hat{p}_e$, so as to calculate changes of spectral-line intensities with $\hat{T}$ and $\hat{p}_e$ as well as would be inferred from an exact treatment in terms of the radial distributions. It must not be overlooked, however, that using effective values $\hat{T}$ and $\hat{p}_e$ involves definitely assuming an effective radius $\hat{R}$.

Mean values of log $\hat{R}_s$ for each effective temperature have been indicated on the left of the diagrams in Fig. 8.8, along with the range of the individual values that were used for evaluating the averages. The mean values of $\hat{R}_s$ are 3·4, 3·2, 2·9, and 2·65 mm.

As long as effective values are appropriate, it makes no difference whether we consider the arc to be imaged on the slit or on the collimator lens of the spectrograph, provided that we employ values of $\hat{T}$ and $\hat{p}_e$ that have been determined from correspondingly integrated radial emission profiles of spectral lines. So, for instance, when imaging the arc on the slit, the upper values of $\hat{T}$ and $\hat{p}_e$ in each row of Table 8.2 apply, whereas with the arc focused on the collimator lens the lower values must be used.

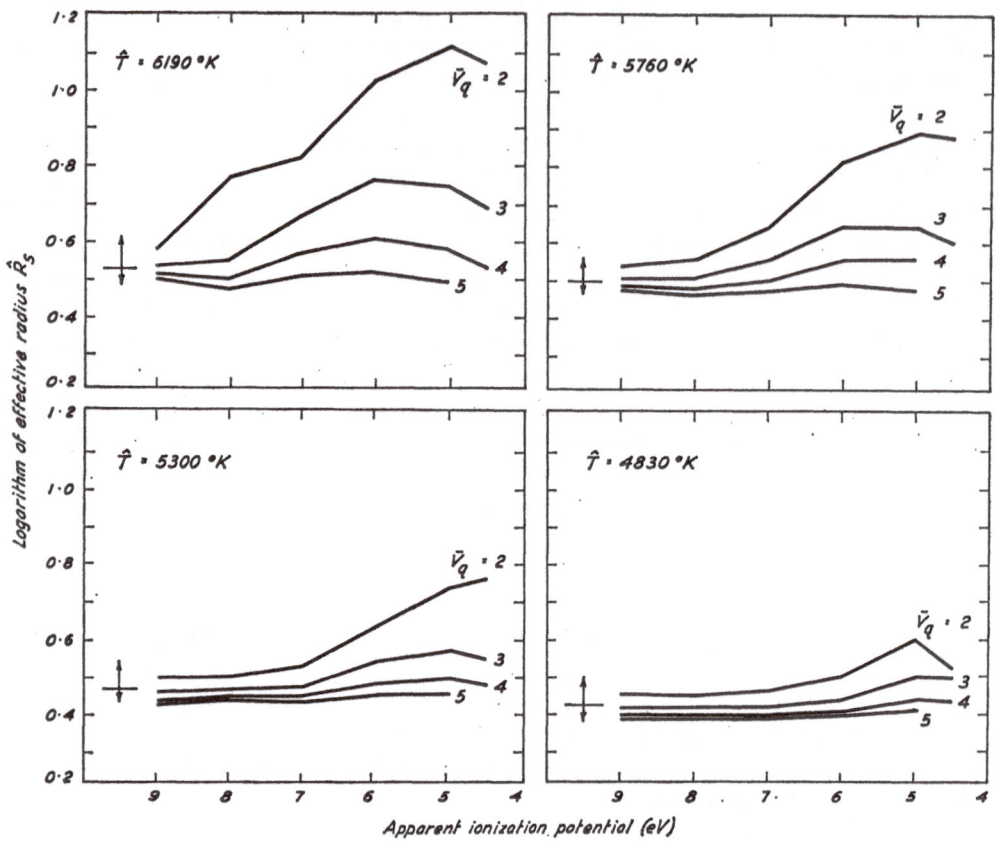

FIG. 8.8. Trends of the effective arc radius $\hat{R}_s$ with the apparent ionization potential $\bar{V}_{ij}$ of the element (abscissa) and the apparent excitation potential $\bar{V}_q$ of the atom line (parameter).

The figure shows results of calculations for the four temperature distributions comprised in the model arc of § 8.1 and characterized by the temperatures at the arc axis $T_0 = 6500$, 6000, 5500, and 5000°K, which correspond with the effective temperatures $\hat{T}$ inserted in the diagrams ($\hat{T}$ was measured from atom lines of zinc with the arc focused on the spectrograph slit).

At the left in each diagram a range of log $\hat{R}_s$ has been indicated within which $\hat{R}_s$ may be considered constant in view of the limitations set to precision in an actual experiment. The horizontal lines in the double-headed arrows indicate the averages computed from the individual values located in the accepted interval. The corresponding mean values of $\hat{R}_s$ are 3·4, 3·2, 2·9, and 2·65 mm, if we proceed from high to low temperature.

★ § 8.6   *Effective temperatures and electron pressures measured in the model arc with the aid of different sets of spectral lines*

The calculations for finding the dependence of $\hat{R}_s$ upon $\bar{V}_q$, $\bar{V}_{ij}$, and $\hat{T}$ provided us also with a large variety of data for computing effective temperatures from other line pairs than that of zinc; for, the primarily computed quantities are the

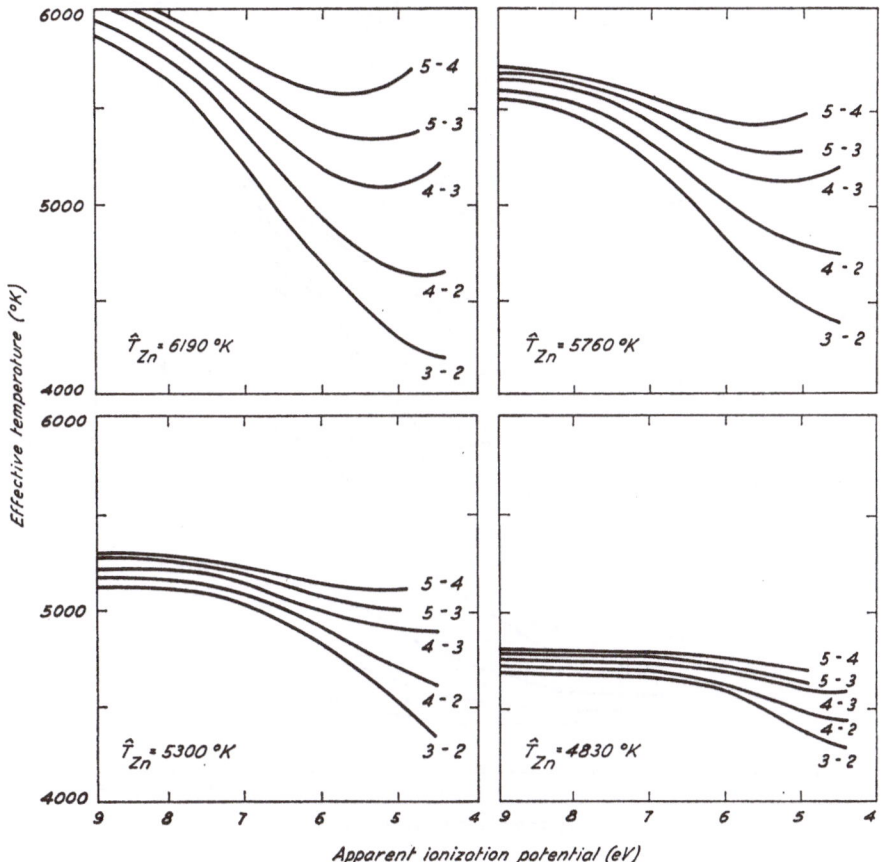

FIG. 8.9. Effective temperatures pertaining to the four temperature distributions of the model arc, when measured with the aid of different atom–atom line pairs of different elements. The diagrams show the trend with the apparent ionization potential $\bar{V}_{ij}$ of the element (abscissa) and with the apparent excitation potentials of the lines (parameter). The considered combinations of the latter ones have been indicated (in eV) at the right in the diagrams. At the left, the effective temperature $\hat{T}_{Zn}$, determined by means of an atom–atom line pair of zinc, is shown.

line intensities I_s, obtained by integrating the emission J in compliance with equation (8.12). Values of I_s were available for lines having an apparent excitation potential of 5, 4, 3, or 2 eV, and belonging to elements having an apparent ionization potential of 9, 8, 7, 6, 5, or 4·5 eV. If we remember our remark in a note at the beginning of § 8.3, that for the model arc, relative gA/λ values of thermometric lines need not be known for deriving effective temperatures, we see that for each of the above elements an effective temperature can be computed from each of six line pairs when combining excitation potentials as follows: 5–4, 5–3, 5–2, 4–3, 4–2, and 3–2.

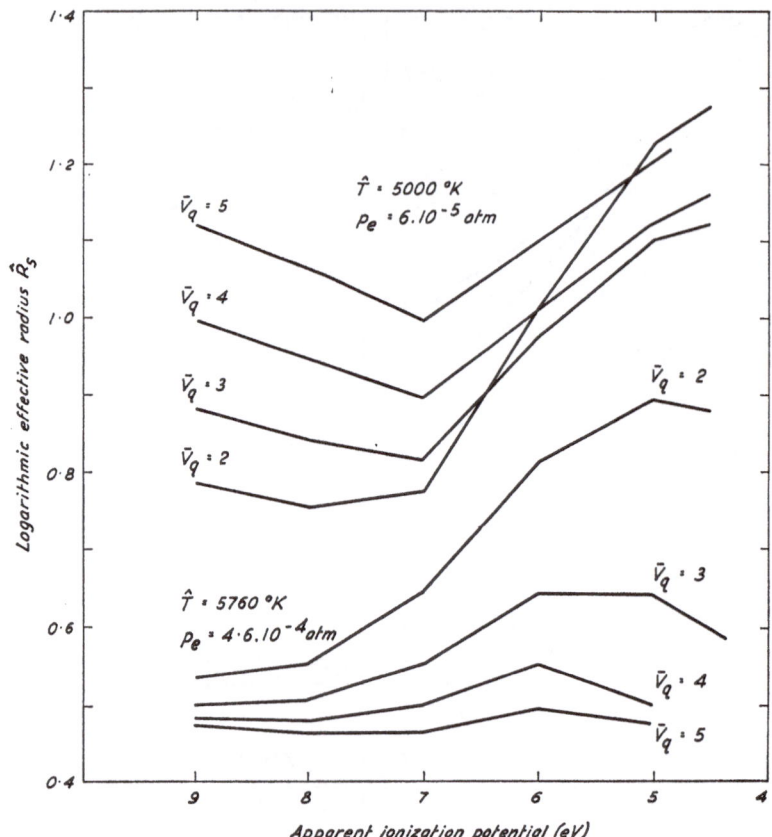

FIG. 8.10. Trends of the effective arc radius $\hat{R}_s$ with the apparent ionization potential $\bar{V}_{ij}$ of the element and with the apparent excitation potential $\bar{V}_q$ of the atom line. Both sets of curves pertain to one and the same case of the model arc, namely that characterized by $T_0 = 6000°K$ and $p_e(0) = 0.6 \cdot 10^{-3}$ atm.

For the lower set, the effective values $\hat{T} = 5760°K$ and $\hat{p}_e = 4.6 \cdot 10^{-4}$ atm, determined with zinc and magnesium, apply; the curves are identical with those produced in the corresponding diagram of Fig. 8.8.

The upper set is based on an effective temperature $\hat{T} = 5000°K$ and an effective electron pressure $\hat{p}_e = 6 \cdot 10^{-5}$ atm, determined with a thermometric and manometric species having an apparent ionization potential of 6 eV.

Results for five pairs have been given in Fig. 8.9 for each of the pertinent zinc temperatures $(\hat{T}_{Zn})$ considered throughout this discussion. In agreement with the picture of effective radii given in the preceding section, Fig. 8.9 shows the ever-growing discrepancies between the temperatures indicated by different line pairs of different elements as the temperature in the arc core (represented by $\hat{T}_{Zn}$) increases. We also see that the effective zinc temperature indicates poorly the excitation conditions for low-level lines of easily-ionized elements. For high-level lines of easily-ionized elements, the correlation with $\hat{T}_{Zn}$ is better, of course, though it is still far from ideal.

The present exposition suggests a remedy for dealing with effective excitation conditions of elements of low ionization potential, namely the application of effective values of temperature and electron pressure that have been derived from spectral lines of those very elements. We have worked out this principle for $\hat{T}_{\mathrm{Zn}} = 5760°\mathrm{K}$, using an element with an apparent ionization potential of 6 eV as both the thermometric and the manometric species. The effective temperature computed from an atom–atom 4–2 line pair was found to be 5000°K. Using a pair of ion–atom lines, both with an apparent excitation potential of 4 eV, we calculated the effective electron pressure to be $6 . 10^{-5}$ atm. Proceeding in a similar way as in §8.5, we calculated the effective radii $\hat{R}_s$, on the basis of $\hat{T} = 5000°\mathrm{K}$ and $\hat{p}_e = 6 . 10^{-5}$ atm. In Fig. 8.10 results are compared with those derived previously on the strength of $\hat{T} = 5760°\mathrm{K}$ (zinc) and $\hat{p}_e = 4 \cdot 6 . 10^{-4}$ atm (magnesium). Obviously, the appropriateness of the newer effective values of T and p_e is poor for elements of high ionization potential, whereas the situation is markedly improved for easily-ionized elements. Still, conditions are less ideal than for the elements of high ionization potential with the earlier effective values; a fairly strong dependence of $\hat{R}_s$ on the ionization potential remains. So, it appears that the range of excitation conditions for easily-ionized elements can be covered adequately only when several thermometric and manometric species are used.

★ §8.7 *Spectral-line intensity as a function of the effective temperature*

Again using the data underlying the calculation of effective radii (Fig. 8.8), we consider the dependence of the atom-line intensity I_s on the effective temperature $\hat{T}$ (measured with zinc as the thermometric species). Fig. 8.11 gives plots of log intensity against $\hat{T}$ for lines of different apparent excitation potential ($\bar{V}_q$) of elements of different apparent ionization potential ($\bar{V}_{ij}$). The continuous curves represent log I_s derived by integrating radial emission profiles [equation (8.12)].

The dashed curves depict the logarithm of $\hat{I}_s$ according to equation (8.13), that is

$$\log \hat{I}_s = \log 2C_{\mathrm{qp}} n_j \frac{(1 - \hat{\alpha}_j) \exp(-\epsilon_q/k\hat{T})}{\hat{Z}_{aj}} \hat{R}_s \qquad (8.18)$$

with a constant value for $\hat{R}_s$. They illustrate what dependence the intensity has on the temperature $\hat{T}$, if we assume the appropriateness of the effective temperatures and electron pressures while omitting to consider changes of the effective radius $\hat{R}_s$. Although for each separate spectral line an arbitrary value of $\hat{R}_s$ can be chosen, we applied different values only for different ionization potentials, thus we have one value for all the curves in one diagram of Fig. 8.11. By using the separate mean values of $\hat{R}_s$ computed for each ionization potential from the data underlying Fig. 8.8, we obtained, on the average, a fair adaptation of the continuous and dashed curves in each of the diagrams of Fig. 8.11.

Finally, we also considered log $\hat{I}_s$ according to equation (8.18), taking into

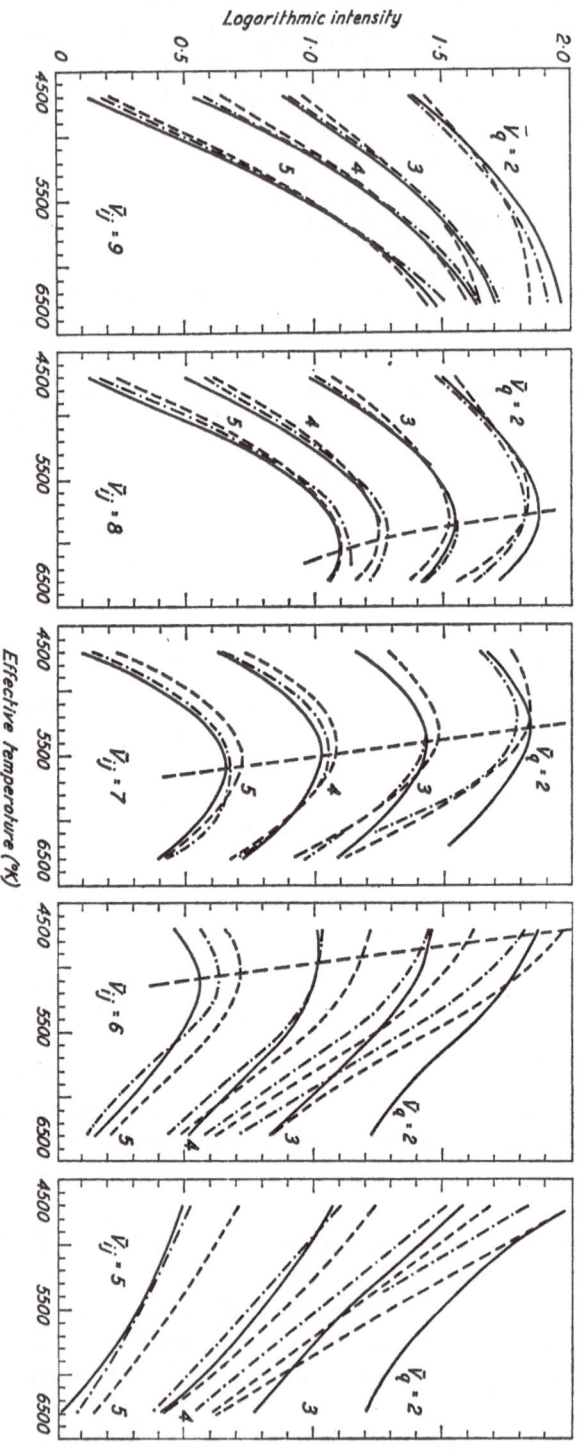

Logarithmic intensity (vertical axis: 0, 0·5, 1·0, 1·5, 2·0)

Effective temperature (°K) (horizontal axis: 4500, 5500, 6500)

Panels labelled: $\bar{V}_{ij} = 9$, $\bar{V}_{ij} = 8$, $\bar{V}_{ij} = 7$, $\bar{V}_{ij} = 6$, $\bar{V}_{ij} = 5$, each with $\bar{V}_a = 2$ and curves 3, 4, 5.

Fig. 8.11. Spectral-line intensity as a function of the effective temperature. Successive diagrams show plots of atom line intensity versus effective (zinc) temperature for different apparent ionization potentials (V_{ij}) of test elements. Each diagram contains the functions for four values of the apparent excitation potential (V_a).

Continuous curves pertain to intensities obtained by integrating radial emission profiles over r. They represent the intensities that can be found experimentally when the arc is focused on the spectrograph slit.

Dashed curves depict intensities computed from effective temperatures and electron pressures, but, with constant effective radius $\hat{R}_s$ for the four curves in each of the five diagrams.

Dashed-dotted curves depict similar intensity functions as the dashed ones, involving, however, a variation of the effective radius R_s with the effective temperature $\hat{T}$. The underlying relation between R_s and $\hat{T}$ is that indicated in Fig. 8.8, i.e. $\hat{T} = 6190\,°K$ and $\hat{R}_s = 3\cdot4$ mm, $\hat{T} = 5760\,°K$ and $\hat{R}_s = 3\cdot2$ mm, $\hat{T} = 5300\,°K$ and $\hat{R}_s = 2\cdot9$ mm, and $\hat{T} = 4830°$ and $\hat{R}_s = 2\cdot65$ mm.

Accordingly, it considers high-level lines only.

The diagrams also demonstrate how the temperature T_M at which atom line intensity is at a maximum shifts with V_i and V_a (see particularly § 8.8).

account a variation of $\hat{R}_s$ with temperature. Employing the mean values of log $\hat{R}_s$ defined in Fig. 8.8, we arrived at the dashed–dotted curves shown in Fig. 8.11.

As anticipated, we find only small discrepancies between the continuous, the dashed, and the dashed–dotted curves for the higher ionization and excitation potentials. When, on the contrary, $\bar{V}_q \leqslant 3$ eV and $\bar{V}_{ij} \leqslant 6$ eV, intensity calculations based on effective values are seen to yield poor results.

Fig. 8.11 also reveals clearly how the temperature T_M at which the atom line intensity is at a maximum shifts with the apparent ionization potential ($\bar{V}_{ij}$) of the element and with the apparent excitation potential ($\bar{V}_q$) of the line. With the assumed correlation between electron pressure and temperature (§ 8.1) we see that, given $\bar{V}_q$, a difference of 1 eV in $\bar{V}_{ij}$ results in a displacement of T_M by 400–500°K, while, given $\bar{V}_{ij}$, temperature T_M is displaced by about 100°K per eV difference in $\bar{V}_q$. So, for instance, if $\bar{V}_q = 4$ eV, we deduce from the figure the following numerical values:

$$T_M > 6300°\text{K for } \bar{V}_{ij} \geqslant 9 \text{ eV}$$

$$T_M \approx 6000°\text{K for } \bar{V}_{ij} = 8 \text{ eV}$$

$$T_M \approx 5500°\text{K for } \bar{V}_{ij} = 7 \text{ eV}$$

$$T_M \approx 5000°\text{K for } \bar{V}_{ij} = 6 \text{ eV}$$

$$T_M < 4700°\text{K for } \bar{V}_{ij} \leqslant 5 \text{ eV}$$

The figure also demonstrates that for finding approximate values of T_M in the range of effective temperatures between 4500 and 6500°K, it makes no great difference whether we base our conclusions on the continuous curves or on the dashed ones. Therefore, we feel justified in examining this matter somewhat more closely in terms of intensity $\hat{I}_s$, effective temperature $\hat{T}$, and effective electron pressure $\hat{p}_e$, neglecting, however, variations of the effective radius $\hat{R}_s$. This approximate consideration, presented in the following section, thus links with the dashed curves of Fig. 8.11.

★ § 8.8 *Optimum temperatures for the emission of different atom lines*

In conjunction with the exposition in the preceding section we now deal in a more general way with the calculation of optimum temperatures (T_M) for the emission of spectral lines. The problem has been considered for a single gas by Larenz (1951); for the composite gas in the arc, Semenova and Durkina (1957) have made calculations. The treatment below follows a similar line of argument to that developed in Semenova and Durkina's theory, although the way of presenting the material is different.

Prior to taking up the pertinent mathematical equations we first note that in the temperature range at present under discussion, say 4000–7000°K, the concept of optimum temperature (*Normtemperatur* after Larenz) applies to atom lines only, because the second stage of ionization is still negligible here. Secondly, we recall the qualitative picture of optimum temperatures produced in § 6.12, which can be summarized as follows. Given the sum of the concentrations of the neutral atoms and the singly-charged ions of an element in a gas mixture, the absolute level population of an excited state of the atom usually varies with temperature according to a function with a maximum. This is the result of two opposing effects: the relative population in the level tends to increase according to Boltzmann's law, whereas the absolute number of the neutral atoms tends to decrease in favour of that of the ions after Saha's relationship.

Mathematically, we discuss the problem first in terms of the emission J of an atom line, which is proportional to the absolute level population. Remembering equation (8.5) we write

$$\log J = \log C + \log n + \log (1 - \alpha) - \frac{5040}{T} V_q \qquad (8.19)$$

where, for the sake of convenience, we have dropped the subscript qp, denoting the pertinence of J and C to a particular transition q → p, and the subscript j, designating that n and α pertain to a particular element; in addition we have considered the partition function Z_a using the apparent excitation potential V_q (cf. § 6.3).

Applying the condition for a maximum or minimum

$$\frac{d \log J}{dT} = 0 \qquad (8.20)$$

we find

$$\frac{d}{dT} \log (1 - \alpha) = \frac{d}{dT} \left(\frac{5040}{T} V_q \right) \qquad (8.21)$$

or

$$\frac{1}{1 - \alpha} \frac{d\alpha}{dT} = m \frac{5040}{T^2} V_q \qquad (8.22)$$

where $m = 2.303$.

Equation (8.22) expresses that the relative rate at which the relative number of neutral atoms decreases with temperature equals the relative rate at which the relative population of the upper level q increases with temperature. The fulfilment of this condition usually gives rise to a principal maximum of the intensity function $J(T)$.

Considering that

$$d \log(1 - \alpha) \equiv \frac{1}{m} \frac{d\alpha}{1 - \alpha} = \frac{\partial \log(1 - \alpha)}{\partial T} dT + \frac{\partial \log(1 - \alpha)}{\partial \log p_e} \frac{\partial \log p_e}{\partial T} dT \qquad (8.23)$$

and remembering equations (7.29) and (7.31) we find

$$\frac{1}{1 - \alpha} \frac{d\alpha}{dT} = \frac{\alpha}{T} \left(\frac{5}{2} + m \frac{5040}{T} V_i - mT \frac{\partial \log p_e}{\partial T} \right) \qquad (8.24)$$

Equating the right-hand sides of (8.22) and (8.24) eventually yields a transcendental equation which expresses the optimum temperature T_M as a function of the apparent ionization potential $\bar{V}_i$ of the element and the apparent excitation potential $\bar{V}_q$ of the line. In order to keep closely in line with the physical meaning of the mathematical expressions, we consider a graphical solution of the equation, involving the plotting of two sets of curves, viz.

(a) $m\dfrac{5040}{T^2}\bar{V}_q$ versus T, with parameter $\bar{V}_q$,

(b) $\dfrac{\alpha}{T}\left(\dfrac{5}{2}+m\dfrac{5040}{T}\bar{V}_i-mT\dfrac{\partial \log p_e}{\partial T}\right)$ versus T, with parameter $\bar{V}_i$.

A point of intersection of a curve of type (a) and a curve of type (b) represents the optimum temperature that corresponds with a particular pair of $\bar{V}_i$ and $\bar{V}_q$ values.

Clearly, the behaviour of the function that expresses the relative rate at which the atomic population is depleted by ionization [depicted by curve (b)] depends largely on the interrelation between the electron pressure and the temperature. As an example, Fig. 8.12 shows for an element with $\bar{V}_i = 7$ eV the trend of curve (b) in the following cases:

1 p_e = constant = 4.10^{-4} atm.
2 p_e = constant = 10^{-3} atm.
3 p_e = constant = 4.10^{-3} atm.
4 p_e is taken to increase from 4.10^{-4} atm at 4500°K to 4.10^{-3} atm at 6500°K, at a rate conforming with $\partial \log p_e/\partial T = 0{\cdot}05$ per 100°K.
5 p_e is taken to decrease from 4.10^{-3} atm at 4500°K to 4.10^{-4} atm at 6500°K, at a rate conforming with $\partial \log p_e/\partial T = -0{\cdot}05$ per 100°K.

Curves of type (a) for different excitation potentials $\bar{V}_q$ have also been included in the diagram to show the optimum temperatures T_M for different spectral lines. For comparison Fig. 8.13 plots the degree of ionization versus the temperature for the five cases considered in Fig. 8.12.

When examining the diagrams we notice first the influence of the absolute value of the electron pressure upon the location of the optimum temperature T_M for cases where p_e is constant throughout. Obviously, T_M shifts to higher values as p_e becomes larger. We readily understand this if we consider that the larger p_e is, the more the ionization of individual vapour components is depressed. Remember, of course, that p_e is principally a function of the plasma composition (cf. §7.9) and that consequently a larger value of p_e ensues as the concentration of metallic constituents in the plasma increases, resulting in smaller degrees of ionization for all plasma constituents. This effect is plainly illustrated by the curves representing the degree of ionization α as a function of temperature in Fig. 8.13. Mark that it is the relative slope of the α versus T curves, i.e. $[1/(1-\alpha)]d\alpha/dT$, which eventually controls the location of the optima. These appear when the relative slope has assumed a fairly substantial value; not until then is the ionization effect capable of competing with and balancing the excitation effect.

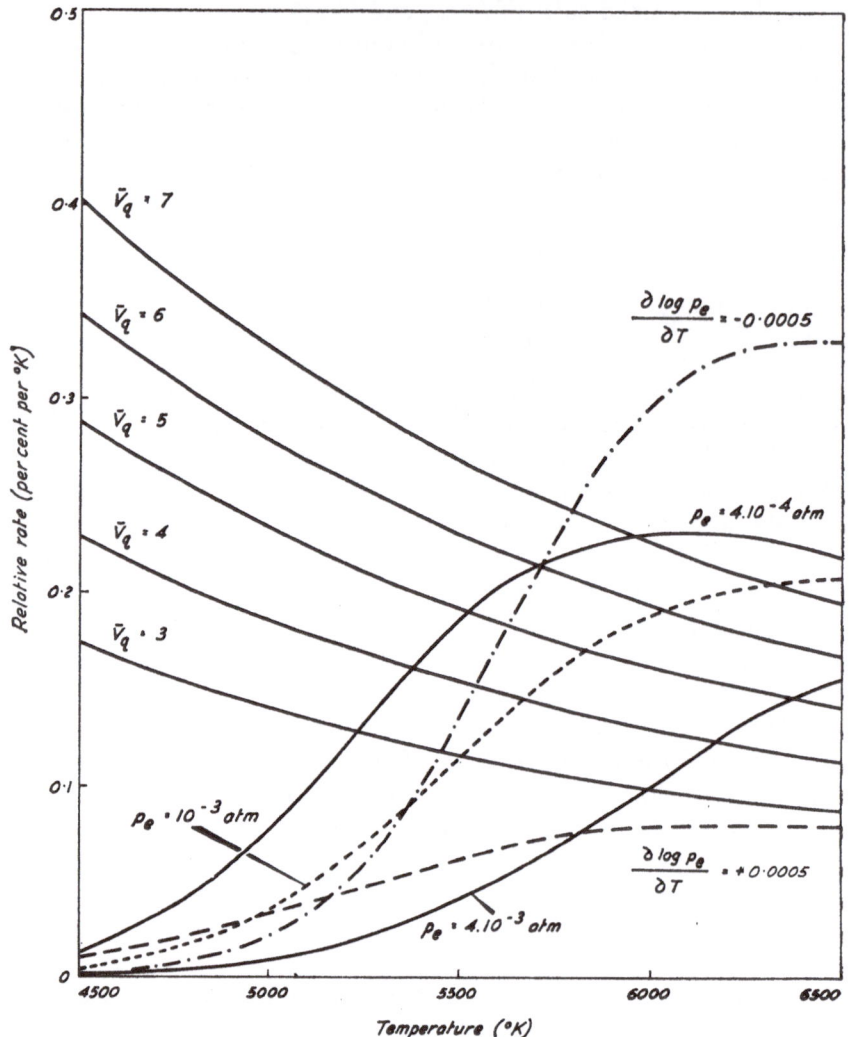

FIG. 8.12. An illustration of the influence of the behaviour of electron pressure (p_e) upon the location of optimum temperatures (T_M) for the emission of atom lines of an element with an apparent ionization potential $\bar{V}_i = 7$ eV.

The curves descending from left to right depict the function $2\cdot303(5040/T^2)\bar{V}_q$, i.e. the relative rate at which the relative population of an atomic level q with an apparent excitation potential $\bar{V}_q$ increases with temperature. The ascending curves represent the function $[1/(1-\alpha)]d\alpha/dT$, i.e. the relative rate at which the relative number of neutral atoms decreases with temperature. The abscissa of a point of intersection of a curve of the former category with one of the latter denotes the optimum temperature for the emission of the relevant atom line.

The function $[1/(1-\alpha)]d\alpha/dt$ is considered for three constant values of p_e (4.10^{-4}, 10^{-3}, and 4.10^{-3} atm) and for when a correlation between p_e and T exists, either positive or negative.

It is interesting that when a strongly positive correlation exists between p_e and T, ionization of vapours proceeds much more slowly with temperature than for constant p_e (see dashed curve in Fig. 8.13). The reason is that the increase of α caused by the rise of T alone is counteracted by the effect of increasing p_e, which, inherently, tends to repress ionization. For the case depicted in Fig. 8.12 (dashed curve) we see that an optimum temperature is not found unless the apparent excitation potential of the spectral line is below about 3 eV.

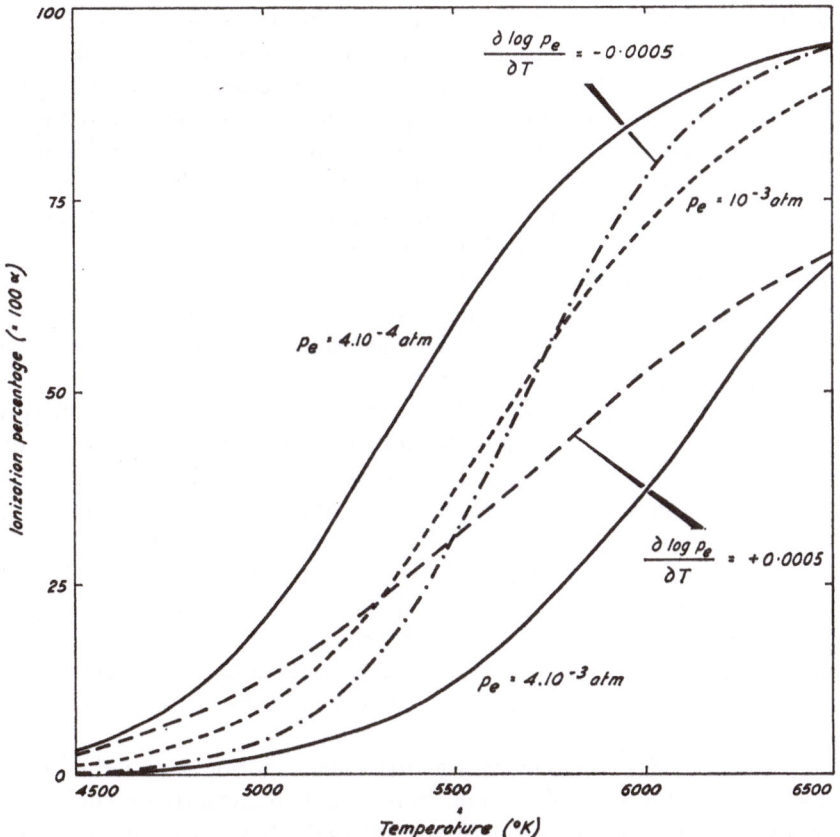

FIG. 8.13. Ionization percentage for an element of apparent ionization potential $V_i = 7$ eV plotted against temperature. The diagram shows the influence both of the magnitude of the electron pressure (p_e) and of its dependence on temperature. The cases considered are similar to those referred to in Fig. 8.12.

A strongly positive correlation between p_e and T is displayed radially in the arc. It is not surprising, therefore, that we did not find pronounced intensity maxima at some distance from the arc axis in radial emission profiles, except for some low-level lines of elements of low ionization potential (see Fig. 8.7). In general, atom line emission is rather smeared out over the arc cross-section (see Fig. 8.5, for instance).

The reverse case, i.e. the one with a strongly negative correlation between p_e and T, is exemplified by the dashed–dotted line in Figs. 8.12 and 8.13. Such a correlation entails that a variation of T and the corresponding variation of p_e shift ionization equilibria in the same direction, causing the ionization to occur faster with increasing temperature than if p_e were constant.

Summarizing, Fig. 8.12 shows that to predict exactly optimum temperatures the behaviour of electron pressure as a function of temperature should be accurately known. The correlation between p_e and T that applies in a particular case depends to a considerable extent on the arc conditions, as has been mentioned in § 7.9. This prevents us, unfortunately, from drawing definite conclusions. However, some generalized inference is of use for special cases. Let us attempt, therefore, to find out which temperatures are generally most favourable for producing low *detection limits* for elements. Evidently, when treating this problem, we use effective temperatures and electron pressures, although we shall not express them explicitly. Before calculating the optimum temperatures, we shall consider which means are at our disposal to control the temperature in the arc.

We have seen in §§ 7.9 and 7.12 that when the correct buffer is added to a sample a definite arc temperature is reached, the magnitude of which depends chiefly on the ionization potential of the buffer element. So, for example, in a 10-amp arc a buffer of KF–graphite yields $T = 5000$–$5200°K$, of LiF–graphite $T = 5500$–$5800°K$, and of Al_2O_3–graphite $T = 5900$–$6100°K$. The electron pressure associated with each of these buffers depends largely on the vapour concentration of the buffer element and thus on the rate of entry, which is controlled by the type of buffer compound, the electrode shape, and the ratio of buffer compound to graphite. In general, we may expect the electron pressure to be about 10^{-3} atm, provided that in unit time a sufficient amount of buffer element is released from the electrode. Therefore, we feel justified in calculating optimum temperatures for $p_e =$ constant $= 10^{-3}$ atm. So we have a reasonable compromise that may well be typical of the general case. Results of such a calculation are given in Fig. 8.14, where the pertinent curves for $\bar{V}_q = 3, 4, 5, 6, 7$, and 8 eV, and for $\bar{V}_i = 5, 6, 7, 8$, and 9 eV, have been plotted.

Before drawing final conclusions we turn our attention to another important point that must not be overlooked here. In our mathematical treatment of optimum temperatures founded in equation (8.19), all deductions pertain to a source of unit volume with a given concentration n of the relevant element in the vapour phase. In spectrochemical analysis, however, it is not the concentration in the vapour phase, but that in the solid sample which is given; and consequently we should incorporate the transport mechanism in our considerations. Anticipating a thorough discussion of this topic in Chapter 9, we mention here an experimental result regarding the transport of vapours through the discharge zone (see particularly § 9.4).

When samples, diluted in various buffer admixtures covering the temperature range between 5000 and 6000°K, are evaporated from the anode into a free-burning 10-amp d.c. arc of 10-mm gap width, the particle concentration n $(= n_a + n_i)$ in the middle of the gap is uniformly distributed in the arc cross-section at least up to 0·4 cm from the arc axis. The average concentration can be related to the rate of evaporation Q (number per sec) by a transport parameter ψ (cm³/sec) such that

$$\overline{\log n} = \log Q - \log \psi \qquad (8.25)$$

where $\log \psi$ is given by the empirical relationship

$$\log \psi = 2·5 + \alpha \qquad (8.26)$$

The equation expresses that the more strongly a component of the vapour is ionized, the more rapidly it passes through the discharge channel. This transport is caused by the longitudinal electric field.

TABLE 8.5

$\bar{V}_i$ (eV)	$\bar{V}_q$ (eV)	$T_M(°K)$ without transport function	$T_M(°K)$ with transport function
5	3	<4500	≪4500
	4	≈4500	≪4500
	5	4750	<4500
·6	3	4900	4500
	5	5250	4700
7	3	5500	5100
	5	5850	5300
8	3	6100	5650
	5	6400	5850
9	3	>6500	6200
	5	>6500	6450

Optimum temperatures (T_M) for the emission of atom lines of apparent excitation potential $\bar{V}_q$ of elements of apparent ionization potential $\bar{V}_i$. Column three gives data for a plasma of given concentration (= number of atoms and ions per unit volume) of the relevant element (cf. Fig. 8.14). Column four presents data pertaining to a 10-amp d.c. arc, where vapour transport through the plasma must be taken into account (cf. Fig. 8.15). Calculations in the latter case are based on an empirically established transport function expressed by equations (8.25) and (8.26) (see § 9.4 for details).

Identifying $\overline{\log n}$ in equation (8.26) with $\log n$ in equation (8.19), we insert (8.25) and (8.26) into (8.19) and again apply the condition for a limit (8.20), obtaining

$$\frac{1}{1-\alpha}\frac{d\alpha}{dT} + m\frac{d\alpha}{dT} = m\frac{5040}{T^2}\bar{V}_q \qquad (8.27)$$

The left-hand side of this equation is readily computed from (8.24) by multiplying the latter by $m(1·434 - \alpha)$.

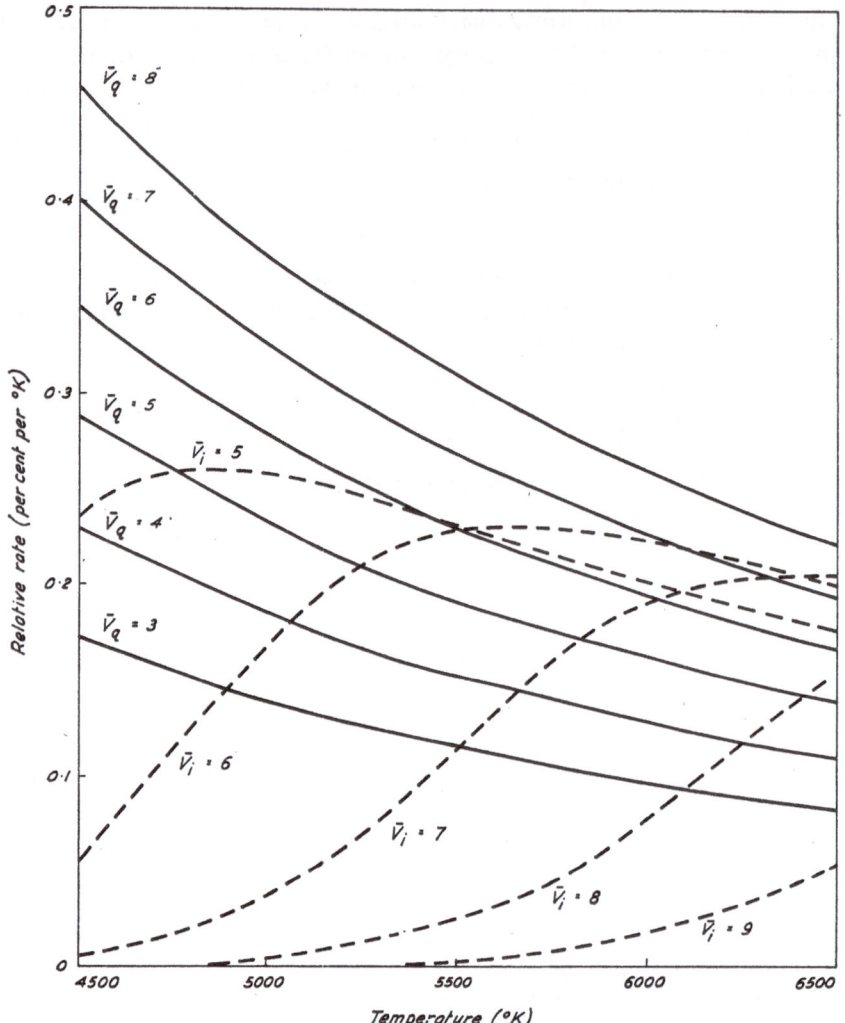

FIG. 8.14. Curves for deriving optimum temperatures (T_M) of atom lines of
different elements. The diagram pertains to a source of unit volume with a
given concentration n of the relevant element in the vapour phase. The electron pressure has been assumed constant and equal to 10^{-3} atm.

Continuous curves depict the function $2 \cdot 303(5040/T^2)\bar{V}_q$, i.e. the relative rate
at which the relative population of an atomic level q with apparent excitation
potential $\bar{V}_q$ increases with temperature. Broken curves represent the function
$[1/(1-\alpha)]d\alpha/dT$, i.e. the relative rate at which the relative number of neutral
atoms decreases with temperature. The abscissa of a point of intersection of a
continuous and a broken line denotes the optimum temperature for the emission
of the relevant line [cf. equation (8.22)].

For finding optimum temperatures valid for the arc the *transport* of vapours
through the plasma must be also considered (see Fig. 8.15).

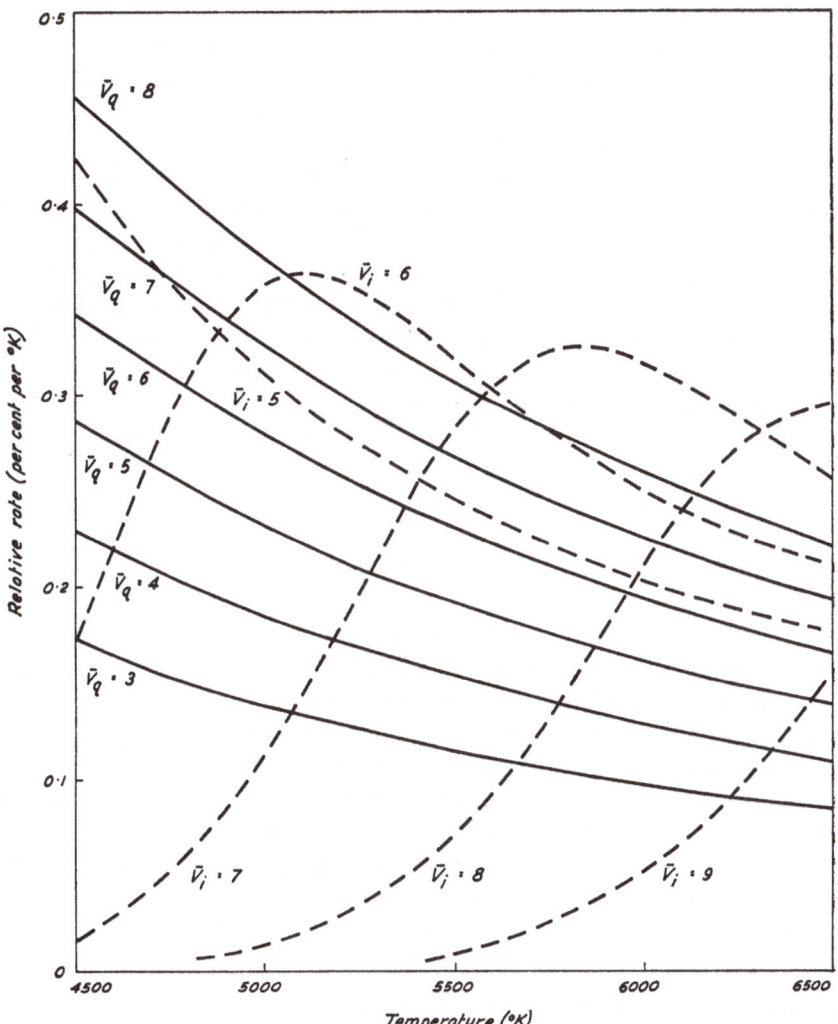

FIG. 8.15. Curves for deriving optimum temperatures (T_M) of atom lines of different elements. In contrast with Fig. 8.14, the presentation here fits in with the actual situation in the arc, as it also takes the vapour transport into account. This appears from the broken curves, which now depict the function $[1/(1-\alpha)]d\alpha/dT+2\cdot303 d\alpha/dT$, the second term being a transport term originating from an empirical relationship established by de Galan [equation (8.26)]. The effect of the transport is to enhance the slope of the dashed curves markedly, with the result that optimum temperatures shift to lower values. The fact is that the more rapidly a vapour passes through the discharge column the more strongly it is ionized. So, a high concentration is favoured by a relatively low temperature.

As previously, the continuous curves depict the function $2\cdot303(5040/T^2)\bar{V}_q$. Strictly speaking, the dependence of the effective arc radius on temperature (§§ 8.4 and 8.5) were to be incorporated in the picture. As its influence is so small that optimum temperatures would be displaced only by about $+50°$K, it has been left out of consideration here (see § 8.9, however).

Curves that take this transport mechanism into account have been plotted in Fig. 8.15. Comparison of this figure with Fig. 8.14 shows that the optimum temperatures have shifted to lower values; evidently, since the transport equations (8.25) and (8.26) involve that a high concentration of a metal vapour in the plasma is favoured by *suppressing* the ionization of that vapour.

Numerical values of T_M for different apparent ionization potentials of elements and for different excitation potentials of lines have been listed in Table 8.5. Both cases, one including the transport function and the other ignoring it, have been considered. The transport is seen to entail a decrease of T_M by 400–550°K. Also, we infer that spectral lines of equal excitation potential of elements differing by 1 eV in ionization potential have optimum temperatures that differ by about 600°K. In turn, a difference of 1 eV in excitation potential for lines of elements of equal ionization potential corresponds to a shift of T_M by about 100°K.

The eventual conclusions for spectrochemical analysis founded on Fig. 8.15 and Table 8.5 will be discussed in the following section.

★ §8.9   *Some conclusions on spectrochemical analysis concerning detection limits and internal standards*

Let us now state, in connection with the previous discussion, a few conclusions concerning spectrochemical analysis. We shall first deal with the problem of choosing excitation conditions that are favourable for attaining low detection limits,* and then with the selection of internal standards.

Studying Fig. 8.15 and considering the arrangement of elements according to apparent ionization potential in Table 8.6 (cf. Table 7.1), we see that the optimum temperatures for the emission of atom lines ($\bar{V}_q = 3$–5 eV) cover a very wide range. We conclude from this that we must excite separate portions of a sample under different conditions (as created by different admixtures) and that the elements must be analysed according to spectrochemical grouping. In doing this we should not neglect to take into account two other points, viz.

1 The intensity and structure of the background.
2 The use of ion lines.

Naturally, it is not the spectral-line intensity alone, but the contrast between line and background (signal-to-noise ratio) which determines whether a line can be detected or not. Whatever instrumental factors influence the selection of optimum conditions for determining lines of given intensity against a given background, with photographic recording attempts must always be made to enhance

---

* A noteworthy discussion on the concept of the detection limit has recently been given by Kaiser (1965). In this connection I mention further Kaiser (1964, 1947), Kaiser and Specker (1956), Kaiser, Massmann, and Hagenah (1962), Nalimov, Nedler, and Men'shova (1961), Zaidel, Malyshev, and Schreider (1964), Burmistrov, Nalimov, and Nedler (1964), Gerbatsch and Scholze (1963), Hubaux and Smiriga–Snoeck (1964), Hobbs and Smith (1966), Mandelshtam and Nedler (1961), Zaidel, Kaliteevskii, Lipis, and Chaika (1960), Fratkin (1960), and the Proceedings of a Conference on Limitations of Detection in Spectrochemical Analysis (see Exeter, 1964).

line intensity, to reduce the level of the background, and to diminish its fluctuations.* Therefore, when searching for the most favourable excitation conditions, we have to consider, in addition to the factors mentioned in the preceding section, the behaviour of the background including the contribution from the blanks. Some aspects of this question are considered below, others will be discussed elsewhere (see Boumans, 1966; and Boumans and Maessen, 1966).†

As has been mentioned in § 7.10, I found in the temperature range 5000–6000°K a minimum background intensity in a 10-amp LiF arc with a moderate lithium concentration in the plasma ($T = 5700$–5800°K). With a greater lithium concentration, the background becomes more intense but shows less band structure ($T \approx 5500$°K). Higher temperatures (Al_2O_3 arc, $T = 6000$°K, or graphite arc, $T = 6200$°K) give rise to a very intense background with dominating band structure; many spectral lines that are clearly visible in the LiF arc are now obscured. For this reason and in view of Fig. 8.15 and Tables 8.5 and 8.6 we recommend the use of a buffer such as LiF or Li_2CO_3 (see § 7.12, however) for establishing conditions in the plasma where a high overall detectability can be guaranteed. Obviously, better detection limits for specific elements can be pursued. The alkali metals require the lowest possible temperature, while the volatile elements of high ionization potential, Cd, Sb, Zn, Te, As, Hg, and P, tend to have better detection limits when high-temperature excitation is chosen, particularly if selective volatilization is applied to avoid excess background. In the latter group of elements, boron must be included if it is fluoridized (cf. §§ 7.12 and 10.4); for this element, high-temperature excitation must be favoured as it also promotes the dissociation of the stable BO molecule (see § 11.3).

At the same time I point out that for a number of elements ion lines can be advantageously used for detection (see Table 8.6).‡ Ion line emission is stimulated by high-temperature excitation. We should bear in mind, however, that ion line intensities are often very sensitive to variations in temperature and electron

* Of course, the recording level of the background on the plate should be sufficiently high, that is, the density must be at least 0·1 to 0·2 (cf. Kaiser, 1947).

† I must remark here that the conclusions constitute only one aspect in the vast number of questions concerning the improvement of detection limits. In addition to instrumental factors, I mention the problem of the blanks and the importance of enrichment and separation procedures, either chemical or physical, among which should be especially noted the selective volatilization of impurity elements in the arc and the preconcentration of such elements by evaporating them, prior to arcing, from the (refractory) base on to a cooled collector electrode (cf. § 10.4).

Of references on trace and ultra-trace analysis I cite but a few that may give the reader further access to this important branch of analytical chemistry, viz. Zaidel, Kaliteevskii, Lipis, and Chaika (1960), Grimaldi and Helz (1961), Knížek and Provazník (1961), Morrison and Rupp (1961), Czakow and Minczewski (1962b), Minczewski (1962a and b, 1963), Cook, Crespi, and Minczewski (1963), Alimarin (1963), Moenke–Blankenburg (1964), Ehrlich (1964), Ehrlich and Rexer (1964), Koch and Koch–Dedic (1964), and Mitteldorf (1965).

‡ The easily-ionized alkali metals are missing in this group because the ions have the same electronic structure as inert gases. The ground state of the alkali metals ions is thus very stable and excitation to higher levels does not occur in the arc. Likewise in virtue of their electron configurations, ions of B, Al, Ga, In, and Tl are not easily excited. A general discussion on relations between electronic structure and line intensity has been provided by Meggers (1941a and b).

pressure, especially if an element is ionized only to a small degree. Let us consider the intensity function for ion lines and compare it with that for atom lines. Written

TABLE 8.6

$\bar{V}_i < 5.5$	4.4 Cs	4.6 Rb	4.7 K	5.0 Ba(+)	5.5 Sr(+)	5.5 Na		
$5.5 < \bar{V}_i < 6.5$	5.5 La(+)	5.7 Li	5.8 Ca(+)	6.2 Sc(+)	6.3 Y(+)			
$6.5 < \bar{V}_i < 7.5$	6.5 In	6.5 Ti(+)	6.5 Tl	6.7 Zr(+)	6.8 Ga	6.8 V	6.8 Al	6.9 Cr
	7.0 Nb	7.2 Mo	7.3 Pb	7.3 Mg(+)	7.3 Mn(+)			
$7.5 < \bar{V}_i < 8.5$	7.5 Ru	7.6 Sn	7.7 Fe	7.7 Rh	7.8 Ta	7.9 Re	7.9 Bi	7.9 Co
	7.9 W	7.9 Pd	8.0 Ag	8.1 Ni	8.1 Cu	8.2 Ge	8.4 Si	
$8.5 < \bar{V}_i$	8.6 Os	8.7 Cd	8.8 Sb	9.0 Be(+)	9.1 Zn	9.1 B	9.2 Te	
	9.3 Pt	9.6 Au	9.7 As	10.0 Hg	10.2 P			

Elements arranged according to their apparent ionization potential $\bar{V}_i$ (cf. § 7.3). Elements marked with a (+) have ion lines that are often used advantageously for detection.

in the form of equation (8.19), the intensity function (8.7) for an ion line reads

$$\log J^+ = \log C^+ + \log n + \log \alpha - \frac{5040}{T} \bar{V}_q^+ \qquad (8.28)$$

Substitution of the transport functions (8.25) and (8.26), and differentiation with respect to T yields

$$\frac{d \log J^+}{dT} = -\frac{d\alpha}{dT} + \frac{0.434}{\alpha} \frac{d\alpha}{dT} + \frac{5040}{T^2} \bar{V}_q^+ \qquad (8.29)$$

This function consists of three terms:

1   The transport term, $-\dfrac{d\alpha}{dT}$.

2   The ionization term, $\dfrac{0.434}{\alpha} \dfrac{d\alpha}{dT}$.

3   The excitation term, $\dfrac{5040}{T^2} \bar{V}_q^+$.

The ionization and excitation influences (2 and 3) act in the same sense and are partly compensated by the transport effect, which is capable of balancing the ionization effect when α is greater than 0·434. Although the sum of 1 and 2 for $\alpha > 0\cdot434$ is negative, it does not assume values that would balance the excitation effect; for, the derivative $d\alpha/dT$ becomes small again when α is large. So, a limit does not occur.

Table 8.7 gives the magnitude of the change of ion line intensity (expressed as the factor by which intensity is increased) for a rise in temperature of 100°K at

TABLE 8.7

$\bar{V}_q$	Ion lines $\bar{V}_i$					Atom lines $\bar{V}_i$				
	5	6	7	8	9	5	6	7	8	9
3	1·11	1·08	1·12	1·43	1·58	0·88	0·82	0·82	1·02	1·10
4	1·15	1·12	1·16	1·48	1·67	0·92	0·86	0·86	1·05	1·14
5	1·20	1·16	1·20	1·54	1·73	0·95	0·89	0·89	1·10	1·19

Sensitivity of ion and atom line intensity for temperature variations according to equations (8.29) and (8.30).

Sensitivity has been expressed as the factor by which the intensity changes if T increases by 100°K. Results pertain to the specific case of an arc with a temperature of 5600°K and an electron pressure of 10^{-3} atm.

$\bar{V}_i$ and $\bar{V}_q$ are the apparent ionization potential of the element and the apparent excitation potential of the line respectively.

A glance at Fig. 8.16 shows that the results represent only a specific case, since the sensitivity depends largely on the temperature under consideration. So, for example, at the temperature chosen here the sensitivity values for atom line intensities of elements with apparent ionization potentials of 6 and 7 eV happen to coincide. The reason is that the values of

$$\frac{d\alpha}{dT} + \frac{0\cdot434}{1-\alpha}\frac{d\alpha}{dT}$$

for $\bar{V}_i = 6$ and $\bar{V}_i = 7$ eV just match at $T = 5600$°K (cf. Fig. 8.15, where the broken lines for $\bar{V}_i = 6$ and $\bar{V}_i = 7$ are seen to intersect at $T = 5600$°K).

$T = 5600$°K. Electron pressure has been taken as constant ($= 10^{-3}$ atm). For comparison, numerical values for atom lines have been also listed. Here the effect of temperature fluctuations on line intensity, i.e.

$$\frac{d\log J}{dT} = -\frac{d\alpha}{dT} - \frac{0\cdot434}{1-\alpha}\frac{d\alpha}{dT} + \frac{5040}{T^2}\bar{V}_q \tag{8.30}$$

is zero in the optimum (T_M) and positive or negative depending on whether the temperature is below or above T_M.

The plots of Fig. 8.16 give an overall picture of spectral-line intensity as a function of temperature. They show the behaviour of the intensity of atom and ion lines with apparent excitation potentials ($\bar{V}_q$) of 3, 5, and 7 eV, and belonging to elements with an apparent ionization potential ($\bar{V}_i$) of 5, 6, 7, 8, and 9 eV. The figure demonstrates the cumulative effect of separate contributions to the logarithmic intensity Y. Evaluation was conducted as follows.

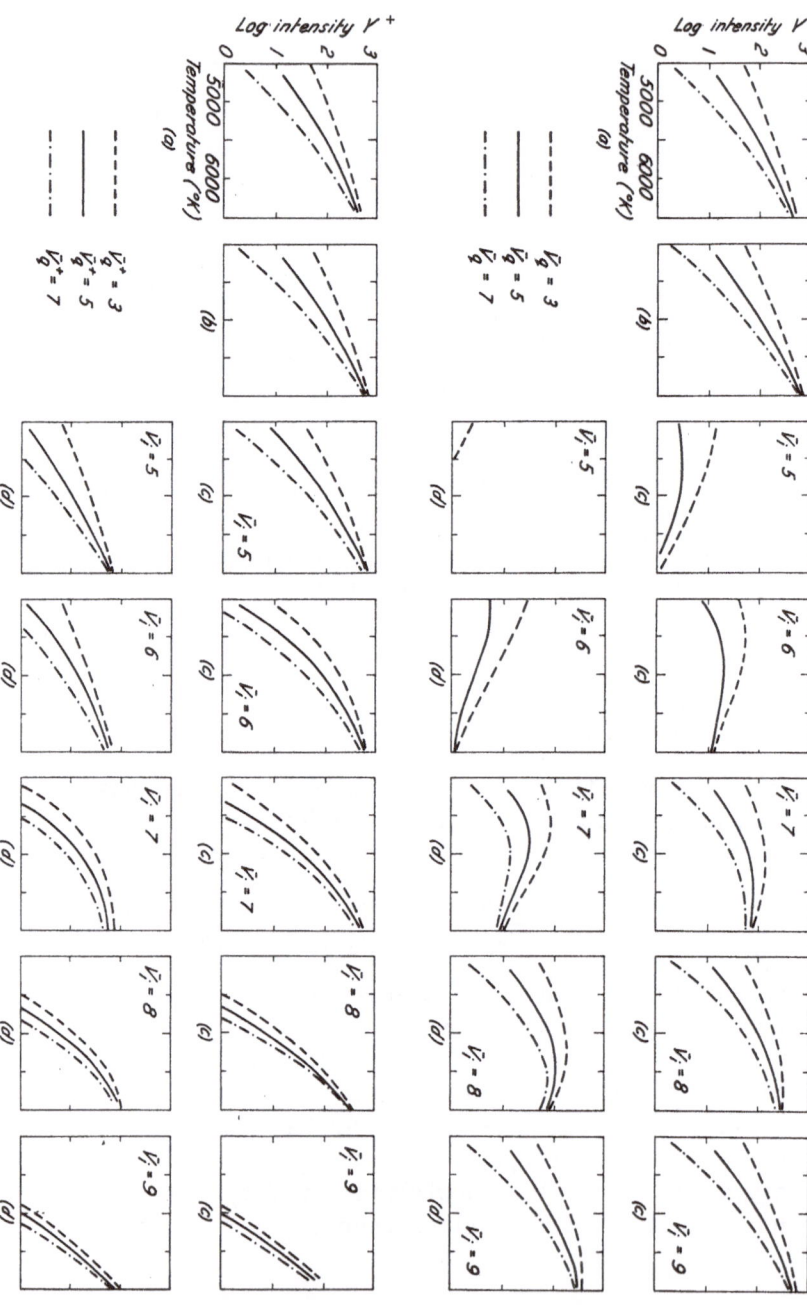

FIG. 8.16. Survey of the behaviour of atom and ion line intensities (expressed as logarithms Y and Y⁺) in the arc, as predicted by cumulating the effects of (1) the Boltzmann exponential factor, (2) the effective arc radius ($\hat{R}$), (3) the ionization factor, (4) the transport factor. Successive diagrams thus represent the functions (a) $Y_a = \log K - (5040/T)V_a$, (b) $Y_b = \log K - (5040/T)V_a + \log \hat{R} + \log(1 - \alpha)$, (c) $Y_c = \log K - (5040/T)V_a + \log \hat{R} + \log(1 - \alpha)$, (d) $Y_d = \log K - (5040/T)V_a + \log \hat{R} + \log(1 - \alpha)$ for apparent ionization potentials (V_i) ranging from 5 to 9 eV. The of the apparent excitation potential ($V_a = 3, 5,$ and 7 eV) and for apparent ionization potentials (V_i) ranging from 5 to 9 eV. The constant log K has been arbitrarily assigned the values 5, 6·5, and 8 for $V_a = 3, 5,$ and 7 eV successively.

Rows one and two refer to the neutral atom, rows three and four to the singly-charged ion, in which case log(1 − α) in the functions Y_c and Y_d has been replaced by log α. Evidently, diagrams (a) and (b) are similar for atoms and ions.

Assuming the arc to be focused on the spectrograph slit* we write for the log relative intensity (Y) of an atom line

$$Y \equiv \log I = \log J + \log \hat{R} + \log F_i \qquad (8.31)$$

where J is the emission per unit volume per unit solid angle, as given by (8.19) in combination with the transport functions (8.25) and (8.26), $\hat{R}$ is the effective arc radius, and F_i is an instrumental factor whose magnitude is independent of the characteristics of the source and the nature of the spectral line.

The concept of effective arc radius was considered at length in §§ 8.4 and 8.5. Accordingly, this parameter defines the thickness of a homogeneous gaseous layer having a temperature and electron pressure equal to the effective temperature and electron pressure of the arc and yielding the same radiation output as the arc. The value of this parameter was seen to be independent, to a first approximation, of the type of line, except for low-level atom lines of easily-ionized elements. On the basis of the applied model, $\hat{R}$ can be stated to change on the average by a factor of 1·3 to 1·4 when the effective temperature changes from 4800 to 6200°K. This order of magnitude of the variation agrees well with experimental values recently determined by de Galan (1965a). To take such a variation into account when constructing Fig. 8.16, we assumed, for convenience, that $\log \hat{R}$ increases at a constant rate of 0·007 per 100°K, which is equivalent to a factor of 1·38 if T changes from 4500 to 6500°K.†

Inserting equations (8.19), (8.25), and (8.26) into the intensity function (8.31) leads to

$$Y = \log K - \frac{5040}{T} \bar{V}_q + \log \hat{R} + \log(1 - \alpha) - \alpha \qquad (8.32)$$

where

$$\log K = \log C + \log Q - 2 \cdot 5 + \log F_i \qquad (8.33)$$

Subsequent diagrams in the first two rows of Fig. 8.16 depict the trend of the functions

(a) $Y_a = \log K - \dfrac{5040}{T} \bar{V}_q,$

(b) $Y_b = \log K - \dfrac{5040}{T} \bar{V}_q + \log \hat{R},$

(c) $Y_c = \log K - \dfrac{5040}{T} \bar{V}_q + \log \hat{R} + \log(1 - \alpha),$

(d) $Y_d = \log K - \dfrac{5040}{T} \bar{V}_q + \log \hat{R} + \log(1 - \alpha) - \alpha \equiv Y.$

* When the arc is imaged on the collimator lens, $\hat{R}$ should be replaced by $\hat{R}^2$, while F_i must be similarly adapted.

† Strictly speaking, we should have also considered the variation of the effective radius with temperature when computing optimum temperatures in the preceding section. As its influence on T_M is only about +50°K, we have ignored it.

Similarly, in rows three and four, diagrams are produced for corresponding ion lines. The plots (a) and (b) are identical with those for atom lines, whereas plots (c) and (d) differ in that the term $\log(1 - \alpha)$ has been replaced by $\log \alpha$.

In summary, Fig. 8.16 displays the cumulative effect of

1 The Boltzmann term, $-\dfrac{5040}{T}V_{\mathrm{q}}$,

2 The effective arc radius, $\log \hat{R}$,

3 The ionization term, $\log(1 - \alpha)$ or $\log \alpha$,

4 The transport term, $-\alpha$,

when we pass from an element of low ionization potential ($V_i = 5$ eV) to one of high ionization potential ($V_i = 9$ eV). Clearly, the influence of the effective arc radius is very small. The direct effect of ionization, on the contrary, changes the picture substantially, while its indirect effect via the transport process brings about a gradual change that quantitatively is far from negligible. In accordance with the exposition on optimum temperatures given in § 8.8 (see especially Figs. 8.14 and 8.15) the diagrams of Fig. 8.16 demonstrate the displacement of the intensity maxima with the ionization potential of the element and with the excitation potential of the line, as well as the shift of the optima to lower temperatures caused by the transport. Fig. 8.16 also discloses, in a self-explanatory way, that atom and ion line intensities exhibit an entirely different behaviour, as discussed above. [Parenthetically we must note here that the plots for the low-level atom line with $V_{\mathrm{q}} = 3$ eV do not depict the true situation for the easily-ionized elements having an apparent ionization potential of 5 or 6 eV (see §§ 8.5 to 8.7).]

Let us finally see what can be inferred from the preceding exposition to help in the choice of an internal standard. We use an internal standard to eliminate the influence of fluctuations in the excitation process upon spectro-analytical results. In the ideal case, these fluctuations should affect both the analysis line and the internal-standard line to the same extent, so that their ratio does not depend on the excitation conditions. An ideal internal standard would thus be an isotope of the analysis element. This being normally impracticable, one must resort to other combinations of analysis elements and internal standards, which inevitably entail the disadvantages of compromise. Compromises are necessary also because practical factors, originating from the nature of the sample and the needs of the analysis, restrict the choice of the internal standard (see e.g. Ahrens and Taylor, 1961, pp. 91 ff.).

When searching for a satisfactory internal standard we should take two major factors into consideration: how does the element volatilize and how does it behave in the plasma? Obviously, an internal standard must evaporate in a similar manner to the analysis element, so that both are simultaneously present in the plasma. Practice has shown that volatilization characteristics usually outweigh the excitation characteristics, probably because with undiluted samples uncontrolled losses of material occur in the evaporation process. Then, by selecting an internal standard that vaporizes in a similar manner to the analysis element, we attempt to eliminate the influence of such losses upon analytical results. At the same time we want the internal standard to obviate the influence of variations of the excitation conditions in the plasma.

The fulfilment of all *desiderata* must necessarily make high demands upon the pairing of internal standard and analysis element. From the discussion on line intensity in this and in the preceding section we conclude that for an internal standard to be efficacious when large temperature fluctuations occur, it should have the same apparent ionization potential as the analysis element, while at the same time the lines should have equal excitation potentials. Self-evidently, it would seem indispensable that either an atom–atom or an ion–ion line pair be used. There are more possibilities, however, provided that the temperature range to be covered is not too large. The key to a rational choice of line pairs is provided by equations (8.29) and (8.30), and by Figs. 8.15 and 8.16.

For two spectral lines to meet the conditions we should demand that

$$\frac{d \log J_1}{dT} = \frac{d \log J_2}{dT} \tag{8.34}$$

which for an atom–atom line pair can be expressed as a transcendental equation between the respective apparent ionization potentials (V_{i1} and V_{i2}) of the elements and the apparent excitation potentials (V_{q1} and V_{q2}) of the lines, namely

$$-\frac{d\alpha_1}{dT} - \frac{0.434}{1-\alpha_1}\frac{d\alpha_1}{dT} + \frac{5040}{T^2}V_{q1} = -\frac{d\alpha_2}{dT} - \frac{0.434}{1-\alpha_2}\frac{d\alpha_2}{dT} + \frac{5040}{T^2}V_{q2} \tag{8.35}$$

where α_1 and α_2 are functions of V_{i1} and V_{i2} respectively.

Naturally, condition (8.35) can be fulfilled only for all temperatures if $V_{i1} = V_{i2}$ and $V_{q1} = V_{q2}$. However, if the range of temperature is moderate, the effect of a difference between ionization potentials of elements might be sometimes compensated for by an opposite difference of excitation potentials of lines. Let us demonstrate this first for a definite value of T and subsequently extend it to a range of T. Fig. 8.15 enables us to solve equation (8.35) graphically, since continuous lines depict the function

$$f(T, V_q) = m\frac{5040}{T^2}V_q$$

and broken ones the function

$$f_\alpha(T, V_i) = m\left(\frac{d\alpha}{dT} + \frac{0.434}{1-\alpha}\frac{d\alpha}{dT}\right)$$

So, for example, at $T = 5600°\text{K}$, the following line pair satisfies equation (8.35):

$$\begin{cases} V_{q1} = 5 \\ V_{i1} = 9 \end{cases} \quad \text{and} \quad \begin{cases} V_{q2} = 6.5 \\ V_{i2} = 8 \end{cases}$$

since

$$\begin{cases} f(T, V_{q1}) = 0 \cdot 19 \\ f_\alpha(T, V_{i1}) = 0 \cdot 01 \end{cases} \quad \text{and} \quad \begin{cases} f(T, V_{q2}) = 0 \cdot 25 \\ f_\alpha(T, V_{i2}) = 0 \cdot 07 \end{cases}$$

The line pair considered in the example is useful also for a larger temperature range; for, from the last two diagrams in row two of Fig. 8.16 we see that lines characterized by $V_{i1} = 9$, $V_{q1} = 5$ (continuous curve) and $V_{i2} = 8$, $V_{q2} = 7$ (dashed–dotted curve) behave very similarly in the temperature range 5000–6000°K.

The difference in excitation potentials is seen to be advantageous for balancing the difference of ionization potentials. Therefore, we should in general select pairs of lines whose intensity curves are similar in the working range of temperature. Even a glance at the diagrams in rows two and four of Fig. 8.16 shows, however, that this condition is much more easily stated than achieved. Surely, it is possible to find other striking combinations, which, at first sight, show little promise. We mention the atom–atom line pairs

$$V_{i1} = 6, \ V_{q1} = V_{q2} \quad \text{and} \quad V_{i2} = 7, \ V_{q2} = V_{q1}$$

in the temperature range > 5400°K. Conversely, it is impossible to find a proper combination of atom lines if $V_{i1} = 7$ and $V_{i2} = 8$.

Even, efficaceous ion–atom line pairs can be found, if the ion line belongs to an element of low ionization potential and the atom line to one of high ionization potential, e.g. the pair $V_{i1} = 5$, $V_{q1}^+ = 3$ and $V_{i2} = 9$, $V_{q2} = 3$, in the temperature range 5000–6000°K.

To summarize, there exist good analytical line pairs for restricted temperature ranges that do not satisfy the condition $V_{i1} = V_{i2}$ and $V_{q1} = V_{q2}$. Their selection is rather critical, whence we must first see whether a combination that does meet this condition can be selected. A very profitable alternative, however, is to limit the range of excitation conditions by using a proper buffer diluent (see § 7.12). Then rigorous demands upon the excitation characteristics of the internal standard can be dropped and its function is mainly to meet with irregularities in the evaporation and with plate errors. Under such circumstances only a few internal standards, or preferably a single one, will suffice to cover the entire set of analysis elements. Examples of recent applications of this principle are the procedures described by the following:

	Buffer	Internal standard
Frisque (1957)	GeO_2 + graphite	Ge
Joensuu and Suhr (1962)	$LiBO_2$ or $Na_2B_4O_7$	Co
Kroonen and Vader (1963)	Li_2CO_3 + graphite	Li
Marinković (1963)	Li_2CO_3 + graphite	Ge

Earlier published methods are included in a survey by Kemp (1958), who lists in tabular form the operating conditions of some fifty routine methods.

CHAPTER NINE

TRANSPORTATION PHENOMENA

★ § 9.1 *Introduction*

In the general outline of variables in spectrochemistry (§ 1.1) we considered briefly the main features of sample excitation. When the vaporizing sample is released from the electrode, fractions of its constituents enter into the discharge zone, where they dissociate, ionize, and become excited. Most of the elements are present in the plasma partly as neutral atoms, partly as singly-charged ions; only a few form molecules or radicals that are stable at the high temperatures prevailing in the core of the arc (cf. § 11.3). The plasma of the free-burning arc, consisting mainly of the elements of the surrounding atmosphere (see § 7.9), flows continuously upwards by convection (see § 2.3). The sample vapours enter into the upward-flowing gas and are transported by diffusion, by migration under the influence of the axial electric field (§ 3.1 and 3.2), and by the gas flow.

These three processes, *diffusion, migration by the action of the electric field, and convection*, govern the transport of metal vapours through the arc. Although the mechanism of this transport has long since attracted the attention of investigators, it was not until rather recently that the importance of transportation phenomena for spectrochemical inference was fully recognized.

When, in accordance with (1.11), (7.9), (8.3), and (8.4), we write for the intensity of an atom line of element j

$$J_{qp} = C_{qp}(1 - \alpha_j)n_j \frac{\exp(-\epsilon_q/kT)}{Z_a} \qquad (9.1)$$

we must regard the equation as expressing the proportionality between line intensity J_{qp} and number of particles n_j (atoms and ions) in the vapour phase. The constant of proportionality is made up of three factors, viz.

1 C_{qp}, a line constant.
2 $[\exp(-\epsilon_q/kT)]/Z_a$, the Boltzmann exponential factor.
3 $(1 - \alpha_j)$, the ionization factor.

For a uniform source factors 2 and 3 are determined by the temperature and the electron pressure of the gas. Now the crucial point is that we assume the vapour concentration n_j to be proportional to the concentration of the relevant element in the sample. This assumption is the basic premise of spectrochemical analysis; indeed, when we do achieve constant excitation conditions, this proportionality is verified and the constant of proportionality enters along with other constants into the working curves. However, when we examine the influence of excitation conditions upon spectral-line intensities, we should not forget to consider the connection between the concentration n_j of a particular element j in the *source* and its concentration G_j in the *sample*, that is the factor of proportionality K_j in the equation

$$n_j = K_j \times G_j \tag{9.2}$$

It must be recognized that the factor K_j is *not* a constant (like the transition probability), but depends on both the excitation conditions (temperature, electron pressure, electric field strength, current strength, polarity, electrode shapes, etc.) and the nature of the element (ionization potential, diffusivity, and volatility). It is preferable that we should consider two aspects of this matter separately, namely

1 The volatilization of the sample.
2 The passage of the material through the discharge channel.

Tackling the second problem prior to considering what effects are to be expected from volatilization phenomena (Chapter 10) seems to be the best approach.

★ § 9.2   *The passage of material through the discharge zone*
A general outline

We might consider that in total energy procedures the vapour transport through the discharge has no effect on the radiation output, provided that the temperature and the electron pressure are constant from one sample to another. Indeed, if constant excitation conditions in the plasma are secured by a correctly chosen buffer admixture, there can be no objection to our assuming that recorded spectral-line intensities (apart from self-absorption influences which are not considered here, see § 10.2) are proportional to the concentrations of elements in samples. Thus, as long as discussion is confined to the analysis of specimens under the conditions produced by the relevant buffer admixture, we do not get entangled in transport problems. Should we want, on the other hand, to compare the radiation efficiency either among different constituents of a sample in a particular buffer, or for each separate constituent in different buffers, we cannot ignore the mechanism that governs the passage of material through the arc. It is therefore a vital link in the chain of matrix effects.

To get an idea of transport processes, let us visualize the following simplified model. Let the arc be a cylinder having uniform temperature and electron pressure and let the base of the cylinder exactly cover the surface of the filled crater of radius a in the lower electrode (anode). The specimen enters the cylinder and is uniformly distributed over the entire base. The rate of entry of an element labelled j is Q_j atoms per sec. Let the particles of this element travel vertically upwards at a constant velocity v_j. Then, if diffusion could be fully ignored and the arc were infinitely long, Q_j particles would spread over a volume of $\pi a^2 v_j$, in other words, we would have in a steady state the vapour concentration

$$n_j = \frac{Q_j}{\pi a^2 v_j} \tag{9.3}$$

That an actual arc has the finite length l should not prevent us from applying equation (9.3) to an actual case, since vapours, once they have migrated from anode to cathode, escape sidewards and are carried away, so that they no longer influence the transport of fresh material through the arc.* Equation (9.3) needs further refinement, however, to allow for the process of lateral diffusion. Moreover, we must consider the assumption that the specimen enters the cylindrical column completely. It seems that sideward losses can occur when the sample is released from the electrode. The effect will depend largely on the electrode shape and on the admixture of the sample. We shall deal with this problem more extensively in Chapter 10 (see also the end of § 9.5). For the present, let us take such losses into account by assuming that a fraction β_j of an element j enters into the discharge zone, whence we write formally

$$n_j = \beta_j \frac{Q_j}{\pi a^2 v_j} \tag{9.4}$$

To relate spectral-line intensity to the concentration of an element in the sample, we need only one more link, namely that between the rate of entry Q_j and the total number N_j of atoms of the element j in the electrode crater. If the sample is burned to completion, we have

$$N_j = \int_0^{t_c} Q_j \, dt \tag{9.5}$$

where the integration extends over the total arcing period t_c. Naturally, it suffices that the relevant *element* has been completely volatilized. To avoid inelegant circumlocution we have adopted the term 'burning to completion' also for where only the impurity elements are completely evaporated from the sample (carrier technique).

In the ideal case of even volatilization, equation (9.5) can be written as

$$Q_j = \frac{N_j}{t_c} \tag{9.6}$$

---

* The influence of the cathode on the transport process is not negligible if the electrode distance is small. Nickel (1965a) demonstrated such an effect at an electrode distance of 4 mm for arcs burning in an argon–oxygen atmosphere.

We are able now, for the simplified model, to relate the time-integrated signal

$$E_{\mathrm{qp}} = \int_0^{t_c} J_{\mathrm{qp}}\, dt \tag{9.7}$$

with the number N_j of atoms of the relevant element in the electrode cavity. Assuming constant temperature and electron pressure during the burning time t_c we find from (9.1), (9.4), and (9.5):

$$E_{\mathrm{qp}} = C_{\mathrm{qp}}(1-\alpha_j)\frac{\exp(-\epsilon_{\mathrm{q}}/kT)}{Z_a}\frac{\beta_j}{\pi a^2 v_j}\, N_j \tag{9.8}$$

Although equation (9.8) must be refined to match entirely the real situation in the arc, it illustrates what influence is exerted on spectral-line intensity by the transport of the metal vapours. To see this more clearly we shall investigate the nature of the transport velocity v_j.

The plasma as a whole travels upwards as a result of convection (see § 2.3). The speed v_c of this motion is of the order of 100 cm/sec. It is an increasing function of the current; it also increases with distance above the lower electrode. We shall assume a constant value over the entire length of the short arcs ($l \leqslant 1$ cm) commonly dealt with in spectrochemical analysis. At a current strength of 10 amp a value of about 150 cm/sec is appropriate. As a first approximation we consider the convection current to be a laminar vertical flow. Evidently, the presence of the electrodes must disturb, to some extent, this ideal picture. This complication is not so serious, since convection is not the only feature and for the majority of elements not the principal feature of the axial vapour transport; for, ions migrate from anode to cathode by the action of the longitudinal field. The mean speed of ions due to the field is substantially beyond the convection velocity, as is shown by the following examples.

According to § 3.7 the drift velocity v_i of ions is calculated to be

$$v_i = \mu_i E \tag{9.9}$$

where v_i is in cm/sec if mobility μ_i is expressed as $\mathrm{cm^2\, V^{-1}\, sec^{-1}}$ and field strength E as V/cm. Taking μ_i to be 70 $\mathrm{cm^2\, V^{-1}\, sec^{-1}}$ (see the end of § 3.7) and recalling the experimental values of E in graphite arcs burning in air alone and in air with different metal vapours (§ 3.2), we find for the drift velocity the values summarized in Table 9.1. Obviously, v_i varies appreciably, namely from 1750 to 735 cm/sec, when we pass from a pure graphite arc to an arc with potassium vapour. Also, the magnitude of this speed is well beyond that of the convection velocity v_c. This comparison might be misleading, however, for judging the part played by ion migration in the mechanism of transport of elements through the arc. Inherently, we must take into account the fractions of the elements that are ionized, i.e. the degrees of ionization α_j, and the fractions that are unionized, i.e. $(1-\alpha_j)$. This is easily done if we consider the following.

The arc plasma is in local thermal equilibrium (Chapter 5), which implies that at a given temperature and electron pressure the degree of ionization of each plasma constituent is constant. This does not mean that each individual particle, either an atom or an ion, travels over the entire path through the arc in the very form in which it arrived. Naturally, we must think of ionization in terms of a dynamic process of continuous ionization and recombination. Stating the degree

TABLE 9.1

Arc	T (°K)	E (V/cm)	v_i (cm/sec)	v_j [$\alpha_j = 1$] (cm/sec)
Graphite	6200	25	1750	1900
Aluminium	6000	15	1050	1200
Lithium	5600	12	840	990
Potassium	5100	10·5	735	885

Drift velocity of ions (v_i) and axial transport velocity (v_j) of completely ionized elements in the discharge column of various d.c. arcs. The axial transport velocity of an element labelled j is given by

$$v_j = v_c + \alpha_j \mu_i E$$

where v_c = convection velocity ($\approx$ 150 cm/sec)
α_j = degree of ionization
μ_i = ion mobility ($\approx$ 70 cm^2 V^{-1} sec^{-1})
E = axial electrical field strength (cf. § 3.2)

of ionization of a particular element j to be α_j means that each separate particle of this element exists, on the average, for a fraction α_j of the time as an ion and for a fraction $(1 - \alpha_j)$ of the time as a neutral atom. Therefore, the problem of computing the speed at which an element as a whole is transported upwards in the arc is at once resolved by considering that a particle travels, on the average, for $(1 - \alpha_j)\tau_j$ sec as an atom and for a $\alpha_j\tau_j$ sec as an ion, where τ_j is the mean time spent by the element j in the arc. So, the mean speed v_j is composed of the atomic speed times $(1 - \alpha_j)$ and of the ionic speed times α_j, viz.

$$v_j = (1 - \alpha_j)v_c + \alpha_j(v_i + v_c) \qquad (9.10)$$

since both neutral atoms and ions travel at the speed of the flowing gas, i.e. the convection velocity v_c, whereas the ions get an additional velocity component v_i imparted by the field. Inserting (9.9) into (9.10) and rearranging we obtain

$$v_j = v_c + \alpha_j \mu_i E \qquad (9.11)$$

Table 9.1 gives numerical values of the quantities involved in equation (9.11) for a graphite arc and for arcs buffered with aluminium, lithium, or potassium. Fig. 9.1 gives in the lower diagram the trend of the axial transport velocity v_j in the four arcs of Table 9.1 for elements having an apparent ionization potential V_{ij} of 5, 6, 7, 8, 9, and 10 eV. For comparison the trend of the degree of ionization α_j has been given in the upper half of the figure. Obviously, neither for elements

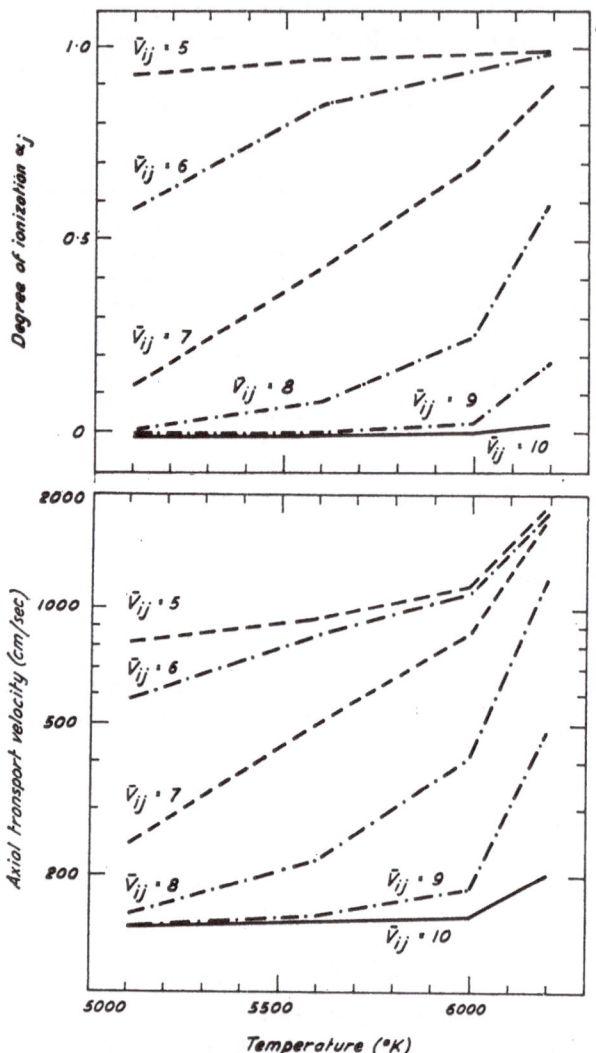

FIG. 9.1. Upper diagram: the degree of ionization (α_j) of elements of different apparent ionization potential ($\bar{V}_{ij}$) plotted against arc temperature. Lower diagram: axial transport velocity (v_j) of different elements plotted versus temperature [formula (9.11)].

In both diagrams points have been inserted for the potassium arc ($T =$ 5100°K), the lithium arc ($T = 5600$°K), the aluminium arc ($T = 6000$°K) and the graphite arc ($T = 6200$°K). Electron pressure has been assumed to be 10^{-3} atm in all cases. The numerical values of electric field strength, ion mobility, and convection velocity used for computing v_j conform with those listed in Table 9.1.

Note that α_j has been plotted on a normal linear scale and v_j on a log scale.

of very low nor for elements of very high ionization potential do large changes of v_j occur when we pass from the arc buffered with potassium to one buffered with aluminium. With a pure graphite arc there is an additional increase of v_j because the field strength is appreciably enhanced, while the electron pressure tends to be substantially lower. Of course, the conditions of the undisturbed graphite arc can be realized only if a very small quantity of the test material is introduced into the discharge.

To summarize, we must expect the transport process to influence spectral-line intensities considerably. Its influence is not exactly such that the intensity E_{qp} is inversely proportional to the axial transport velocity v_j, as predicted by equation (9.8). The fact is that vapours do not only move vertically but also spread laterally by diffusion. The eventual concentration distribution of metal vapour is the net result of the competition between the vertical and the lateral dilution of the vapour released from the electrode. The faster an element is carried vertically through the discharge zone, the less it is allowed to spread radially, and vice versa: the slower it moves axially, the more it spreads out. Before we go into details of a model that quantitatively deals with both effects as well as with the experimental evidence furnished in favour of it by a recent study by de Galan (1965a, 1966a), it appears useful to review the literature on this subject.

★ § 9.3   *Vapour transport and mean time spent by atoms in the excitation zone*
         A review of literature

Among the oldest studies on the transport of metal vapours in the arc I mention the investigations by Lenard (1903, 1905), which I have already referred to in conjunction with the radial emission distribution (§ 6.12). I further mention the studies by Mannkopff and Peters (1931), Mannkopff (1932), Ginsel (1933), Mandelshtam (1938), Lauwerier (1956), and Boumans (1957).

The topic of mass transport to and through the discharge zone has regained interest in spectrochemical analysis, particularly in the study of the carrier effect. The term 'carrier distillation' was coined by Scribner and Mullin (1946) to denote the effect of an admixture, such as Ga_2O_3, used in the analysis of uranium oxide samples. The carrier depressed the uranium spectrum and the background and promoted the selective volatilization of impurity elements from the base, resulting in a marked improvement of detection limits for a number of elements. The nature of the carrier effect has been and still is a subject of investigation and discussion. The carrier is considered either to influence the volatilization, by thermochemical reactions, or to act upon the conditions of the plasma. Both effects have been proved to occur and arguments in favour of one view or the other have been made (see also Chapter 10).

Interesting studies are provided by Zaidel, Kaliteevskii, Kund, and Fratkin (1957), Gol'dfarb and Il'ina (1961), Il'ina and Gol'dfarb (1962), Raikhbaum and Malykh (1960, 1961), Vukanović (1960, 1964), Samsonova (1962), Engel'sht and Spektorov (1962), and Malykh and Serd (1964). They introduced the concept of mean time spent by atoms in the discharge zone (transit time). It is considered that the number $\mathscr{N}_j$ of particles of an element j present in the arc is governed by an equation of the form

$$\frac{d\mathscr{N}_j}{dt} = Q_j - \lambda_j \mathscr{N}_j \qquad (9.12)$$

where Q_j is the number entering the arc in unit time and λ_j is a constant that determines the rate at which the particles leave the discharge. Then, in the steady state, we have

$$\mathscr{N}_j = \frac{Q_j}{\lambda_j} = Q_j \tau_j \qquad (9.13)$$

where τ_j is the mean time spent by particles of species j in the arc.

Mark that the mean time τ_j is related to the mean concentration $\bar{n}_j$ of particles in the plasma as follows

$$\bar{n}_j = \frac{\mathscr{N}_j}{V_{\text{arc}}} = \frac{Q_j \tau_j}{V_{\text{arc}}} \qquad (9.14)$$

where V_{arc} is the discharge volume.

To ascertain experimentally what influence additions exert on the transport of material through the arc, we can either measure simultaneously the absolute concentration n_j of particles in the arc and their rate of entry Q_j, or determine their mean transit time τ_j and the arc volume V_{arc}. The former method was recently chosen by de Galan (1965a, 1966a) and will be discussed in § 9.4. Mean transit times have been measured by several Russian investigators. In both methods we must be aware that the experimental results are indirect results of the transport mechanism and that the very processes which lead to the observed values, say of n_j/Q_j or τ_j/V_{arc}, require a separate explanation. The main difficulty is to incorporate all the details of the relevant physical picture of the transport process in an adequate mathematical model. Another difficulty is to insert proper values for atomic constants, such as the constant of diffusion, into the equations.

Gol'dfarb and Il'ina (1961) (see also Il'ina and Gol'dfarb, 1962) took high-speed motion pictures to study the time spent by Li, Na, and Ba in a 5-amp d.c. carbon arc with an 8-mm electrode separation. They introduced compounds of the test elements in the form of a granule on the point of a quartz needle, which was swept at a constant velocity through the discharge channel, and followed the evaporation of the sample by observing the appearance and dissipation of the luminous cloud with the aid of motion pictures, using only radiation in narrow bands adjacent to suitable spectral lines.

From optical density measurements made on the photographs, Gol'dfarb and Il'ina were able to estimate the mean transit time τ for the test elements, which was found to be 2, 3, and 5 msec for Li, Na, and Ba respectively. The basic principle is that the intensity decay of a spectral line of a test element, caused by the change of the vapour concentration in the source after cessation of the inflow of atoms, follows an exponential law [equation (9.12)]. The values of τ quoted demonstrate the order of magnitude of the mean time spent by atoms in the discharge zone; they agree well with estimates from measurements of absolute concentrations and volatilization rates (see § 9.4).

Gol'dfarb and Il'ina derived, on the basis of a mathematical model that considers the spherical diffusion of the test material in the discharge region, numerical values for diffusion coefficients (D) from their observations. Results tend to be somewhat lower than the values I estimated from recent literature values (see § 3.7). In addition, the dependence on the atomic mass seems to be more pronounced than would be expected. Altogether, the results obtained by Gol'dfarb and Il'ina are highly interesting.

Experiments conducted by Raikhbaum and Malykh (1960, 1961) and by Malykh and Serd (1964) show a close similarity with those mentioned above. These authors introduced the test substance into the arc on moving probes attached to the shaft of a synchronous motor. The probes were connected to the cathode and had a negative potential with respect to it to decrease the perturbation of the arc plasma by the probes. The elements present in the form of chlorides on the probe's surfaces evaporated in the arc when it was traversed by the probe. Simultaneously with the probe current, intensities of selected spectral lines were photoelectrically recorded. From the former signal the moment was ascertained at which the probe left the discharge region, i.e. the instance at which the entry of vapours was stopped; from a second signal provided by the intensity decay the authors deduced the time it took the atoms to leave the discharge.

Initial investigations were made with Li, Na, Ca, Zn, Ag, Cd, and Tl as the test elements; a 10-amp d.c. carbon arc with a 5-mm gap was used. The mean time spent by the atoms of different elements in the arc varied from 1 to 2 msec and showed a close, positive correlation with atomic mass.* This dependence on mass led the authors to consider the exit of atoms to be completely controlled by the diffusion process. Assuming a cylindrical diffusion zone they calculated diffusion coefficient values. The absolute values are lower than those predicted from recent literature data (see § 3.7).

Raikhbaum and Malykh (1961) also studied the influence of carriers upon the mean time spent by atoms in the discharge zone. The carriers were supplied from

---

* Corliss (1962c) (see also Corliss and Bozman, 1962) fitted the following equation to Raikhbaum and Malykh's results:

$$\tau = 0.440 + 0.714 \log A$$

where τ is mean transit time and A is atomic weight. The experimental points are stated to deviate by less than one per cent from the line.

Corliss used this relation for estimating the persistence of atoms in the N.B.S. copper arc (see § 1.5). The objection can be raised that even in this relatively low-temperature arc $(T = 5100°K)$ ionization of elements of low and medium ionization potential $(V_i < 7 \text{ eV})$ is still too large and so we cannot ignore the transport by the electric field, the more so, as, in the small gap used by Corliss, the field is mainly controlled by the anode and cathode falls; its strength might thus be appreciable (cf. §§ 3.1 and 3.2).

the electrodes, while the test elements again were evaporated from rotating probes. The authors found as a result of the introduction of the carrier a substantial increase of the transit time τ for elements of low ionization potential. The effect diminished with increasing ionization potential and disappeared completely for mercury. Raikhbaum and Malykh conclude that the enhancement of τ in the presence of a carrier is caused not only by changes in the thermal parameters, but also in the electrical parameters in the arc column. This, indeed, agrees with the picture given in § 9.2 (see also §§ 9.4 and 9.5): the introduction of the carrier reduces the temperature and may increase the electron concentration, so that the degrees of ionization (α_j) of individual vapour components are lowered, which, in turn, results in decreased axial transport velocities (v_j). At the same time the inflow of the carrier weakens the electric field and thus further lowers the axial transport velocities (cf. Table 9.1). Using the probe method Raikhbaum and Malykh also observed a substantial decline of the field strength. Their ascribing this to the formation of negative ions is, in my opinion, at least a speculative supposition.

Highly interesting observations on the mean time spent by atoms in different sources have been recently reported by Malykh and Serd (1964). Measurements were conducted in the flame, in d.c. and a.c. arcs, and in the plasma jet. Their method again involved the pulse-like introduction of vapours of a substance into the luminescent region of the source and the observation of the change of the vapour luminescence during the pulse. The results are reproduced in Table 9.2. Malykh and Serd state that 'It is known that the escape of vapours of a substance from the excitation region is due either to diffusion of atoms from the luminescent cloud, or to the transfer by a gas stream as occurs in convection and in the flame of various types of burner. The diffusion transfer of the atoms of a substance is characterized by the dependence of the diffusion rate on the mass of the diffusing atoms. The rate of escape of atoms from the luminescent region when the transfer is due to gas streams does not depend on the properties of the atoms and is determined only by the velocity of the gas flow. It has been shown (Raikhbaum and Malykh, 1960) that diffusion processes are essential in the exit of atoms from the d.c. arc. As the table (Table 9.2) shows, external factors (e.g. blowing air into the arc) do not change the characteristic relationship.'

Evidently, the absence of a correlation between the transit time and the nature of the element in the plasma jet and the air-acetylene flame show the transport of vapours in these excitation sources to be entirely due to transfer by a gas stream. Strangely enough, Malykh and Serd do not even mention the action of the electric field upon vapour transport in the d.c. arc, although their results plainly demonstrate that atomic mass is not the only factor to govern the time atoms spend in the arc (compare the τ-values for Tl and Hg which differ by a factor of 2). In addition, the enhanced transit times found in the a.c. arc again illustrate, in my opinion, the influence of the electric field; for, in the d.c. arc the field acts continuously in one direction and stimulates the escape of partially ionized substances in this direction; in the a.c. arc, on the contrary, the alternating field counteracts the unidirectional exit of vapours. Indeed, when we compare τ-values in the a.c. arc with those in the d.c. arc, we see that the relative increments of the former are large for the easily-ionized elements Li, Na, and Tl, and small for Hg. Also, there is no difference between τ for Tl and Hg in an a.c. arc, indicating that here the resultant influence

TABLE 9.2

Source	Conditions	τ (msec)			
		Li $A = 6\cdot94$ $V_i = 5\cdot39$	Na $A = 23\cdot0$ $V_i = 5\cdot38$	Tl $A = 204\cdot4$ $V_i = 6\cdot1$	Hg $A = 200\cdot7$ $V_i = 10\cdot48$
D.C. arc	Free-burning arc, current 10 amp	1·2	1·7	2·5	5·0
	Arc with air blasts (Stallwood-jet), current 10 amp	0·7	—	1·4	3·0
A.C. arc	Free-burning arc, current 10 amp	3·2	4·1	6·9	6·7
Plasma jet	Argon consumption 2 l/min, current 15 amp	0·48	—	0·47	0·45
	Argon consumption 8 l/min, current 15 amp	0·25	—	0·23	0·24
Air-acetylene flame	Fuel mixture consumption 2 l/min	9·2	9·6	9·7	—
	Fuel mixture consumption 5 l/min	3·2	3·4	3·3	—

Mean transit time (τ) of atoms in different excitation sources (after Malykh and Serd, 1964).

A = atomic weight V_i = ionization potential (in eV)

of the alternating field is to allow diffusion alone to dominate the transport process. Finally, we note that for the reduced τ-values in a d.c. arc with air blasts several reasons can be considered:

1 A direct influence of the increased streaming velocity of the arc gas upon the transport of the vapours,

2 Stimulated radial diffusion of atoms in consequence of the peripheral gas flow, which purifies the fringe and causes the concentration of atoms to be low at the edges, and perhaps

3 An indirect influence on the axial vapour transport as a result of increased field strength.

Vainshtein and his co-workers (see Vainshtein and Belyaev, 1958, 1959; Belyaev, Vainshtein, and Korolev, 1959; and Vainshtein, 1962) used radioactive tracers for studying the spatial distribution of elements in the arc. The active isotope distribution of elements in the plasma was recorded in γ-rays by means of a pin-hole camera. They investigated the distribution in the conventional d.c. arc, a.c. arc, and spark, as well as in a so-called pulsed arc (see Korolev and Vainshtein, 1958, 1960; Belyaev, Vainshtein, and Korolev, 1959; and Vainshtein, 1962). Optical density measurements obtained from the photographs provided information on the particle distribution in the gap between the electrodes. Most results are given in the form of curves describing the longitudinal distribution from anode to cathode. The shape of the curves is the principal criterion according to which the distributions must be judged. A serious drawback of the presentation is the arbitrariness of the scales, which hampers a quantitative interpretation.

Engel'sht and Spektorov (1962) express an opinion on the transport processes in a d.c. arc which is similar to the ideas expressed in § 9.2. Their paper cited here seems to be only a synopsis of a more detailed work that, unfortunately, I was unable to obtain.

An excellent approach to unravelling the various factors encompassed in the carrier effect has been made by Samsonova (1962). She examined the influence of the carriers, NaCl, Ga_2O_3, AgCl, powdered metallic silver, H_3BO_3, and S, upon the intensities of atom and ion lines of some twenty elements that were added in concentrations varying between 0·1 to 0·01 per cent to a beryllium oxide base. To distinguish between changes of spectral-line intensity caused by the effect of the carrier on the kinetics of evaporation of the added constituents and its effect on the conditions in the excitation zone, she compared the spectra obtained when the carrier and the sample were mixed with those obtained during the combustion of equivalent quantities of sample and carrier isolated from one another in a double electrode. The results of her experiments showed that generally the carriers did not influence the intensity via the mechanism of evaporation of the constituents. Only for a small number of elements was this additional effect found.

The main line of Samsonova's investigation was to study the effect of the carriers on line intensities through their influence on the conditions prevailing in the plasma. To that end she compared the intensities of spectrograms taken during the evaporation of the sample to which the carrier had *not* been added with those taken when the sample and the carrier were isolated from one another in a double electrode. The evaporation conditions were identical and the change in intensity was entirely due to the presence of the carrier in the discharge zone.

Samsonova was able to analyse three factors that affect line intensities through the excitation conditions in the plasma:

1 Excitation in the strictest sense (Boltzmann's law).

2 Ionization (Saha' relationship).

3 A 'rest factor', interpreted in terms of the mean transit time of particles in the discharge zone.

The rest factor is shown to play a substantial part in the carrier effect. As the influence on the mean transit times was found to be crudely related to the change in degrees of ionization, Samsonova notes that the electric field is likely to play a significant role. An attempt to elucidate the transport mechanism is not made, however.

Vukanović (1960) analysed the changes of spectral-line intensities caused by the presence of a carrier (Ga_2O_3). He considered the influence of changes of the temperature and the electron pressure upon line intensities and showed that the observed intensity variations could not be entirely explained by excitation and ionization effects. Later, Vukanović (1964) approached the problem by considering excitation and ionization phenomena in terms of radial distributions, thereby avoiding errors inherent in the use of effective temperatures and electron pressures (cf. Chapter 8). This analysis again led to the conclusion that the carrier also acts upon the mean transit time of atoms in the discharge gap. The effect was found to be the more pronounced the lower the ionization potential of the relevant element.

★ § 9.4 *Measurements of absolute particle concentrations in the arc*
Establishment of a transport parameter

Recently, de Galan (1965a, 1966a) conducted a long series of experiments to collect decisive experimental evidence as a basis for a theory of transport of metal vapours through the arc. The principle underlying his study was suggested by the following considerations. Let an element volatilize from the supporting electrode at a rate of Q atoms per sec. The vapour will spread over the discharge zone under the influence of transport forces, and in a steady state a definite concentration distribution $n(z, r)$ will ensue. The ratio

$$\psi(z, r) = \frac{Q}{n(z, r)} \tag{9.15}$$

defines a transport parameter ψ whose numerical value can be determined if both Q and n are measured simultaneously in absolute units. In the equation, z and r denote the axial and radial cylindrical coordinates respectively.

To elucidate the meaning of the transport parameter and the significance of determining it experimentally we shall first consider a simple example. Let the arc be a uniform, vertical cylinder with a 1 cm^2 cross-sectional area. Vapours enter the plasma evenly distributed over the base of the cylinder at a rate Q and are transported longitudinally at constant speed v. If the rate of entry is 10^{15} atoms per sec and the absolute particle concentration within the cylinder is 10^{12} per cm^3, the transport parameter ψ is 10^3 cm^3/sec. This value depicts the volume occupied by the particles that evaporate per sec, in other words, the rate of flow of the vapour. As we assumed the particles to be contained in a cylinder with a 1 cm^2 cross-sectional area, the numerical value of ψ also equals the axial transport velocity v, which therefore is found to be 10^3 cm/sec.

The transport mechanism of an actual arc is far more involved than is pictured by the simple model, which was primarily included for instruction purposes. It demonstrates clearly, however, what kind of information is inferred from measurements of the transport parameter ψ. The determination of numerical values of this quantity for some ten elements under widely differing excitation conditions (covering the temperature range 5000–6000°K) has been the object of de Galan's investigation. His remarkable results have enabled us to elucidate the transport mechanism satisfactorily, although there remain several questions to be resolved (see § 9.5).

Concentration distributions and absolute concentrations of particles in the discharge zone have been measured by de Galan from intensities of spectral lines emitted by atoms or ions. The principle of the method, which has been applied previously by Huldt (1948b), Eberhagen (1955a), and Belousova (1962), is seen from equations (8.3) and (8.5). In terms of concentration n_j they read

$$n_j = \frac{4\pi J_{qp}}{g_q A_{qp} h\nu_{qp}} \frac{Z_{aj}}{1 - \alpha_j} \exp(-\epsilon_q/kT) \tag{9.16}$$

and

$$n_j = \frac{4\pi J_{qp}^+}{g_q^+ A_{qp}^+ h\nu_{qp}^+} \frac{Z_{ij}}{\alpha_j} \exp(-\epsilon_q^+/kT) \tag{9.17}$$

where n_j is the concentration of the element labelled j, that is the sum of atoms and ions per unit volume, J is the emission per unit volume per unit solid angle, g_q the statistical weight of the upper level q, A_{qp} the probability of the transition q → p, h the Planck constant, ν_{qp} the frequency of the transition, Z_{aj} the partition function of the atom, α_j the degree of ionization of the relevant element, ϵ_q the excitation energy of the upper level, k the Boltzmann constant, and T the absolute temperature; quantities provided with distinguishing labels i or + refer to ions.

Evidently, either an atom line or an ion line can be used. For determining absolute concentrations the procedure comprises the following steps.

1 Emission J of a spectral line is measured in absolute units. To that end the spectroscopic equipment is previously calibrated with the help of a standardized lamp, so that relative intensity readings can be converted to an absolute scale (see the end of § 1.2).

2 A reliable gA-value for the pertinent spectral line is selected from available literature data. The first factor at the right of the equation can thus be computed.

3 The temperature of the plasma is measured so that the Boltzmann exponential factor, the partition function, and the degree of ionization can be calculated; the evaluation of the degree of ionization with the aid of Saha's relationship requires a concomitant determination of the electron pressure (cf. § 7.5 and 7.6).

De Galan employed the lines Zn 3076 and Zn 3282 for temperature measurement (see §§ 6.5 to 6.8) and Mg II 2796, Mg I 2852, and Mg I 2780 for determining the electron pressure (see §§ 7.6 and 7.7).

When applying this procedure to the arc we must consider the spatial in-homogeneity of the source. Consequently, spatial resolution of intensities should precede the calculation of concentrations with either of the equations (9.16) or (9.17), since these pertain to a unit volume of a uniform source. So, for examining the radial distributions of particle concentration, one of the procedures involving the Abel inversion of edge-on intensity distributions, outlined in § 6.14, must be applied for the two zinc lines, for two magnesium lines, and for the line λ_{qp} of the element j whose concentration distribution is determined. From the radial emission profiles of the zinc and magnesium lines we compute the radial distributions of temperature and electron pressure (§ 7.17), and eventually from these distributions and the radial emission profile of the line λ_{qp} the distribution of the particle concentration of the element j.

To collect a great many data in a comparatively short time, de Galan used the two-disk method described at the end of § 6.14. The radial distributions thus are obtained in four steps only, which proved to be adequate for the purpose. We refer the reader to de Galan's work for practical details regarding the calibration of the arrangement and the elaboration of intensity readings. Let us confine our discussion to the main lines of the investigation and to the final results.

Measurements were made in a vertical, free-burning 10-amp d.c. arc with 10-mm electrode separation. The supporting graphite electrode, the anode, was of the type shown in Fig. 7.14. The cathode was of carbon. Test mixtures containing the elements In, Ga, Al, Tl, Mg, Pb, Sn, B, Be, Zn, and Sb as oxides in graphite powder were prepared and suitable portions of them were diluted with each of the following buffer admixtures: Al_2O_3 + graphite, LiF + graphite, and KF + graphite. These buffers (or matrices if related to the entire electrode charge) were decided upon after preliminary studies of the excitation conditions (temperature, electron pressure) in the arc and of the volatilization characteristics of the test elements had been made. The temperatures achieved in the central region of the arc were 6000°K with the Al_2O_3 matrix, 5600°K with the LiF matrix, and 5100°K with the KF matrix.

Four-step radial distributions were determined at three different heights in the arc, namely in the middle section and in the neighbourhood of each electrode. The region investigated extended usually to 0·4 cm and occasionally to 0·7 cm from the arc axis. Samples were burnt to completion and exposure times were taken over the entire arcing period.

Relative concentration distributions: Results for the middle section of the arc showed the particle concentrations of the test elements to be uniform up to 0·4 cm from the axis. At least, no systematic variation with radial distance to the axis could be shown to exist within experimental error. The magnitude of the latter was estimated to be such that a systematic decrease of the concentration over the entire distance of 0·4 cm by a factor exceeding 1·5 would have been detected.

There was one exception, boron, which exhibited a strong radial decline of particle concentration. This behaviour was explained quantitatively on the basis of formation of BO molecules. When determining particle concentrations only atoms and ions are considered; so, low concentration values are found if an element forms molecules that are stable in the temperature range of the arc. For the test elements used by de Galan only boron has this effect, since its oxide remains stable up to 6000°K. The concentration of boron atoms decreases rapidly with an increase in temperature in favour of BO molecules. The effect is noticeable if we proceed radially from the arc axis outward and also if we consider a definite region, say the vicinity of the axis, in arcs with different temperatures. De Galan (1965a and b, 1966a) demonstrated that a dissociation energy of 8·3 eV for BO was the best value to fit his experimental results; it is intermediate between the literature values of 7·5 and 9·0 eV (see also §11.3).

For the regions near the electrodes, observations of thin slices of plasma at a distance of 0·2 cm from either electrode revealed an even radial distribution of particle concentration at the upper electrode (the cathode) and a pronounced radial decline at the lower electrode. The radial gradient of concentration at the anode increased rapidly with decreasing arc temperature, as is seen from values of the difference $\Delta = \log n(r = 0) - \log n(r = 0.4)$ averaged over all test elements for each matrix:

matrix	Al_2O_3	LiF	KF
temperature (°K)	6000	5600	5100
Δ	0·3	0·5	0·8

In summary, particle concentrations exhibit a sharp radial variation in the region adjacent to the lower electrode and become more uniform as the distance from it increases. In the middle of the gap an even distribution has ensued and this situation extends nearly to the upper electrode.

The available results also enabled de Galan to compare the particle concentrations at three different heights in the arc, in other words, to draw conclusions about the *axial* distributions. These were studied, moreover, in a separate experiment, which involved examining the longitudinal variation of temperature, electron concentration, and particle concentrations from spectral-line intensities that were not resolved into radial emission profiles.

Thus, calculations were carried out with intensities that represented integrals of radiation taken along the line of sight. The experiment yielded effective values of the temperature and the electron concentrations, in the sense defined in §8.3. Effective concentrations derived by formally applying equation (9.16) or (9.17) in fact represent integrals over the depth of the source. When the radial distribution is uniform, the measured quantity is simply the product of the concentration and the effective arc diameter, or, in terms of equation (8.16), the product $2n_j\hat{R}_s$, where $\hat{R}_s$ is the effective arc radius. On the strength of the measured radial distributions referred to above, de Galan showed that the effective radius does not change appreciably over the arc length. Hence, an observed longitudinal variation of the effective particle concentration represents well the axial distribution of the actual particle concentration n_j.

Results, plotted as concentration versus axial distance (z) to the lower electrode, could be related to the degrees of ionization of the elements. Those elements having a low degree of ionization displayed a marked decrease of concentration with increasing z. This occurred for elements of high ionization potential irrespective of the matrix and for nearly all elements in the potassium fluoride matrix. Completely ionized elements showed a somewhat smaller, though still significant enhancement of concentration near the upper electrode (the cathode). The effects were more pronounced in the low temperature KF arc than in the high temperature Al_2O_3 arc; apparently, the axial concentration distribution becomes more uniform as the arc temperature increases.

These conclusions, based on the observed longitudinal variation of effective concentrations, were substantiated by considering actual concentrations at the arc axis derived from radial emission profiles. We mentioned above that such determinations were conducted at three different levels in the arc. Comparison of the concentrations at the anode and the cathode again showed the concentration gradient to diminish as the degree of ionization of element increased. The maximum logarithmic difference, $\log n$ (anode) $- \log n$ (cathode), was found to be 0·5, which is equivalent to a factor of 3 in the concentration.

Absolute concentrations: Hitherto we have dealt with the relative concentration distributions only; now we shall be engaged in the absolute values. A careful choice of spectral lines with reliably known absolute transition probabilities permitted de Galan to put results of concentration measurement on an absolute scale. In addition, as known amounts of the test elements were burnt to completion and as volatilization was known to proceed evenly, he was able to compute the average rate of volatilization for each test element in each arcing. Consequently, connecting absolute values of observed concentrations $n(z, r)$ with rates of evaporation Q by a transport parameter ψ, as defined in equation (9.15), could be attempted. For that purpose the concentrations determined in the middle section of the arc were chosen. Further discussion will thus be restricted to these results.

Since the radial distribution of particle concentration in the middle section was shown to be uniform within the zone of observation, that is, up to 0·4 cm from the arc axis, it was permissible to average the results of the four steps. Mean values obtained after performing the whole experiment repeatedly were finally used to connect transport parameter, rate of evaporation, and concentration, by the equation

$$\log \psi = \log Q - \overline{\log n} \qquad (9.18)$$

For a detailed discussion of the results, including an exposition on precision and accuracy, we refer the reader to de Galan's original work.

By arranging the observed values of $\log \psi$ in sets according to the ionization potential of the relevant element and the temperature of the pertinent arc, a pronounced trend of $\log \psi$ with the degree of ionization of the element became apparent. If details are disregarded, the aggregate of $\log \psi$ values suggested a final representation of results as produced in Table 9.3. The values in the solid frame pertain to completely ionized elements, those in the dotted frame to very slightly ionized elements, and the remaining ones to partly ionized elements. The values 4 and 3 in the first row refer to lithium and potassium in the lithium and potassium arc respectively and are uncertain, owing to the unreliability of the absolute transition probabilities of the lines used and to interference by self-absorption.

The results listed in Table 9.3 reflect the general trend of the experimental results and show their most striking characteristics, i.e. the dependence of $\log \psi$ on the degree of ionization; it agrees with the trends of the observed axial distributions

TABLE 9.3

Ionization potential of element (in eV)	Matrix		
	Al_2O_3 (6000°K)	LiF (5600°K)	KF (5100°K)
4-5	—	4 ?	3 ?
6	3·5	3·0	2·7
7·5	3·0	2·7	2·5
9	2·6	2·7	2·4

Values of the transport parameter (expressed as $\log \psi$) as suggested by the complete set of experimental results for elements of different ionization potential.

Values in the solid frame pertain to completely ionized elements, those in the dotted frame to very slightly ionized elements, and the remaining values refer to partly ionized elements.

(cf. § 9.5). We note, however, that some outlying values were found too. So, results for gallium (in LiF and KF), for lead (in Al_2O_3 and LiF) and for thallium (in all three matrices) deviated by far too much from the suggested scheme for the discrepancies to be attributed to random or systematic errors; accordingly, other influences on the transport parameter, for instance atomic mass, should not be excluded from consideration.

On the whole, the results compiled in Table 9.3 can be covered by the approximate relationship

$$\log \psi = 2\cdot5 + \alpha \qquad (9.19)$$

where α is degree of ionization. It expresses that the particle concentration decreases by a factor of 10 when the degree of ionization changes from 0 to 1, provided, of course, that the rate of evaporation remains the same. Consequently, reduction of the arc temperature tends to enhance particle concentrations. This effect must be taken into account as a substantial contribution when the dependence of spectral-line intensity on arc temperature is considered. We discussed this matter at length in §§ 8.8 and 8.9, to which the reader is referred.

The order of magnitude of the transport parameter, that is, $\psi \approx 10^3$ cm³/sec, fits in with that of transit times (τ) determined by Russian authors (see § 9.3). Since ψ is the volume occupied by the particles evaporated per sec, we can estimate τ by dividing the arc volume by ψ [cf. equation (9.14)]. Assuming a value of 1 cm³ for this volume, we compute τ to be of the order of 1 msec.*

Values of τ reported for the free-burning d.c. arc generally range from one to several milliseconds (see e.g. Table 9.2). From de Galan's maximum value of $\log \psi$, that is $\log \psi = 3.5$, we derive a minimum value of $\tau \approx 0.3$ msec, which therefore appears low. Although the value to be taken for the arc volume makes estimates of τ from ψ somewhat uncertain, there arises a discrepancy that cannot be lightly attributed to random errors. Possibly, de Galan's entire set of results is subject to a systematic error, that has caused *all* values of $\log \psi$ to be too high by a constant amount. It does not affect the main feature of the results, that is, the strong correlation between $\log \psi$ and α as expressed by (9.19), as it is merely the value of the constant, i.e. 2.5, that is in doubt. Conclusions for spectro-chemical analysis, such as the calculation of optimum temperatures for the emission of atom lines (§§ 8.8 and 8.9) are therefore unchanged.

Bringing the value of the constant in (9.19) more into line with values found by Russian authors changes the equation to

$$\log \psi = 2.0 + \alpha \qquad (9.20)$$

The upper limit of $\log \psi$, i.e. 3, now also agrees better with estimates of the particle concentration from the electron concentration. On this basis we expect $\log n$ to be about 15 for an easily-ionized element that carries the discharge (see § 7.9). The same value follows from de Galan's measurement of the evaporation velocity, namely $\log Q = 18$, if we take $\log \psi = 3$, in conformity with (9.20). Using (9.19) we arrive at $\log n = 14.5$.

Perhaps the coefficient of α in (9.19) or (9.20) should not be exactly equal to unity but somewhat, though not essentially, smaller, say 0.9, expressing that the total range of ψ would be covered by a factor of 8 instead of 10.

In addition to the conclusions for spectrochemical analysis produced in §§ 8.8 and 8.9, we point out the consequences of the transport when we relate the composition of the vapour cloud to the sample composition. Clearly, if the constituents of the sample are only slightly ionized in the arc, relative concentrations

---

* In connection with the order of magnitude of the mean time spent by atoms in the discharge zone, we may ask how many quanta a particle is likely to emit during its passage. Of course a particle is excited and de-excited a great many times, but only a few excitations lead to radiation (see Chapter 5). To calculate the radiation output of one particle for a particular transition we consider the following example.

Let the upper level of the transition be located at 4 eV above the ground level and have a statistical weight equal to the partition function of the atom. Thus at about 5000°K the population of the upper level is computed to be $10^{-4} N$ [equation (6.6)], where N is the total number of atoms available. This is equivalent to saying that each atom exists, on the average, during the proportion 10^{-4} of the time in the pertinent excited state. Thus, with a transit time $\tau \approx 1$ msec, we find the atom during 10^{-7} sec in that state. The natural mean lifetime being of the order of 10^{-8} to 10^{-7} sec (§ 1.3), we expect 1 to 10 radiative transitions during the passage of the atom through the arc. In the case when a transition to only one lower level is involved, there will be emitted 1 to 10 quanta of the same frequency.

of elements in the vapour phase will be close to those in the solid sample. Deviations will be caused by different rates of diffusion (cf. § 9.5). Conversely, if a sample is excited at a relatively high temperature and it contains elements of widely differing ionization potential, we can expect a discrepancy between the ratio of the concentrations of two elements in the vapour phase and in the sample as large as a factor of 10. As a matter of course we must always realize that the vapours of the sample as a whole constitute about 1 per cent of the arc plasma at most (see § 7.9).

Let us finally cite a few studies on absolute concentrations in the arc that are not based on the emission method discussed above. We mention determinations from profiles of lines subject to self-absorption by Dvornikova and Nagibina (1958), Nagibina (1958, 1959, 1963), Burakov and Naumenko (1963), and Rukosneva (1964), and measurements by the method of anomalous dispersion by Nikonova and Prokof'ev (1959). It is difficult to compare the absolute values of concentrations found by these authors with those measured by de Galan, since operating conditions differ too much for that. Only the measurements by Nikonova and Prokof'ev lend themselves for comparison and they show indeed the same order of magnitude for the concentration of a major constituent of a sample, that is, $n = 10^{15}$ per cm^3. Hefferlin (1965) estimated the concentration of Mn I in an atmospheric manganese arc using different methods. Taking into account the fairly large dispersion of the results, he concluded that the concentration of Mn I in the centre of his arc was, to a factor of 2 or so, 5.10^{14} per cm^3.

★ § 9.5   *An approximate model for the vapour transport through the arc*

We shall now consider an approximate model for the vapour transport through the arc that may account for the general trends observed experimentally by de Galan (see § 9.4). The main problem is the incorporation of all relevant physical processes into a model that remains mathematically manageable.

Starting from the elementary considerations of § 9.2 we consider that:

1 At the lower electrode (the anode) a stream of vapour of element j enters the discharge zone at a rate of Q_j particles per sec.

2 The vapour spreads axially and radially by diffusion (coefficient of diffusion $= D_j$).

3 Directed transport forces, originating from the electric field and the convection flow, are superimposed on the diffusion forces. This causes an additional axial velocity component v_j to act upward.

The axial transport velocity is given by (9.11), i.e.

$$v_j = v_c + \alpha_j \mu_i E$$

where v_c is the convection velocity, α_j the degree of ionization of species j, μ_i the ion mobility, and E the electric field strength. For the sake of convenience we shall omit in the following the label j of v, α, and D.

Prior to dealing with the mathematical outline of the model we shall estimate numerical values for the parameters that will appear in the equations. It will be shown below that the governing parameters of the transport process are the ratio $v/2D$ and the coefficient of diffusion D. The former, defining the balance between the axial motion and the lateral expansion of a vapour, determines preponderantly the relative distribution of particle concentration, whereas the latter (D) dominates the absolute magnitude of the concentration. We suppose, in compliance with considerations brought out in § 3.7, the validity of the kinetic theory relation (3.26), which in practical form reads

$$\frac{\mu_i}{D_i} = \frac{1 \cdot 17 \cdot 10^4}{T} \qquad (9.21)$$

Moreover, we premise the diffusion coefficient of ions and atoms to be of equal magnitude (see also note on p. 59). The parameter $v/2D$ can thus be written as

$$\frac{v}{2D} = \frac{v_c}{2D} + \frac{5850}{T}\alpha E \qquad (9.22)$$

For the convection velocity v_c we take the constant value of 150 cm/sec (see §§ 9.2 and 2.3) and for the electric field strength E the value of 10 V/cm at 5000°K and 16 V/cm at 6000°K (cf. Table 3.1). The coefficient of diffusion D is assumed to be in a range between 25 and 50 cm²/sec at 5000°K and in a range between 35 and 70 cm²/sec at 6000°K (see § 3.7). Thus, when the degree of ionization α is taken to vary from 0 to 1, the numerical values of $v/2D$ and v listed in Table 9.4

TABLE 9.4

$T = 5000°K$	$E = 10$ V/cm			$T = 6000°K$	$E = 16$ V/cm		
$v/2D$	v	D	α	$v/2D$	v	D	α
3·0	150	25	0	2·1	150	35	0
1·5	150	50	0	1·1	150	70	0
8·8	440	25	0·5	10·0	700	35	0·5
7·3	735	50	0·5	8·9	1240	70	0·5
14·7	735	25	1	17·7	1240	35	1
13·2	1320	50	1	16·7	2330	70	1

Basic numerical values of parameters involved in the vapour transport through the d.c. arc.
v = directed transport velocity in cm/sec [equation (9.11)]
D = coefficient of diffusion in cm²/sec (cf. § 3.7)
$v/2D$ = parameter that determines the shape of the concentration distribution (expressed in cm⁻¹)
α = degree of ionization of element
T = absolute temperature
E = electric field strength (see Table 3.1)

are obtained. Evidently, $v/2D$ is in a range between 1 and 18 cm⁻¹, while v varies from 150 to 2400 cm/sec. We notice that $v/2D$ depends little on D when α is large or moderately large. Therefore, relative concentration distributions should be principally controlled by α, E, and T, and only to a small extent by the coefficient of diffusion and thus by the atomic mass of the relevant element.

We now proceed to a transport model that takes the diffusion and the directed axial transport into account. We define a cylindrical coordinate system with its z axis along the arc axis and its origin ($z = 0$, $r = 0$) at the centre of the surface of the lower electrode (the anode). We first suppose that the directed transport acts in the entire space $0 < r < \infty$, $-\infty < z < +\infty$, and imposes on the particles a constant velocity v in the positive z direction. Let the coefficient of diffusion D be also constant in the whole space and let particles be produced at a rate Q by a point source at the origin.

By considering the flow through an infinitesimal element of volume (see, e.g. Crank, 1956, for general principles) the following partial differential equation can be derived, if a steady state is assumed:

$$\frac{dn}{dt} = D\left[\frac{1}{r}\frac{\partial}{\partial r}\left(r\frac{\partial n}{\partial r}\right) + \frac{\partial^2 n}{\partial z^2}\right] - v\frac{\partial n}{\partial z} = 0 \qquad (9.23)$$

A well-known solution of this equation is

$$n(z, r) = \frac{Q}{4\pi D\sqrt{(z^2+r^2)}} \exp\left\{-\frac{v}{2D}[\sqrt{(z^2+r^2)} - z]\right\} \qquad (9.24)$$

It satisfies the boundary conditions

$$n = 0 \qquad z = \infty$$
$$n = 0 \qquad r = \infty \qquad (9.25)$$

and

$$\lim_{\rho \to 0} 4\pi\rho^2 D\frac{\partial n}{\partial \rho} = Q \qquad (9.26)$$

where $\rho = \sqrt{(z^2+r^2)}$. The former condition expresses that n is very small at great distances from the origin; the latter one is a normalization condition expressing that the diffusive flux through a small sphere around the origin equals Q.*

---

* Equations (9.23) and (9.24) appeared in an article by Wilson (1912) in connection with studies on diffusion in flames. They were also used, for example, by Ginsel (1933) and by Ginsel and Ornstein (1933) for determining coefficients of diffusion in flames (cf. § 3.7). The point source was a bead of salt put in the flame on a platinum wire. This example resembles a point source in an infinitely large medium subjected to a laminar flow. The conditions permitting the use of this approximation are also satisfied in the arrangements for measuring coefficients of diffusion at high temperatures as employed by Walker and Westenberg (1958a) and by Ember, Ferron, and Wohl (1962). In all those cases v is the velocity of a laminar gas flow. Ginsel (1933) suggested the adaptation of equation (9.24) to the transport in the arc, v then being given by (9.11). This application was made later by van Stekelenburg (1943).

Solution (9.24) does not describe the concentration distribution in the arc, as it predicts, for example, the concentration at the axis ($r = o$) to be independent of the axial velocity v, which is in contradiction of experimental evidence (see § 9.4). To find fair agreement between theory and experiment we considered, in co-operation with mathematicians at the Mathematical Centre at Amsterdam, new boundary conditions, that I shall explain below (see also Boumans and de Galan, 1966, and Bavinck, 1965).

A rigorous treatment of the problem in terms of radial distribution functions $D(r)$ and $v(r)$ would require a differential equation that can be solved only by numerical methods. To keep the mathematics as simple as possible we therefore retained equation (9.23) and sought a solution that is significant only in a region where D and v can be regarded constant. This provides sufficient information, since our chief interest is in the central core of the arc, where D and v are approximately constant. Besides, the present object is to show some general trends rather than to give a minute account of concentration distributions.

In the new model we consider the transport through two boundaries, viz. (1) a cylinder of radius $r = R$ that envelops the arc, (2) the basis of this cylinder in the plane $z = o$, through which the intake of material occurs.

The presence of the cathode is ignored; in fact, we assume the arc to be infinitely long. This is permitted if the vapours, once they have reached the cathode, do not return to the column. As ions travel to the cathode, metal vapours tend to accumulate in front of it. From this cathode layer, atoms diffuse mainly sidewards and are carried away along the cathode by the upward convection flow. So, the effect of the cathode as an obstacle will be perceptible only in a region extending over a few millimetres below the cathode and, consequently, we can ignore its presence when considering the transport processes in the column of a long arc, for example, the 1-cm arc used by de Galan. The smaller the gap, however, the larger the influence of the cathode. Experiments with radioactive isotopes carried out by Nickel (1965a and b) indicate an appreciable influence of the upper electrode on the vapour transport in a 4-mm arc burning in an argon-oxygen atmosphere.

Boundary Conditions:
1 A boundary condition that takes the radial variation of D and v approximately into account is the following:

$$r = R \qquad D\frac{\partial n}{\partial r} = kn \qquad\qquad (9.27)$$

where $o \leqslant k \leqslant \infty$. If the constant k (in cm/sec) is zero, equation (9.27) expresses that no diffusive transport through the boundary takes place. This special case, $k = o$, appears to be a good approximation if v is large or moderately large in the arc core, that is for elements with a high or medium degree of ionization. We argue this as follows.

The rather steep radial decline of the temperature, which follows approximately a parabolic profile (see §§ 6.15 and 8.1), causes the diffusion coefficient of free atoms and ions to drop rapidly at the edge of the core. In addition, in the arc fringe free atoms combine into molecules, which have a larger mass and therefore a smaller diffusion coefficient. To conclude, in the arc fringe the rate of diffusion is noticeably smaller than in the more or less uniform column; the fringe thus acts as a wall and restricts outward diffusion from the core.

Now consider the boundary condition $\partial n/\partial r = 0$ at $r = R$. It implies the preclusion of diffusive transport through the boundary. It seems a contradiction in terms, however; for the arc flame to act as an impenetrable wall, it should contain atoms and molecules of the element whose diffusion is considered. This difficulty is removed if discussion pertains to a steady state; then, we may conceive that particles have diffused outwards prior to the establishment of the steady state. We must then make an additional assumption, i.e. that the particles remain in the fringe and are not carried away. This assumption is justified if the axial velocity v is large inside the cylinder $r = R$ and small outside it, which applies for elements having a high or medium degree of ionization in the core. For then the degree of ionization drops almost abruptly to very low values at the edge (see Fig. 9.2) and so does the axial transport velocity v; the latter is likely to decline even more sharply, owing to the vanishing of the electric field outside the core, and, to a lesser extent, because the ion mobility drops radially with the temperature. Only for weakly ionized elements does v change gradually, as it now virtually equals the speed of convection v_c. Consequently, the appropriateness of the condition $\partial n/\partial r = 0$ at $r = R$ becomes worse as v decreases.

To fit in with the entire range of v, we state the boundary condition in the general form of equation (9.27). So, provision is made for a leakage of particles through the wall. This is stringent, particularly for small v.

2 The second boundary condition concerns the supply of particles from the source, i.e. the electrode. We assume a disk source of radius a in the plane $z = 0$. Particles enter the discharge space through the area πa^2 of the disk at a rate of Q per sec. The condition

$$z = 0 \qquad -D\frac{\partial n}{\partial z} + vn = \frac{Q}{\pi a^2}\vartheta(a-r) \qquad\qquad (9.28)$$

where $\vartheta(a-r)$ is the Heaviside step function

$$\vartheta(a-r) = 0 \qquad r > a$$
$$\vartheta(a-r) = 1 \qquad r < a$$

expresses that no mass transport takes place through the plane $z = 0$, except in the disk source.

3 In addition to the conditions (9.27) and (9.28), two elementary boundary conditions must be satisfied:

$$r = 0 \qquad z \neq 0 \qquad n \text{ regular} \qquad (9.29)$$

and

$$z = \infty \qquad\qquad n \text{ bounded} \qquad (9.30)$$

Solution: With reference to a detailed mathematical treatment by Bavinck (1965) I state the solution of (9.23) that satisfies the boundary conditions (9.27) to (9.30), viz.

$$n(z/R, r/R) = \frac{2Q}{\pi a D} \sum_{n=1}^{\infty} \frac{1}{[\sqrt{(w^2 + \xi_n^2)} + w]} \times \frac{J_1(\xi_n a/R)}{\xi_n [J_0^2(\xi_n) + J_1^2(\xi_n)]}$$

$$\times J_0(\xi_n r/R) \exp\{-[\sqrt{(w^2 + \xi_n^2)} - w]z/R\} \qquad (9.31)$$

where $J_0(x)$ and $J_1(x)$ are the Bessel functions of the zero and the first order respectively, w is the reduced dimensionless transport velocity $vR/2D$, and ξ is a positive number satisfying

$$\xi J_1(\xi) - K J_0(\xi) = 0 \qquad (9.32)$$

For each positive value of $K (= k/D)$ there is an infinite series of positive numbers ξ_n that satisfy (9.32). These numbers can be taken from a table.

General feature of the solution: A remarkable feature of the solution (9.31) is the following. For given values of the parameters a, R, K, and $v/2D$ the absolute magnitude of the concentration n is inversely proportional with D. The shape of the distribution is determined by the sum function and depends on the values of the parameters, among which $v/2D$. We have already argued that for large and moderately large values of v, the ratio $v/2D$ depends only slightly on D [cf. equation (9.23)]. Hence elements of equal (apparent) ionization potential, when strongly or moderately ionized, are similarly distributed in the arc, while the absolute values of their concentrations decrease inversely proportional with the coefficient of diffusion. Thus, the difference of transit times for lithium and sodium in a free-burning d.c. arc observed by Malykh and Serd (see Table 9.2) can be accounted for quantitatively, if D is related to atomic mass by (3.28) with an average molecular weight of 24 for the arc gas.

Results of numerical calculations: Numerical calculations of the distribution $n(z, r)$ have been carried out with an electronic computor at the Mathematical Centre in Amsterdam. The calculation was conducted with different values of the parameters D, v, R, a, and K. As an example, Fig. 9.3 presents radial distributions of n/Q at five heights in the arc for $v = 600$ cm/sec, $D = 30$ cm²/sec, and $K = 0$.*

* If the absolute magnitude of n/Q is decreased by half, the plots represent $v = 1200$ cm/sec, $D = 60$ cm²/sec, $K = 0$.

It represents a moderately ionized element with a medium coefficient of diffusion (cf. Table 9.4). Results are given for a point source ($a = 0$), and for disk sources with a radius $a = 0{\cdot}2$, $0{\cdot}3$, and $0{\cdot}4$ cm, in each case for three values of R, namely $R = 0{\cdot}7$, $0{\cdot}5$, and $0{\cdot}4$ cm.

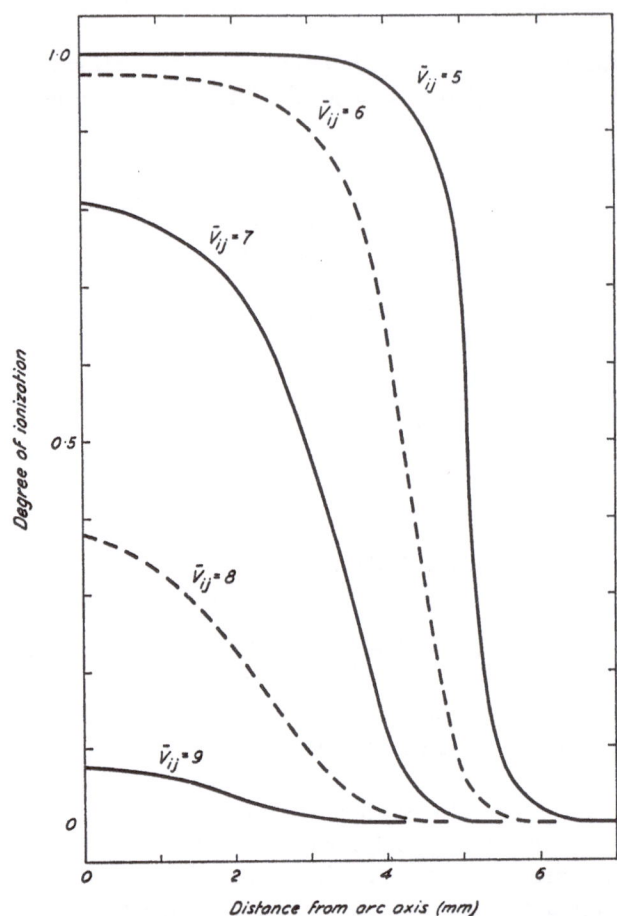

FIG. 9.2. The degree of ionization (α_j) of elements of different apparent ionization potential (V_{ij}) plotted as a function of the radial distance from the arc axis. The curves are for the model arc of Chapter 8 and correspond to the radial profiles of the temperature and the electron concentration shown in Figs. 8.1 and 8.3 for $T_0 = 6000°$K.

We have considered various values of R for formal reasons chiefly; physically the value $R = 0{\cdot}4$ seems the most realistic one. Confining discussion to this case (continuous curves in the figure) we first recognize the flattening of the *radial* distribution pattern achieved by replacing the point source by a disk source of increasing radius. As a matter of course, the effect is more pronounced in the vicinity of the source than at some distance from it. Secondly, we see the *axial* distribution grow more uniform in the disk source approximation. Note that a

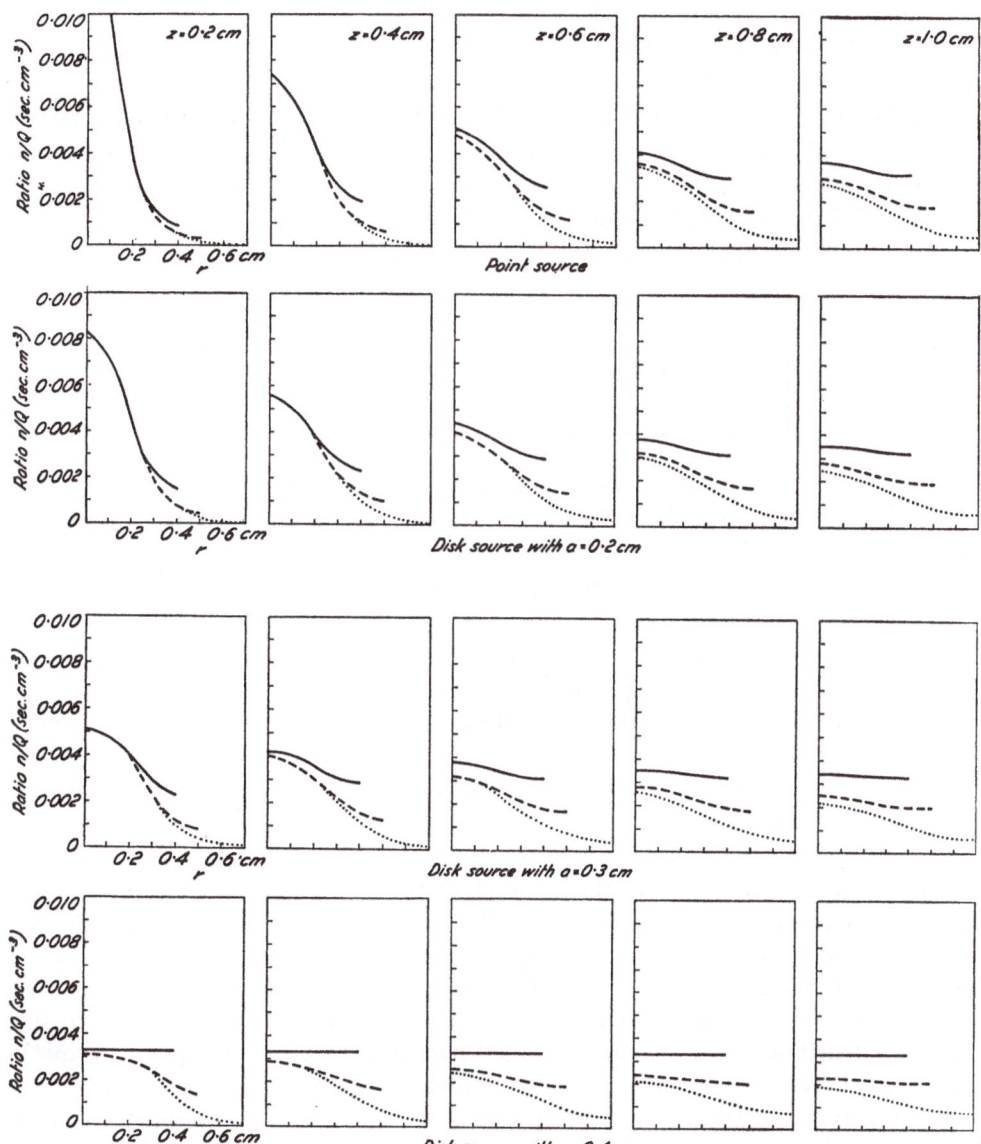

FIG. 9.3. Radial distributions of concentration computed from (9.31) with $v = 600$ cm/sec, $D = 30$ cm²/sec, and $K = 0$.

In each diagram the ratio n/Q of concentration (in cm⁻³) and volatilization rate (in sec⁻¹) has been plotted versus the radial distance r from the arc axis. Each row presents radial distributions for five sections z through the arc, at 0·2, 0·4, 0·6, 0·8, and 1·0 cm above the lower electrode. Successive rows correspond to different values of the radius a of the source. Different curves pertain to different locations of the boundary $r = R$.

———— $R = 0.4$ cm   ------ $R = 0.5$ cm   .... $R = 0.7$ cm perfectly homogeneous distribution is reached if $a = R$, since then particles are not allowed to spread outside a cylinder having the disk source as a base. The results for this trivial case link up with the simplified model of § 9.2.

To examine how the present model fits in with the experimental results obtained by de Galan (§ 9.4) we calculated concentration distributions for a highly ionized element, a moderately ionized element, and a weakly ionized one, characterized by the following sets of values (cf. Table 9.4):

	α		
	1	0·5	0
v (cm/sec)	1200	600	150
D (cm²/sec)	30	30	30
$v/2D$ (cm^{-1})	20	10	2·5

With $R = 0\cdot4$ we varied the parameter values of the 'leak' constant K and that of the disk radius a, and observed which values of K and a yielded the best agreement with the trends found experimentally, viz.

1 A marked radial decline of the concentration just above the lower electrode, which increases as the degree of ionization falls.

2 A virtually uniform radial distribution at a height of more than 0·5 cm above the lower electrode, independent of the degree of ionization.

3 A virtually uniform axial distribution for highly ionized elements and a decline from the lower electrode to the upper one for weakly ionized elements.

4 A very strong dependence of the absolute concentration, for example, at about 0·5 cm above the source on the degree of ionization (an increase of about a factor of 10 if α falls from 1 to 0).

Independent of the choice of the parameter values we found the first trend in the theoretical curves, as is demonstrated by the example in Fig. 9.3. The high degree of uniformity of the experimental radial distributions (trend 2) was not covered by the theoretical results. We must bear in mind, however, that the actual arc studied was a free-burning, non-stabilized arc, which tends to wander; so, the observed distributions, that were averaged over a fairly long time (about 5 min), are spread more evenly than the calculated ones.

When discussing the agreement between theoretical and experimental results for the axial distribution and the absolute concentration (trends 3 and 4), R, a, and K must be specified precisely. As $R = 0\cdot4$ cm is physically the most realistic value, we decided upon this one. The values of a and K were varied and calculations were conducted for the three representative velocities mentioned above. Conclusions can be summarized as follows.

1 It turned out that the radius of the disk source is a vital parameter. At high axial transport velocities the diffusion process has little time to spread the vapours laterally during their short stay in the discharge zone. So, to obtain uniform distributions the disk radius must be assigned a rather large value, say 0·3 cm. This

value is larger than the bore radius and even larger than the radius of the lower electrode used in the experiments.

It seems therefore that the processes occurring near the electrode do not permit us to approximate the electrode surface as a simple disk source. Jet-like vapour streams from the electrode, diffusion through the electrode wall, and the spatial distribution of the electric field in the vicinity of the electrode might be points to consider in this respect.

Probably, we must conceive the disk source as a *virtual* source at a small distance above the lower electrode, say at 1 to 2 mm above it, and apply the disk source model from thereon. The vapour transport from the electrode to the disk source should be studied separately.

2 If K is taken to be smaller than 1 cm^{-1}, its exact value is immaterial for highly and moderately ionized elements; consequently, we can base our choice on the behaviour of weakly ionized elements and use one value to cover the entire range of degree of ionization.

3 When discussing the best value of K, we have to consider simultaneously its effect on the absolute magnitude of the concentration and its influence on the axial variation of the concentration. These effects are always correlated, because an axial fall in the concentration from the lower electrode to the upper one results from losses by radial diffusion: the more particles allowed to escape from the core by lateral diffusion the more pronounced the axial fall in the concentration and the lower the absolute concentration. For a given value of the diffusion constant D, the magnitude of losses by radial diffusion depends on the two parameters v and K.

The influence of v when the transport is not hampered by a wall at $r = R$ is elucidated in Fig. 9.4. The left-hand diagram refers to a small axial velocity, $v = 150$ cm/sec, and the right-hand one to a moderate velocity, $v = 600$ cm/sec. In the latter, the axial transport is four times as great as that in the former and the concentration distribution is qualitatively derived by spreading out the particles in the shaded region of the left-hand diagram over an area of the same width but four times the length. We see that with fast axial transport a vapour is more spread out longitudinally than with slow axial transport, whereas lateral diffusion is much more pronounced for small v than for large v. To conclude, the influence of the directed axial transport and the radial diffusion on the absolute concentration in the core balance each other to some extent. In other words, the radial diffusion causes the absolute concentration to vary less steeply with v than it does in the tube model discussed in § 9.2.

The experimental results, particularly the strong dependence of the absolute concentration on α [equations (9.15) and (9.19)], led us to the conclusion that the influence of the axial transport dominates. This result can be understood only if we assume the presence of a wall, which for other reasons is also a plausible assumption, as was indicated above. The permeability of the wall is expressed numerically by the value of K. The smaller we make it, the less the sideward losses and the

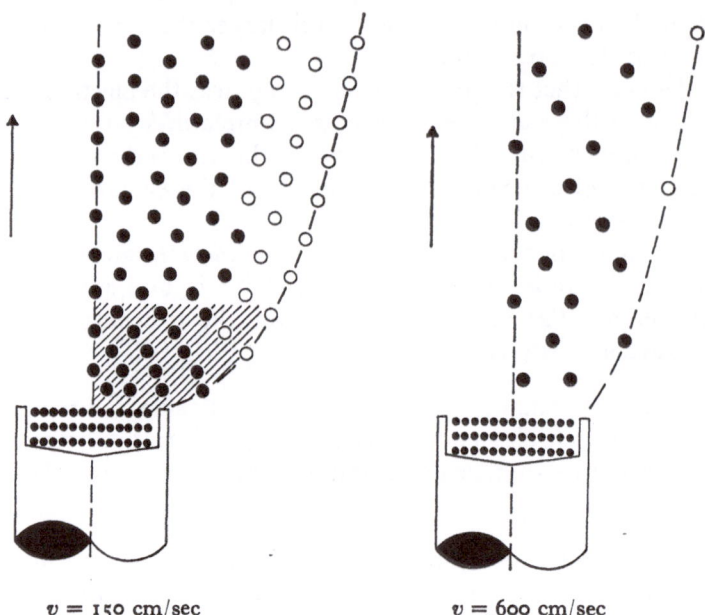

$$v = 150 \text{ cm/sec} \qquad v = 600 \text{ cm/sec}$$

FIG. 9.4 Schematic representation of the influences of the diffusion and the axial transport velocity upon the concentration distribution in the arc.

Increasing the axial transport velocity v by a factor of four gives rise to a concentration distribution that is spread out axially on a fourfold magnified scale (compare the distribution in the shaded region at the left with the distribution at the right). The figure shows that at low values of v a higher concentration ensues in the arc core, although a relatively greater fraction of the particles (o) escapes sidewards from the excitation region than at high axial transport velocities.

stronger the dependence of the concentration on α. By making $K \approx 0$, it would be possible to find the strong correlation [equation (9.19)] between the concentration and the degree of ionization quantitatively. However, this would preclude the existence of a marked axial fall in the concentration for weakly ionized elements, which was also found experimentally. Consequently, we must come to a compromise.

Satisfactory agreement with the experimental trends is obtained if $K = 0.5 \text{ cm}^{-1}$. Results for $R = 0.4 \text{ cm}$, $a = 0.3 \text{ cm}$, $D = 30 \text{ cm}^2/\text{sec}$, and $v = 1200$, 600, and 150 cm/sec are shown in Fig. 9.5. The diagram presents the ratio $\bar{n}/Q$ at different heights above the disk source; the average $\bar{n}$ has been taken over a circular section with a 0.3 cm radius. Accordingly, a change of v from 1200 to 150 cm/sec (corresponding to a change in α from 1 to 0) causes an increase in the absolute concentration by a factor of four at a height of 0.5 cm above the source. The largest fall in the axial concentration, occurring when $v = 150 \text{ cm/sec}$, amounts to a factor of nearly three. Considering the approximate character of the model and the assumptions made about the numerical values of the constants, we conclude that the theory agrees well with experiment.

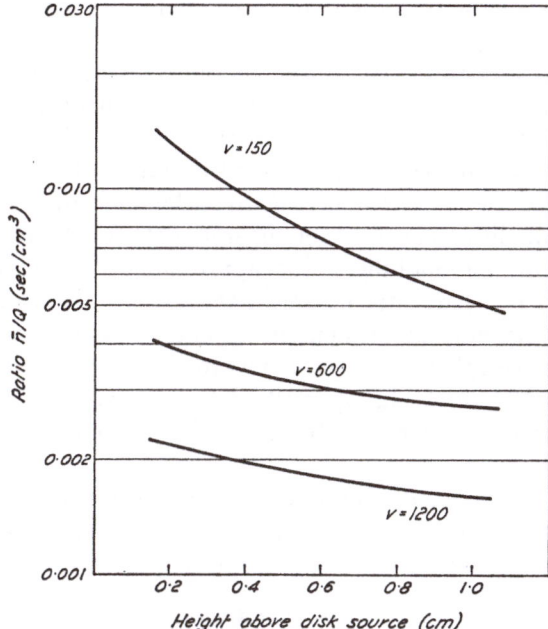

FIG. 9.5. Axial distribution of the mean concentration in the arc core (expressed as $\bar{n}/Q$) for three values of the axial transport velocity v (cm/sec).

The average $\bar{n}$ has been taken over a circular section with a 0·3-cm radius. The results refer to the following parameter values: radius of enveloping cylinder $R = 0·4$ cm, radius of disk source $a = 0·3$ cm, diffusion coefficient $D = 30$ cm²/sec, 'leak' constant $K = 0·5$ cm⁻¹.

Yet we should point out some discrepancies. For $z = 0·5$ cm, theory gives for $\bar{n}/Q$ a range from 0·002 to 0·008 when α runs from 1 to 0. Equations (9.15) and (9.19), on the other hand, give a range of $\bar{n}/Q$ between 0·0003 and 0·003. So the theoretical range encompasses a factor of four, the experimental one a factor of ten. At the same time the entire theoretical range is systematically higher than the empirical one. For this systematic difference there are several possible causes. First, there may be losses of material owing to the mechanism of entry, so that the Q-value for the virtual disk source located at some millimetres above the lower electrode need not coincide with the experimental Q-value, computed as the quotient of the number of atoms in the electrode and the evaporation time. Second, the experimental values of the absolute concentrations may be systematically low as a result of bias in the basic data, e.g. the measurement of the absolute temperature and the electron concentration. Third, our estimate of the diffusion coefficient may be too low. In this connection we also mention a possible difference in the diffusion coefficient for neutral atoms and singly-charged ions. If $D_i > D_a$, the theoretical range of $\bar{n}/Q$ would extend to lower values, as the absolute concentration is inversely proportional with D, and D must be written now as

$$D = (1 - \alpha)D_a + \alpha D_i \tag{9.33}$$

It is not likely that, owing to the radial electric field, the radial diffusion proceeds more rapidly for strongly-ionized elements; for, the enhanced radial diffusion would necessarily cause the axial concentration gradient to become large, which contradicts experimental evidence. Only an overall enhancement of the diffusion rate of ions can thus be considered and not that of the radial diffusion alone.

The approximate mathematical model for the transport of vapours through the discharge channel of the d.c. arc allows us to understand the various experimental trends more than qualitatively. Lack of accurate data for the transport parameters (diffusion coefficient, ion mobility), the approximations underlying the model, and the possibility of systematic errors in the experimental data do not permit us to compare theory and experiment in an entirely quantitative way. It seems beyond doubt, however, that the essential features of the transport mechanism have been clarified. Further research, especially into the nature of the mechanism of entry of material from the electrode to the discharge zone, should be undertaken.

I note finally that an entirely different transport problem arises when the lower electrode is made the cathode. Then, the resultant transport velocity for strongly and moderately ionized elements is directed in the negative z direction, i.e.

$$v = v_c - \alpha \mu_i E \qquad (9.34)$$

is negative. The volatilizing vapours, being retained by the field, collect near the cathode. This condition is commonly referred to as the typical cathode layer effect (Mannkopff and Peters, 1931, and Mannkopff, 1932). We must visualize that the particles escape from the cathode layer by sideward diffusion to the cooler fringe, where ionization is negligible so that electric forces cease to act on the transport. Diffused particles are then carried upward by the convection flow. In the higher layers of the arc, we may find a low concentration at the axis and a comparatively high concentration at the edges of the arc, resulting in an inward diffusion of material. We made an attempt to have the typical cathode layer effect described mathematically (Boumans, 1957, Lauwerier, 1956). The mathematical model was not wholly adequate, and the work was not continued.

CHAPTER TEN

VOLATILIZATION PHENOMENA
AND CHEMICAL REACTIONS

★ § 10.1 *Introduction*

In § 1.1 we introduced the term 'sample excitation' to denote the group of processes that determine the radiation output of the source for a given sample. In theory, the aggregate of processes can be treated in separate steps; in practice, the processes take place jointly, however, and observed intensities therefore reflect overall effects. Only by means of ingenious experiments are we able to decide upon the roles played by the individual steps. In general we attempt to distinguish between the part played by

1 The thermal processes in the plasma (dissociation, ionization, excitation).
2 The transport mechanism in the plasma.
3 The volatilization phenomena and the chemical reactions in the electrode cavity.

The majority of processes occurring in the plasma, including the transport of vapours through the discharge zone, have been discussed in the preceding chapters: only dissociation phenomena remain to be dealt with (see Chapter 11). All these processes pertain to the gaseous phase. Samples, on the contrary, are either solids or liquids and must be converted into the gaseous state to obtain the desired radiation. This conversion process, the evaporation or volatilization of the sample, also is a crucial link in the chain connecting the composition of the sample with the radiation output of the source.

At any given moment the composition of the plasma is determined by the evaporation and by the rates at which the various components leave the discharge zone. Intake and escape of material are not independent of one another; for, the instantaneous composition of the discharge cloud, which depends on the rate of entry of the material, principally controls the excitation conditions, viz. the temperature, the electron pressure, and the transport parameters in the plasma, so that the exit is indirectly influenced by the entry. Conversely, the rate of entry depends on the temperature distribution in the electrode, which in turn is affected by the conditions in the plasma. The complete interrelation is too complex by far to permit a quantitative account to be given, particularly when we consider the volatilization phenomena, where a multiplicity of chemical aspects is encountered; they make the picture less straightforward than that for the plasma processes, which can be dealt with in terms of only a few physical parameters.

At any rate, it is held that the evaporation of the samples is a vital point in spectrochemical analysis. Irregular volatilization is felt to influence results of analysis unfavourably. Since in this respect much depends on the type of the sample and its composition, the evaporation process is considered to contribute pre-eminently to the matrix effects. Unfortunately, little detailed work has been carried out on the nature of the volatilization processes, particularly on their role in the sequence of steps comprised in excitation. Most studies concern the overall effects of matrices on spectral-line intensities and so there is no analysis of the observations in terms of sub-effects due to variations of temperature, electron pressure, spatial distributions, and transport parameters. Inherently, conclusions merely sum up the established facts, and possible explanations are speculative. Except for a few cases (e.g. de Galan's study discussed in § 9.4), an essential link is always missing, so that an adequate and decisive explanation cannot be arrived at.

Now the reader must thoroughly realize which aspects of volatilization phenomena I am talking about, otherwise I run the risk of being entirely misunderstood. We must, in my opinion, discern three distinct points when discussing the importance of the evaporation process in spectrochemical analysis:

1 By which mechanism do volatilization characteristics of sample components affect precision and accuracy of spectrochemical results in total energy procedures, that is, when samples are burnt to completion?

2 How are volatilization characteristics related to chemical reactions in the electrode crater?

3 By what means can we promote selective volatilization of definite constituents, present as impurities in the samples, to create circumstances that are most favourable for trace analysis?

Clearly, the rather pessimistic view expressed above about our inadequate knowledge of the evaporation process pertains primarily to the understanding of the influence of the volatilization characteristics upon analytical results in total energy procedures. Our knowledge about volatilization and its enhancement or depression by additions is far greater. Systematic studies have been made and some insight into the nature of the volatilization process, particularly its interrelation with chemical reactions, has been provided (see § 10.4).

★ § 10.2   *Volatilization phenomena and total energy procedures*

Let us examine the reasons that explain the influence of the volatilization characteristics upon time-integrated intensities of spectral lines, such as those recorded when samples are burnt to completion. It is preferable that we distinguish between three categories of volatilization influences, namely

1 *Obvious effects*, that inherently do not set unsurmountable problems to our imagination and can be discussed clearly.

2 *Trivial effects*, such as spluttering and spattering of material from the electrode, that are obvious and of course highly important in common practice, but uninteresting for theoretical considerations.

3 *Obscure effects*, that have not yet been analysed and can be discussed mainly in a speculative manner.

Obvious effects are the following.

1 *Influence of the rate of volatilization of the sample upon excitation and transport parameters*

After the thorough exposition of how the composition of the arc plasma controls the excitation and transport parameters (see particularly §§ 3.9, 7.7 to 7.9, 8.8, 8.9, 9.2, 9.4, and 9.5), it is evident that the rate at which a sample enters the discharge column strongly influences spectral-line intensities, especially if the sample contains large amounts of easily-ionized elements.

The influence of the plasma composition on the excitation conditions of the vapour is so great that no experiment can exclude it. Only when the arc is adequately buffered are we justified in basing our conclusions on constant excitation conditions in the plasma. A gradual change in the temperature over a few hundred degrees during the burning period is not harmful; time-integrated intensities can still be interpreted then with reasonable accuracy in terms of mean temperatures, as has been shown by de Galan (1965a). The crucial point is that the constancy of such mean temperatures is maintained from one arcing to another.

From the literature concerning influences on spectral-line intensities, numerous examples can be cited of investigations in which the nature of the excitation phenomena in the plasma has been ignored. Unfortunately, the failure to consider fundamental processes has led authors to conduct laborious experiments and to produce ingenious theories that seem to be inappropriate and only express a personal view. However meritorious such investigations might have been at the time they were carried out, they, in my opinion, should now belong to the history of spectroscopy. What we really need today are critical studies of the fundamental processes and well-designed experiments that definitely demarcate the field of non-resolved problems of spectrochemical physics. Examples of such studies are those of Vukanović (1960, 1964), Samsonova (1961) and de Galan (1965a, 1966a) (see §§ 9.3 and 9.4).

2 *Self-absorption of spectral lines*

The instantaneous concentrations of elements in the discharge zone are proportional to their rates of evaporation, provided that excitation conditions in the plasma are constant. Since the degree of self-absorption depends on the instantaneous concentration of particles in the vapour phase, we can visualize how the rate of evaporation may affect line intensities via this mechanism. If we suppose equation (5.12) to be appropriate for describing self-absorption in the cooler arc fringe, we can conclude that it is the absolute magnitude of variations of the vapour concentration which predetermines the relative magnitude of intensity changes. Therefore, relatively small changes in the vapour concentration in the arc may have large repercussions on the degree of self-absorption, and consequently, time-integrated intensities depend on the rate of evaporization of a given amount of element (cf. Preuss, 1956).

The presence or absence of self-absorption can be checked by measuring the relative intensities of component lines of a multiplet. If an observed intensity ratio of two multiplet components deviates from the theoretical value, self-absorption is proved. If we do not know the theoretical ratio, we should vary the concentration and determine whether the line-intensity ratio depends on the concentration. The procedure is exemplified by Ahrens and Taylor (1961, p. 100). They also point out the poor reproducibility of line intensities when self-absorption occurs.

When no multiplet component line is available to ascertain whether the line under investigation is free from interfering self-absorption or not, we can resort to the method of concentration variation and observe the slope of a working curve, i.e. a plot of the log intensity of a line versus the log concentration of an element. Unit slope may serve as an indication that self-absorption is absent. This check is less convincing, however, than that using multiplet component lines.

An alternative method is to install a concave mirror behind the source, which in the absence of self-absorption should redouble line intensity, provided that it reflects perfectly. The instability of the d.c. arc is a serious drawback for this experiment to be conducted precisely.

3 *Schwarzschildt effect*

When a photographic plate is used for recording spectra, we must consider the reciprocity law failure (Schwarzschildt effect) as a possible cause of interference in the study of volatilization effects. We recall equation (1.20), i.e. $E = Ht_e^p$, where E is the exposure, H the irradiance, t_e the exposure time, and p the Schwarzschildt exponent. If $p \neq 1$ and $Ht_e = $ constant, different exposures result when a line is either emitted at high intensity for a short time or at low intensity for a long time. So, for example, when $p = 0.85$, the ratio E_1/E_2 corresponding to $H_1 = 2H_2$ and $t_{e1} = \frac{1}{2}t_{e2}$ is computed to be 1.11. When $H_1 = 3H_2$ and $t_{e1} = \frac{1}{3}t_{e2}$, $E_1/E_2 = 1.18$.

The role of the Schwarzschildt effect in connection with volatilization phenomena has been clearly demonstrated by Beintema (1957).

4 *Ignition losses*
See p. 310.

It appears probable that many anomalies described in the literature could have been accounted for if the proper additional measurements had been made. Yet, it has not been established with certainty whether some effects still escape attention. So, doubt remains about the mechanism that regulates the passage of the material to the base of the discharge column. We have seen in § 9.5 that a vapour once it has arrived at the base of the column is subjected to a transport process that seems to be sufficiently understood. However, defining accurately the cross-sectional area of this virtual base and stating the distribution of the vapour in this area presented difficulties in the model we considered. The quintessence of the problem is that we do not know exactly how the substance leaves the electrode. Is there a jet-like flow, which may cause vapours to be blown sidewards out of the discharge volume until they are grasped by the longitudinal transport? To what extent does material diffuse through the electrode walls and vaporize at the exterior of the electrode, thus being prevented from passing properly into the discharge zone? What role does the incomplete dissociation of substances leaving the electrode play?

The significance of these obscure effects will depend largely on the electrode shape and on the composition of the specimen in the electrode cavity. Now suppose this shape to be given, what can we expect then if a sample is carefully diluted with graphite and a suitable buffer compound, so that the excitation conditions in the plasma are stabilized? Should matrix effects be fully annihilated or do, for example, differences in chemical structure between natural samples and synthetic standards prepared by grinding oxides remain to offset accuracy of analysis? Arguments in favour of either possibility can be found in the spectrochemical literature. Recent examples of the latter group are studies by Nickel and Pflugmacher (1961a and b) who analysed fly ashes, iron ores, and related materials. Their investigation, which we can consider as representative of many others of a similar tenor, can be summarized as follows. To determine the major constituents of fly ashes the authors developed a method that uses pelleted anodes (one part of sample or standard to nineteen parts of graphite) which are burnt at 5 amp in an argon atmosphere. They established that a fly ash and a synthetic standard of mixed oxides having approximately equal chemical composition yield markedly differing evaporation curves for the analysis elements and also for the internal standard (Be), which was added by impregnating the graphite powder prior to mixing and pelleting. Analyses made with reference to calibration curves for synthetic standards showed systematic relative errors of 26 per cent for Fe, of about 100 per cent for SiO_2, and of about 30 per cent for Al_2O_3. Pretreatment of samples and standards, involving heating for two hours in an oxygen stream at 800°C and subsequent fusion with excess boric acid, resulted in identical volatilization characteristics for both fly ashes and synthetic standard substances, which, in turn, removed the systematic errors in the analysis.

The general line of this investigation gives rise to the following considerations. We are able to ascertain experimentally that different volatilization characteristics between standards and samples affect accuracy of analysis unfavourably. We are also able to establish that different volatilization characteristics are caused by differences in the chemical structure of the standards and the samples. Further, by carefully studying the course of chemical reactions in the electrode cavity, we are even able to explain the pattern that evaporation follows with time (see Marinković, 1959, 1960; 1965;* Pavlyuchenko and Dubovik, 1963; and Nickel, 1965b).

* These papers are principally quoted as examples of studies that relate the dependence of volatilization on time with the proceeding of chemical reactions in the electrode. Marinković explains on this strength why impurities in uranium oxides (UO_2 and U_3O_8) evaporate either completely or incompletely within a prefixed time. Inference does not touch the problem connected with burning to completion at present under discussion.

11

However, we are not in a position to indicate why different volatilization characteristics give rise to inaccurate analysis results, in other words, why different time-integrated intensities or intensity ratios can result from different volatilization patterns, if indeed, as is commonly assumed and emphasized in such experiments, excitation conditions in the plasma have been constant throughout the burning period. The crucial and fundamental question we are thus concerned with is the reason why different time-integrated intensities originate from different volatilization curves. Are excitation conditions as constant as we believe them to be, and should we seek a hitherto hidden factor in the mechanism of entry of the material? Or should we doubt the constancy of excitation?

In any case, events occurring in the initial arcing period are likely to produce marked effects. Not only are excitation conditions irregular and badly defined then—which primarily would effect precision—but the proportions of sample constituents that evaporate during this time interval may vary according to the state of chemical bonding of elements, giving rise to different ignition losses* between samples and synthetic standards. By ignition losses we denote all those influences on time-integrated line intensities that originate from the irregular and variable escape of material from the electrode when the arc is struck; perhaps vapours are blown partly at high velocity through the discharge volume, partly sidewards out of the excitation zone. Anyhow, the influence of ignition losses should be reduced by covering the specimen with a layer of pure diluent (buffer + graphite) and by using a long supporting electrode with a small outer diameter and a deep, narrow bore (say with an internal diameter of 2–3 mm).

In addition to ignition losses we must consider the effects of diffusion of substances through the electrode walls (cf. Leuchs, 1950) and of incomplete dissociation of vapours that enter the discharge zone (cf. Pavlyuchenko and Dubovik, 1963). Further experiments are needed, however, before we can consider the passage of material into the arc in terms other than questions and speculative suppositions. The problem to be solved is not that volatilization characteristics and the chemical behaviour of elements in the electrode *do* affect time-integrated intensities, but rather *why* they do this.

§ 10.3   *Graphite and carbon electrodes*

It has become customary in carbon arc spectrochemical analysis to distinguish roughly between two grades of electrode material, amorphous carbon and graphite. In fact, both show crystallinity and there is a gradual transition from one form to the other. In compliance with general need, manufacturers of spectroscopic electrode materials usually supply material that represents the extremes of the

---

* A lucid description of how the evaporation proceeds, in general, from the moment of ignition to the time the sample has completely volatilized is given by Hegemann and Schöntag (1953). They distinguish three stages: first, glowing particles of the sample are sprayed into the arc and volatile components distil selectively; then, evaporation ceases for a few seconds while the arc consumes the remaining electrode wall and burns almost as a pure carbon arc; finally, a more or less regular evaporation of the test substance sets in, which reaches a final maximum just before the very end.

entire region of possibilities. So, there are available various grades of graphite and various grades of carbon. They differ in degree of graphitization, which is indicated by the average linear diameter of crystallites, by the interlayer spacing of the graphitic planes, and by the extent of orientation of the crystallites. In carbon the average linear diameter of the crystallites is smaller than in graphite, whereas the interlayer spacing of the graphitic planes is larger, and the orientation of the crystallites is less specific. Various structural aspects have been discussed at length by Rüssmann (1958) and concisely by Mellichamp and Finnegan (1959).

Carbon and graphite as well as various grades of either form differ in physical properties, such as density, hardness, electrical resistivity, and thermal conductivity. Numerical values of physical properties of some twenty commercially available grades have been summarized by Rüssmann (1958). The most important features for spectrochemical analysis are thermal conductivity, electrical resistivity, and ease of machining into desired electrode shapes. The last property particularly has been a decisive factor that led spectrochemists to prefer graphite rods to carbon ones, since the latter are harder to shape into electrodes. Nowadays this objection does not really apply, as preforms and electrodes of special shape made from carbon can be purchased. The choice of either graphite or carbon in a particular case can be made therefore by considering a more essential property, the thermal conductivity. It is related to the electrical resistivity, which itself has no direct bearing for spectrochemical analysis, as the heat generated by the electrical resistance of the electrodes is negligible in the energy balance. On the other hand, numerical values of electrical resistivity may be used as an indication of the grade of the material. According to measurements by Rüssmann the resistivity of various grades of graphite at room temperature is in a range between 700 and 1750 $\mu\Omega$ cm (most frequently it amounts to 800–1100 $\mu\Omega$ cm), whereas carbon grades have a resistivity over 4500 $\mu\Omega$ cm. We should note that the resistivity depends largely on temperature, particularly that of carbon which decreases rapidly with increasing temperature. The *Handbook of Chemistry and Physics* (Hodgman, Weast, and Selby, 1957, pp. 2384–2386) reports for carbon a decrease from 3500 $\mu\Omega$ cm at 0°C to 900 $\mu\Omega$ cm at 2500°C; for graphite, a gradual rise from 800 to 1100 $\mu\Omega$ cm is reported for this temperature range.

Although data on electrical resistivity reflect what difference must be expected between thermal conductivities of carbon and graphite, it is not easy to state accurately numerical values of thermal conductivity. Available data (see Mellichamp and Finnegan, 1959; and Euler, 1956) are scarce, especially for the temperature range 1500–3400°K, and scatter widely. From Euler's measurements and from data reproduced by Mellichamp and Finnegan (1959) after Castle (1956) we can infer that the thermal conductivity of carbon varies only slightly with temperature up to about 3400°K, its numerical value being some 0·1 W/cm degree. The thermal conductivity of graphite is of the order of 1 W/cm degree at room temperature, sinks with increasing temperature, and becomes equal to that of carbon at about 3400°K. From then on carbon and graphite behave similarly and thermal conductivity decreases fairly rapidly with increasing temperature, showing a transition at 3700°K, where it falls abruptly to a very low value.

The dependence of thermal conductivity on temperature controls the longi-tudinal temperature distribution in the electrode. The temperature of the anode spot has been measured to be 4000°K (Euler, 1956, 1956/57). The temperature sinks very rapidly to 3700°K across a thin layer of about 0·05 mm thickness, at current strengths between 8 and 10 amp. Then, the temperature falls somewhat less steeply to about 3400°K, which is reached at about 0·5 mm below the electrode surface (Euler, 1956). The transition at 3700°K substantiates the presumption that the melting-point of carbon is located at this temperature.

Some useful estimates of the temperature distribution in the region below 3400°K have been made by Leuchs (1950). He obtained them by observing the thermochemical behaviour of different substances and studying diffusion and volatilization phenomena in the cavity of the anode of a carbon arc. Results obtained by optical pyrometry pertaining to both carbon and graphite cathodes have been reported by Hoens and Smit (1957). Recently, Nickel (1956a, b, and c) published highly interesting studies on the temperature distribution in pelleted graphite anodes that were burnt in an argon–oxygen atmosphere. His results are founded on the investigation of thermochemical reactions and on micropyrometric measure-ments.

Leuchs' estimates, referring to an anode 3 mm in diameter and an arc 1·3 mm in length, burning with a current of 5·5–8 amp, are as follows:

Distance from electrode top (mm)	0	1	2	3	4	5
Temperature (°K)	3800	2800	2400	2200	2000	1900

Knowledge of the longitudinal distribution of temperature is essential for under-standing the complex of chemical processes occurring in the electrode cavity, and unravelling these processes would, in turn, greatly facilitate further insight into the nature of volatilization phenomena (cf. §§ 10.2 and 10.4).

It is of interest in this connection to consider the energy balance at the anode. Euler (1956) points out that in the electrode energy is emitted on five counts and consumed on five. The energy emitted encompasses mainly the electric energy dissipated in the anode fall, the energy released by the entry of electrons into the anode surface, and the heat of combustion of the electrode material; the two other sources, the energy transferred from the column to the electrode either by radiation or by conduction, are virtually negligible. Numerical values of the main contributions are estimated by Euler for a carbon arc in air to be

1 Energy dissipated in the anode fall = 270 W
   (anode fall = 27 V, current strength = 10 amp)
2 Energy released by entry of electrons = 45 W
   (potential drop = 4·5 V)
3 Heat of combustion = 13 W
   (rate of combustion of carbon = 1·5 mg/sec, heat of combustion in the reaction $C + \frac{1}{2}O_2 \rightarrow CO$ = 26·64 kcal/mol = 9·4 kWsec/g)

Energy is consumed by three major processes: radiation, convection, and thermal conduction in the electrode, for which numerical values have been estimated by Euler to be 136, 85 ± 20, and 108 ± 20 W respectively. The last value was calculated from the complete energy balance.

We conclude from these data that in the carbon arc in air the energy dissipated in the anode fall contributes by far the greatest portion of the total power input at the electrode. Consequently, when the anode fall is changed by the evaporation of a substance or by burning the arc in a controlled atmosphere (inert gas), we must expect noticeable repercussions on the amount of energy that flows into the electrode. It is difficult to predict quantitatively how the magnitudes of the energy consuming processes are affected separately when the power input changes. The heat inflow by thermal conduction is the one principally of interest when considering the evaporation of a substance, as the evaporation usually does not take place at the anode spot but somewhat lower in the electrode crater. Interesting observations on the rate of evaporation of samples in different atmospheres have been reported by Vallee and co-workers (see review paper by Thiers and Vallee, 1957).

A parameter related to thermal conductivity is the rate of consumption of the electrode material; it is easy to measure. According to Mellichamp and Finnegan (1959) a carbon anode having a diameter of 1/8 inch and burning in a 10-amp d.c. arc with a 1-mm gap is consumed about 1·4 times faster than a graphite anode; by contrast, carbon cathodes are consumed at a slightly slower rate than graphite cathodes. Mellichamp and Buder (1963) extended the investigation of the burning properties of carbon and graphite to include the dependence of consumption rates on current strength and on cross-sectional area. The test showed that the three variables, current strength, cross-sectional area, and electrode grade, are so interrelated that, to some extent, changes in one variable can be partially compensated for by an equivalent change in another. So, they found that a change of 0·01 inch in anode diameter is equivalent to a current change of approximately 1 amp; also, the difference between carbon and graphite grades could be expressed as equivalent to a current change of about 5 amp. These data apply to the anode without sample material; evidently the presence of a specimen greatly modifies the relationship between the three variables.

Mellichamp and Buder also noted, as did Euler (1956/57) too, that carbon cannot withstand as high a current density as graphite and is more easily overloaded. (See note on p. 35 for remarks on overloading and hissing.)

In conclusion, in electrodes of similar shape, burnt under identical conditions, temperature distributions are different for graphite and carbon. The temperature gradient in a graphite electrode is smaller than that in a carbon electrode. This results in a marked difference in rates of consumption and in rates of volatilization of a substance. Ahrens and Taylor (1961, p. 57) state that the period for complete volatilization of a silicate sample may be shortened by a factor of two or three when a carbon anode is used instead of a graphite anode. They also note that selective volatilization is more marked when graphite electrodes are used (see also Mellichamp and Finnegan, 1957). Clearly, we must bear in mind that the volatilization characteristics of constituents of a sample are strongly influenced by the electrode shape and by the addition of admixtures (see § 10.4). For comments on different types of electrodes we refer the reader to Ahrens and Taylor (1961, p. 57). These authors recommend the use of carbon as the upper electrode, irrespective of whether the lower one is graphite or carbon, because it usually decreases wandering of the arc over the upper electrode surface. This recommendation is also made by Rüssmann (1958) and by Mellichamp and Finnegan (1957).

The question may be raised whether the temperature in the arc depends on the electrode material. It is improbable that a large temperature difference between the graphite and the carbon arc will be found, since the energy balance of the column is virtually independent of the electrode characteristics. When a sample is arced, the electrode grade might have a secondary influence on the arc temperature, resulting from different rates of evaporation and consequently different plasma compositions in a graphite arc and a carbon arc. Measurements made by Rüssmann (1958) in arcs without added material indicate a difference of 700°K between carbon and graphite arcs. Probably, these results are in error, for the following reason. He measured the temperature with the aid of the CN band sequences o → o and 1 → 1 (3883 and 3872 Å); self-absorption is likely to have influenced this determination and made the results too high. De Galan (1965a) showed this source of error to be critical, particularly if measurements are made within a few millimetres of the electrodes. As Rüssmann used an electrode separation of only 4 mm, the higher temperature value established in the carbon arc might well have arisen from self-absorption, which, as a matter of fact, should be more pronounced in the carbon arc in view of the greater consumption rate.

Concluding this section we point out a new development in spectrographic electrode material, the so-called pyrolytic graphite, which exhibits high thermal and electrical anisotropy (see Mooney, Schoder, and Garbini, 1964). We note an observation by Scholze and Gerbatsch (1964) (see also Ehrlich, 1964), who established that consumption of graphite at the side surface of the anode proceeds stepwise; during the first 2 min the length over which the exterior is consumed increases regularly; then the region of consumption ceases to grow longitudinally and the part of the electrode thus demarcated is consumed radially. After 6 min the entire process is repeated on a subsequent part of the electrode. The phenomenon is of interest for distinguishing between impurities present at the outer surface of spectroscopic electrodes and those in the interior. Finally, we mention a paper by Pepper, Pardi, and Atwell (1963) on the effects of various grades of graphite electrodes on the carrier distillation of impurities in uranium oxide, an article by Spindler (1961) on the effects of replacing graphite powder with carbon powder (cf. § 7.12), and an article by Mitteldorf (1965a) on spectroscopic electrodes.

★ § 10.4   *Selective volatilization and chemical reactions in the electrode cavity*

The term 'selective volatilization' refers to the phenomenon of successive volatilization of separate elements of a specimen containing several components. The phenomenon is also referred to as fractional distillation or selective vaporization.

As there are excellent discussions* on this subject available, I shall confine myself to a few comments.

The order in which elements and their compounds volatilize in a d.c. arc has been systematically studied by Rusanov (1941, 1943, 1945) and by Gorfinkle and Ahrens (see Ahrens and Taylor, 1961, p. 82). They established volatility series for elements, oxides, and sulphides, which may serve generally as a helpful guide in spectrochemical analysis. We should be aware, however, that the volatility of an element is not an intrinsic property, but depends to a considerable degree on the environment, that is, on the presence of other substances and on the thermal conditions in the electrode cavity. We note here only the reduction of oxides and the formation of involatile carbides in the carbon electrode (see comments by Zaidel *et al.*, 1960; Leuchs, 1950; and Schroll, 1963). Other striking examples are the conversion of the volatile As_2O_3 into a substantially less volatile compound Fe_2As in an iron oxide base (Leuchs) and the conversion of the non-volatile boron carbide into the volatile boron fluoride BF_3 by the addition of a mixture of AlF_3 and NaF (Brandenstein, Janda, and Schroll, 1960). In conclusion, we must fully appreciate that the volatilization characteristics of an element can be markedly altered by changing the composition of the specimen in the electrode crater. For that reason admixtures play an important part in spectrochemical analysis. We have already considered some other aspects of this matter in §§ 1.1 and 7.12, where emphasis was on the diluting and buffering action of admixtures when samples of widely different composition are analysed and matrix effects have to be annulled.

An entirely different but highly important aspect of selective volatilization is the deliberate promotion of this phenomenon to improve the limits of detection for impurity elements present in a base, which, when volatilized to completion, would obscure the spectrum by the multiplicity of lines and the enhanced intensity of the continuous background (cf. § 7.11). It is especially in this field that knowledge of the chemical reactions occurring in the electrode cavity serves as a valuable and indispensable guide in the selection of suitable admixtures. Here, the question of the carrier once more enters into the discussion. We have shown in § 9.3 that one contribution to the carrier effect arises from the influence exerted by the presence of the carrier in the plasma upon the excitation conditions, that is on the temperature, the electron pressure, and the transport parameters. Carriers acting in this way must be identified with spectroscopic buffers. The relation between the transport of vapours through the plasma and spectral-line intensities was ignored until fairly recently and it was believed therefore that substances denoted as carriers had a peculiar role in this respect. However, as de Galan's experiments (§ 9.4) affirm, a change of the transport parameters is inherently associated with the changes of temperature and electron pressure that can be brought about by the mere addition of a substance to the arc plasma.

On the other hand, there are many examples of investigations where carriers

* A useful introduction is given by Ahrens and Taylor (1961, Chapter 7); a more elaborate treatment is found in the book by Zaidel, Kaliteevskii, Lipis, and Chaika (1960, Chapter 7 and 8); see also Leuchs (1950) and Schroll (1963).

promote the selective volatilization of elements in the sample through the mechanism of chemical reaction and the formation of volatile compounds in the electrode cavity. In a survey of literature, Schroll (1963) emphasizes this aspect of carriers and at the same time produces interesting results and conclusions from his own systematic study of selective volatilization in relation to thermochemical reactions. Of investigations concerned with the chemical aspects of the carrier we mention particularly those by Pszonicki (1960, 1962), by Pszonicki and Minczewski (1962), by Kung-Soo Chang (1963), by Janda, Schausberger, and Schroll (1963), by Daniel (1960), and by Marinković (1959, 1960). Other articles of interest in this connection are those by Rusanov, Alekseeva, and Il'yasova (1961), Zakhariya, Turulina, Karpenko, and Voloshchenko (1963), and Schroll and Weniger (1965) on sulphiding agents, the papers by Nickel (1963, 1965a, b, and c) on chemical reactions of various boron compounds in graphite pellets heated in a d.c. arc in an argon–oxygen atmosphere, and the paper by Tymchuck, Russel, and Berman (1965) on the effect of halide carriers on the volatilization of trace impurities from copper.

Special attention must be drawn to a procedure for analysing refractory materials. It is known as the evaporation method. This method, mainly explored by Russians, separates completely the excitation of the spectrum from the process of volatilization of the impurities from the sample. The sample is heated in a graphite cup by resistance heating. The evaporated volatile elements are condensed on a cooled collector electrode, which is subsequently arced or sparked in the usual way. By the use of this enrichment procedure extremely low detection limits can be reached. For detailed information on this technique we refer the reader to the book by Zaidel, Kaliteevskii, Lipis, and Chaika (1960). A new apparatus for the vaporization-concentration method has been described by O'Connell (1964).

CHAPTER ELEVEN

DISSOCIATION PHENOMENA
AND BAND EMISSION

§ 11.1 *Thermal dissociation and the plasma composition of the carbon arc in air*

The plasma of the carbon arc in air is composed of the elements nitrogen, oxygen and carbon. Unless the discharge takes place in a controlled atmosphere of dried air, hydrogen must also be considered. The molecular composition of the plasma is complex at the elevated temperatures prevailing in the arc. We are primarily concerned with the following dissociation and ionization equilibria:

$$
\begin{array}{llll}
N_2 \leftrightharpoons 2N & (a) & N \leftrightharpoons N^+ + e^- & (g) \\
O_2 \leftrightharpoons 2O & (b) & O \leftrightharpoons O^+ + e^- & (h) \\
CO \leftrightharpoons C + O & (c) & C \leftrightharpoons C^+ + e^- & (i) \\
NO \leftrightharpoons N + O & (d) & N_2 \leftrightharpoons N_2{}^+ + e^- & (j) \\
CN \leftrightharpoons C + N & (e) & O_2 \leftrightharpoons O_2{}^+ + e^- & (k) \\
C_2 \leftrightharpoons 2C & (f) & NO \leftrightharpoons NO^+ + e^- & (l)
\end{array}
\qquad (11.1)
$$

For computing the main molecular composition the ionization equilibria (g) to (l) need not be considered, as the total degree of ionization is small ($\approx 10^{-4}$–10^{-3}, cf. § 3.9). References to calculations of the thermal dissociation of nitrogen and air in different temperature ranges are given in § 3.9. We shall now examine some results of Roes' calculations (1962) for the carbon arc plasma, that is, for air with atomic carbon and carbon compounds (C_2, CN, and CO). First of all, let us consider a few fundamental concepts of thermal dissociation.

For the dissociation equilibrium

$$ XY \leftrightharpoons X + Y \qquad (11.2) $$

of a diatomic molecule XY the constant of dissociation K_n is defined as

$$ K_n = \frac{n_X n_Y}{n_{XY}} \qquad (11.3) $$

where n is concentration expressed as number per unit volume. The constant K_n is a function of temperature (cf. Saha's relationship in § 7.1), viz.

$$ K_n = \left(\frac{2\pi}{h^2} \frac{m_X m_Y}{m_{XY}} kT \right)^{\frac{3}{2}} \frac{Z_X Z_Y}{Z_{XY}} \exp(-\epsilon_d/kT) \qquad (11.4) $$

where m is the mass, T the absolute temperature, Z the partition function, ϵ_d the energy of dissociation, h Planck's constant, and k Boltzmann's constant.

When concentration is expressed as number per cm³ the practical form of (11.4) reads (cf. Smit, 1950, p. 105):

$$K_n = 1 \cdot 88 \cdot 10^{20} \left(\frac{M_X M_Y}{M_{XY}}\right)^{\tfrac{3}{2}} T^{\tfrac{3}{2}} \frac{Z_X Z_Y}{Z_{XY}} 10^{-(5040/T)V_d} \qquad (11.5)$$

where M is the molecular weight in atomic units and V_d the dissociation potential in electron volts.

In partial pressures we have

$$K_p = \frac{p_X p_Y}{p_{XY}} = 2 \cdot 56 \cdot 10^{-2} \left(\frac{M_X M_Y}{M_{XY}}\right)^{\tfrac{3}{2}} T^{\tfrac{5}{2}} \frac{Z_X Z_Y}{Z_{XY}} 10^{-(5040/T)V_d} \qquad (11.6)$$

where p is in atm. The partition functions Z_X, Z_Y, and Z_{XY} are defined by (6.4); for molecules, electronic, vibrational, and rotational levels must be included in the summation. Generally, it suffices to consider the lowest electronic level only. The partition function for one electronic state of a diatomic molecule can be approximated by (see Herzberg, 1950, pp. 123, 125, 467; and Smit, 1950, p. 160):

$$Z = \frac{kT}{hcB} \frac{g \exp(-\epsilon/kT)}{1 - \exp(-hc\omega/kT)} \qquad (11.7)$$

where h is the Planck constant, k the Boltzmann constant, c the velocity of light, T the absolute temperature, ϵ the electronic energy, B the rotational constant, ω the vibrational constant, and g the statistical weight of the electronic state. Numerical values of B and ω for diatomic molecules can be taken from Herzberg (1950, p. 501 ff.). The value of g depends on the type of electronic level, for example

$$g = 1 \quad \text{for } {}^1\Sigma$$
$$g = 2 \quad \text{for } {}^2\Sigma,\ {}^1\Pi,\ {}^2\Pi_{\tfrac{1}{2}},\ {}^2\Pi_{\tfrac{3}{2}},\ {}^3\Pi_0,\ {}^3\Pi_1,\ {}^3\Pi_2,\ {}^1\Delta$$
$$g = 3 \quad \text{for } {}^3\Sigma$$
$$g = 4 \quad \text{for } {}^2\Pi\ (\equiv {}^2\Pi_{\tfrac{1}{2}+\tfrac{3}{2}})$$
$$g = 6 \quad \text{for } {}^3\Pi\ (\equiv {}^3\Pi_{0+1+2})$$

The dissociation constant K_n is readily computed from the practical logarithmic formula

$$\log K_n = 20 \cdot 432 + \tfrac{3}{2} \log \frac{M_X M_Y}{M_{XY}} + \log \frac{Z_X Z_Y}{Z_{XY}} + \log B - \log g$$

$$+ \tfrac{5}{2} \log T + \log(1 - 10^{-0 \cdot 625\omega/T}) - \frac{5040}{T} V_d \qquad (11.8)$$

where B and ω are in cm^{-1} and V_d is in electron volts. It has been assumed that only the lowest electronic level contributes substantially to the partition function of the molecule.

For calculating the plasma composition of the carbon arc in air, one must consider the six equilibria (a) to (f) indicated in (11.1). The concentrations of the nine species involved are then determined by the following set of equations:

1 Six equilibrium conditions of the type (11.3).

2 The condition that the sum of the concentrations of all species equals the total concentration, that is, at a pressure of 1 atm, $7\cdot340\cdot10^{21}/T$ per cm^3.

3 An equation that specifies the total carbon concentration, i.e.

$$n_C + 2n_{C_2} + n_{CN} + n_{CO} = \mathcal{N}_C$$

4 An equation expressing the ratio of the concentrations of oxygen and nitrogen in air, i.e.

$$n_N + 2n_{N_2} + n_{CN} + n_{NO} = 3\cdot71(n_O + 2n_{O_2} + n_{CO} + n_{NO})$$

When estimates are made about the locations of the equilibria (a) to (f), equations (2) to (4) can be considerably simplified:

$$n_N + n_O + n_{N_2} + n_{CO} = 7\cdot340\cdot10^{21}/T \qquad (2')$$

$$n_C + n_{CN} + n_{CO} = \mathcal{N}_C \qquad (3')$$

$$n_N + n_{N_2} = 3\cdot71(n_O + n_{CO}) \qquad (4')$$

The set of equations comprised in (1), $(2')$, $(3')$, and $(4')$ can be solved for the concentrations of the nine species, N$_2$, N, O$_2$, O, NO, C, CO, CN, and C$_2$, if appropriate values of the constants are inserted and the total carbon concentration $\mathcal{N}_C$ is known.

The calculations carried out by Roes (1962), results of which are given in Fig. 11.1, are based on the following data. Constants needed for computing the partition functions of the molecules were taken from Herzberg (1950). The numerical values of dissociation energies as well as their origins are summarized in Table 11.1, along with values of the dissociation constants at different temperatures. A critical point is the estimation of the total carbon concentration $\mathcal{N}_C$. Roes based his estimate on the subsequent considerations. Carbon vapour consisting chiefly of CO originates from the electrode surface, which, to a first approximation, can be regarded as a disk source. The vapour is carried upwards by convection and spreads laterally by diffusion. To find the vapour concentration at the arc axis, Roes applied the first transport model outlined in § 9.5 [equations (9.23) to (9.26)], however, using a disk source approximation instead of a point source approximation. He determined the rate of entry Q at different currents and computed from the measured values of Q, and $v = 100$ cm/sec and $D = 38$ cm^2/sec, the total carbon

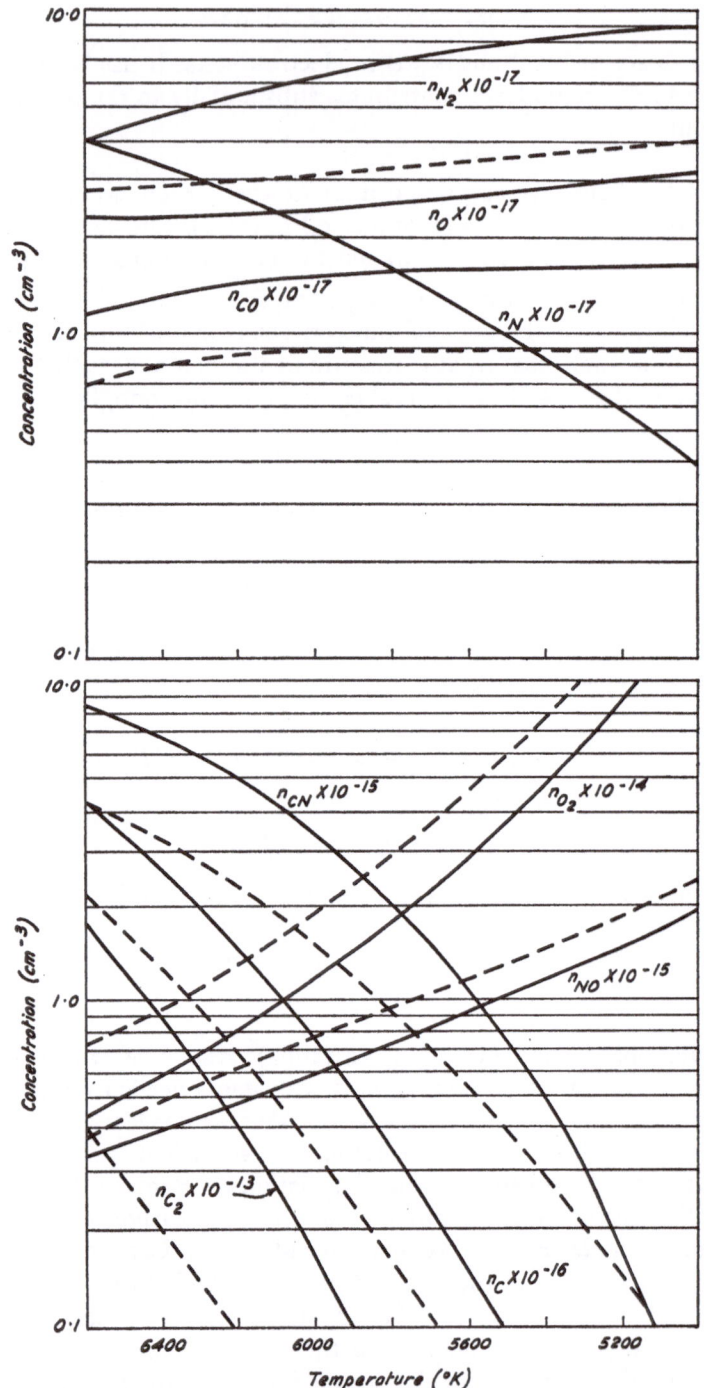

Fig. 11.1. Concentrations of components of a gas mixture consisting of air and carbon vapours plotted as functions of the temperature (after Roes, 1962, pp. 102–103). (Legend continues on facing page.)

TABLE 11.1

Molecule	V_d	Reference	K_n		
			5500°K	6000°K	6500°K
N_2	9·756	A	$1·2.10^{16}$	$7·1.10^{16}$	$3·0.10^{17}$
C_2	6·5	B	$1·7.10^{19}$	$4·8.10^{19}$	$1·2.10^{20}$
O_2	5·114	A	$2·9.10^{20}$	$5·2.10^{20}$	$9·8.10^{20}$
CN	8·2	C	$1·3.10^{17}$	$5·2.10^{17}$	$1·7.10^{18}$
CO	11·09	A	$1·8.10^{15}$	$1·2.10^{16}$	$6·0.10^{16}$
NO	6·503	A	$2·7.10^{18}$	$8·5.10^{18}$	$2·2.10^{19}$

Dissociation potentials V_d (in eV) and dissociation constants K_n (in cm^{-3}) of molecular species at different temperatures (after Roes, 1962, p. 137).

A The values for N_2, O_2, CO, and NO listed here are generally accepted as the correct ones (see e.g. Wilkinson, 1963).

B The value for C_2 given here has been taken from Chupka and Inghram (1955). Wilkinson (1963) lists a value of $6·25 \pm 0·2$ eV after recent measurements by Brewer, Hicks, and Krikorian (1962).

C The value of 8·2 eV listed here originates from a Birge–Sponer extrapolation by Carroll (1956). Mass spectrometric measurements by Berkowitz (1962) yielded a value of $7·5 \pm 0·1$ eV in close agreement with one obtained from X-ray densitometry measurements of shock waves in cyanogen by Knight and Rink (1961) (cf. Wilkinson, 1963). Roes (1962) considered the values of 8·2, 6·9, and 6·2 eV listed by Douglas and Routly (1955) and he decided upon 8·2 eV in view of Carroll's value. Roes states that the results of his calculations agree better with experiment if the last mentioned value (instead of 6·9 or 6·2 eV) is used.

concentration at $z = 0·5$ cm to be $1·24.10^{11}$, $1·65.10^{17}$, and $2·16.10^{17}$ per cm^3 at current strengths of 3, 5, and 7 amp respectively. The value $\mathcal{N}_C = 1·6.10^{17}$ per cm^3 was finally used for computing the composition of the arc plasma at different temperatures. Besides he conducted calculations with a lower value of $\mathcal{N}_C$, namely $0·9.10^{17}$ per cm^3. Results are shown in self-explanatory way in Fig. 11.1.

★ § 11.2 *Band spectra emitted by the carbon arc in air*

In view of its molecular nature the plasma of the arc in air emits a large number of band spectra.* A useful aid for identifying band heads is the compilation by Pearse and Gaydon (1963). More detailed information on band heads of C_2, CH, CN, CO, NH, NO, O_2, OH, and their ions has been summarized by Wallace

* A concise, general discussion on band spectra has been given in § 6.10.

The diagrams pertain to a total pressure of 1 atm. Continuous curves are for a total carbon concentration $\mathcal{N}_C = 1·6.10^{17}$ per cm^3, broken ones for $\mathcal{N}_C = 0·9.10^{17}$ per cm^3. These total carbon concentrations correspond to about $1/11$ and $1/20$ respectively of the total number of nitrogen atoms (either free or bound ones) per cm^3 at 5000°K. The former should depict the actual situation in a 5-amp carbon arc in air at about 0·5 cm above the lower electrode, made the anode (see text).

Appendix 1 has tables for the conversion of concentrations into partial pressures.

(1962). Roes (1962) (see also Roes and Smit, 1957) investigated the band spectra emitted by the carbon arc in nitrogen and in air, particularly in the wavelength range 2300–3500 Å. The following bands were identified.*

N_2 bands: Intense N_2 bands that cover the wavelength range 3372–2814 Å and belong to the second positive system (see Pearse and Gaydon, p. 211). The band heads are:

λ(Å)	v'–v"	λ(Å)	v'–v"	λ(Å)	v'–v"	λ(Å)	v'–v"
3371	0–0	3159	1–0	2977	2–0	2820	3–0
		3136	2–1	2962	3–1	2814	4–1
		3117	3–2	2953	4–2		

C_2 bands: Weak bands that have two intensity maxima in the vicinity of 2310 Å which are headless C_2 bands with origins at 2312·6, 2314·0, and 2315·4 Å and belong to Mulliken's system (see Pearse and Gaydon, p. 97).

CN bands: Tail bands that cover the wavelength range 3465–3127 Å (see Pearce and Gaydon, p. 113). The sequences $\Delta v = 2$ and $\Delta v = 3$ reported by Feast (1949) were found. The band heads are:

λ(Å)	v'–v"	λ(Å)	v'–v"
3465	12–10	3232	11–8
3433	11–9	3204	10–7
3405	10–8	3180	9–6
3380	9–7	3160	8–5
3359	8–6	3143	7–4
3340	7–5	3128	6–3

NO bands: Bands that belong to the γ system of NO and contribute the main portion of the spectrum below about 3000 Å (see Pearse and Gaydon, p. 226). Six double-headed bands were found:

λ(Å)	v'–v"	λ(Å)	v'–v"	λ(Å)	v'–v"
3009	0–6	2722	0–4	2479	0–2
2998		2713		2472	
2860	0–5	2595	0–3	2370	0–1
2850		2588		2363	

O_2 bands: Bands that cover the wavelength range 3000–3400 Å and belong to the Schumann–Runge system of O_2 described by Feast (1949). The band structure of the spectrum is poor. Lines are very weak in the carbon arc in air and

* Wavelengths given below have been taken from Pearse and Gaydon (1963). They differ sometimes by 1 Å from those reported by Roes. We have listed only those heads that were detected by Roes in the spectrum of the carbon arc in air.

become more intense when the arc temperature is lowered by the addition of an alkali metal. The bands identified are:

λ(Å)	v'–v"	λ(Å)	v'–v"	λ(Å)	v'–v"
3370	0–14	3232	0–13	3104	0–12

OH band: The 3064 Å system of OH which is degraded towards longer wavelengths and extends from 3064 to about 3250 Å (see Pearse and Gaydon, p. 31).

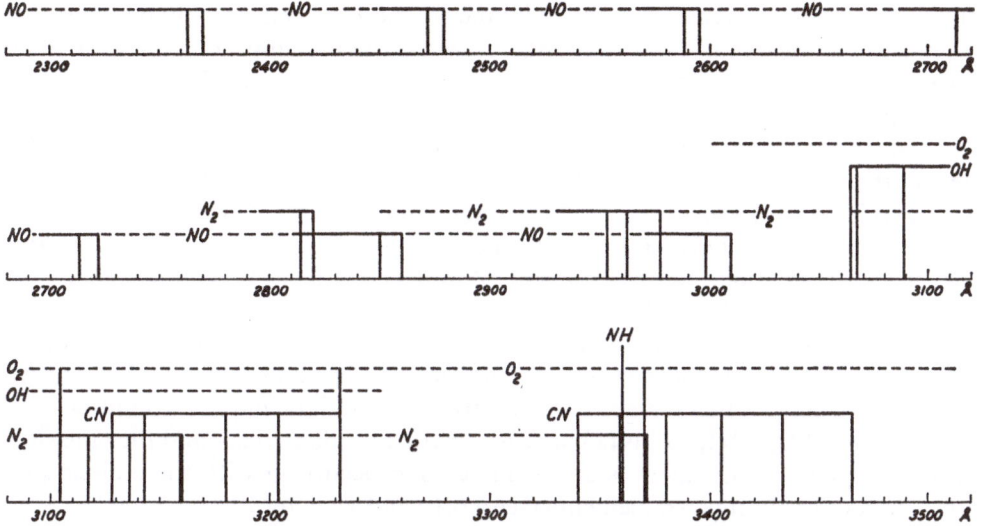

FIG. 11.2. Schematical representation of band spectra occurring in the wavelength region 2300–3500 Å in the carbon arc in air at atmospheric pressure (after data reported by Roes, 1962).

Horizontal lines indicate approximately the extent of the bands. The N_2, CN, and NO bands are degraded towards the violet, the OH is faded towards the red. Vibrational transitions and wavelengths of band heads are mentioned in the text.

The OH band is intense and extended if the arc temperature is lowered by the addition of an easily-ionized substance. The OH band occurs when water vapour is present in air, so it can be suppressed by using a controlled atmosphere of dry air. The band heads are:

λ(Å)	v'–v"
3064	0–0
3067	0–0
3089	0–0

NH band: A headless band which has an intensity maximum at 3360 Å and belongs to the 3360 Å system (see Pearse and Gaydon, p. 222).

The aggregate of band spectra occurring in the wavelength range 2300–3500 Å is shown schematically in Fig. 11.2.

The relative and absolute intensities of the various bands are influenced by the current strength and by the addition of substances that affect arc temperature. In the arc in air alone, the intensity of the CN tail bands increases (as well as that of the main violet CN bands) rapidly with the current (see Roes and Smit, 1957). It is caused by both the enhanced consumption of the carbon electrodes and the increased excitation at higher current. The same effect occurs with the C_2 bands. The intensity of the N_2 and O_2 bands depends only slightly on the current strength; the rise for the N_2 band can be virtually explained from the increase of the radius of the core. The emission per unit volume of the core remains almost constant when the current is varied in the range between 2 and 9 amp. The reason is that the enhanced excitation at higher temperature is compensated for by the increasing dissociation of the molecules.

When the arc temperature is lowered by the addition of an easily-ionized substance, the emission of CN, C_2, N_2, and NO bands is greatly reduced. The bands of O_2, OH, and NH remain at about the same intensity level, however, which must be attributed to the balancing of reduced excitation by an enhancement of the concentration of molecules as a result of repelled dissociation. The energy of dissociation of O_2, OH, and NH is relatively small, namely 5·1, 4·3, and 4·2 eV, so that in the arc in air alone the bands of these molecules are dominantly emitted from the arc fringe, in contrast to the bands of CN, N_2, and NO, which originate from the arc core. When the arc temperature is lowered, favourable conditions for the emission of O_2, OH, and NH bands are created in the arc core and the emission zone is displaced. The effect is analogous to that occurring with the emission of atom lines of easily-ionized elements (see § 6.12 and Chapter 8).

★ § 11.3   *Molecular species in the plasma originating from constituents of the sample*

At the high temperature prevailing in the interior of the arc discharge most bonds between components of samples are broken and specimens desintegrate almost completely into free atoms; however, some elements capable of forming diatomic compounds with oxygen are present partly in molecular form, even in the arc core. The proportion of molecules increases when we proceed radially from the arc core to the flame. In the latter region other molecular species than oxides become important, for example halides, when the sample contains halogens. We shall consider this question more extensively below. Let us first briefly state the significance of the formation of molecules in spectrochemical analysis and in spectrochemical physics.

1 The emission of molecular bands contributes to the background and deteriorates detection limits. Well-known bands causing interference are the AlO bands, 4374–5424 Å, the SiO bands, 2414–2925 Å, and the CaO bands, 5983–6362 and 5473–5560 Å (cf. Pearse and Gaydon, 1963; and Ahrens and Taylor, 1961, p. 172).

In addition to AlO, SiO, and CaO, Pearse and Gaydon list the following molecules of which bands have been observed in arcs:

AsO, BO, BaBr, BaCl, BaF, BaO, BeF, BeO, BiCl, BiO, CaCl, CaF, CeO, CrO, CuO, FeO, GeO, HfO, InO, LaO, LuO, MgCl, MgF, MgO, MnO, NiO, PO, PbO, PrO, SO, SbO, ScO, SnO, SrCl, SrF, SrO, TaO, TiO, VO, WO, YO, ZrO.

We must not conclude that band spectra of all these species listed are readily found under common operating conditions.

2 Bands of CaF, SrF, and CaCl can be used for detecting and determining the elements fluorine and chlorine in the arc (see Ahrens and Taylor, 1961, p. 277). Also bands of diatomic oxides, e.g. LaO bands, can sometimes be advantageously used in analysis, in preference to atom or ion lines (see Ahrens and Taylor, 1961, p. 226).

3 When an element forms molecules in the arc fringe, interfering self-absorption of atom lines in the outer regions is reduced.

4 For elements whose oxides are stable at the high temperature of the arc core, that is, for temperatures beyond 5000°K, matrix effects can occur as a result of the displacement of dissociation equilibria with increasing temperature. Particularly in view of these effects we shall now give some numerical examples of dissociation equilibria.

General principles of thermal dissociation have been summarized in § 11.1. We recall especially formulae (11.2) to (11.8) pertaining to a dissociation equilibrium

$$XY \leftrightharpoons X + Y$$

of a diatomic molecule. Bringing equation (11.3) into the form

$$\frac{n_{XY}}{n_X} = \frac{n_Y}{K_n} \qquad (11.9)$$

and identifying X with a metal and Y with oxygen, we have expressed the ratio of the concentrations of oxide molecules and free metal atoms into the equilibrium constant K_n and the concentration of free oxygen atoms. The numerical value of the ratio n_{XY}/n_X at a given temperature indicates whether the formation of molecules is important or not. Clearly, both the concentration of free oxygen (n_Y) and the dissociation constant (K_n) must be considered. For computing K_n at a given temperature, we must know the numerical value of the dissociation energy of the molecule and the numerical values of the partition functions of the free atoms and the molecule [cf. equations (11.4) to (11.8)].

Values of dissociation energy of diatomic molecules have been reviewed by Wilkinson (1963), Herzberg (1957, 1950), Inghram, Chupka, and Berkowitz (1957), and Gaydon (1953). Table 11.2 gives available data for oxides and halides of interest in the arc.

The procedure for computing the partition function of the molecule has been outlined in § 11.1 [see particularly equation (11.7)]. Numerical values of the rotational and vibrational constants, B and ω, respectively, as well as the designation of ground states for finding the statistical weight g can be taken from Herzberg (1950).*

TABLE 11.2

Molecule	V_d	Ref.	Molecule	V_d	Ref.	Molecule	V_d	Ref.
AgO	(1·8)	A B	MgO	3·9	A	AlF	6·7	A
AlO	≤5·0	A	MnO	(4·4)	B	BaCl	(2·7)	B
AsO	5·0	B	PO	(6·2)	A	BaF	5·8	A
BO	8·3	C	PbO	(4·3)	B	CaCl	≤2·8	A B
BaO	5·0	D	SO	5·4	A	CaF	5·4	A
BeO	4·6	A	SbO	(3·8)	B	GaF	6·2	A
BiO	(2·9)	A B	ScO	6·0	A	InF	5·5	A
CaO	3·9	D	SiO	(8·0)	A	KF	5·0	A
CrO	5·3	A	SnO	5·4	A	MgF	4·5	A
CuO	4·9	A	SrO	4·1	D	MnF	(3·9)	A B
FeO	4·3	A	TiO	6·8	A B	NiCl	(7·3)	B
GaO	(2·9)	B	VO	6·4	A	SiCl	(4·0)	A B
GeO	6·8	A	WO	(7·2)	A	SiF	5·4	A
InO	(1·3)	A B	YO	(9)	A B	SnF	(3·9)	B
LaO	8·2	A	ZrO	7·8	A B	SrF	5·4	A

Dissociation potentials V_d (in eV) of diatomic oxides and halides of interest in the arc. Values given in parentheses are uncertain.

A As listed by Wilkinson (1963).

B As listed by Herzberg (1950).

C Wilkinson gives ≤ 7·4 eV; de Galan (1965a and b) found a value of 8·3 eV to give good agreement with his measurements of absolute concentrations in the arc (cf. § 9.4).

D After Hollander (1964), Hollander, Kalff, and Alkemade (1964), and Kalff, Hollander, and Alkemade (1965).

Data on partition functions of free atoms at 5000 and 6000°K have been included in § 6.2.† Values for lower temperatures (say ≤4000°K) are easily calculated according to (6.4) from the data compiled by Moore (1949, 1952, 1958); only a few terms have to be considered at these low temperatures.‡

We now compute the dissociation constant K_n for the oxides BO, TiO, SnO, and AlO at $T = 6000, 5000$, and $4000°K$, using the values of the dissociation energy given in Table 11.2. Calculations are conducted with the aid of the logarithmic formula (11.8). For the sake of illustration, Table 11.3 shows the numerical values of separate terms of the equation for $T = 5000°K$. Linked to this, Table 11.4

* Uncertainty exists about the ground state designation of alkaline earth oxides. Hollander, Kalff, and Alkemade (1964) (see also Hollander, 1964) bring out arguments in favour of the ground state being a triplet state, most probably a $^3\Pi$ state.

† The partition function of O, which is not listed in § 6.2, practically equals the statistical weight of the ground term, i.e. 9·0.

‡ A practical form of (6.4), to be used in conjunction with a table of the exponential function e^{-z}, is

$$ Z = \sum_0^j g_m \exp(-1\cdot439\sigma_m/T) $$

where σ_m is the term value expressed as cm^{-1} (or kaysers).

TABLE 11.3

Term in equation (11.8)	BO	TiO	SnO	AlO
log constant	20·432	20·432	20·432	20·432
$\frac{2}{3}\log M_X M_Y / M_{XY}$	1·221	1·612	1·724	1·503
$\log Z_X Z_Y$	1·732	2·426	1·666	1·725
$\log B$	0·250	−0·272	−0·451	−0·193
$-\log g$	−0·301	−0·778	0	−0·301
$\frac{1}{2}\log T$	1·849	1·849	1·849	1·849
$\log(1 - 10^{-0.625\omega/T})$	−0·378	−0·599	−0·676	−0·611
$-(5040/T)\,V_d$	−8·366	−6·854	−5·443	−5·040
$\log K_n$	16·439	17·816	19·101	19·364

Illustration of the calculation of the dissociation constant K_n of diatomic molecules, BO, TiO, SnO, and AlO, at $T = 5000°K$.

The table gives the values of separate terms in the logarithmic equation (11.8) for computing K_n. Although individual terms may differ considerably among molecules, their joint effect is such that $-(5040/T)V_d$ is the only essential contribution [see Table 11.4 and equation (11.10)].

brings out numerical values of the three main factors of which K_n is composed, namely

1 The constant $2·70 . 10^{20}$

2 The function

$$Cf(T) = \frac{B}{g}\left(\frac{M_X M_Y}{M_{XY}}\right)^{\frac{2}{3}} Z_X Z_Y \, T^{\frac{1}{2}}(1 - 10^{-0.625\omega/T})$$

3 The exponential factor

$$10^{-(5040/T)V_d}$$

Evidently, however different the separate quantities making up $Cf(T)$ may be (see Table 11.3), the entire function $Cf(T)$ depends only slightly on the type of molecule and does not vary largely with temperature. This circumstance is highly convenient, as it generally allows estimates of K_n to be based on the dissociation energy only, that is from the equation

$$K_n \approx 5.10^{24} . 10^{-(5040/T)V_d} \tag{11.10}$$

where the coefficient must be assigned an uncertainty factor, say, of $1·5$. Mark that the energy of dissociation is likely to have a much larger uncertainty; for instance, an error of $0·2$ eV in V_d corresponds to a factor of $1·6$ in K_n at $T = 5000°K$.

Table 11.4 contains values of n_Y, that is of the concentration of free oxygen atoms, at the three temperatures under consideration. These values, based on Roes' calculations (see § 11.1), eventually enable us to find the ratio n_{XY}/n_X, i.e. the ratio of concentrations of oxide molecules and free metal atoms. The data show in a self-explanatory way that dissociation equilibria of oxides tend to become important in the arc core only when the energy of dissociation is beyond 6 eV. Matrix effects

Table 11.4

	BO ($V_d = 8{\cdot}3$ eV)			TiO ($V_d = 6{\cdot}8$ eV)		
	6000°K	5000°K	4000°K	6000°K	5000°K	4000°K
Constant	$2{\cdot}70.10^{20}$	$2{\cdot}70.10^{20}$	$2{\cdot}70.10^{20}$	$2{\cdot}70.10^{20}$	$2{\cdot}70.10^{20}$	$2{\cdot}70.10^{20}$
$Cf(T)$	$2{\cdot}25.10^{4}$	$2{\cdot}36.10^{4}$	$2{\cdot}49.10^{4}$	$2{\cdot}00.10^{4}$	$1{\cdot}73.10^{4}$	$1{\cdot}54.10^{4}$
$10^{-(5040/T)V_d}$	$1{\cdot}07.10^{-7}$	$4{\cdot}30.10^{-9}$	$3{\cdot}48.10^{-11}$	$1{\cdot}94.10^{-6}$	$1{\cdot}40.10^{-7}$	$2{\cdot}71.10^{-9}$
K_n (cm^{-3})	$6{\cdot}50.10^{17}$	$2{\cdot}74.10^{16}$	$2{\cdot}34.10^{14}$	$1{\cdot}05.10^{19}$	$6{\cdot}54.10^{17}$	$1{\cdot}13.10^{16}$
n_Y (cm^{-3})	2.10^{17}	$3{\cdot}5.10^{17}$	5.10^{17}	2.10^{17}	$3{\cdot}5.10^{17}$	5.10^{17}
n_{XY}/n_X	$0{\cdot}3$	$12{\cdot}8$	2140	$0{\cdot}02$	$0{\cdot}53$	$44{\cdot}2$

	SnO ($V_d = 5{\cdot}4$ eV)			AlO ($V_d = 5{\cdot}0$ eV)		
	6000°K	5000°K	4000°K	6000°K	5000°K	4000°K
Constant	$2{\cdot}70.10^{20}$	$2{\cdot}70.10^{20}$	$2{\cdot}70.10^{20}$	$2{\cdot}70.10^{20}$	$2{\cdot}70.10^{20}$	$2{\cdot}70.10^{20}$
$Cf(T)$	$1{\cdot}36.10^{4}$	$1{\cdot}29.10^{4}$	$1{\cdot}18.10^{4}$	$0{\cdot}88.10^{4}$	$0{\cdot}94.10^{4}$	$1{\cdot}02.10^{4}$
$10^{-(5040/T)V_d}$	$2{\cdot}91.10^{-5}$	$3{\cdot}61.10^{-6}$	$1{\cdot}57.10^{-7}$	$6{\cdot}31.10^{-5}$	$9{\cdot}12.10^{-6}$	$5{\cdot}01.10^{-7}$
K_n (cm^{-3})	$1{\cdot}07.10^{20}$	$1{\cdot}26.10^{19}$	$5{\cdot}00.10^{17}$	$1{\cdot}50.10^{20}$	$2{\cdot}31.10^{19}$	$1{\cdot}38.10^{18}$
n_Y (cm^{-3})	2.10^{17}	$3{\cdot}5.10^{17}$	5.10^{17}	2.10^{17}	$3{\cdot}5.10^{17}$	5.10^{17}
n_{XY}/n_X	$0{\cdot}002$	$0{\cdot}03$	$1{\cdot}0$	$0{\cdot}001$	$0{\cdot}015$	$0{\cdot}36$

Calculation of the extent of oxide formation in the arc for the elements boron, titanium, tin, and aluminium.

The table shows three separate factors making up the constant of dissociation, viz.

1 A constant, $2{\cdot}70.10^{20}$.
2 A function $Cf(T) = (B/g)(M_X M_Y/M_{XY})^{\frac{3}{2}} Z_X Z_Y T^{\frac{5}{2}}(1 - 10^{-0{\cdot}625\omega/T})$, which altogether varies only slightly.
3 The exponential factor $10^{(5040/T)V_d}$, which essentially controls the value of the dissociation constant.

In addition there have been included in the table the concentration (n_Y) of free oxygen atoms after Roes' calculations (§ 11.1) and the ratio (n_{XY}/n_X) of concentrations of oxide molecules and free metal atoms computed as n_Y/K_n [equation (11.9)].

will show for relevant elements by an indirect influence, namely via the arc temperature: when the temperature changes with the matrix, dissociation equilibria of slightly and moderately dissociated oxides will shift and hence the concentrations of free atoms and ions are altered, given the total concentration of the element in the arc.

As an example we consider titanium. Using equation (7.28) and remembering equations (7.7) and (7.8) we compute the ratio n_i/n_a of concentrations of ions and atoms at 6000 and 5000°K. If we assume the electron pressure to be constant and equal to 10^{-3} atm, we calculate with $\bar{V}_{ij} = 6{\cdot}55$ eV (after Table 7.1):

$$n_i/n_a = 5{\cdot}77 \text{ at } T = 6000°K \qquad \text{and} \qquad n_i/n_a = 0{\cdot}29 \text{ at } T = 5000°K$$

When the total concentration is given and no molecules are formed, we have

$$n_i + n_a = n = \text{constant}$$

and thus

$$\left. \begin{array}{l} n_a = 0.15n \\ n_i = 0.85n \end{array} \right\} T = 6000°K \qquad \left. \begin{array}{l} n_a = 0.77n \\ n_i = 0.23n \end{array} \right\} T = 5000°K$$

so that the change of the concentration of atoms and ions with temperature is found to be

$$\frac{n_a(6000°K)}{n_a(5000°K)} = 0.195 \qquad \text{and} \qquad \frac{n_i(6000°K)}{n_i(5000°K)} = 3.7$$

These results are markedly affected if the formation of TiO has to be taken into account. Now we have

$$n_i + n_a + n_m = n = \text{constant}$$

where n_m is the concentration of TiO molecules. In view of the data given in Table 11.4, viz.

$$n_m/n_a = 0.02 \text{ at } T = 6000°K \qquad \text{and} \qquad n_m/n_a = 0.53 \text{ at } T = 5000°K$$

we compute

$$\left. \begin{array}{l} n_a = 0.147n \\ n_i = 0.850n \\ n_m = 0.003n \end{array} \right\} T = 6000°K \qquad \text{and} \qquad \left. \begin{array}{l} n_a = 0.55n \\ n_i = 0.16n \\ n_m = 0.29n \end{array} \right\} T = 5000°K$$

The change of the concentration of atoms and ions with temperature turns out to be

$$\frac{n_a(6000°K)}{n_a(5000°K)} = 0.27 \qquad \text{and} \qquad \frac{n_i(6000°K)}{n_i(5000°K)} = 5.3$$

which values must be compared with 0.195 and 3.7 respectively, calculated above. The discrepancy, which in practice will show as a matrix effect, is due to the displacement of the dissociation equilibrium TiO $\leftrightharpoons$ Ti + O with increasing temperature.

An extreme example is, of course, furnished by boron. The element is only slightly ionized, even at 6000°K. For the present objective we may ignore ionization altogether and compute the change of the concentration of atoms with temperature while assuming $n_a + n_m = n = \text{constant}$. From Table 11.4 we find

$$n_a = 0.77n \text{ at } T = 6000°K \qquad \text{and} \qquad n_a = 0.072n \text{ at } T = 5000°K$$

Accordingly, owing to the formation of BO, the atom concentration decreases by more than a factor of 10 when the temperature is reduced from 6000 to 5000°K.

A last question to be considered is whether matrix effects are likely to arise from the formation of molecules between elements of the sample, for instance by the formation of halides, especially of fluorides. Table 11.2 shows that some fluorides have a fairly high dissociation energy and should be stable in the arc core when the temperature is not too elevated. So, for AlF we expect the dissociation constant to be of about the same magnitude as that given in Table 11.2 for TiO. The ratio n_{XY}/n_X for AlF, however, will be smaller at least by a factor of 100 than for TiO, because it is not feasible for the concentration of fluorine in the arc to become much larger than 10^{15} per cm^3 (cf. § 7.9). This representative example therefore demonstrates that in general we must not expect matrix effects to arise from the formation of molecular compounds between sample components in the arc core. In the flamy fringe, on the other hand, the situation is very different and the concentration of free atoms may depend largely upon dissociation equilibria. However, it rarely happens that atom line emission is virtually confined to the fringe, only when low-level atom lines of easily-ionized elements are concerned (see Chapter 8), and so we must expect in those cases merely a pronounced influence of dissociation equilibria in the fringe on atom line intensities.

MATRIX EFFECTS IN RETROSPECT

> We have not succeeded in answering all our problems—indeed we sometimes
> feel we have not completely answered any of them. The answers we have found
> only served to raise a whole set of new questions. In some ways we feel that we
> are as confused as ever, but we think we are confused on a higher level and about
> more important things.
>
> (From *The Workshop Ways of Learning* by Earl C. Kelley)

At the end of this book I want to return to the fascinating topic of matrix effects (cf. § 1.1). My purpose for doing so is not to mention aspects that have not already been considered in the text, but merely to produce a synopsis that may point out to the reader how the thread of the exposition on this subject runs through the work.

We discussed the matter in the light of the working curve

$$I = K \times G^m \tag{1}$$

or

$$\log I = m \log G + \log K \tag{2}$$

which expresses the proportionality between the spectral-line intensity I and G^m, i.e. the concentration of the analysis element raised to a power m to allow for the self-absorption, or, more generally, for the line profile (see §§ 1.1 and 5.7).* I note that the matrix can influence spectral-line intensities through volatilization effects that cause a change in self-absorption (see § 10.2).

If we confine further discussion to total energy procedures and photographic recording, intensity I is identified with exposure *E*. Then, in consequence of reciprocity law failures of the photographic emulsion (Schwarzschildt effect), matrix effects originating from the influence of volatilization rate upon the recorded densities on the plates must sometimes be considered (see the end of § 1.2 and § 10.2).

---

* I recall my remark in § 1.2 that the intensity of a spectral line should be measured by integrating over the line profile. As we usually take, in spectrochemistry, the peak value of the line to be representative of the intensity, we must expect that I is not always proportional to G, even if self-absorption is absent.

The value of m in an actual working curve might be influenced by instrumental factors, for example, by the calibration of an emulsion. The use of an incorrect calibration curve causes, for instance, the background correction to be inaccurate, which has its repercussion on m.

The majority of matrix effects, however, are classed among influences on the excitation conditions (in the widest sense), ignition losses, and obscure volatilization effects that are not yet clearly understood (see § 10.2).

Unravelling the network of interrelations between the nature of the matrix, excitation conditions, and spectral-line intensities can be conveniently accomplished in the following stages. Firstly, we consider what influences the matrix exerts on the excitation and transport parameters, and then we can relate spectral-line intensities to these quantities.

The parameters that chiefly control both excitation and transportation are the *temperature* and the *electron pressure*. Their values in an arc depend to a large extent on the composition of the plasma, particularly on the ionization potential and the concentration of the element that carries the discharge (see §§ 3.8, 3.9, 7.7 to 7.10, 7.12, and 7.14).

Temperature and electron pressure govern the ionization equilibria of individual vapour components and predetermine the degrees of ionization (α_j) of all species present (see §§ 7.1 to 7.5). Degrees of ionization indicate the proportions of neutral atoms, $(1 - \alpha_j)$, and singly-charged ions, α_j, which, in turn, determine the fractions of the elements available either for atom line emission or for ion line emission [equations (7.9) and (7.10)]. The degree of ionization is also the governing feature of the transport process, as was established experimentally (§ 9.4) and expounded theoretically (§§ 9.1, 9.2, and 9.5). The greater the extent to which an element is ionized, the larger the speed at which it migrates through the discharge zone and the lower its concentration at a given rate of evaporation.

The influence of the rate of volatilization upon time-integrated intensities as recorded in total energy procedures should be an indirect one only, owing to its effect upon the plasma composition and hence its influence on temperature and electron pressure. If the values of these parameters are stabilized by a buffer admixture (§ 7.12), the rate of volatilization is not likely to have a particular significance when samples are burnt to completion (see § 10.2, however). For then, it is only the total number of atoms of species j present in the sample and the transport velocity in the discharge zone which determine the time-integrated concentration of the element in the plasma [see equation (9.8)], the rate of entry being immaterial, apart from the secondary influences on line intensities exerted by self-absorption and the Schwarzschildt effect.

For a number of elements we must consider as a possible matrix effect the formation of stable molecules with oxygen. The influence of the matrix again is a secondary one, namely that occurring via the temperature of the arc. A change of the arc temperature causes shifts of dissociation equilibria and acts upon the concentrations of free atoms and ions in the plasma (see § 11.3). The effect should be noticeable only for the few elements whose oxides have a dissociation potential exceeding 6 eV (cf. Table 11.2). We therefore ignore it in the following.

In compliance with de Galan's measurements discussed in § 9.4, we write for the average particle concentration ($\bar{n}_j$) of an element j in the middle of the gap of not too short an arc, with the lower electrode made the anode,

$$\bar{n}_j = cQ_j 10^{-\hat{\alpha}_j} \tag{3}$$

where Q_j is the rate of volatilization, $\hat{\alpha}_j$ the effective degree of ionization, and c a constant; $c \approx 0\cdot003$ sec/cm^3, if n is expressed as cm^{-3} and Q as sec^{-1} [cf. equations (9.15), (9.19), and (9.20)].

The concentrations of atoms and ions are given separately by

$$n_{aj} = cQ_j(1 - \alpha_j)10^{-\hat{\alpha}_j} \tag{4}$$

and

$$n_{ij} = cQ_j\alpha_j 10^{-\hat{\alpha}_j} \tag{5}$$

These formulae express implicitly how temperature and electron pressure effect the concentrations of atoms and ions in the arc, viz. via the degree of ionization α_j (see §7.1 to 7.5). To find the dependence of spectral-line intensities on these parameters we must add one more factor, i.e. the Boltzmann exponential factor divided by the partition function (see §§ 6.1 to 6.3).* For atom lines we thus have

$$J = C \cdot cQ_j(1 - \alpha_j)10^{-\hat{\alpha}_j} \frac{10^{-(5040/T)V_q}}{Z_{aj}} \tag{6}$$

and for ion lines

$$J^+ = C^+ \cdot cQ_j\alpha_j 10^{-\hat{\alpha}_j} \frac{10^{-(5040/T)V_q^+}}{Z_{ij}} \tag{7}$$

where J is the emission per unit volume per unit solid angle, C is a line constant [see equations (8.4) and (8.8)] containing the statistical weight (g_q) of the upper level of the transition, the probability (A) of the transition, and its frequency (ν); V_q and V_q^+ are the excitation potentials of the lines (in eV), and Z_{aj} and Z_{ij} are the partition functions of the atom and ion respectively.

Actually we do not observe the quantities J and J^+ immediately, but the volume integrals, viz.

$$I = \iiint J(x, y, z)\,dx\,dy\,dz \tag{8}$$

as was shown in §§ 1.2 and 6.13.

In the arc we have to take into account particularly the radial distribution of the emission. If the zone of observation is a horizontal slice through the arc located at some millimetres above the lower electrode, we are justified in assuming that the radial distribution of particle concentration n_j is uniform within the region observed. This fact was established experimentally and underlies equation (3),

---

* It is very useful in this connection to consider by what mechanism the Boltzmann distribution of energy levels is established (see §§ 5.1 to 5.6).

where $\bar{n}_j$ represents the average concentration in the arc cross-section (cf. § 9.4).

In contrast with particle concentration n_j, both temperature and electron pressure exhibit a pronounced radial decline, and consequently the functions J and J^+ follow radial distributions similar to those of the temperature and the electron pressure (see §§ 6.12, 6.13, 8.1, 8.2, and 8.3). For the majority of spectral lines the complications arising from the radial structure of the arc can be by-passed by applying effective values of temperature and electron pressure, provided that we observe the effective arc radius $\hat{R}$ (see §§ 8.3 to 8.7), whose influence is small, however, in comparison with other effects (see § 8.9). Only for atom lines of easily-ionized elements do difficulties concerning the use of effective quantities occur (see §§ 8.6 and 8.7).

In terms of effective values $\hat{T}$, $\hat{a}$, $\hat{Z}$, and $\hat{R}$, we have for the intensity of an atom line

$$\hat{I}_s = 2C . cQ_j(1 - \hat{a}_j)10^{-\hat{a}_j} \frac{10^{-(5040/\hat{T})V_q}}{\hat{Z}_{aj}} \hat{R}_s \tag{9}$$

if the arc is focused on the slit, and

$$\hat{I}_c = 2\pi C . cQ_j(1 - \hat{a}_j)10^{-\hat{a}_j} \frac{10^{-(5040/\hat{T})V_q}}{\hat{Z}_{aj}} \hat{R}_c^2 \tag{10}$$

if the arc is imaged on the collimator lens (cf. § 8.4). Similar expressions hold for ion lines.

We recall that the partition function can be included, to a first approximation, in the excitation potential (§§ 6.3 and 8.4), so that, for instance, (9) simplifies to

$$\hat{I}_s = 2 C' . cQ_j(1 - \hat{a}_j)10^{-\hat{a}_j} . 10^{-(5040/\hat{T})\bar{V}_q} \hat{R}_s \tag{11}$$

The dependence of this function on temperature and electron pressure has been considered at length in §§ 8.8 and 8.9 (see particularly Fig. 8.16).

Let us finally consider also the working curve (2) for when photographic detection is applied. The signal to which the detector responds is the irradiance H at the wavelength of the spectral line, which is related to the intensity $\hat{I}_s$ by an instrumental factor f_i, viz.

$$H = f_i \hat{I}_s \tag{12}$$

If the sample is burnt to completion, we integrate H over the arcing period t_c and adopt the exposure [defined by equation (1.20)], viz.

$$E = \int_{t_c} Hpt^{p-1} dt \tag{13}$$

as the intensity I that must be inserted in the equation for the working curve. Supposing Q_j to be the average rate of volatilization and assuming the excitation conditions to be constant during the burning period we have from (11), (12), and (13):

$$I = 2f_iC'.cQ_jt_c{}^{p}(1 - \hat{\alpha}_j)10^{-\hat{\alpha}_j}.10^{-(5040/\hat{T})\bar{V}_a}\,\hat{R}_s \qquad (14)$$

Q_jt_c is the total number of particles j present in the sample; it is proportional to the concentration G_j of the relevant element, viz.

$$Q_jt_c :: G_j \qquad (15)$$

Now combining all the proportionality constants we can put equation (14) into the form

$$\log I = m\log G_j + \log(1 - \hat{\alpha}_j) - \hat{\alpha}_j - \frac{5040}{\hat{T}}\,\bar{V}_a + \log \hat{R}_s - (1 - p)\log t_c + \log \Gamma \quad (16)$$

where we have included the empirical factor m (see above).

Likewise we have for an ion line

$$\log I^{+} = m\log G_j + \log \hat{\alpha}_j - \hat{\alpha}_j - \frac{5040}{\hat{T}}\,\bar{V}_q^{+} + \log \hat{R}_s - (1 - p)\log t_c + \log \Gamma^{+} \quad (17)$$

When the arc is imaged on the collimator lens, $\hat{R}_s$ must be replaced by $\hat{R}_c{}^2$ and the constant Γ must be accordingly adapted (cf. § 8.4).

In conclusion, when samples are burnt to completion in the d.c. arc, the majority of matrix effects can be discussed on the strength of equations (16) and (17). From them the equations for working curves with an internal standard are readily derived (cf. § 8.9). Additional matrix effects may be caused by volatilization 'defects', which have not yet been satisfactorily clarified (see §§ 10.1 and 10.2).

APPENDIX 1

Table for the conversion of concentration n (expressed as number per cm^3) into partial pressure p (expressed as atm) at different temperatures T.

Formula: $p = nT/7 \cdot 340 \cdot 10^{21}$
Entry: $T/7 \cdot 340 \cdot 10^3$
Conversion: for obtaining p multiply n by entry $\times 10^{-18}$

T	000	100	200	300	400	500	600	700	800	900
4	0·545	0·559	0·572	0·586	0·600	0·613	0·627	0·640	0·654	0·668
5	0·681	0·695	0·708	0·722	0·736	0·749	0·763	0·777	0·790	0·804
6	0·817	0·831	0·845	0·858	0·872	0·886	0·899	0·913	0·926	0·940

Table for the conversion of partial pressure p (expressed as atm) into concentration n (expressed as number per cm^3) at different temperatures T.

Formula: $n = (7 \cdot 340 \cdot 10^{21}/T)p$
Entry: $7 \cdot 340 \cdot 10^3/T$
Conversion: for obtaining n multiply p by entry $\times 10^{18}$

T	000	100	200	300	400	500	600	700	800	900
4	1·835	1·790	1·748	1·707	1·668	1·631	1·596	1·562	1·529	1·498
5	1·468	1·439	1·411	1·385	1·359	1·334	1·312	1·288	1·265	1·244
6	1·223	1·203	1·184	1·165	1·147	1·129	1·112	1·096	1·080	1·064

APPENDIX 2

Concise table for rapid conversion of reciprocal centimetres (cm^{-1}) or kaysers (K) into electron volts (eV).

Numbers within the frame represent energy values (σ) in 10^3 cm^{-1} or kilo-kaysers (kK). Marginal numbers refer to values (V) in electron volts: the upper row gives the whole number of eV, the column on the left the decimal fraction beyond the whole number. The table has been arranged so that no interpolation is required. Accuracy is within 0·03 eV.

Examples: all values of σ in the range 20·0 kK $< \sigma \leqslant$ 20·4 kK are converted to be V = 2·50 eV; those in the range 64·3 kK $< \sigma \leqslant$ 64·7 kK are read as V = 8·00 eV.

	0	1	2	3	4	5	6	7	8	9
00	0·2	8·3	16·3	24·4	32·5	40·5	48·6	56·7	64·7	72·8
05	0·6	8·7	16·7	24·8	32·9	40·9	49·0	57·1	65·1	73·2
10	1·0	9·1	17·1	25·2	33·3	41·3	49·4	57·5	65·6	73·6
15	1·4	9·5	17·5	25·6	33·7	41·7	49·8	57·9	66·0	74·0
20	1·8	9·9	18·0	26·0	34·1	42·2	50·2	58·3	66·4	74·4
25	2·2	10·3	18·4	26·4	34·5	42·6	50·6	58·7	66·8	74·8
30	2·6	10·7	18·8	26·8	34·9	43·0	51·0	59·1	67·2	75·2
35	3·0	11·1	19·2	27·2	35·3	43·4	51·4	59·5	67·6	75·6
40	3·4	11·5	19·6	27·6	35·7	43·8	51·8	59·9	68·0	76·0
45	3·8	11·9	20·0	28·0	36·1	44·2	52·2	60·3	68·4	76·4
50	4·2	12·3	20·4	28·4	36·5	44·6	52·6	60·7	68·8	76·8
55	4·6	12·7	20·8	28·8	36·9	45·0	53·0	61·1	69·2	77·2
60	5·0	13·1	21·2	29·2	37·3	45·4	53·4	61·5	69·6	77·6
65	5·4	13·5	21·6	29·6	37·7	45·8	53·9	61·9	70·0	78·1
70	5·8	13·9	22·0	30·1	38·1	46·2	54·3	62·3	70·4	78·5
75	6·3	14·3	22·4	30·5	38·5	46·6	54·7	62·7	70·8	78·9
80	6·7	14·7	22·8	30·9	38·9	47·0	55·1	63·1	71·2	79·3
85	7·1	15·1	23·2	31·3	39·3	47·4	55·5	63·5	71·6	79·7
90	7·5	15·5	23·6	31·7	39·7	47·8	55·9	63·9	72·0	80·1
95	7·9	15·9	24·0	32·1	40·1	48·2	56·3	64·3	72·4	80·5

APPENDIX 3

Table of 5040/T for temperatures in the range 4000–8000°K.

	00	10	20	30	40	50	60	70	80	90
40	1·2600	1·2569	1·2537	1·2506	1·2475	1·2444	1·2414	1·2383	1·2353	1·2323
41	1·2293	1·2263	1·2233	1·2203	1·2174	1·2145	1·2115	1·2086	1·2057	1·2029
42	1·2000	1·1971	1·1943	1·1915	1·1887	1·1859	1·1831	1·1803	1·1776	1·1748
43	1·1721	1·1694	1·1667	1·1640	1·1613	1·1586	1·1560	1·1533	1·1507	1·1481
44	1·1455	1·1429	1·1403	1·1377	1·1351	1·1326	1·1300	1·1275	1·1250	1·1225
45	1·1200	1·1175	1·1150	1·1126	1·1101	1·1077	1·1053	1·1028	1·1004	1·0980
46	1·0957	1·0933	1·0909	1·0886	1·0862	1·0839	1·0815	1·0792	1·0769	1·0746
47	1·0723	1·0701	1·0678	1·0655	1·0633	1·0611	1·0588	1·0566	1·0544	1·0522
48	1·0500	1·0478	1·0456	1·0435	1·0413	1·0392	1·0370	1·0349	1·0328	1·0307
49	1·0286	1·0265	1·0244	1·0223	1·0202	1·0182	1·0161	1·0141	1·0120	1·0100
50	1·0080	1·0060	1·0040	1·0020	1·0000	0·9980	0·9960	0·9941	0·9921	0·9902
51	0·9882	0·9863	0·9844	0·9825	0·9805	0·9786	0·9767	0·9749	0·9730	0·9711
52	0·9692	0·9674	0·9655	0·9637	0·9618	0·9600	0·9582	0·9564	0·9545	0·9527
53	0·9509	0·9492	0·9474	0·9456	0·9438	0·9421	0·9403	0·9385	0·9368	0·9351
54	0·9333	0·9316	0·9299	0·9282	0·9265	0·9248	0·9231	0·9214	0·9197	0·9180
55	0·9164	0·9147	0·9130	0·9114	0·9097	0·9081	0·9065	0·9048	0·9032	0·9016
56	0·9000	0·8984	0·8968	0·8952	0·8936	0·8920	0·8905	0·8889	0·8873	0·8858
57	0·8842	0·8827	0·8811	0·8796	0·8780	0·8765	0·8750	0·8735	0·8720	0·8705
58	0·8690	0·8675	0·8660	0·8645	0·8630	0·8615	0·8601	0·8586	0·8571	0·8557
59	0·8542	0·8528	0·8514	0·8499	0·8485	0·8471	0·8456	0·8442	0·8428	0·8414
60	0·8400	0·8386	0·8372	0·8358	0·8344	0·8331	0·8317	0·8303	0·8289	0·8276
61	0·8262	0·8249	0·8235	0·8222	0·8208	0·8195	0·8182	0·8169	0·8155	0·8142
62	0·8129	0·8116	0·8103	0·8090	0·8077	0·8064	0·8051	0·8038	0·8025	0·8013
63	0·8000	0·7987	0·7975	0·7962	0·7950	0·7937	0·7925	0·7912	0·7900	0·7887
64	0·7875	0·7863	0·7850	0·7838	0·7826	0·7814	0·7802	0·7790	0·7778	0·7766
65	0·7754	0·7742	0·7730	0·7718	0·7706	0·7695	0·7683	0·7671	0·7660	0·7648
66	0·7636	0·7625	0·7613	0·7602	0·7590	0·7579	0·7568	0·7556	0·7545	0·7534
67	0·7522	0·7511	0·7500	0·7489	0·7478	0·7467	0·7456	0·7445	0·7434	0·7423
68	0·7412	0·7401	0·7390	0·7379	0·7368	0·7358	0·7347	0·7336	0·7326	0·7315
69	0·7304	0·7294	0·7283	0·7273	0·7262	0·7252	0·7241	0·7231	0·7221	0·7210
70	0·7200	0·7190	0·7179	0·7169	0·7159	0·7149	0·7139	0·7129	0·7119	0·7109
71	0·7099	0·7089	0·7079	0·7069	0·7059	0·7049	0·7039	0·7029	0·7019	0·7010
72	0·7000	0·6990	0·6981	0·6971	0·6961	0·6952	0·6942	0·6933	0·6923	0·6914
73	0·6904	0·6895	0·6885	0·6876	0·6866	0·6857	0·6848	0·6839	0·6829	0·6820
74	0·6811	0·6802	0·6792	0·6783	0·6774	0·6765	0·6756	0·6747	0·6738	0·6729
75	0·6720	0·6711	0·6702	0·6693	0·6684	0·6675	0·6667	0·6658	0·6649	0·6640
76	0·6632	0·6623	0·6614	0·6606	0·6597	0·6588	0·6580	0·6571	0·6563	·06554
77	0·6545	0·6537	0·6528	0·6520	0·6512	0·6503	0·6495	0·6486	0·6478	0·6470
78	0·6462	0·6453	0·6445	0·6437	0·6429	0·6420	0·6412	0·6404	0·6396	0·6388
79	0·6380	0·6372	0·6364	0·6356	0·6348	0·6340	0·6332	0·6324	0·6316	0·6308

APPENDIX 4

Table of 5/2 log T for temperatures in the range 4000–8000°K.

	00	10	20	30	40	50	60	70	80	90
40	9·005	9·008	9·011	9·013	9·016	9·019	9·021	9·024	9·027	9·029
41	9·032	0·035	9·037	9·040	9·043	9·045	9·048	9·050	9·053	9·056
42	9·058	9·061	9·063	9·066	9·069	9·071	9·074	9·076	9·079	9·081
43	9·084	9·086	9·089	9·091	9·094	9·096	9·099	9·101	9·104	9·106
44	9·109	9·111	9·114	9·116	9·119	9·121	9·123	9·126	9·128	9·131
45	9·133	9·136	9·138	9·140	9·143	9·145	9·148	9·150	9·152	9·155
46	9·157	9·159	9·162	9·164	9·166	9·169	9·171	9·173	9·176	9·178
47	9·180	9·183	9·185	9·187	9·190	9·192	9·194	9·196	9·199	9·201
48	9·203	9·205	9·208	9·210	9·212	9·214	9·217	9·219	9·221	9·223
49	9·226	9·228	9·230	9·232	9·234	9·237	9·239	9·241	9·243	9·245
50	9·248	9·250	9·252	9·254	9·256	9·258	9·261	9·263	9·265	9·267
51	9·269	9·271	9·273	9·275	9·278	9·280	9·282	9·284	9·286	9·288
52	9·290	9·292	9·294	9·296	9·298	9·301	9·303	9·305	9·307	9·309
53	9·311	9·313	9·315	9·317	9·319	9·321	9·323	9·325	9·327	9·329
54	9·331	9·333	9·335	9·337	9·339	9·341	9·343	9·345	9·347	9·349
55	9·351	9·353	9·355	9·357	9·359	9·361	9·363	9·365	9·367	9·369
56	9·371	9·373		9·376	9·378	9·380	9·382	9·384	9·386	9·388
57	9·390	9·392	9·394	9·396	9·397	9·399	9·401	9·403	9·405	9·407
58	9·409	9·411	9·412	9·414	9·416	9·418	9·420	9·422	9·424	9·425
59	9·427	9·429	9·431	9·433	9·435	9·436	9·438	9·440	9·442	9·444
60	9·446	9·447	9·449	9·451	9·453	9·455	9·456	9·458	9·460	9·462
61	9·463	9·465	9·467	9·469	9·471	9·472	9·474	9·476	9·478	9·479
62	9·481	9·483	9·485	9·486	9·488	9·490	9·492	9·493	9·495	9·497
63	9·498	9·500	9·502	9·504	9·505	9·507	9·509	9·510	9·512	9·514
64	9·516	9·517	9·519	9·521	9·522	9·524	9·526	9·527	9·529	9·531
65	9·532	9·534	9·536	9·537	9·539	9·541	9·542	9·544	9·546	9·547
66	9·549	9·551	9·552	9·554	9·556	9·557	9·559	9·560	9·562	9·564
67	9·565	9·567	9·569	9·570	9·572	9·573	9·575	9·577	9·578	9·580
68	9·581	9·583	9·585	9·586	9·588	9·589	9·591	9·593	9·594	9·596
69	9·597	9·599	9·600	9·602	9·604	9·605	9·607	9·608	9·610	9·611
70	9·613	9·614	9·616	9·618	9·619	9·621	9·622	9·624	9·625	9·627
71	9·628	9·630	9·631	9·633	9·634	9·636	9·637	9·639	9·640	9·642
72	9·643	9·645	9·646	9·648	9·649	9·651	9·652	9·654	9·655	9·657
73	9·658	9·660	9·661	9·663	9·664	9·666	9·667	9·669	9·670	9·672
74	9·673	9·675	9·676	9·678	9·679	9·681	9·682	9·683	9·685	9·686
75	9·688	9·689	9·691	9·692	9·694	9·695	9·696	9·698	9·699	9·701
76	9·702	9·704	9·705	9·706	9·708	9·709	9·711	9·712	9·714	9·715
77	9·716	9·718	9·719	9·721	9·722	9·723	9·725	9·726	9·728	9·729
78	9·730	9·732	9·733	9·735	9·736	9·737	9·739	9·740	9·741	9·743
79	9·744	9·746	9·747	9·748	9·750	9·751	9·752	9·754	9·755	9·756

APPENDIX 5

Table of 3/2 log T for temperatures in the range 4000–8000°K.

	00	10	20	30	40	50	60	70	80	90
40	5·403	5·405	5·406	5·408	5·410	5·411	5·413	5·414	5·416	5·417
41	5·419	5·421	5·422	5·424	5·426	5·427	5·429	5·430	5·432	5·433
42	5·435	5·436	5·438	5·439	5·441	5·443	5·444	5·446	5·447	5·449
43	5·450	5·452	5·453	5·455	5·456	5·458	5·459	5·461	5·462	5·464
44	5·465	5·467	5·468	5·470	5·471	5·473	5·474	5·475	5·477	5·478
45	5·480	5·481	5·483	5·484	5·486	5·487	5·489	5·490	5·491	5·493
46	5·494	5·495	5·497	5·498	5·500	5·501	5·503	5·504	5·505	5·507
47	5·508	5·510	5·511	5·512	5·514	5·515	5·516	5·518	5·519	5·520
48	5·522	5·523	5·525	5·526	5·527	5·528	5·530	5·531	5·533	5·534
49	5·535	5·537	5·538	5·539	5·541	5·542	5·543	5·545	5·546	5·547
50	5·549	5·550	5·551	5·552	5·554	5·555	5·556	5·557	5·559	5·560
51	5·561	5·563	5·564	5·565	5·567	5·568	5·569	5·570	5·571	5·573
52	5·574	5·575	5·576	5·578	5·579	5·580	5·582	5·583	5·584	5·585
53	5·586	5·588	5·589	5·590	5·591	5·593	5·594	5·595	5·596	5·597
54	5·599	5·600	5·601	5·602	5·603	5·604	5·605	5·607	5·608	5·609
55	5·611	5·612	5·613	5·614	5·615	5·616	5·618	5·619	5·620	5·621
56	5·622	5·624	5·625	5·626	5·627	5·628	5·629	5·630	5·632	5·633
57	5·634	5·635	5·636	5·637	5·638	5·640	5·641	5·642	5·643	5·644
58	5·645	5·646	5·647	5·649	5·650	5·651	5·652	5·653	5·654	5·655
59	5·656	5·657	5·658	5·660	5·661	5·662	5·663	5·664	5·665	5·666
60	5·667	5·668	5·669	5·670	5·672	5·673	5·674	5·675	5·676	5·677
61	5·678	5·679	5·680	5·681	5·682	5·683	5·684	5·685	5·687	5·688
62	5·689	5·690	5·691	5·692	5·693	5·694	5·695	5·696	5·697	5·698
63	5·699	5·700	5·701	5·702	5·703	5·704	5·705	5·706	5·707	5·708
64	5·709	5·710	5·711	5·712	5·713	5·714	5·715	5·716	5·717	5·718
65	5·719	5·720	5·721	5·722	5·723	5·724	5·725	5·726	5·727	5·728
66	5·729	5·730	5·731	5·732	5·733	5·734	5·735	5·736	5·737	5·738
67	5·739	5·740	5·741	5·742	5·743	5·744	5·745	5·746	5·747	5·748
68	5·749	5·750	5·751	5·752	5·753	5·754	5·754	5·756	5·756	5·757
69	5·758	5·759	5·760	5·761	5·762	5·763	5·764	5·765	5·766	5·767
70	5·768	5·769	5·769	5·771	5·771	5·772	5·773	5·774	5·775	5·776
71	5·777	5·778	5·779	5·780	5·781	5·781	5·782	5·783	5·784	5·785
72	5·786	5·787	5·788	5·789	5·790	5·790	5·791	5·792	5·793	5·794
73	5·795	5·796	5·797	5·798	5·799	5·799	5·800	5·801	5·802	5·803
74	5·804	5·805	5·806	5·807	5·807	5·808	5·809	5·810	5·811	5·812
75	5·813	5·813	5·814	5·815	5·816	5·817	5·818	5·819	5·820	5·820
76	5·821	5·822	5·823	5·824	5·825	5·826	5·826	5·827	5·828	5·829
77	5·830	5·831	5·831	5·832	5·833	5·834	5·835	5·836	5·837	5·837
78	5·838	5·839	5·840	5·841	5·841	5·842	5·843	5·844	5·845	5·846
79	5·846	5·847	5·848	5·849	5·850	5·851	5·851	5·852	5·853	5·854

APPENDIX 6

Table of the Saha ionization constant S_{pj} (in atm) as a function of temperature for different values of the apparent ionization potential V_{ij} (cf. §§ 7.2, 7.3, 7.5, and 7.8). V_{ij} is expressed as eV, the temperature as °K. $S_{pj} = p_{ij}p_e/p_{aj} = 6{\cdot}58.10^{-7}\,T^{\frac{5}{2}}\,10^{-(5040/T)\bar{V}_{ij}}$ [cf. equations (7.18), (7.21), and (7.20)].

T

$\bar{V}_{ij}$	4000	4100	4200	4300	4400	4500	4600	4700	4800	4900
4·5	$1{\cdot}43.10^{-3}$	$2{\cdot}10.10^{-3}$	$3{\cdot}01.10^{-3}$	$4{\cdot}27.10^{-3}$	$5{\cdot}94.10^{-3}$	$8{\cdot}18.10^{-3}$	$1{\cdot}11.10^{-2}$	$1{\cdot}50.10^{-2}$	$1{\cdot}99.10^{-2}$	$2{\cdot}61.10^{-2}$
5·0	$3{\cdot}35.10^{-4}$	$5{\cdot}07.10^{-4}$	$7{\cdot}55.10^{-4}$	$1{\cdot}10.10^{-3}$	$1{\cdot}59.10^{-3}$	$2{\cdot}25.10^{-3}$	$3{\cdot}15.10^{-3}$	$4{\cdot}34.10^{-3}$	$5{\cdot}93.10^{-3}$	$8{\cdot}00.10^{-3}$
5·5	$7{\cdot}85.10^{-5}$	$1{\cdot}23.10^{-4}$	$1{\cdot}90.10^{-4}$	$2{\cdot}86.10^{-4}$	$4{\cdot}26.10^{-4}$	$6{\cdot}21.10^{-4}$	$8{\cdot}91.10^{-4}$	$1{\cdot}26.10^{-3}$	$1{\cdot}77.10^{-3}$	$2{\cdot}45.10^{-3}$
6·0	$1{\cdot}84.10^{-5}$	$2{\cdot}99.10^{-5}$	$4{\cdot}76.10^{-5}$	$7{\cdot}43.10^{-5}$	$1{\cdot}14.10^{-4}$	$1{\cdot}71.10^{-4}$	$2{\cdot}53.10^{-4}$	$3{\cdot}68.10^{-4}$	$5{\cdot}28.10^{-4}$	$7{\cdot}48.10^{-4}$
6·5	$4{\cdot}31.10^{-6}$	$7{\cdot}28.10^{-6}$	$1{\cdot}20.10^{-5}$	$1{\cdot}93.10^{-5}$	$3{\cdot}04.10^{-5}$	$4{\cdot}71.10^{-5}$	$7{\cdot}16.10^{-5}$	$1{\cdot}07.10^{-4}$	$1{\cdot}58.10^{-4}$	$2{\cdot}29.10^{-4}$
7·0	$1{\cdot}01.10^{-6}$	$1{\cdot}77.10^{-6}$	$3{\cdot}01.10^{-6}$	$5{\cdot}00.10^{-6}$	$8{\cdot}13.10^{-6}$	$1{\cdot}30.10^{-5}$	$2{\cdot}03.10^{-5}$	$3{\cdot}12.10^{-5}$	$4{\cdot}71.10^{-5}$	$7{\cdot}01.10^{-5}$
7·5	$2{\cdot}37.10^{-7}$	$4{\cdot}28.10^{-7}$	$7{\cdot}55.10^{-7}$	$1{\cdot}30.10^{-6}$	$2{\cdot}18.10^{-6}$	$3{\cdot}57.10^{-6}$	$5{\cdot}74.10^{-6}$	$9{\cdot}08.10^{-6}$	$1{\cdot}41.10^{-5}$	$2{\cdot}14.10^{-5}$
8·0	$5{\cdot}56.10^{-8}$	$1{\cdot}04.10^{-7}$	$1{\cdot}90.10^{-7}$	$3{\cdot}36.10^{-7}$	$5{\cdot}82.10^{-7}$	$9{\cdot}84.10^{-7}$	$1{\cdot}63.10^{-6}$	$2{\cdot}64.10^{-6}$	$4{\cdot}20.10^{-6}$	$6{\cdot}56.10^{-6}$
8·5	$1{\cdot}30.10^{-8}$	$2{\cdot}53.10^{-8}$	$4{\cdot}76.10^{-8}$	$8{\cdot}73.10^{-8}$	$1{\cdot}56.10^{-7}$	$2{\cdot}71.10^{-7}$	$4{\cdot}61.10^{-7}$	$7{\cdot}67.10^{-7}$	$1{\cdot}25.10^{-6}$	$2{\cdot}01.10^{-6}$
9·0	$3{\cdot}05.10^{-9}$	$6{\cdot}14.10^{-9}$	$1{\cdot}20.10^{-8}$	$2{\cdot}26.10^{-8}$	$4{\cdot}17.10^{-8}$	$7{\cdot}46.10^{-8}$	$1{\cdot}31.10^{-7}$	$2{\cdot}23.10^{-7}$	$3{\cdot}74.10^{-7}$	$6{\cdot}15.10^{-7}$
9·5	$7{\cdot}16.10^{-10}$	$1{\cdot}49.10^{-9}$	$3{\cdot}01.10^{-9}$	$5{\cdot}87.10^{-9}$	$1{\cdot}11.10^{-8}$	$2{\cdot}06.10^{-8}$	$3{\cdot}70.10^{-8}$	$6{\cdot}50.10^{-8}$	$1{\cdot}12.10^{-7}$	$1{\cdot}88.10^{-7}$
10·0	$1{\cdot}68.10^{-10}$	$3{\cdot}62.10^{-10}$	$7{\cdot}55.10^{-10}$	$1{\cdot}52.10^{-9}$	$2{\cdot}98.10^{-9}$	$5{\cdot}66.10^{-9}$	$1{\cdot}05.10^{-8}$	$1{\cdot}89.10^{-8}$	$3{\cdot}33.10^{-8}$	$5{\cdot}75.10^{-8}$
11·0	$9{\cdot}23.10^{-12}$	$2{\cdot}14.10^{-11}$	$4{\cdot}76.10^{-11}$	$1{\cdot}03.10^{-10}$	$2{\cdot}13.10^{-10}$	$4{\cdot}29.10^{-10}$	$8{\cdot}39.10^{-10}$	$1{\cdot}60.10^{-9}$	$2{\cdot}97.10^{-9}$	$5{\cdot}38.10^{-9}$
12·0	$5{\cdot}07.10^{-13}$	$1{\cdot}26.10^{-12}$	$3{\cdot}01.10^{-12}$	$6{\cdot}90.10^{-12}$	$1{\cdot}52.10^{-11}$	$3{\cdot}26.10^{-11}$	$6{\cdot}74.10^{-11}$	$1{\cdot}35.10^{-10}$	$2{\cdot}65.10^{-10}$	$5{\cdot}05.10^{-10}$
13·0	$2{\cdot}79.10^{-14}$	$7{\cdot}43.10^{-14}$	$1{\cdot}90.10^{-13}$	$4{\cdot}64.10^{-13}$	$1{\cdot}09.10^{-12}$	$2{\cdot}47.10^{-12}$	$5{\cdot}41.10^{-12}$	$1{\cdot}15.10^{-11}$	$2{\cdot}36.10^{-11}$	$4{\cdot}72.10^{-11}$
14·0	$1{\cdot}53.10^{-15}$	$4{\cdot}38.10^{-15}$	$1{\cdot}20.10^{-14}$	$3{\cdot}13.10^{-14}$	$7{\cdot}80.10^{-14}$	$1{\cdot}87.10^{-13}$	$4{\cdot}33.10^{-13}$	$9{\cdot}73.10^{-13}$	$2{\cdot}10.10^{-12}$	$4{\cdot}43.10^{-12}$
15·0	$8{\cdot}41.10^{-17}$	$2{\cdot}59.10^{-16}$	$7{\cdot}55.10^{-16}$	$2{\cdot}10.10^{-15}$	$5{\cdot}57.10^{-15}$	$1{\cdot}42.10^{-14}$	$3{\cdot}47.10^{-14}$	$8{\cdot}22.10^{-14}$	$1{\cdot}87.10^{-13}$	$4{\cdot}14.10^{-13}$

APPENDIX 6 (continued)

T

$\bar{V}_{ij}$	5000	5100	5200	5300	5400	5500	5600	5700	5800	5900
4·5	$3{\cdot}40{\cdot}10^{-2}$	$4{\cdot}38{\cdot}10^{-2}$	$5{\cdot}61{\cdot}10^{-2}$	$7{\cdot}11{\cdot}10^{-2}$	$8{\cdot}93{\cdot}10^{-2}$	$1{\cdot}11{\cdot}10^{-1}$	$1{\cdot}38{\cdot}10^{-1}$	$1{\cdot}70{\cdot}10^{-1}$	$2{\cdot}08{\cdot}10^{-1}$	$2{\cdot}53{\cdot}10^{-1}$
5·0	$1{\cdot}07{\cdot}10^{-2}$	$1{\cdot}41{\cdot}10^{-2}$	$1{\cdot}84{\cdot}10^{-2}$	$2{\cdot}38{\cdot}10^{-2}$	$3{\cdot}05{\cdot}10^{-2}$	$3{\cdot}88{\cdot}10^{-2}$	$4{\cdot}91{\cdot}10^{-2}$	$6{\cdot}15{\cdot}10^{-2}$	$7{\cdot}66{\cdot}10^{-2}$	$9{\cdot}46{\cdot}10^{-2}$
5·5	$3{\cdot}34{\cdot}10^{-3}$	$4{\cdot}51{\cdot}10^{-3}$	$6{\cdot}01{\cdot}10^{-3}$	$7{\cdot}96{\cdot}10^{-3}$	$1{\cdot}04{\cdot}10^{-2}$	$1{\cdot}35{\cdot}10^{-2}$	$1{\cdot}74{\cdot}10^{-2}$	$2{\cdot}22{\cdot}10^{-2}$	$2{\cdot}81{\cdot}10^{-2}$	$3{\cdot}54{\cdot}10^{-2}$
6·0	$1{\cdot}05{\cdot}10^{-3}$	$1{\cdot}44{\cdot}10^{-3}$	$1{\cdot}97{\cdot}10^{-3}$	$2{\cdot}67{\cdot}10^{-3}$	$3{\cdot}56{\cdot}10^{-3}$	$4{\cdot}71{\cdot}10^{-3}$	$6{\cdot}18{\cdot}10^{-3}$	$8{\cdot}03{\cdot}10^{-3}$	$1{\cdot}03{\cdot}10^{-2}$	$1{\cdot}32{\cdot}10^{-2}$
6·5	$3{\cdot}28{\cdot}10^{-4}$	$4{\cdot}63{\cdot}10^{-4}$	$6{\cdot}46{\cdot}10^{-4}$	$8{\cdot}91{\cdot}10^{-4}$	$1{\cdot}21{\cdot}10^{-3}$	$1{\cdot}64{\cdot}10^{-3}$	$2{\cdot}19{\cdot}10^{-3}$	$2{\cdot}90{\cdot}10^{-3}$	$3{\cdot}80{\cdot}10^{-3}$	$4{\cdot}95{\cdot}10^{-3}$
7·0	$1{\cdot}03{\cdot}10^{-4}$	$1{\cdot}49{\cdot}10^{-4}$	$2{\cdot}12{\cdot}10^{-4}$	$2{\cdot}98{\cdot}10^{-4}$	$4{\cdot}15{\cdot}10^{-4}$	$5{\cdot}70{\cdot}10^{-4}$	$7{\cdot}78{\cdot}10^{-4}$	$1{\cdot}05{\cdot}10^{-3}$	$1{\cdot}40{\cdot}10^{-3}$	$1{\cdot}85{\cdot}10^{-3}$
7·5	$3{\cdot}22{\cdot}10^{-5}$	$4{\cdot}75{\cdot}10^{-5}$	$6{\cdot}93{\cdot}10^{-5}$	$9{\cdot}98{\cdot}10^{-5}$	$1{\cdot}42{\cdot}10^{-4}$	$1{\cdot}99{\cdot}10^{-4}$	$2{\cdot}76{\cdot}10^{-4}$	$3{\cdot}78{\cdot}10^{-4}$	$5{\cdot}14{\cdot}10^{-4}$	$6{\cdot}92{\cdot}10^{-4}$
8·0	$1{\cdot}01{\cdot}10^{-5}$	$1{\cdot}52{\cdot}10^{-5}$	$2{\cdot}27{\cdot}10^{-5}$	$3{\cdot}34{\cdot}10^{-5}$	$4{\cdot}84{\cdot}10^{-5}$	$6{\cdot}92{\cdot}10^{-5}$	$9{\cdot}79{\cdot}10^{-5}$	$1{\cdot}37{\cdot}10^{-4}$	$1{\cdot}89{\cdot}10^{-4}$	$2{\cdot}59{\cdot}10^{-4}$
8·5	$3{\cdot}16{\cdot}10^{-6}$	$4{\cdot}89{\cdot}10^{-6}$	$7{\cdot}45{\cdot}10^{-6}$	$1{\cdot}12{\cdot}10^{-5}$	$1{\cdot}65{\cdot}10^{-5}$	$2{\cdot}41{\cdot}10^{-5}$	$3{\cdot}47{\cdot}10^{-5}$	$4{\cdot}94{\cdot}10^{-5}$	$6{\cdot}95{\cdot}10^{-5}$	$9{\cdot}68{\cdot}10^{-5}$
9·0	$9{\cdot}91{\cdot}10^{-7}$	$1{\cdot}57{\cdot}10^{-6}$	$2{\cdot}44{\cdot}10^{-6}$	$3{\cdot}74{\cdot}10^{-6}$	$5{\cdot}64{\cdot}10^{-6}$	$8{\cdot}37{\cdot}10^{-6}$	$1{\cdot}23{\cdot}10^{-5}$	$1{\cdot}79{\cdot}10^{-5}$	$2{\cdot}56{\cdot}10^{-5}$	$3{\cdot}62{\cdot}10^{-5}$
9·5	$3{\cdot}10{\cdot}10^{-7}$	$5{\cdot}02{\cdot}10^{-7}$	$8{\cdot}00{\cdot}10^{-7}$	$1{\cdot}25{\cdot}10^{-6}$	$1{\cdot}93{\cdot}10^{-6}$	$2{\cdot}92{\cdot}10^{-6}$	$4{\cdot}37{\cdot}10^{-6}$	$6{\cdot}46{\cdot}10^{-6}$	$9{\cdot}40{\cdot}10^{-6}$	$1{\cdot}35{\cdot}10^{-5}$
10·0	$9{\cdot}73{\cdot}10^{-8}$	$1{\cdot}61{\cdot}10^{-7}$	$2{\cdot}62{\cdot}10^{-7}$	$4{\cdot}19{\cdot}10^{-7}$	$6{\cdot}58{\cdot}10^{-7}$	$1{\cdot}02{\cdot}10^{-6}$	$1{\cdot}55{\cdot}10^{-6}$	$2{\cdot}33{\cdot}10^{-6}$	$3{\cdot}46{\cdot}10^{-6}$	$5{\cdot}07{\cdot}10^{-6}$
11·0	$9{\cdot}55{\cdot}10^{-9}$	$1{\cdot}66{\cdot}10^{-8}$	$2{\cdot}81{\cdot}10^{-8}$	$4{\cdot}69{\cdot}10^{-8}$	$7{\cdot}67{\cdot}10^{-8}$	$1{\cdot}23{\cdot}10^{-7}$	$1{\cdot}95{\cdot}10^{-7}$	$3{\cdot}05{\cdot}10^{-7}$	$4{\cdot}68{\cdot}10^{-7}$	$7{\cdot}10{\cdot}10^{-7}$
12·0	$9{\cdot}38{\cdot}10^{-10}$	$1{\cdot}70{\cdot}10^{-9}$	$3{\cdot}02{\cdot}10^{-9}$	$5{\cdot}25{\cdot}10^{-9}$	$8{\cdot}93{\cdot}10^{-9}$	$1{\cdot}49{\cdot}10^{-8}$	$2{\cdot}46{\cdot}10^{-8}$	$3{\cdot}98{\cdot}10^{-8}$	$6{\cdot}32{\cdot}10^{-8}$	$9{\cdot}93{\cdot}10^{-8}$
13·0	$9{\cdot}20{\cdot}10^{-11}$	$1{\cdot}75{\cdot}10^{-10}$	$3{\cdot}24{\cdot}10^{-10}$	$5{\cdot}87{\cdot}10^{-10}$	$1{\cdot}04{\cdot}10^{-9}$	$1{\cdot}81{\cdot}10^{-9}$	$3{\cdot}10{\cdot}10^{-9}$	$5{\cdot}19{\cdot}10^{-9}$	$8{\cdot}55{\cdot}10^{-9}$	$1{\cdot}39{\cdot}10^{-8}$
14·0	$9{\cdot}04{\cdot}10^{-12}$	$1{\cdot}79{\cdot}10^{-11}$	$3{\cdot}47{\cdot}10^{-11}$	$6{\cdot}58{\cdot}10^{-11}$	$1{\cdot}22{\cdot}10^{-10}$	$2{\cdot}19{\cdot}10^{-10}$	$3{\cdot}90{\cdot}10^{-10}$	$6{\cdot}78{\cdot}10^{-10}$	$1{\cdot}16{\cdot}10^{-9}$	$1{\cdot}94{\cdot}10^{-9}$
15·0	$8{\cdot}87{\cdot}10^{-13}$	$1{\cdot}84{\cdot}10^{-12}$	$3{\cdot}73{\cdot}10^{-12}$	$7{\cdot}36{\cdot}10^{-12}$	$1{\cdot}42{\cdot}10^{-11}$	$2{\cdot}66{\cdot}10^{-11}$	$4{\cdot}91{\cdot}10^{-11}$	$8{\cdot}85{\cdot}10^{-11}$	$1{\cdot}56{\cdot}10^{-10}$	$2{\cdot}72{\cdot}10^{-10}$

T

$\bar{V}_{ij}$	6000	6100	6200	6300	6400	6500	6600	6700	6800	6900
4·5	$3{\cdot}06{\cdot}10^{-1}$	$3{\cdot}67{\cdot}10^{-1}$	$4{\cdot}39{\cdot}10^{-1}$	$5{\cdot}22{\cdot}10^{-1}$	$6{\cdot}19{\cdot}10^{-1}$	$7{\cdot}29{\cdot}10^{-1}$	$8{\cdot}57{\cdot}10^{-1}$	$1{\cdot}00{\cdot}10^{0}$	$1{\cdot}16{\cdot}10^{0}$	$1{\cdot}35{\cdot}10^{0}$
5·0	$1{\cdot}16{\cdot}10^{-1}$	$1{\cdot}42{\cdot}10^{-1}$	$1{\cdot}72{\cdot}10^{-1}$	$2{\cdot}08{\cdot}10^{-1}$	$2{\cdot}50{\cdot}10^{-1}$	$2{\cdot}98{\cdot}10^{-1}$	$3{\cdot}56{\cdot}10^{-1}$	$4{\cdot}21{\cdot}10^{-1}$	$4{\cdot}95{\cdot}10^{-1}$	$5{\cdot}82{\cdot}10^{-1}$
5·5	$4{\cdot}43{\cdot}10^{-2}$	$5{\cdot}48{\cdot}10^{-2}$	$6{\cdot}76{\cdot}10^{-2}$	$8{\cdot}28{\cdot}10^{-2}$	$1{\cdot}01{\cdot}10^{-1}$	$1{\cdot}22{\cdot}10^{-1}$	$1{\cdot}48{\cdot}10^{-1}$	$1{\cdot}77{\cdot}10^{-1}$	$2{\cdot}11{\cdot}10^{-1}$	$2{\cdot}51{\cdot}10^{-1}$
6·0	$1{\cdot}68{\cdot}10^{-2}$	$2{\cdot}12{\cdot}10^{-2}$	$2{\cdot}65{\cdot}10^{-2}$	$3{\cdot}30{\cdot}10^{-2}$	$4{\cdot}08{\cdot}10^{-2}$	$5{\cdot}01{\cdot}10^{-2}$	$6{\cdot}12{\cdot}10^{-2}$	$7{\cdot}45{\cdot}10^{-2}$	$8{\cdot}99{\cdot}10^{-2}$	$1{\cdot}08{\cdot}10^{-1}$
6·5	$6{\cdot}40{\cdot}10^{-3}$	$8{\cdot}18{\cdot}10^{-3}$	$1{\cdot}04{\cdot}10^{-2}$	$1{\cdot}31{\cdot}10^{-2}$	$1{\cdot}65{\cdot}10^{-2}$	$2{\cdot}05{\cdot}10^{-2}$	$2{\cdot}55{\cdot}10^{-2}$	$3{\cdot}13{\cdot}10^{-2}$	$3{\cdot}83{\cdot}10^{-2}$	$4{\cdot}67{\cdot}10^{-2}$
7·0	$2{\cdot}43{\cdot}10^{-3}$	$3{\cdot}16{\cdot}10^{-3}$	$4{\cdot}08{\cdot}10^{-3}$	$5{\cdot}22{\cdot}10^{-3}$	$6{\cdot}65{\cdot}10^{-3}$	$8{\cdot}39{\cdot}10^{-3}$	$1{\cdot}06{\cdot}10^{-2}$	$1{\cdot}31{\cdot}10^{-2}$	$1{\cdot}63{\cdot}10^{-2}$	$2{\cdot}01{\cdot}10^{-2}$
7·5	$9{\cdot}25{\cdot}10^{-4}$	$1{\cdot}22{\cdot}10^{-3}$	$1{\cdot}60{\cdot}10^{-3}$	$2{\cdot}08{\cdot}10^{-3}$	$2{\cdot}69{\cdot}10^{-3}$	$3{\cdot}44{\cdot}10^{-3}$	$4{\cdot}38{\cdot}10^{-3}$	$5{\cdot}53{\cdot}10^{-3}$	$6{\cdot}95{\cdot}10^{-3}$	$8{\cdot}69{\cdot}10^{-3}$
8·0	$3{\cdot}52{\cdot}10^{-4}$	$4{\cdot}71{\cdot}10^{-4}$	$6{\cdot}28{\cdot}10^{-4}$	$8{\cdot}28{\cdot}10^{-4}$	$1{\cdot}09{\cdot}10^{-3}$	$1{\cdot}41{\cdot}10^{-3}$	$1{\cdot}82{\cdot}10^{-3}$	$2{\cdot}33{\cdot}10^{-3}$	$2{\cdot}96{\cdot}10^{-3}$	$3{\cdot}75{\cdot}10^{-3}$
8·5	$1{\cdot}34{\cdot}10^{-4}$	$1{\cdot}82{\cdot}10^{-4}$	$2{\cdot}46{\cdot}10^{-4}$	$3{\cdot}30{\cdot}10^{-4}$	$4{\cdot}38{\cdot}10^{-4}$	$5{\cdot}77{\cdot}10^{-4}$	$7{\cdot}55{\cdot}10^{-4}$	$9{\cdot}79{\cdot}10^{-4}$	$1{\cdot}26{\cdot}10^{-3}$	$1{\cdot}62{\cdot}10^{-3}$
9·0	$5{\cdot}08{\cdot}10^{-5}$	$7{\cdot}03{\cdot}10^{-5}$	$9{\cdot}66{\cdot}10^{-5}$	$1{\cdot}31{\cdot}10^{-4}$	$1{\cdot}77{\cdot}10^{-4}$	$2{\cdot}36{\cdot}10^{-4}$	$3{\cdot}14{\cdot}10^{-4}$	$4{\cdot}12{\cdot}10^{-4}$	$5{\cdot}37{\cdot}10^{-4}$	$6{\cdot}97{\cdot}10^{-4}$
9·5	$1{\cdot}93{\cdot}10^{-5}$	$2{\cdot}72{\cdot}10^{-5}$	$3{\cdot}78{\cdot}10^{-5}$	$5{\cdot}22{\cdot}10^{-5}$	$7{\cdot}16{\cdot}10^{-5}$	$9{\cdot}68{\cdot}10^{-5}$	$1{\cdot}30{\cdot}10^{-4}$	$1{\cdot}73{\cdot}10^{-4}$	$2{\cdot}29{\cdot}10^{-4}$	$3{\cdot}01{\cdot}10^{-4}$
10·0	$7{\cdot}34{\cdot}10^{-6}$	$1{\cdot}05{\cdot}10^{-5}$	$1{\cdot}49{\cdot}10^{-5}$	$2{\cdot}08{\cdot}10^{-5}$	$2{\cdot}89{\cdot}10^{-5}$	$3{\cdot}96{\cdot}10^{-5}$	$5{\cdot}41{\cdot}10^{-5}$	$7{\cdot}29{\cdot}10^{-5}$	$9{\cdot}75{\cdot}10^{-5}$	$1{\cdot}30{\cdot}10^{-4}$
11·0	$1{\cdot}06{\cdot}10^{-6}$	$1{\cdot}57{\cdot}10^{-6}$	$2{\cdot}29{\cdot}10^{-6}$	$3{\cdot}30{\cdot}10^{-6}$	$4{\cdot}71{\cdot}10^{-6}$	$6{\cdot}65{\cdot}10^{-6}$	$9{\cdot}31{\cdot}10^{-6}$	$1{\cdot}29{\cdot}10^{-5}$	$1{\cdot}77{\cdot}10^{-5}$	$2{\cdot}41{\cdot}10^{-5}$
12·0	$1{\cdot}53{\cdot}10^{-7}$	$2{\cdot}34{\cdot}10^{-7}$	$3{\cdot}52{\cdot}10^{-7}$	$5{\cdot}22{\cdot}10^{-7}$	$7{\cdot}69{\cdot}10^{-7}$	$1{\cdot}11{\cdot}10^{-6}$	$1{\cdot}61{\cdot}10^{-6}$	$2{\cdot}29{\cdot}10^{-6}$	$3{\cdot}21{\cdot}10^{-6}$	$4{\cdot}49{\cdot}10^{-6}$
13·0	$2{\cdot}22{\cdot}10^{-8}$	$3{\cdot}48{\cdot}10^{-8}$	$5{\cdot}41{\cdot}10^{-8}$	$8{\cdot}28{\cdot}10^{-8}$	$1{\cdot}25{\cdot}10^{-7}$	$1{\cdot}87{\cdot}10^{-7}$	$2{\cdot}77{\cdot}10^{-7}$	$4{\cdot}04{\cdot}10^{-7}$	$5{\cdot}82{\cdot}10^{-7}$	$8{\cdot}36{\cdot}10^{-7}$
14·0	$3{\cdot}21{\cdot}10^{-9}$	$5{\cdot}20{\cdot}10^{-9}$	$8{\cdot}32{\cdot}10^{-9}$	$1{\cdot}31{\cdot}10^{-8}$	$2{\cdot}05{\cdot}10^{-8}$	$3{\cdot}13{\cdot}10^{-8}$	$4{\cdot}77{\cdot}10^{-8}$	$7{\cdot}14{\cdot}10^{-8}$	$1{\cdot}06{\cdot}10^{-7}$	$1{\cdot}55{\cdot}10^{-7}$
15·0	$4{\cdot}63{\cdot}10^{-10}$	$7{\cdot}76{\cdot}10^{-10}$	$1{\cdot}28{\cdot}10^{-9}$	$2{\cdot}08{\cdot}10^{-9}$	$3{\cdot}33{\cdot}10^{-9}$	$5{\cdot}26{\cdot}10^{-9}$	$8{\cdot}22{\cdot}10^{-9}$	$1{\cdot}26{\cdot}10^{-8}$	$1{\cdot}92{\cdot}10^{-8}$	$2{\cdot}89{\cdot}10^{-8}$

APPENDIX 7

Abridged table for converting $\log[\alpha/(1-\alpha)]$ to $\log \alpha$ and $\log(1-\alpha)$ (cf. § 7.5).

Marginal numbers: $|\log[\alpha/(1-\alpha)]|$

Entry: $-\log(1-\alpha)$ in upper rows and $-\log \alpha$ in lower rows if $\log[\alpha/(1-\alpha)] > 0$
$-\log \alpha$ in upper rows and $-\log(1-\alpha)$ in lower rows if $\log[\alpha/(1-\alpha)] < 0$

Examples: $\log[\alpha/(1-\alpha)] = 0.93 \rightarrow -\log \alpha = 0.050$ and $-\log(1-\alpha) = 0.980$
$\log[\alpha/(1-\alpha)] = -0.93 \rightarrow -\log \alpha = 0.980$ and $-\log(1-\alpha) = 0.050$

	0	1	2	3	4	5	6	7	8	9
2.0	2.005	2.015	2.025	2.035	2.045	2.055	2.065	2.075	2.085	2.095
	0.005	0.005	0.005	0.005	0.005	0.005	0.005	0.005	0.005	0.005
1.9	1.905	1.915	1.925	1.935	1.945	1.955	1.965	1.975	1.985	1.995
	0.005	0.005	0.005	0.005	0.005	0.005	0.005	0.005	0.005	0.005
1.8	1.805	1.815	1.825	1.835	1.845	1.855	1.865	1.875	1.885	1.895
	0.005	0.005	0.005	0.005	0.005	0.005	0.005	0.005	0.005	0.005
1.7	1.710	1.720	1.730	1.740	1.750	1.760	1.765	1.775	1.785	1.795
	0.010	0.010	0.010	0.010	0.010	0.010	0.005	0.005	0.005	0.005
1.6	1.610	1.620	1.630	1.640	1.650	1.660	1.670	1.680	1.690	1.700
	0.010	0.010	0.010	0.010	0.010	0.010	0.010	0.010	0.010	0.010
1.5	1.515	1.525	1.535	1.545	1.550	1.560	1.570	1.580	1.590	1.600
	0.015	0.015	0.015	0.015	0.010	0.010	0.010	0.010	0.010	0.010
1.4	1.415	1.425	1.435	1.445	1.455	1.465	1.475	1.485	1.495	1.505
	0.015	0.015	0.015	0.015	0.015	0.015	0.015	0.015	0.015	0.015
1.3	1.320	1.330	1.340	1.350	1.360	1.370	1.380	1.390	1.400	1.405
	0.020	0.020	0.020	0.020	0.020	0.020	0.020	0.020	0.020	0.015
1.2	1.225	1.235	1.245	1.255	1.265	1.275	1.285	1.295	1.300	1.310
	0.025	0.025	0.025	0.025	0.025	0.025	0.025	0.025	0.020	0.020
1.1	1.135	1.145	1.150	1.160	1.170	1.180	1.190	1.200	1.210	1.215
	0.035	0.035	0.030	0.030	0.030	0.030	0.030	0.030	0.030	0.025
1.0	1.040	1.050	1.060	1.070	1.075	1.085	1.095	1.105	1.115	1.125
	0.040	0.040	0.040	0.040	0.035	0.035	0.035	0.035	0.035	0.035
0.9	0.950	0.960	0.970	0.980	0.985	0.995	1.005	1.015	1.025	1.030
	0.050	0.050	0.050	0.050	0.045	0.045	0.045	0.045	0.045	0.040
0.8	0.865	0.875	0.880	0.890	0.900	0.905	0.915	0.925	0.935	0.945
	0.065	0.065	0.060	0.060	0.060	0.055	0.055	0.055	0.055	0.055
0.7	0.780	0.790	0.795	0.805	0.810	0.820	0.830	0.840	0.845	0.855
	0.080	0.080	0.075	0.075	0.070	0.070	0.070	0.070	0.065	0.065
0.6	0.695	0.705	0.715	0.720	0.730	0.740	0.745	0.755	0.760	0.770
	0.095	0.095	0.095	0.090	0.090	0.090	0.085	0.085	0.080	0.080
0.5	0.620	0.625	0.635	0.640	0.650	0.660	0.665	0.675	0.680	0.690
	0.120	0.115	0.115	0.110	0.110	0.110	0.105	0.105	0.100	0.100
0.4	0.545	0.555	0.560	0.565	0.575	0.580	0.590	0.595	0.605	0.610
	0.145	0.145	0.140	0.135	0.135	0.130	0.130	0.125	0.125	0.120
0.3	0.475	0.485	0.490	0.495	0.505	0.510	0.515	0.525	0.530	0.540
	0.175	0.175	0.170	0.165	0.165	0.160	0.155	0.155	0.150	0.150
0.2	0.410	0.420	0.425	0.430	0.435	0.445	0.450	0.455	0.465	0.470
	0.210	0.210	0.205	0.200	0.195	0.195	0.190	0.185	0.185	0.180
0.1	0.355	0.360	0.365	0.370	0.375	0.380	0.390	0.395	0.400	0.405
	0.255	0.250	0.245	0.240	0.235	0.230	0.230	0.225	0.220	0.215
0.0	0.300	0.305	0.310	0.315	0.320	0.325	0.330	0.335	0.345	0.350
	0.300	0.295	0.290	0.285	0.280	0.275	0.270	0.265	0.265	0.260

APPENDIX 8

Basic numerical data of radial distribution functions underlying the arc model discussed in §§ 8.1 to 8.7.

T_0 = temperature at arc axis

$\log A = \frac{3}{4} \log T - \log \exp[-c(d - T_0)r^2]$

Note that $\log[\alpha_j/(1 - \alpha_j)]$ is computed as [cf. (8.6)]:

$$\log[\alpha_j/(1 - \alpha_j)] = 15 \cdot 684 - \log n_e(0) + \log A - (5040/T)V_{ij} + \log Z_{ij}/Z_{aj}.$$

Values of $\log \alpha_j$ and $\log(1 - \alpha_j)$ are found from $\log[\alpha_j/(1 - \alpha_j)]$ with the aid of the table in Appendix 7 (cf. § 7.5).

r (mm)	$T_0 = 6500°K$			$T_0 = 6000°K$			$T_0 = 5500°K$			$T_0 = 5000°K$		
	$T(°K)$	$5040/T$	$\log A$	$T(°K)$	$5040/T$	$\log A$	$T(°K)$	$5040/T$	$\log A$	$T(°K)$	$5040/T$	$\log A$
0	6500	0·7754	5·719	6000	0·8400	5·667	5500	0·9164	5·611	5000	1·0080	5·548
0·5	6470	0·7790	5·727	5970	0·8442	5·680	5470	0·9214	5·629	4970	1·0141	5·572
1·0	6400	0·7875	5·752	5890	0·8557	5·720	5390	0·9351	5·684	4880	1·0328	5·642
1·5	6270	0·8038	5·794	5760	0·8750	5·787	5250	0·9600	5·775	4740	1·0633	5·758
2·0	6100	0·8262	5·852	5580	0·9032	5·881	5060	0·9960	5·903	4540	1·1101	5·920
2·5	5870	0·8586	5·924	5340	0·9438	5·998	4810	1·0478	6·066	4270	1·1803	6·123
3·0	5590	0·9016	6·012	5050	0·9980	6·141	4500	1·1200	6·262	3960	1·2727	6·374
3·5	5260	0·9582	6·113	4700	1·0723	6·306	4140	1·2174	6·490	3580	1·4078	6·660
4·0	4880	1·0328	6·228	4300	1·1721	6·492	3720	1·3548	6·746	3140	1·6051	6·982
4·5	4450	1·1326	6·352	3850	1·3091	6·698	3250	1·5508	7·027	2650	1·9020	
5·0	3970	1·2695	6·484	3350	1·5045	6·916	2720	1·8529		2100	2·4000	
5·5	3440	1·4651	6·619	2790	1·8064	7·141	2140	2·3551				
6·0	2860	1·7622	6·748	2180	2·3119							

APPENDIX 9
Physical constants and conversion factors*

N	Avogadro's number	$= 6 \cdot 02252 . 10^{23}\,\mathrm{mol}^{-1}$
c	Speed of light *in vacuo*	$= 2 \cdot 997925 . 10^{8}\,\mathrm{m\ sec}^{-1}$
		$= 2 \cdot 997925 . 10^{10}\,\mathrm{cm\ sec}^{-1}$
e	Electronic charge	$= 1 \cdot 60210 . 10^{-19}\,\mathrm{C}$
		$= 4 \cdot 80298 . 10^{-10}\,\mathrm{esu}$
m_e	Electron rest mass	$= 9 \cdot 1091 . 10^{-31}\,\mathrm{kg}$
		$= 9 \cdot 1091 . 10^{-28}\,\mathrm{g}$
h	Planck constant	$= 6 \cdot 6256 . 10^{-27}\,\mathrm{erg\ sec}$
		$= 6 \cdot 6256 . 10^{-34}\,\mathrm{J\ sec}$
R	Gas constant	$= 8 \cdot 3143\,\mathrm{J\ {}^\circ K^{-1}\,mol^{-1}}$
		$= 8 \cdot 3143 . 10^{7}\,\mathrm{erg\ {}^\circ K^{-1}\,mol^{-1}}$
k	Boltzmann constant	$= 1 \cdot 38054 . 10^{-23}\,\mathrm{J\ {}^\circ K^{-1}}$
		$= 1 \cdot 38054 . 10^{-16}\,\mathrm{erg\ {}^\circ K^{-1}}$
V_0	Normal volume of perfect gas	$= 2 \cdot 24136 . 10^{-2}\,\mathrm{m^3\,mol^{-1}}$
		$= 2 \cdot 24136 . 10^{4}\,\mathrm{cm^3\,mol^{-1}}$
n_0	Loschmidt's number ($= N/V_0$)	$= 2 \cdot 6870 . 10^{25}\,\mathrm{m}^{-3}$
		$= 2 \cdot 6870 . 10^{19}\,\mathrm{cm}^{-3}$
c_1	First radiation constant ($2\,hc^2$)	$= 3 \cdot 7405 . 10^{-16}\,\mathrm{W\ m^2}$
		$= 3 \cdot 7405 . 10^{-5}\,\mathrm{erg\ cm^2\,sec^{-1}}$
c_2	Second radiation constant (hcN/R)	$= 1 \cdot 43879 . 10^{-2}\,\mathrm{m\ {}^\circ K}$
		$= 1 \cdot 43879\,\mathrm{cm\ {}^\circ K}$
b	Wien displacement constant	$= 2 \cdot 8978 . 10^{-3}\,\mathrm{m\ {}^\circ K}$
		$= 2 \cdot 8978 . 10^{-1}\,\mathrm{cm\ {}^\circ K}$
σ	Stefan–Boltzmann constant	$= 5 \cdot 6697 . 10^{-8}\,\mathrm{W\ m^{-2}\,{}^\circ K^{-4}}$
		$= 5 \cdot 6697 . 10^{-5}\,\mathrm{erg\ cm^{-2}\,sec^{-1}\,{}^\circ K^{-4}}$
amu	Atomic mass unit (= unit of mass equal to one-sixteenth the mass of the atom of oxygen of mass number 16)	$= 1 \cdot 65979 . 10^{-24}\,\mathrm{g\ (phys.)}$
		$= 1 \cdot 66024 . 10^{-24}\,\mathrm{g\ (chem.)}$
eV	Electron volt	$= 1 \cdot 60210 . 10^{-19}\,\mathrm{J\ (eV)^{-1}}$
		$= 1 \cdot 60210 . 10^{-12}\,\mathrm{erg\ (eV)^{-1}}$
J	Joule	$= 6 \cdot 2418 . 10^{18}\,\mathrm{eV\ J^{-1}}$
erg	Erg	$= 6 \cdot 2418 . 10^{11}\,\mathrm{eV\ erg^{-1}}$
ν	Frequency associated with 1 eV	$= 2 \cdot 41804 . 10^{14}\,\mathrm{sec^{-1}\,(eV)^{-1}}$
λ	Wavelength associated with 1 eV	$= 1 \cdot 23981 . 10^{-4}\,\mathrm{cm\ eV}$
σ	Wave-number associated with 1 eV	$= 8 \cdot 06573 . 10^{3}\,\mathrm{cm^{-1}\,(eV)^{-1}}$
T	Temperature associated with 1 eV ($=$ energy/k)	$= 1 \cdot 16049 . 10^{4}\,\mathrm{{}^\circ K\ (eV)^{-1}}$
atm	Normal atmospheric pressure	$= 101325\,\mathrm{N\ m^{-2}}$
		$= 1013250\,\mathrm{dyn\ cm^{-2}}$
cal	Thermochemical calorie	$= 4 \cdot 1840\,\mathrm{J}$
		$= 4 \cdot 1840 . 10^{7}\,\mathrm{erg}$
	Partial pressure p (in atm) associated with concentration n (in cm^{-3}) at temperature T (in °K)	$= 1 \cdot 3625 . 10^{-22} . Tn\dagger$
	Concentration n (in cm^{-3}) associated with partial pressure p (in atm) at temperature T (in °K)	$= 7 \cdot 340 . 10^{21} . p/T\dagger$

* As recommended by the National Academy of Science and the National Research Council and are published in *J. Opt. Soc. Amer.*, 1964, **54**, 281.

† See Appendix 1.

REFERENCES TO LITERATURE COMPILATIONS

Reviews in Analytical Chemistry:

MEGGERS, W. F. (1949) Emission Spectroscopy, *Anal. Chem.*, **21**, 29–31, 64 refs.

MEGGERS, W. F. (1950) Emission Spectroscopy, *Anal. Chem.*, **22**, 18–23, 104 refs.

MEGGERS, W. F. (1952) Emission Spectroscopy, *Anal. Chem.*, **24**, 23–27, 139 refs.

MEGGERS, W. F. (1954) Emission Spectroscopy, *Anal. Chem.*, **26**, 54–58, 147 refs.

MEGGERS, W. F. (1956) Emission Spectroscopy, *Anal. Chem.*, **28**, 616–621, 99 refs.

SCRIBNER, B. F. (1958) Emission Spectroscopy, *Anal. Chem.*, **30**, 596–604, 263 refs.

SCRIBNER, B. F. (1960) Emission Spectroscopy, *Anal. Chem.*, **32**, 229R–237R, 282 refs.

SCRIBNER, B. F. (1962) Emission Spectrometry, *Anal. Chem.*, **34**, 200R–209R, 339 refs.

MARGOSHES, M. (1962) Emission Flame Photometry, *Anal. Chem.*, **34**, 221R–224R, 128 refs.

SCRIBNER, B. F., and MARGOSHES, M. (1964) Emission Spectrometry, *Anal. Chem.*, **36**, 329R–343R, 565 refs.

Index to the Literature on Spectrochemical Analysis (Am. Soc. Testing Materials, Philadelphia, Pa.)

MEGGERS, W. F., and SCRIBNER, B. F. (1940) Part I, 1920–1939, 1446 refs.

SCRIBNER, B. F., and MEGGERS, W. F. (1947)

Part II, 1940–1945, 1053 refs.

SCRIBNER, B. F., and MEGGERS, W. F. (1954) Part III, 1946–1950, 1231 refs.

SCRIBNER, B. F., and MEGGERS, W. F. (1959) Part IV, 1951–1955, 1878 refs.

Spectrochemical Abstracts (Hilger and Watts Ltd, London)

TWYMAN, F. (1938) Vol. I, 1933–1937, 228 refs.

VAN SOMEREN, E. H. S. (1941) Vol. II, 1938–1939, 166 refs.

VAN SOMEREN, E. H. S. (1947) Vol. III, 1940–1945, 468 refs.

VAN SOMEREN, E. H. S., and LACHMAN, F. (1955) Vol. IV, 1946–1951, 827 refs.

VAN SOMEREN, E. H. S., and LACHMAN, F. (1958) Vol. V, 1952–1953, 258 refs.

VAN SOMEREN, E. H. S., and LACHMAN, F. (1960) Vol. VI, 1954–1955, 361 refs.

VAN SOMEREN, E. H. S., and LACHMAN, F. (1962) Vol. VII, 1956–1959, 484 refs.

VAN SOMEREN, E. H. S., LACHMAN, F. and BIRKS, F. T. (1963) Vol. VIII, 1958–1961, 424 refs.

VAN SOMEREN, E. H. S., LACHMAN, F. and BIRKS, F. T. (1964) Vol. IX, 1962–1963, 397 refs.

VAN SOMEREN, E. H. S., LACHMAN, F. and BIRKS, F. T. (1965) Vol. X, 1963–1964, 495 refs.

REFERENCES CITED

(Numbers in square brackets refer to pages where references are cited)

AARTS, C. J. M. (1952) Investigation of an arc discharge which is stabilized with a screw-shaped stream of air. (In Dutch) Ph.D. Thesis, University of Utrecht. [39, 41, 43, 144, 148]

ADCOCK, B. D., and PLUMTREE, W. E. G. (1964) On excitation temperature measurements in a plasma-jet and transition probabilities for argon lines, *J. Quant. Spectry Radiative Transfer*, **4**, 29. [5]

ADDINK, N. W. H. (1959) Determination of transition probability (reciprocal lifetime) of excited atoms and ions from spectrochemical data and the importance of lifetime values in spectrochemistry, *Spectrochim. Acta*, **15**, 349. [20]

ADELSTEIN, S. J., and VALLEE, B. L. (1954) The effect of argon atmospheres on the intensity of certain spectral lines, *Spectrochim. Acta*, **6**, 134. [87]

AHRENS, L. H., and TAYLOR, S. R. (1961) *Spectrochemical Analysis* (Addison-Wesley Publ. Comp., Reading, Mass.). [7, 87, 112, 153, 154, 270, 308, 313–315, 324, 325]

ALIMARIN, I. P. (1963) Modern progress and problems in the determination of trace elements in pure substances, *Zh. Analit. Khim.*, **18**, 1412; *J. Anal. Chem. USSR*, **18**, 1229. [265]

ALLEN, C. W. (1955) *Astrophysical Quantities* (Athlone Press, London). [24, 174]

ANDERMANN, G., and KEMP, J. W. (1959) Effect of gaseous atmospheres on excitation. Symposium on Spectroscopic Excitation, *A.S.T.M. Spec. Tech. Publ.* No. 259, 23 (Am. Soc. Testing Materials, Philadelphia, Pa.). [87]

ANNELL, C. S., and HELZ, A. W. (1960) Spectrochemical analysis using controlled atmospheres with a simple gas jet, *U.S. Geol. Surv. Profess. Papers* **400-B**, 497. [87]

ANNELL, C. S., and HELZ, A. W. (1961) A constant-feed direct-current arc, *U.S. Geol. Surv. Bull.*, **1084-J**, 231. [87]

ARNOLD, J. H. (1930) Studies in diffusion. I, Estimation of diffusivities in gaseous systems, *Ind. Eng. Chem.*, **22**, 1091 [58]

A.S.T.M. (1964) *Methods for Emission Spectrochemical Analysis*, 4th ed. (Am. Soc. Testing Materials, Philadelphia, Pa.). [6, 7]

AYRTON, H. (1902) *The Electric Arc* (Electrician Printing and Publ. Comp., London). [41, 43]

BAER, W. K., and HODGE, E. S. (1960) The spectrochemical analysis of solutions: a comparison of five techniques, *Appl. Spectry*, **14**, 141. [5]

BAKER, S. C. (1964) Spectroscopic estimation of d.c. arc electrode vapor-jet velocities, *J. Opt. Soc. Am.*, **54**, 1204. [35]

BARDÓCZ, Á., VÖRÖS, T., and VANYEK, U. M. (1961) The course in time of the field strength and the ion concentration in the spark discharge, *Spectrochim. Acta*, **17**, 642. [60]

BARDÓCZ, Á., VÖRÖS, T., and VANYEK, U. M. (1962) Zeitlicher Verlauf der Verbreiterung von Spektrallinien und der Ionenkonzentration in Funkenentladungen, *Z. Angew. Phys.*, **14**, 581. [60]

BARR, W. L. (1962) Method for computing the radial distribution of emitters in a cylindrical source, *J. Opt. Soc. Am.*, **52**, 885. [140]

BARTELS, H., and ZWICKER, H. (1962) Die Bestimmung der Linienform aus der Linienkontur bei Linien mit Selbstumkehr. I, Theorie der Methodik, *Z. Physik*, **166**, 148. [89]

BASS, S. T., and SOULATI, J. (1963) Use of a vacuum cup electrode for spark emission spectrographic analysis of plant tissue. *Developments in Applied Spectroscopy* **2**, 211 (Plenum Press, New York, N.Y.). [5]

BAUER, A., and SCHULZ, P. (1954) Elektrodenfälle und Bogengradienten in Hochdruckentladungen, insbesondere bei Xenon, *Z. Physik* **139**, 197. [43]

BAUER, G. (1962) *Strahlungsmessung im Optischen Bereich* (Friedr. Vieweg und Sohn, Braunschweig); English Translation: *Measurement of Optical Radiation* (Focal Press, London, New York, 1965). [8]

BAVINCK, H. (1965) Mass transport through arcs and flames, *Rept TW* 98 (Mathematical Centre Amsterdam). [295, 297]

BEINTEMA, J. (1957) Volatilization phenomena in the direct current arc used as a spectrochemical light source for the analysis of powders (Proc. VI Colloq. Spectros. Intern., Amsterdam 1956), *Spectrochim. Acta* **11**, 186. [207, 308]

BEINTEMA, J., and KROONEN, J. (1955) Eine allgemeine Methode zur spektrochemischen Analyse von nichtleitenden Stoffen (Proc. V Colloq. Spectros. Intern., Gmunden 1954), *Mikrochim. Acta*, 345. [206, 207, 209]

BELOUSOVA, I. M. (1962) Mechanism of formation of the equilibrium concentration of the electrode substance in the plasma of an arc discharge, *Opt. i Spektroskopiya*, **13**, 12; *Opt. Spectry (USSR)*, **13**, 6. [286]

BELYAEV, YU. I., VAINSHTEIN, E. E., and KOROLEV, V. V. (1959) A comparative study of the spatial distribution of elements in a d.c. arc and in an impulse arc by means of radioactive isotopes, *Zh. Analit. Khim.*, **14**, 147; *J. Anal. Chem. USSR*, **14**, 161. [283, 284]

BELYAKOV-BODIN, I. B., and MANDELSHTAM, S. L. (1944) Stabilizing the conditions of excitation of spectrum in the volta arc, *Zh. Tekhn. Fiz.*, **14**, 400 (in Russian). [39]

BERGSTEDT, K., FERGUSON, E., SCHLÜTER, H., and WULFF, H. (1962) Experimentelle Beiträge zur Theorie der Spektrallinienverbreiterung. *Proc. 5th Intern. Conf. Ionization Phenomena Gases, Munich 1961*, **1**, 437. (North-Holland Publ. Comp., Amsterdam). [60]

BERKOWITZ, J. (1962) Heat of formation of the CN radical, *J. Chem. Phys.*, **36**, 2533. [321]

BERTHOLD, R. G. (1922) Spektroheliographische Untersuchungen am elektrischen Lichtbogen, *Physikal. Z.*, **23**, 178. [126]

BEZ, W., and HÖCKER, K. H. (1956) Theorie des Anodenfalls. V, Das Zischen des Homogenkohle-Hochstrombogens in Luft, *Z. Naturforsch.*, **11a**, 192. [35]

BIERMANN, L., and LÜBECK, K. (1948) Übergangswahrscheinlichkeiten von Spektrallinien verschiedener Atome, *Z. Astrophys.*, **25**, 325. [24, 25, 174]

BOCKASTEN, K. (1961) Transformation of observed radiances into radial distribution of the emission of a plasma, *J. Opt. Soc. Am.*, **51**, 943. [140]

BOESCHOTEN, F. (1953) Specimen of an apparatus for the direct photoelectrical measurement of spectral relative intensities as an aid for the spectrochemical determination of trace elements. (In Dutch) Ph.D. Thesis, University of Utrecht. [152]

VAN DEN BOLD, H. J. (1945) The flame of atomic hydrogen and its application to the determination of transition probabilities. Ph.D. Thesis, University of Utrecht. [116]

VAN DEN BOLD, H. J., and SMIT, J. A. (1946) Optical investigation of an atomic hydrogen flame, *Physica*, 12, 475. [116]

BOUMANS, P. W. J. M. (1957) Konzentrationsverteilung des Metalldampfes im Glimmschichtbogen (Proc. VI. Colloq. Spectros. Intern., Amsterdam 1956), *Spectrochim. Acta*, 11, 146. [112, 279, 304]

BOUMANS, P. W. J. M. (1959) Eine neue Anordnung zur Messung der achsialen Intensitätsverteilung im Bogen (Proc. VII Colloq. Spectros. Intern., Liege 1958), *Rev. Universelle Mines*, 15, 396. [142]

BOUMANS, P. W. J. M. (1961) Some fundamental aspects of d.c. arc spectrochemical analysis. (In Dutch) Ph.D. Thesis, University of Amsterdam. [46, 93, 94, 101, 108, 110, 161, 173, 174, 208]

BOUMANS, P. W. J. M. (1962) Temperatur- und Elektronendruckmessungen im Gleichstromkohlebogen und ihre Deutung für die spektrochemische Analyse. *Proc. IX Colloq. Spectros. Intern.*, Lyons 1961, 2, 84 (Groupement pour l'Avancement des Méthodes Spectrographiques, Paris). [46, 93, 94, 101, 108, 110, 161, 173, 174, 208]

BOUMANS, P. W. J. M. (1962a) Lithium fluoride vs lithium carbonate as a buffer, *The Spex Speaker* 7 (4), 6 (Spex Industries, Inc., Metuchen, N.J.). [208]

BOUMANS, P. W. J. M. (1966) Zur Verbesserung der spektralanalytischen Nachweisgrenzen durch Zusätze. *Proc. 2nd Intern. Symp. High-Purity Materials, Dresden* 1965, 1 (in press) (Akademie-Verlag, Berlin) [265]

BOUMANS, P. W. J. M., and DE GALAN, L. (1966) A theoretical calculation of the transport of metal vapors through the discharge column of a d.c. arc, *Anal. Chem.*, 38 (in press). [295]

BOUMANS, P. W. J. M., and MAESSEN, F. M. J. M. (1966) The evaluation of detection limits in emission spectroscopy, *Z. Anal. Chem.* (in press). [265]

BOUMANS, P. W. J. M., and ROUWS, C. J. J. (1965) Fortgesetzte Untersuchungen über Anregungstemperatur, Elektronendichte und Untergrundintensität in Metalldampfgraphitbögen. Ein Studium des LiF-Graphit-Bogens, *Collection Czech. Chem. Commun.*, 30, 1268. [108,110, 173, 174, 191, 198]

BRACEWELL, R. N. (1956) Strip integration in radio astronomy, *Australian J. Phys.*, 9, 198. [135]

BRANDENSTEIN, M., JANDA, I., and SCHROLL, E. (1960) Emissionsspektrographische Methode zur Bestimmung geringster Borgehalte in Reaktorgraphiten, *Miktrochim. Acta*, 935. [208, 315]

BRÄUER, E. (1919) Zur Kenntnis des zischenden Lichtbogens, *Physikal. Z.*, 20, 409. [35]

BREENE, R. G. (1961) *The Shift and Shape of Spectral Lines* (Pergamon Press, London, New York, Paris). [89]

BREWER, L., HICKS, W. T., and KRIKORIAN, O. H. (1962) Heat of sublimation and dissociation energy of gaseous C_2, *J. Chem. Phys.*, 36, 182. [321]

BRINKMAN, H. (1937) Optical study of the electric arc. (In Dutch) Ph.D. Thesis, University of Utrecht. [63, 69, 72, 73, 80, 102–104, 122, 123, 127, 135, 144, 145, 152, 154, 183]

BRODE, R. B. (1929) The absorption coefficient for slow electrons in alkali metal vapors, *Phys. Rev.*, 34, 673. [52]

BURAKOV, V. S., and NAUMENKOV, P. A. (1963) Investigation of the gas cloud composition of an a.c. arc, *Opt. i Spektroskopiya*, 15, 818; *Opt. Spectry (USSR)*, 15, 442. [292]

BURHORN, F. (1955) Temperatur und thermisches Gleichgewicht in Eisenbogen, *Z. Physik* 140, 440. [82, 103]

BURHORN, F. (1959), Berechnung und Messung der Wärmeleitfähigkeit von Stickstoff bis 13 000°K, *Z. Physik*, 155, 42. [52, 63, 69, 72]

BURKHARDT, G. (1948) Zusammensetzung, innere Energie und spezifische Wärme der Luft bei hohen Temperaturen, *Z. Naturforsch.*, 3a, 603. [63]

BURMISTROV, M. P., NALIMOV, V. V., and NEDLER, V. V. (1964) Selecting the optimum conditions for measuring weak spectral lines, *Zavodsk. Lab.*, 30, 544; *Ind. Lab. (USSR)*, 30, 681. [264]

BUSZ-PEUCKERT, G., and FINKELNBURG, W. (1955) Die Abhängigkeit des Anodenfalles von Stromstärke und Bogenlänge bei Hochtemperaturbögen, *Z. Physik*, 140, 540. [42]

VAN CALKER, J. (1962) Untersuchungen zur spektralen Anregung in hochfrequenten Plasmaflammen. *Proc. IX Colloq. Spectros. Intern.*, Lyons 1961, 2, 48 (Groupement pour l'Avancement des Méthodes Spectrographiques, Paris). [5]

VAN CALKER, J., and BRAUNISCH, H. (1956) Über spektroskopische Temperaturmessungen an kondensierten Funkenentladungen, *Z. Naturforsch.*, 11a, 612. [150]

VAN CALKER, J., and TAPPE, W. (1963) Die Anwendung hochfrequenter Plasmaflammen als spectrochemische Lichtquelle, *Arch. Eisenhüttenw.*, 34, 679. [5]

CARROLL, P. K. (1956) The spectrum of CN in the vacuum ultraviolet, *Can. J. Phys.* 34, 83. [321]

CASTLE, J. G., Jr. (1956) *Proc. 1st and 2nd Conf. Carbon, Buffalo, N.Y.*, p. 13 (The Waverly Press, Inc., Baltimore, Maryland). [311]

CHAMP, W. H. (1962) An account of the development of the Stallwood jet, *The Spex Speaker* 7(3), 1 (Spex Industries, Inc., Metuchen, N.J.). [30]

CHAMPION, K. S. W. (1952) The theory of gaseous arcs. I, The fundamental relations for the positive columns. II, The energy balance equation for the positive columns, *Proc. Phys. Soc. (London)*, 65B, 329, 345. [69]

CHAMPION, K. S. W. (1953) The energy balance equation for the positive columns of high pressure arcs, *Proc. Phys. Soc. (London)*, 66B, 169. [69]

CHAMPION, K. S. W. (1956) Radiation from gaseous arcs, *Proc. Phys. Soc. (London)*, 69B, 120. [69]

CH'EN SHANG-YI, and MAKOTO TAKEO (1957) Broadening and shift of spectral lines due to the presence of foreign gases, *Rev. Mod. Phys,* 29, 20. [89]

CHUPKA, W. A., and INGHRAM, M. G. (1955) Direct determination of the heat of sublimation of carbon with the mass spectrometer, *J. Phys. Chem.*, 59, 100. [321]

CLAAS, W. J. (1949) The partition function for the elements in the solar armosphere, *Proc. Koninkl. Ned. Akad. Wetenschap.*, 52, 518. [93, 101]

CLAAS, W. J. (1951) The composition of the solar atmosphere, *Rech. Astron. Obs. Utrecht*, 12(1), 49. [93, 100, 101]

CLARK JONES, R. (1963) Terminology in photometry and radiometry, *J. Opt. Soc. Am.*, 53, 1314 [10]

COHEN, R. S., SPITZER, L., and ROUTLY, McR. P. (1950) The electrical conductivity of an ionized gas, *Phys. Rev.*, 80, 230. [53]

COHEUR, F. P. (1942) Sur la mesure des températures au moyen des bandes moléculaires, *Bull. Classe Sci. Acad. Roy. Belg.*, 28, 569. [121]

COHEUR, P. (1942) Modifications des spectres d'étencelle en fonction du temps (Abfunkeffekt): Mesure des températures. *Mem. Soc. Roy. Sci. Liege*, 5, Ser. 4, 367. [121]

COLLINS, A. G., and PEARSON, C. A. (1964) Emission spectrometric determination of beryllium in oilfield water using plasma arc, *Anal. Chem.*, 36, 787. [5]

COMPTON, K. T. (1923) Theory of the electric arc, *Phys. Rev.* 21, 266, 382. The cause of ionization in the carbon arc, *Phys. Rev.*, 21, 476. [59, 79]

COMPTON, K. T., and LANGMUIR, I. (1930) Electrical discharges in gases. Part I, Survey of fundamental processes, *Rev. Mod. Phys.*, 2, 123. [47, 49, 50]

COOK, G. B., CRESPI, M. B. A., and MINCZEWSKI, J. (1963) International comparison of analytical methods for nuclear materials. I, Accuracy and precision of some techniques in routine trace analysis, *Talanta*, 10, 917. [265]

CORLISS, C. H. (1962a) Temperature of a copper arc, *J. Res. Nat. Bur. Std. A. Physics and Chemistry*, 66A, 5. [26, 105, 106]

CORLISS, C. H. (1962b) Ionization in the plasma of a copper arc, *J. Res. Nat. Bur. Std. A. Physics and Chemistry*, 66A, 169. [26, 28, 93, 94, 163, 164, 171, 173]

CORLISS, C. H. (1962c) Experimental gf-values for seventy elements, *Astrophys. J.*, 136, 916. [26, 93, 94, 163, 164, 171, 173, 281]

CORLISS, C. H., and BOZMAN, W. R. (1962). *Experimental Transition Probabilities for Spectral Lines of Seventy Elements. Derived from the N.B.S. tables of spectral-line intensities. The wavelength, energy levels, transition probability, and oscillator strength of 25 000 lines between 2000 and 9000 Å for 112 spectra of 70 elements.* (N.B.S. Monograph 53, Washington, D.C.). [20, 26, 28, 85, 151, 281]

COWAN, R. D., and DIEKE, G. H. (1948) Self-absorption of spectrum lines, *Rev. Mod. Phys.*, 20, 418. [89, 90]

CRANK, J. (1956) *The Mathematics of Diffusion* (Clarendon Press, Oxford). [294]

CURTIS, C. D. (1962) Cyanogen band suppression in direct-current spectrographic analysis, *Nature*, 196, 1087. [87]

CZAKOW, J. (1960) Über verschiedenartige Einflüsse bei der Spektralanalyse nach dem Schüttverfahren im Bogen. *Proc. VIII Colloq. Spectros. Intern.*, Luzern 1959, p. 114 (Verlag H. R. Sauerländer und Co., Aarau, Switzerland). [5]

CZAKOW, J. (1962) Spectral analysis by the powder sifting method. IV, Sifting technique by means of an electrode with an exchangeable pressed sieved bottom, *Polish Rept. PAN/IBJ* 370/VIII. [Cf. *Spectrochem. Abstracts*, Vol. 9, 3443, (Hilger and Watts Ltd, London). [5]

CZAKOW, J., and MINCZEWSKI, J. (1960) Spectral analysis with a sifter electrode. II, The influence of alkali metal halides on excitation, *Chem. Anal. (Warsaw)*, 5, 863 (Cf. *Anal. Abstr.*, 1961, 8, 3527). [5]

CZAKOW, J., and MINCZEWSKI, J. (1962a) Spectral analysis by the powder sifting method. III, Addition and division method, *Polish Rept. PAN/IBJ* 312/VIII. [Cf. *Spectrochem. Abstracts*, Vol. 9, 3505. (Hilger and Watts Ltd, London)]. [5]

CZAKOW, J., and MINCZEWSKI, J. (1962b) Überblick über verschiedene spektrographische Methoden zur Untersuchung von Nuklear-materialien, *Acta Chim. Acad. Sci. Hung.*, **30**, 395. [265]

CZAKOW, J., and GRZELAK, R. (1964) Die Anwendung einer kupfernen Siebelelektrode mit gepresstem Boden zur Spektralanalyse von Blei. *Emissionsspektroskopie*, p. 231 (Akademie-Verlag, Berlin). [5]

DANIEL, J. L. (1960) *A detailed study of the carrier concentration method of spectrochemical analysis* (Unclassified Research and Development Report U.S. A.E.C., HW-64299). [316]

DANIELSSON, A., SUNDKVIST, G., and LUND-GREN, F. (1959) The tape machine and some of its applications (Proc. VII Colloq. Spectros. Intern., Liege 1958), *Rev. Universelle Mines*, **15**, 358. [5]

DANIELSSON, A., LUNDGREN, F., and SUND-KVIST, G. (1959) The tape machine. I, A new tool for spectrochemical analysis. II, Applications using different kinds of isoformation. III, Notes on useful corrections in spectrochemical analysis with the tape technique, *Spectrochim. Acta*, **15**, 122, 126, 134. [5]

DAVIS, G. E. (1924) Coefficients of diffusion of certain alkali salt vapors in the Bunsen flame, *Phys. Rev.*, **24**, 383. [57]

DEMTRÖDER, W. (1962) Bestimmung von Oszillatorstärken durch Lebensdauermessungen der ersten angeregten Niveaus für die Elemente Ga, Al, Mg, Tl, und Na, *Z. Physik*, **166**, 42. [21, 24, 25, 174]

DIKHOFF, J. A. M. (1957). Temperaturbestimmungen im Gleichstromkohlebogen (Proc. VI Colloq. Spectros. Intern., Amsterdam 1956), *Spectrochim. Acta*, **11**, 162. [108, 110, 112]

DOERFFEL, K., and GEYER, R. (1964) Untersuchungen zur spektrochemischen Anregung pulverförmiger Stoffe. *Z. Anal. Chem.*, **200**, 411. [46, 224]

DOUGLAS, A. E., and ROUTLY, P. M. (1955) The spectrum of the CN molecule, *Astrophys. J., Suppl. Ser.*, **1**, 295. [321]

DRAWIN, H. W., and FELENBOK, P. (1965) *Data for Plasmas in Local Thermal Equilibrium* (Gauthier-Villars, Paris). [93, 162]

DUNKEN, H., MIKKELEIT, W., and KNIESCHE, W. (1962) Verwendung einer Hochfrequenz-Fackelentladung als Anregungsquelle für die Lösungsspektralanalyse, *Acta Chim. Acad. Sci. Hung.*, **33**, 67. [5]

DVORNIKA, I. V. and NAGIBINA, I. M. (1958) Determination of the degree of inhomogeneity in d.c. and a.c. arc discharges. *Opt. i Spektroskopiya*, **4**, 421 (in Russian). Translation in:

Optical Transition Probabilities, A Representative Collection of Russian Articles, 1932–1962 (Israel Program for Scientific Translations, Jerusalem, 1963. Published for the U.S. Department of Commerce and the National Science Foundation, Washington, D.C., O.T.S.-63-11135). [292]

EATHER, R. H. (1963) The silent and hissing d.c. electric arc, *Australian J. Phys.*, **16**, 228. [35]

EBERHAGEN, A. (1955a) Eine quantitative Untersuchung der Lenardschen Hohlflammen, *Z. Physik*, **143**, 312. [44, 127, 135, 144, 148, 173, 286]

EBERHAGEN, A. (1955b) Eine Methode zur experimentellen Bestimmung von Übergangs-wahrscheinlichkeiten, *Z. Physik*, **143**, 392. [21, 173, 232]

ECKER, G. (1961) Electrode components of the arc discharge, *Ergeb. Exakt. Naturw.*, **33**, 1. [35, 41, 48, 69, 77]

ECKER, G., and WEIZEL, W. (1956) Zustandssumme und effektive Ionisierungsspannung eines Atoms im Inneren des Plasmas, *Ann. Physik*, **17**(6), 126. [93, 162]

EDELS, H., HEARNE, K., and YOUNG, A. (1962) Numerical solutions of the Abel integral equation, *J. Math. Phys.*, **41**, 62. [140]

EHRLICH, G. (1964) Entwicklungstendenzen in der Spektralanalyse von Reinststoffen. *Emissionsspektroskopie*, p. 21 (Akademie-Verlag, Berlin). [265, 314]

EHRLICH, G., and REXER, E. (1964) Probleme bei der Analyse von Reinststoffe, *Wiss. Z. Techn. Hochsch. Chem., Leuna-Merseburg*, **6**, 207. [265]

EICHHOFF, H. J., and ADDINK, N. W. H. (1960). Untersuchungen zur Übertragbarkeit spektrochemischer Verfahren mit vollständiger Verdampfung. *Proc. VIII Colloq. Spectros. Intern., Luzern* 1959, p. 89 (Verlag H.R. Sauerländer und Co., Aarau, Switzerland). [112]

EICKE, H. F. (1962). Die Verbesserung einer Methode zur experimentellen Bestimmung von Übergangswahrscheinlichkeiten und ihre Anwendung auf das Spektrum des Barium, *Z. Physik*, **168**, 227. [21, 24, 25, 135, 173, 232)

EINSTEIN, A. (1917) Zur Quantentheorie der Strahlung, *Physikal. Z.*, **18**, 121. [17]

ELENBAAS, W. (1934) Die Quecksilber-Hochdruckentladung, *Physica*, **1**, 673. [69, 71]

ELENBAAS, W. (1951) *The High Pressure Mercury Vapour Discharge* (North-Holland Publ. Co., Amsterdam). [69]

EMBER, G., FERRON, J. R., and WOHL, K. (1962) Self-diffusion coefficients of carbon dioxide at 1180–1680°K, *J. Chem. Phys.*, **37**, 891. [56, 294]

EMBER, G., FERRON, J. R., and WOHL, K. (1964) A flow method for measuring transport properties at flame temperatures, *Am. Inst. Chem. Engrs J.*, **10**, 68. [56]

VON ENGEL, A. (1955) *Ionized Gases* (Clarendon Press, Oxford). [47,49,50,55,68,86,87,102]

VON ENGEL, A. and STEENBECK, M. (1934) *Elektrische Gasentladungen*, I, II (Springer-Verlag, Berlin). [76]

ENGEL'SHT, V. S., and SPEKTOROV, L. A. (1962) Transport processes in a d.c. arc, *Izv. Akad. Nauk SSSR, Ser. Fiz.*, **26**, 887; *Bull. Acad. Sci. USSR, Phys. Ser.*, **26**, 890. [279, 284]

EULER, J. (1956) Die axiale Temperaturverteilung im inneren der Anode des Kohlebogens und das Wärmeleitvermögen von Graphit bei hohen Temperaturen, *Ann. Physik* **18**(6), 345. [311, 312]

EULER, J. (1956/57) Der Grafit-Kohlebogen als Strahldichtestandard. *Sitzber. Heidelb. Akad. Wiss., Math.-Naturw.Kl.*, Abhandl. 4, 61. [15, 312, 313]

EXETER (1964) *Limitations of Detection in Spectrochemical Analysis. Papers presented at a conference held at the University of Exeter, July 1964* (Hilger and Watts Ltd, London). [264]

FARSKÝ, R., (1960) The use of special atmospheres in spectrographic analysis, *Hutnicke Listy*, **15**, 548 (in Czech), *Metallurgia, Manchr.*, 1963, **67**, 51 (in English). (Cf. *Anal. Abstr.* 1963, **10**, 5006; 1961, **8**, 868.) [87]

FEAST, M. W. (1949) On the Schumann-Runge O_2 bands emitted at atmospheric pressure, *Proc. Phys. Soc. (London)*, **62A**, 114. [322]

FELDMANN, C., and WITTELS, M. K. (1957) Sample transport and temperature studies in porous-cup discharges, *Spectrochim. Acta*, **9**, 19. [149]

FINKELNBURG, W., and MAECKER, H. (1956) Elektrische Bögen und thermisches Plasma. *Handbuch Physik*, Vol. 22, pp. 254–444 (Springer-Verlag, Berlin, Göttingen, Heidelberg). [30, 35–39, 41, 43, 48, 53, 59, 60, 69, 72, 73, 79, 80, 82, 84, 86, 102, 103, 202]

FISHMAN, I. S. (1962) Relationship between the self-absorption of spectral lines and their total intensities, and the *b* values for several lines excited in the vapour jet of a spark discharge, *Opt. i Spektroskopiya*, **13**, 630; *Opt. Spectry (USSR)*, **13**, 360. [89]

FISHMAN, I. S., SHAIMANOV, I. Sh., and ILIN, G. G. (1963) Some experimental relations of the integral characteristics of radiation in an arc, *Opt. i Spektroskopiya*, **15**, 595; *Opt. Spectry (USSR)*, **15**, 323. [89]

FISQ, J. B. (1936) Theory of the scattering of slow electrons by diatomic molecules, *Phys. Rev.*, **49**, 167. [51]

FISQ, J. B., (1937) On the cross-sections of Cl_2 and N_2 for slow electrons, *Phys. Rev.*, **51**, 25. [51]

FOSTER, E. W. (1964) The measurement of oscillator strengths in atomic spectra, *Rept Progr. Phys.*, **27**, 469. [18]

FOWLER, R. G. (1956) Radiation from low pressure discharges. *Handbuch Physik*, Vol. 22, pp. 209–253. (Springer-Verlag, Berlin, Göttingen, Heidelberg). [87]

FRANCIS, G. (1956) The glow discharge at low pressure. *Handbuch Physik*, Vol. 22, pp. 53–208 (Springer-Verlag, Berlin, Göttingen, Heidelberg). [39]

FRANK, P., and VON MISES, R. (1961) *Die Differential- und Integralgleichungen der Mechanik und Physik*, Vol. 1, p. 482 (Friedr. Vieweg und Sohn, Braunschweig). [135]

FRATKIN, Z. G. (1960) Connection between sensitivity of the determination and parameters of spectrographic apparatus, *Zavodsk. Lab.*, **26**, 971; *Ind. Lab. (USSR)*, **26**, 1034. [264]

FRIE, W. (1963a) Entmischungseffekte bei Gemischen ionisierender Atomgase, *Z. Physik*, **172**, 99. [107]

FRIE, W. (1963b) Zur Auswertung der Abelschen Integralgleichung, *Ann. Physik.* **10**(7), 332. [139, 140]

FRIE, W., and MAECKER, H. (1961) Massentrennung durch Diffusion reagierender Gase, *Z. Physik*, **162**, 69. [107]

FRIEDRICH, J. (1959) Zur Auswertung seitlicher Beobachtungen an zylindrischen Bögen, *Ann. Physik.* **3**(7), 327. [135]

FRISQUE, A. J. (1957) Internally standardized general spectrographic method, *Anal. Chem.*, **29**, 1277. [272]

FRISQUE, A. (1960) Causes and control of matrix effects in spectrographic discharges, *Anal. Chem.*, **32**, 1484. [108]

FUCKS, W., BOHN, W. L., HEINRICH, G., and PLATZ, P. (1962) Elektronendichten und Elektronentemperaturen bei Verdichtungsstössen, *Z. Physik*, **170**, 409. [60]

DE GALAN, L. (1965a) Particle distribution in the d.c. carbon arc. Ph.D. Thesis, University of Amsterdam. [16, 44, 45, 108, 111, 143, 144, 154, 155, 173, 193, 194, 232, 237, 269, 279, 280, 285, 288, 307, 314, 326]

DE GALAN, L. (1965b) An experimental value for the dissociation energy of boron oxide, *Physica*, **31**, 1286. [288, 326]

DE GALAN, L. (1966a) Distribution of elements in the d.c. carbon arc, *J. Quant. Spectry Radiative Transfer*, **5**, 735, 1965. [108, 143, 144, 154, 155, 173, 193, 194, 232, 237, 279, 280, 285, 288, 307]

DE GALAN, L. (1966b) The possibility of a truly absolute method of spectrographic analysis, *Anal. Chim. Acta.*, **34**, 2. [16, 108]

DE GALAN, L., and BOUMANS, P. W. J. M. (1965) Optimum excitation conditions in spectrographic analysis with the d.c. carbon arc, *Z. Anal. Chem.*, **214**, 161. [193]

GARRETT, W. R., and MANN, R. A. (1964) Elastic scattering of slow electrons from alkali atoms, *Phys. Rev.*, **135**, A580. [52]

GARSTANG, R. H. (1962) Transition probabilities in Mg I, Zn I, Cd I, and Hg I, *J. Opt. Soc. Am.*, **52**, 845. [175]

GAYDON, A. G. (1953) *Dissociation Energies and Spectra of Diatomic Molecules* (Chapman and Hall, London). [325]

GEGUS, E. (1964a) Eine spektrochemische Lösungsmethode zur Analyse von Einschlüssen in Stählen. *Emissionsspektroskopie*, p. 307 (Akademie-Verlag, Berlin). [5]

GEGUS, E. (1964b). Interelementareffekte bei der Lösungsspektrochemischen Analyse nichtleitender Stoffe, *Mikrochim. Ichnoanal. Acta*, 807. [5]

GERBATSCH, R., and SCHOLZE, H. (1963) Einfluss des Auflösungsvermögens auf die Nachweisempfindlichkeit in der Spektralanalyse, *Jenaer Rundschau*, **8**, 89. [264]

GERLACH, W. (1925) Zur Frage der richtigen Ausführung und Deutung der 'quantitativen Spektralanalyse', *Z. Anorg. Allg. Chem.*, **142**, 383. [4]

GERLACH, W., and SCHWEITZER, E. (1930) *Die Chemische Emissionsspektralanalyse* (Leopold Voss, Leipzig). [4, 209]

GINSEL, L. A. (1933) Massentransport in Lichtbögen und Flammen. Optische Bestimmung der Alkali-Atomradien. Ph.D. Thesis, University of Utrecht. [57, 279, 294]

GINSEL, L. A., and ORNSTEIN, L. S. (1933) Die optische Bestimmung der Diffusionskonstante für Natrium, *Z. Physik*, **84**, 276. [57, 294]

GINZBURG, V. L., and GLUKHOVETSKAYA, N. P. (1962a) The dependence of spectral line intensities on the effective ionization potential of the arc, *Opt. i Spektroskopiya*, **12**, 344; *Opt. Spectry (USSR)*, **12**, 190. [228, 231]

GINZBURG, V. L., and GLUKHOVETSKAYA, N. P. (1962b) Notes on the article by O. P. Semenova and M. A. Levchenko 'Dependence of the effective ionization potential on the concentration of easily ionized impurities in the arc discharge', *Opt. i Spektroskopiya*, **13**, 881; *Opt. Spectry (USSR)*, **13**, 507. [228, 231]

GINZBURG, V. L., GLUKHOVETSKAYA, N. P., and LERNER, L. A. (1963) Fluoridation of samples in spectral analysis, *Zavodsk. Lab.* **29**, 684; *Ind. Lab. (USSR)*, **29**, 729. [208]

GLENNON, B. M., and WIESE, W. L. (1962) *Bibliography on Atomic Transition Probabilities* (N.B.S. Monograph 50. Washington, D.C.). [18, 19, 23]

GOL'DFARB, B. H., and IL'INA, E. V. (1961) Determination of the mean time spent by the atoms, and of the diffusion coefficient in the plasma of the arc discharge, *Opt. i Spektroskopiya*, **11**, 445; *Opt. Spectry (USSR)*, **11**, 243. [34, 58, 279, 280]

GOL'DFARB, V. M., and IL'INA, E. V. (1964) The broadening of cesium lines in the plasma of a direct-current arc, *Opt. i Spektroskopiya* **17**, 302; *Opt. Spectry (USSR)*, **17**, 159. [60]

GOLLING, E. (1957) Über den Untergrund im Spektrum des Kohlebogens und seine Bedeutung für die Spektralanalyse (Proc. VI Colloq. Spectros. Intern., Amsterdam 1956), *Spectrochim. Acta*, **11**, 125 [46, 202]

GORBACHEVA, N. S., and PREOBRAZHENSKII, N. G. (1963). Asymmetric self-reversible spectral line profile as a source of information on plasma properties, *Opt. i Spektroskopiya*, **15**, 453; *Opt. Spectry (USSR)*, **15**, 244. [89]

GRAY, W. T. (1935) Thermal equilibrium of the gas in the d.c. carbon arc, *Phys. Rev.*, **48**, 474. [121]

GRECHIKHIN, L. I. (1962) Broadening of the 4982·8 Å sodium line in the direct-current arc jet, *Opt. i Spektroskopiya*, **13**, 578; *Opt. Spectry (USSR)*, **13**, 325. [89]

GRECHIKHIN, L. I. (1963) Effect of polarity on the physical characteristics of the plasma stream of an arc generator, *Zh. Analit. Khim.*, **18**, 20, *J. Anal. Chem. USSR*, **18**, 16. [5]

GRECHIKHIN, L. I., and SHIMANOVITCH, V. D. (1962) Investigation of the jet of a plasma generator, *Opt. i Spektroskopiya*, **13**, 626; *Opt. Spectry (USSR)*, **13**, 358. [5]

GREENFIELD, S., JONES, I. L. and BERRY, C. T. (1964) High-pressure plasmas as spectroscopic emission sources, *Analyst*, **89**, 713. [5]

GRIEM, H. R. (1962) Plasma spectroscopy, *Proc. 5th Intern. Conf. Ionization Phenomena Gases*, Munich 1961, **2**, 1857 (North-Holland Publ. Co., Amsterdam). [60, 93, 102, 162, 202]

GRIEM, H. R. (1964a) Stark broadening calculations, *J. Quant. Spectry Radiative Transfer*, **4**, 669. [60]

GRIEM, H. R. (1964b) *Plasma Spectroscopy* (McGraw-Hill Book Publ. Comp., New York, Toronto, London). [93, 102, 162, 202]

GRIEM, H. R., KOLB, A. S., and SHEN, K. J. (1959) Stark broadening of hydrogen lines in plasma, *Phys. Rev.*, **116**, 4 (Cf. Naval Res. Rept 5455, 1960, U.S.N.R.L., Washington, D.C.). [60]

13

GRIMALDI, F. S., and HELZ, A. W. (1961) Trace element sensitivities, *U.S. Geol. Surv. Profess. Papers* **424-D**, 388. [265]

GUREVICH, D. B., and PODMOSHENSKII, I. V. (1963) The relationship between the excitation temperature and the gas temperature in the positive column of an arc discharge, *Opt. i Spektroskopiya*, **15**, 587, *Opt. Spectry (USSR)*, **15**, 319. [82, 83, 87, 88]

GVOSDOVER, S. D. (1937) The mobility and mean free path of electrons in the positive column, *Physikal. Z. USSR*, **12**, 164. [53]

HAGENAH, W. (1950) Das Strömungsfeld im freien Kohlebogen, *Z. Physik*, **128**, 279. [34, 72]

HAHN-WEINHEIMER, P. (1962) Quantitative spektrochemische Bestimmung von Spurenelementen in Karbonatgesteinen nach der Doppelbogenmethode (Shaw, Joensuu und Ahrens) und der Verdampfung aus der Ringrinne (Holdt). *Proc. IX Colloq. Spectros. Intern.*, Lyons 1961, **2**, 594 (Groupement pour l'Avancement des Méthodes Spectrographiques, Paris). [112]

HAMEL, G. (1949) *Integralgleichungen* (Springer-Verlag, Berlin, Göttingen, Heidelberg). [135]

HANSEN, C. F. (1964) Combined Stark and Doppler line broadening, *J. Opt. Soc. Am.*, **54**, 1198. [60]

HARRINGTON, F. D. (1963) Time-resolution spectroscopy. *Developments in Appl. Spectry* **2**, 162 (Plenum Press, New York, N.Y.). [151]

HEFFERLIN, R. (1965) Seven density determinations in an atmospheric manganese arc, *J. Quant. Spectry Radiative Transfer*, **5**, 425. [292]

HEFFERLIN, R., and GEARHART, J. (1964) Laboratory high-excitation relative *f*-values for manganese, *J. Quant. Spectry Radiative Transfer*, **4**, 9. [21, 139, 233]

HEGEMANN, F., and SCHÖNTAG, A. (1953) Fehlerquellen bei der quantitativen spektrochemischen Mineralanalyse im Kohlebogen, ihre Ursachen und ihre Beseitigung, *Z. Wiss. Phot. Photophysik Photochem.*, **48**, 170. [46, 224, 310]

VAN DER HELD, E. F. M., and MIESOWICZ, M. (1937) Messung der Diffusion von Metallatomen in Gasen bei Zimmertemperatur auf optischem Wege, *Physica*, **4**, 559. [57]

HELLER, G. (1935) Dynamical similarity laws of the mercury high pressure discharge, *Physics*, **6**, 389. [69, 71]

VAN HENGSTUM, J. P. A. (1955) Spectral measurement of transition probabilities in the cadmium atom. (In Dutch) Ph.D. Thesis, University of Utrecht. [19, 20]

VAN HENGSTUM, J. P. A., and SMIT, J. A. (1956) Measurement of 'optical' transition probabilities of Cd, *Physica*, **22**, 86. [9, 19, 20]

HERZBERG, G. (1944) *Atomic Spectra and Atomic Structure* (Dover Publications, Inc., New York, N.Y.). [25]

HERZBERG, G. (1950) *Molecular Spectra and Molecular Structure. I, Spectra of Diatomic Molecules* (D. van Nostrand Company, Inc., Princeton, N.J.). [117, 120, 318, 325, 326]

HERZBERG, G. (1957) Recent laboratory investigations on molecules of astronomical interest, *Mem. Soc. Roy. Sci. Liege*, **18**, Ser. 4, 397. [325]

HESS, H., KLOSS, H.-G., RADEMACHER, K. and SELIGER, K. (1962a and b) Vergleich zwischen einem Verfahren zur Bestimmung von Bogentemperaturen mit Hilfe von Stosswellen und einer spektroskopischen Methode, *Beitr. Plasmaphysik*, **2**, 171; *Monatsber. Deut. Akad. Wiss. Berlin*, **4**, 510. [117]

HILL, R. A. (1964) Tables of electron density as a function of the half-width of Stark broadened hydrogen lines, *J. Quant. Spectry Radiative Transfer*, **4**, 857. [60]

HIRSCHFELDER, J. O., CURTISS, C. F., and BIRD, R. B. (1954) *Molecular Theory of Gases and Liquids* (John Wiley and Sons, Inc., New York; Chapman and Hall Ltd, London). [56]

HOBBS, D. J., and SMITH, D. M. (1966) The evaluation and application of the 'limit of detection' in quantitative spectrographic analysis, *Can. Spectry*, **11**, 5. [264]

HÖCKER, K. H. (1946) Die Bedeutung des Stickoxydes für die Trägerdichte in einer Niederstrombogensäule in Luft, *Z. Naturforsch.*, **1**, 384. [63]

HÖCKER, K. H., and BEZ, W. (1955) Theorie des Anodenfalls. II, Möglichkeiten und Grenzen der Feldionisierung. III, Äquipotential-flächen vor der Lichtbogenanode, *Z. Naturforsch.*, **10a**, 706, 714. [35]

HÖCKER, K. H., and FINKELNBURG, W. (1946) Theorie der Hochstrombogensäule, *Z. Naturforsch.*, **1**, 305. [69, 72]

HÖCKER, K. H., and SCHULZ, P. (1949) Über die Wärmeleitung in der Hochstrombogensäule, *Z. Naturforsch.*, **4a**, 266. [72]

HODGMAN, C. D., WEAST, R. C., and SELBY, S. M. (1957-1958) *Handbook of Chemistry and Physics*, 39th ed. (Chemical Rubber Publishing Comp., Cleveland, Ohio). [311]

HOENS, M. F. A., and SMIT, J. A. (1957). Stabilization of the vaporization of the filled cathode in a d.c. carbon arc (Proc. VI Colloq. Spectros. Intern., Amsterdam 1956), *Spectrochim. Acta*, **11**, 192. [30, 312]

HOLDT, G. (1962) A study of the influence of buffers on the accuracy and sensitivity of spectrochemical results, *Appl. Spectry*, **16**, 96. [6, 208]

HOLDT, G., MARITZ, F., and VAN DER MARK, C. (1962) Zur Wirksamkeit verschiedener Puffer bei schneller und langsamer Verdampfung. *Proc. IX Colloq. Spectros. Intern.*, Lyons 1961, **2**, 572 (Groupement pour l'Avancement des Méthodes Spectrographiques, Paris). [208]

HOLLANDER, TJ. (1964). Self-absorption, ionization and dissociation of metal vapours in flames. Ph.D. Thesis, University of Utrecht. [326]

HOLLANDER, TJ., KALFF, P. J., and ALKEMADE, C. TH. J. (1964) Dissociation energies and excitation levels of alkaline-earth oxides, *J. Quant. Spectry Radiative Transfer*, **4**, 577. [326]

HÖRMANN, H. (1935) Temperaturverteilung und Elektronendichte in frei brennenden Lichtbögen, *Z. Physik*, **97**, 539. [32, 43, 44, 135, 144, 147]

TER HORST, D. T. J., and KRIJGSMAN, C. (1934) Measurement of the arc temperature from C_2 bands. The transition probability of vibrational transitions, *Physica*, **1**, 114. [121]

HUBAUX, A., and SMIRIGA-SNOECK, N. (1964) On the limit of sensitivity and the analytical error, *Geochim. Cosmochim. Acta*, **28**, 1199. [264]

HULDT, L. (1948a) Berechnung des chemischen Gleichgewichts und der Temperaturionisation der Luft im elektrischen Lichtbogen, *Arkiv Math. Astron. Fysik*, **34A**, nr 30. [63]

HULDT, L. (1948b) Eine spektroskopische Untersuchung des elektrischen Lichtbogens und der Azetylen-Luftflamme. Ph.D. Thesis, University of Uppsala. [63, 116, 117, 135, 144, 173, 190, 228, 286]

HULDT, L. (1948c) Spektroskopische Messung der Temperatur und des Elektronendruckes im Kohlelichtbogen, *Arkiv. Mat. Astron. Fysik*, **36**, nr 3. [173, 228]

HULDT, L. (1955) Zur Deutung spektroskopischer Temperaturmessungen, *Spectrochim. Acta*, **7**, 264. [150]

HULPKE, E., PAUL, E., and PAUL, W. (1964) Bestimmung von Oszillatorenstärken durch Lebensdauermessungen der ersten angeregten Niveaus für die Elemente Ba, Sr, Ca, In, und Na, *Z. Physik*, **177**, 257. [21]

HURWITZ, J. K. (1959) Slopes of working curves in emission spectrometric analysis of certain silicates, *Appl. Spectry*, **13**, 113. [87]

IL'INA, E. V., and GOL'DFARB, V. M. (1962) Determination of the mean imprisonment time of atoms and the diffusion coefficient in the plasma of an arc discharge, *Izv. Akad. Nauk SSSR, Ser. Fiz.*, **26**, 939; *Bull. Acad. Sci. USSR, Phys. Ser.*, **26**, 943. [58, 279, 280]

INGHRAM, M. G., CHUPKA, W. A., and BERKOWITZ, J. (1957) Dissociation energies from thermodynamic equilibria studied with a mass spectrometer, *Mem. Soc. Roy. Sci. Liege*, **18**, Ser. 4, 513. [325]

JAHN, R. E. (1962) Spectroscopic measurements of the temperature of plasma jets. *Proc. 5th Intern. Conf. Ionization Phenomena Gases, Munich 1961*, **1**, 955 (North-Holland Publ. Comp., Amsterdam). [5]

JAHN, R. E. (1963) Temperature distribution and thermal efficiency of low power arc-heated plasma jets, *Brit. J. Appl. Phys.*, **14**, 585. [5]

JANDA, I., SCHAUSBERGER, I., and SCHROLL, E. (1963) Beitrag zur emissionsspektrographischen Spurenanalyse in Uranoxyd, *Mikrochim. Ichnoanal. Acta*, **122** [316]

JEANS, J. H. (1925) *The Dynamical Theory of Gases*, 4th ed. (Cambridge University Press, Cambridge). [48, 50]

JECHT, U., and KESSLER, W. (1963) Beobachtungen und Untersuchungen an HF-Plasmaflammen, *Z. Anal. Chem.*, **198**, 27. [5]

JECHT, U., and KESSLER, W. (1964) Über den Anregungsmechanismus einer HF-Fackelentladung bei 2400 MHz, *Z. Physik*. **178**, 133. [5]

JENKINS, F. A., and WHITE, H. E. (1957). *Fundamentals of Optics*, 3rd ed. (McGraw-Hill Book Company, Inc., New York, Toronto, London). [8]

JOENSUU, O. I., and SUHR, N. H. (1962) Spectrochemical analysis of rocks, minerals, and related materials, *Appl. Spectry*, **16**, 101. [272]

JUNKES, J., and SALPETER, E. W. (1958) Effective line widths in photographic photometry, *Ric. Spettroscopiche, Lab. Astrofis. Specola Vaticana*, **2**, 205. [9]

JUNKES, J., and SALPETER, E. W. (1959, 1961, 1963) Linienbreiten in der photographischen Spektralphotometrie. I, Zur Geschichte der photographischen Photometrie in Astronomie und Spektroskopie. II, Das Optische Linienprofil. III, Das photographische Profil und seine Messung, *Ric. Spettroscopiche, Lab. Astrofis. Specola Vaticana*, **2**, 221, 255, 485. [9, 89]

KAISER, H. (1947) Die Berechnung der Nachweisempfindlichkeit, *Spectrochim. Acta*, **3**, 40. [264, 265]

KAISER, H. (1949) Bemerkungen zum Verfahren der homologen Linienpaare, *Z. Naturforsch.*, **4a**, 565. [4]

KAISER, H. (1964) Zur Theorie der Eichfunktion bei der spektrochemischen Analyse, *Optik*, **21**, 309. [16, 264]

KAISER, H. (1965) Zum Problem der Nach-weisgrenze, Z. Anal. Chem., 209, 1. [264]

KAISER, H., LAQUA, K., and SCHIRRMEISTER, H. (1966) Das Verhalten stromführender und stromfreier Plasmen bei der Zufuhr von Aerosolen in Hinblick auf die Spektralanalyse. Proc. XII Colloq. Spectros. Intern., Exeter 1965, p. 205. (Hilger and Watts Ltd, London). [5]

KAISER, H., MASSMANN, H., and HAGENAH, W. D. (1962) Berechnung der Nachweisgrenzen bei photoelektrischen Analysenverfahren (Emission und Röntgenfluoreszenz). Proc. IX Colloq. Spectros. Intern., Lyons 1961, 3, 479 (Groupement pour l'Avancement des Méthodes Spectrographiques, Paris). [264]

KAISER, H., and SPECKER, H. (1956) Bewertung und Vergleich von Analysenverfahren, Z. Anal. Chem., 149, 46. [264]

KALFF, P. J., HOLLANDER, TJ., and ALKEMADE, C. TH. J. (1965) Flamephotometric determination of the dissociation energies of the alkaline-earth oxides, J. Chem. Phys., 43, 2299. [326]

KAUFMANN, W. (1900) Elektrodynamische Eigentümlichkeiten leitender Gase, Ann. Physik 2(4), 158. [36]

KEMP, N. (1958) Experimentally tested procedures useful in a general spectrochemical method of analysis, Belg. Chem. Ind. 23, 615 (in Dutch). [7, 272]

KENNARD, E. H. (1938) Kinetic Theory of Gases (McGraw-Hill Book Comp., Inc., New York, London). [48, 55]

KERSTEN, J. A. H. (1941) The relative transition probabilities in the spectrum of magnesium. (In Dutch) Ph.D. Thesis, University of Utrecht. [144, 151]

KERSTEN, J. A. H., and ORNSTEIN, L. S. (1941) The relative transition probabilities in the spectrum of magnesium, Physica, 8, 1124. [144]

KESSLER, W., and VON DOBENECK, D. (1963) Kritische Betrachtungen zur Bestimmung von Übergangswahrscheinlichkeiten aus der Linienintensität im Kupferbogen, Z. Anal. Chem., 198, 35. [26]

KESSLER, W., JECHT, U., and ZOTTMANN, R. (1965) Einfluss von Korngrösse und chemische Natur einer Probensubstanz auf die Intensität der von einem Funkenplasma emittierten Spektrallinien, Z. Anal. Chem., 208, 161. [5]

KIBISOV, G. I., and ANTROPOV, N. P. (1962) Elimination of the effect of the composition of the material on the results of quantitative spectrographic analysis, Zh. Analit. Khim., 17, 155; J. Anal. Chem. USSR, 17, 154. [5]

KIBISOV, G. I., and KUBASOVA, N. B. (1961) The effect of the degree to which test material is ground on the results of quantitative spectrographic analysis by the blast method, Zh. Analit. Khim., 16, 660; J. Anal. Chem. USSR, 16, 651. [5]

KING, L. A. (1954) The positive column of high- and low-current arcs, Nature, 174, 1008. [69, 72]

KING, L. A. (1956) The positive column of high- and low-current arcs, Appl. Sci. Res., 5, Sect. B, 189. [69, 72, 73]

KING, L. A. (1957) Theoretical calculations of arc temperatures in different gases (Proc. VI Colloq. Spectros. Intern., Amsterdam 1956,) Spectrochim. Acta 11, 152. [69, 72]

KING, L. A. (1962) The voltage gradient of the free burning arc in air or nitrogen. Proc. 5th Intern. Conf. Ionization Phenomena Gases, Munich 1961, 1, 871 (North-Holland Publ. Comp., Amsterdam). [43]

KING, R. B. (1947) Relative gf-values for lines of V I, Astrophys. J., 105, 376. [106]

KING, R. B. (1963) The measurement of absolute oscillator strengths for lines of neutral atoms, J. Quant. Spectry Radiative Transfer, 3, 299. [18, 21]

KISCHL, K. H., and WILHELM, J. (1960) Zur Deutung des Steenbeckschen Minimumprinzips, Monatsber. Deut. Akad. Wiss. Berlin, 2, 322. [77]

KISCHL, K. H., and WILHELM, J. (1960/61) Beitrag zum Problem der Entscheidbarkeit der Gültigkeit eines Prinzips extremalen Energieverbrauchs beim Stromdurchgang durch einen Leiter, Beitr. Plasmaphysik, 1, 11. [77]

KITAYEVA, V. F., KOLESNIKOV, V. N., OBUKHOV-DENISOV, V. V., and SOBOLEV, N. N. (1962) Structure of the column in an argon arc. I, Local electrical characteristics of the column, Zh. Tekhn. Fiz., 32, 1084; Soviet Phys. Tech. Phys., 1963, 7, 796. [60]

KITAYEVA, V. F., OBUKHOV-DENISOV, V. V., and SOBOLEV, N. N. (1962) Concentration of charged particles in the plasma of an arc burning in an atmosphere of argon and helium, Opt. i Spektroskopiya, 12, 178; Opt. Spectry (USSR), 12, 94. [60, 87]

KITAYEVA, V. F., and SOBOLEV, N. N. (1962) On the broadening of hydrogen lines in the plasma of arc and shock tube. Proc. 5th Intern. Conf. Ionization Phenomena Gases, Munich 1961, 2, 1897 (North-Holland Publ. Comp., Amsterdam). [60]

KNIGHT, H. T., and RINK, J. P. (1961) Dissociation energy of cyanogen and related quantities by X-ray densitometry of shock waves, J. Chem. Phys., 35, 199. [321]

KNÍŽECK, M., and PROVAZNÍK, J. (1961) Problem of blanks in the analysis of trace elements, *Chem. Listy*, **55**, 389 (Cf. *Anal. Abstr.*, 1961, **8**, 4040. [265]

KOCH, O. G., and KOCH-DEDIC, G. A. (1964) *Handbuch der Spurenanalyse* (Springer-Verlag, Berlin, Göttingen, Heidelberg). [265]

KOHN, H., and GUCKEL, M. (1924) Untersuchungen am Kohlelichtbogen: Dampfdruckbestimmungen des Kohlenstoffs, *Z. Physik*, **27**, 305. [43]

KOLB, A. C., and GRIEM, H. (1958) Theory of line broadening in multiplet spectra, *Phys. Rev.*, **111**, 514. [60]

KOLESNIKOV, V. N. and SOBOLEV, N. N. (1962) Structure of the column in an argon arc. II, Radius of the column and shape of the radial distributions, *Zh. Tekhn. Fiz.*, **32**, 1090; *Soviet Phys. Tech. Phys.*, 1963, **7**, 801. [62]

KOLESNIKOV, V. N. and SOBOLEV, N. N. (1963) The establishment of the thermal equilibrium for d.c. arc plasma in inert gases. (XI Colloq. Spectros. Intern., Belgrade 1963). [87]

KOLLATH, R. (1930). Übersicht über den Stand der Wirkungsquerschnitt-Forschung, *Physikal. Z.*, **31**, 985. [49, 51, 52]

KOPFERMANN, H., and WESSEL, G. (1951) Die absoluten f-Werte der Fe-I-Resonanzlinien $\lambda = 3070$ Å und $\lambda = 3737$ Å, *Z. Physik*, **130**, 100. [21]

KOROLEV, V. V., and KVARATSKHELI, Yu. K. (1961) The plasmatron (plasma jet) as a light source for spectroscopy, *Opt. i Spektroskopiya*, **10**, 398; *Opt. Spectry (USSR)*, **10**, 200. [5]

KOROLEV, V. V., and VAINSHTEIN, E. E. (1958) The use of an impulse source for spectra excitation during the spectrographic analysis of silicates, *Zh. Analit. Khim.*, **13**, 627; *J. Anal. Chem. USSR*, **13**, 707. [284]

KOROLEV, V. V., and VAINSHTEIN, E. E. (1959) Application of plasma generator as a source of excitation in spectral analysis, *Zh. Analit. Khim.* **14**, 658; *J. Anal. Chem. USSR*, **14**, 731. [5]

KOROLEV, V. V., and VAINSHTEIN, E. E. (1960) Reasons for the enhanced precision of spectrographic analysis on using an impulse excitation source, *Zh. Analit. Khim.*, **15**, 413; *J. Anal. Chem. USSR*, **15**, 473. [284]

KOSHELEVA, M. M., and KUZNETSOVA, T. I. (1959) The use of certain additives in the determination of rare elements by spectrographic means, *Zavodsk. Lab.* **25**, 964; *Ind. Lab. (USSR)*, **25**, 1006. [208]

KÖSTLIN, H. (1964) Ein Beitrag zur experimentellen Bestimmung von Übergangswahrscheinlichkeiten in der Atomhülle und deren Messung im Spektrum von Kalzium, *Z. Physik*, **178**, 200. [21, 51, 128, 144, 148, 173, 232]

KRANZ, E. (1963) Aufbau und Eigenschaften eines verunreinigungsfreien Plasmabrenners für spektroskopische Zwecke, *Monatsber. Deut. Akad. Wiss. Berlin*, **5**, 308. [5]

KRANZ, E. (1964) Aufbau und Eigenschaften eines verunreinigungsfreien Plasmabrenners für spectroskopische Zwecke. *Emissionsspektroskopie*, p. 160 (Akademie-Verlag, Berlin). [5]

KRANZ, E. (1965) Die Eigenschaften und Anwendungen eines neuentwickelten Plasmabrenners für spektroskopische Zwecke. *Proc. XII Colloq. Spectros. Intern.*, *Exeter* 1965, p. 574 Hilger and Watts Ltd, London). [5]

KREMPL, H. (1962) Über das thermodynamische Gleichgewicht in Funkenplasmen, *Z. Physik*, **167**, 302. [79, 150]

KRINBERG, I. A. (1964) On the theory of the electric arc, burning in conditions of natural convection, I, *Zh. Tekhn. Fiz.*, **34**, 888; *Soviet Phys. Tech. Phys.* **9**, [69]

KROONEN, J., and VADER, D. (1963) *Line Interference in Emission Spectroscopic Analysis. A general emission spectrographic method including sensitivities of analytical lines and interfering lines* (Elsevier Publ. Comp., Amsterdam, London, New York). [272]

KRUITHOF, A. M. (1943a) An investigation of the validity of Saha's equation in the electric arc. (In Dutch) Ph.D. Thesis, University of Utrecht. [86, 127, 128, 135–137, 144, 151, 173, 232]

KRUITHOF, A. M. (1943b) The radial course of the population of the energy levels in the column of an electric arc, and some relative transition probabilities in the spectra of Ba I and Ba II, *Physica*, **10**, 493. [127, 135, 144]

KRUITHOF, A. M., and SMIT, J. A. (1944) The appropriateness of the Saha equation in the electric arc, *Physica*, **11**, 129 (in Dutch). [86, 127, 135, 144, 173, 232]

KUHN, H. G. (1962) *Atomic Spectra* (Longmans, London). [25]

KUNG-SOO CHANG (1963). Carrier effect in spectrographic analysis. Part I. Chemical reaction of chloride carrier in the electrode, *Acta Chim. Sinica*, **29**, 414. [316]

LADENBURG, R. (1921) Die quantentheoretische Deutung der Zahl der Dispersionselektronen, *Z. Physik*, **4**, 451. [23]

LANDOLT-BÖRNSTEIN (1950) *Zahlenwerte und Funktionen*, Vol. 1, Part 1, 6th ed. (Springer-Verlag, Berlin, Göttingen, Heidelberg). [49–52]

LAQUA, K., and HAGENAH, W.-D. (1963). Critical review of methods and results of time-resolved spectroscopy. *Proc. X Colloq. Spectros. Intern.*, *Maryland* 1962, p. 91 (Spartan Books, Washington, D.C.). [150, 151]

LARENZ, R. W. (1951) Über ein Verfahren zur Messung sehr hoher Temperaturen in nahezu durchlässigen Bogensäulen, *Z. Physik*, **129**, 327, 343. [150, 255]

LAUWERIER, H. A. (1956) Diffusion from a point source into a space bounded by an impenetrable plane, *Appl. Sci. Res.*, 6, Sect. A, 197. [279, 304]

LAWRENCE, G. M., LINK, J. K., and KING, R. B. (1965) The absolute oscillator strengths of lines in the spectra of ten elements, *Astrophys. J.*, **141**, 293. [21]

LEME, A. B. P. (1938) *Bol. Museu. Nacl.*, **9**, 3. [5]

LENARD, P. (1903) Über den elektrischen Bogen und die Spektren der Metalle, *Ann. Physik*, **11**(4), 636. [33, 126, 279]

LENARD, P. (1905) Über die Lichtemissionen der Alkalimetalldämpfe und Salze, und über die Zentren dieser Emissionen, *Ann. Physik*, **17**(4), 197. [33, 126, 279]

LEUCHS, O. (1950) Chemische Vorgänge in den Kohleelektroden bei spektrochemischen Untersuchungen und ihr Einfluss auf dem Nachweis der einzelnen Elemente, *Spectrochim. Acta*, **4**, 237. [310, 312, 315]

LILEY, P. E. (1959) Survey of recent work on the viscosity, thermal conductivity and diffusion of gases and gas mixtures. *Symp. Thermal Properties, Lafayette, Ind.* 1959 (McGraw-Hill Book Comp., Inc., New York, Toronto, London). [56]

LINCOLN, A. J., and KOHLER, J. C. (1964) Spectrographic determination of trace impurities in high-purity gold. *Developments in Applied Spectroscopy* 3, p. 265 (Plenum Press, New York N.Y.). [5]

VAN LINGEN, D. (1936a) Investigations in the copper arc. (In Dutch) Ph.D. Thesis, University of Utrecht. [116, 117]

VAN LINGEN, D. (1936b) Bestimmung von Übergangswahrscheinlichkeiten im Kupferspektrum und ein Studium des Metallbogens, *Physica*, **9**, 977. [116, 117]

LOCHTE-HOLTGREVEN, W. (1955) Ionization measurements at high temperatures. *Temperature, its Measurement and Control in Science and Industry, Symposium, Washington* 1954, p. 413 (Reinhold Publ. Comp., New York, N.Y.). [78]

LOCHTE-HOLTGREVEN, W. (1956a) Über den auf einer Kohleanode umlaufenden Lichtbogenansatz, *Z. Physik*, **145**, 451. [35]

LOCHTE-HOLTGREVEN, W. (1956b) The thermally excited plasma in the column of an electric arc, *Appl. Sci. Res.*, **5** Sect. B, 182. [60]

LOCHTE-HOLTGREVEN, W. (1957) Die Abhängigkeit der Emission eines Lichtbogens von äusseren Einflüssen (Proc. VI Colloq. Spectros.

Intern., Amsterdam 1956), *Spectrochim. Acta*, **11**, 111. [202]

LOCHTE-HOLTGREVEN, W. (1958) Production and measurement of high temperatures, *Rept Progr. Phys.*, **21**, 312. [60, 81, 82, 102, 103, 202].

LOCHTE-HOLTGREVEN, W. (1960) Spektroskopische Diagnostik in der Physik heisser Plasmen. *Proc. VIII Colloq. Spectros. Intern., Luzern* 1959, p. 9 (Verlag H. R. Sauerländer und Co., Aarau, Switzerland). [60, 202]

LOCHTE-HOLTGREVEN, W. (1963) Über die quantitative spektroskopische Analyse von Plasmen, *Z. Anal. Chem.*, **198**, 1. [60, 93, 107, 162)

LOCHTE-HOLTGREVEN, W., and MAECKER, H. (1937) Temperaturbestimmung an frei brennenden Kohlelichtbögen mit Hilfe der CN-Banden, *Z. Physik*, **105**, 1. [121]

LURIO, A. (1964) Lifetime of the first excited $^1P_1$ state of Mg and Ba; hfs of Ba^{137}, *Phys. Rev.*, **136**, A376. [21, 175]

MACPHERSON, H. G. (1940) The carbon arc as a radiation standard, *J. Opt. Soc. Am.*, **30**, 189. [15]

MACPHERSON, H. G. (1941) The carbon arc as a radiation standard. *Temperature, its Measurement and Control in Science and Industry, Symposium, New York* 1939, p. 1141. (Reinold Publ. Comp., New York, N.Y.). [15]

MAECKER, H. (1950) Zur Prüfung der Bogentheorie. Der Wirkungsquerschnitt der Luft bei höheren Temperaturen, *Z. Physik*, **128**, 289. [52]

MAECKER, H. (1951) Der elektrische Lichtbogen, *Ergeb. Exakt. Naturw.*, **25**, 293. [43, 48, 52, 68, 69, 71, 77, 102, 144]

MAECKER, H. (1953) Elektronendichte und Temperatur in der Säule des Hochstromkohlebogens, *Z. Physik*, **136**, 119. [60, 139]

MAECKER, H. (1956) Ein zylindrischer Bogen für hohe Leistungen, *Z. Naturforsch.*, **11a**, 457. [52]

MAECKER, H. (1959) Über die Charakteristiken zylindrischer Bögen, *Z. Physik*, **157**, 1. [69]

MAECKER, H. (1960) Messung und Auswertung von Bogencharakteristiken (Ar, N₂), *Z. Physik*, **158**, 392. [52, 69]

MAECKER, H. (1962) Fortschritte in der Bogenphysik. *Proc. 5th Intern. Conf. Ionization Phenomena Gases, Munich* 1961, **2**, 1793 (North-Holland Publ. Comp., Amsterdam). [52, 69, 107]

MAECKER, H., and PETERS, T. (1954) Das Elektronenkontinuum in der Säule des Hochstromkohlebogens und in anderen Bögen, *Z. Physik*, **139**, 448. [202]

MAECKER, H., and PETERS, T. (1956) Einheitliche Dynamik und Thermodynamik des thermischen Plasmas, *Z. Physik*, **144**, 586. [69]

MAECKER, H., PETERS, T., and SCHENK, H. (1955) Ionen- und Atomquerschnitte im Plasma verschiedener Gase, *Z. Physik*, **140**, 119. [52, 53]

MAGNUS, W., and OBERHETTINGER, F. (1949) *Special Functions of Mathematical Physics* (Chelsea Publ. Comp., New York, N.Y.).[135]

MAJKOWSKI, R. F., and SCHREIBER, T. P. (1960) Use of controlled atmospheres to minimise matrix effects in spectrographic analysis of tool effects, *Spectrochim. Acta*, **16**, 1200. [87]

MALYKH, V. D., and SERD, M. A. (1964) Measurement of the time spent by atoms in light sources for spectrum analysis, *Opt. i Spektroskopiya*, **16**, 368; *Opt. Spectry (USSR)*, **16**, 203. [279, 281–283]

MANDELSHTAM, S. L. (1938) L'intensité des raies spectrales dans un arc à électrodes de charbon, *Compt. Rend. (Doklady) Acad. Sci. USSR*, **18**, 559. [279]

MANDELSHTAM, S. L. (1957) Über die Spektrumanregung in Funkenentladungen (Proc. VI Colloq. Spectros. Intern., Amsterdam 1956), *Spectrochim. Acta*, **11**, 245. [79]

MANDELSHTAM, S. L. (1962) Some theoretical aspects of spectrographic analysis, *Izv. Akad. Nauk. SSSR, Ser. Fiz.* **26**, 848; *Bull. Acad. Sci. USSR, Phys. Ser.*, **26**, 850 [82]

MANDELSHTAM, S. L., and MAZING, M. A. (1960) On spectral line broadening in the plasma. *Proc. VIII Colloq. Spectros. Intern., Luzern* 1959, p. 184 (Verlag H. R. Sauerländer und Co., Aarau, Switzerland). [60]

MANDELSHTAM, S. L., and NEDLER, V. V. (1961) On the sensitivity of emission spectrochemical analysis, *Spectrochim. Acta*, **17**, 885. [264]

MANNKOPFF, R. (1932) Anregungsvorgänge und Ionenbewegung im Lichtbogen, *Z. Physik*, **76**, 396. [82, 87, 279, 304]

MANNKOPFF, R. (1933) Über Elektronendichte und Elektronentemperatur in frei brennenden Lichtbögen, *Z. Physik*, **86**, 161. [82, 87]

MANNKOPFF, R. (1943) Die Berechnung der Lichtbogentemperatur und das Stabilitätsproblem der Lichtbogensäule, *Z. Physik*, **120**, 228. [43, 69, 144]

MANNKOPFF, R., and PETERS, C. (1931) Über quantitative Spektralanalyse mit Hilfe der negativen Glimmschicht in Lichtbogen, *Z. Physik*, **70**, 444. [279, 304]

MARGENAU, H., and MURPHY, G. M. (1943) *The Mathematics of Physics and Chemistry* (D. van Nostrand Company, Inc., Princeton, N.J.). [135]

MARGOSHES, M. (1960) Some properties of new or modified excitation sources. Symposium on Spectroscopic Excitation. *A.S.T.M. Spec.* Tech. Publ. No. 259, 46 (Am. Soc. Testing Materials, Philadelphia, Pa.). [5, 30]

MARGOSHES, M. (1965) Recent advances in excitation of atomic spectra. *Proc. XII Colloq. Spectros. Intern., Exeter* 1965, p. 26. (Hilger and Watts Ltd, London). [5, 87]

MARGOSHES, M., and SCRIBNER, B. F. (1959) The plasma jet as a spectroscopic source, *Spectrochim. Acta*, **15**, 138. [5]

MARGOSHES, M., and SCRIBNER, B. F. (1963) A study of the gas-stabilized arc as an emission source for the measurement of oscillator strengths. Determination of some relative gf-values for Fe I, *J. Res. Nat. Bur. Std.*, **67A**, 561. [5, 21, 151]

MARGOSHES, M., and SCRIBNER, B. F. (1964) Simple arc devices for spectral excitation in controlled atmospheres, *Appl. Spectry*, **18**, 154. [87]

MARINKOVIC, M. (1963) A universal semiquantitative spectrochemical method for the analysis of powdered samples, *Bull. Boris Kidrić Inst. Nucl. Sci.*, **14**(3), P/305, 111. [272]

MARINKOVIĆ, S. N. (1959) Behaviour of the system uranium oxide + impurities + graphite at higher temperatures with particular regard to the spectrochemical excitation in a d.c. arc Proc. VII Colloq. Spectros. Intern., Liege (1958) *Rev. Universelle Mines*, **15**, 393. [309, 316]

MARINKOVIĆ, S. N. (1960) A study of the electrode processes occurring in the spectrochemical analysis of uranium oxides. *Proc. VIII Colloq. Spectros. Intern., Luzern* 1959, p. 121. (Verlag H. R. Sauerländer und Co., Aarau, Switzerland). [309, 316]

MARINKOVIĆ, S. N. and MIRJANIĆ, M. (1965). Effects of solid-solution formation in the spectrochemical analysis of uranium. *Proc. XII Colloq. Spectros. Intern., Exeter* 1965, p. 505. (Hilger and Watts Ltd, London). [309]

MARITZ, F. R., and STRASHEIM, A. (1964a and b) A study of the efficiency of spectrochemical buffers. I, Influence of buffer cations. II, Influence of buffer anions, *Appl. Spectry*, **18**, 97, 185. [208]

MASON, R. C. (1948) Some properties of gas discharges used as spectral sources. Symposium on Spectroscopic Sources. *A.S.T.M. Spec. Tech. Publ.* No. 76 (Am. Soc. Testing Materials, Philadelphia, Pa.). [32, 39]

MATHERNY, M. (1962) Emissions-Lösungs-Spektrochemie. Eine kritische Wertung des gegenwärtigen Standes auf Grund der Diskussion fundameller Probleme und der Praxis der Lösungs-Spektrochemie, *Chem. Anal. (Warsaw)*, **7**, 75. [5]

MAVRODINEANU, R., and HUGHES, R. C. (1963) Excitation in radio-frequency discharges, *Spectrochim. Acta*, **19**, 1309. [5]

MAVRODINEANU, R., and HUGHES, R. C. (1964) The RF discharge at atmospheric pressure and its use as an excitation source in analytical spectroscopy. *Developments in Applied Spectroscopy*, **3**, 305 (Plenum Press, New York, N.Y.). [5]

MEEK, J. M., and CRAGGS, J. D. (1953) *Electrical Breakdown of Gases* (Clarendon Press, Oxford). [54]

MEGGERS, W. F. (1941a) Notes on the physical basis for spectrographic analysis, *J. Opt. Soc. Am.*, **31**, 39. [265]

MEGGERS, W. F. (1941b) The strongest lines of singly ionized atoms, *J. Opt. Soc. Am.*, **31**, 605. [265]

MEGGERS, W. F., CORLISS, C. H., and SCRIBNER, B. F. (1961a) Relative intensities for the arc spectra of seventy elements, *Spectrochim. Acta*, **17**, 1137. [25]

MEGGERS, W. F., CORLISS, C. H., and SCRIBNER, B. F. (1961b) *Tables of Spectral-Line Intensities. Part I, Arranged by Elements. Part II, Arranged by Wavelengths* (N.B.S. Monograph 32, Washington, D.C.). [25]

MEIXNER, J. (1952) Zur Theorie der Wärmeleitfähigkeit reagierender fluider Mischungen, *Z. Naturforsch.*, **7a**, 553. [72]

MELLICHAMP, J. W. (1965) Characterization of the voltage fluctuation in the direct current arc for spectrochemical analysis, *Anal. Chem.*, **37**, 1211. [46]

MELLICHAMP, J. W., and BUDER, R. K. (1963) Burning properties of electrodes for DC arc analysis, *Appl. Spectry*, **17**, 57. [313]

MELLICHAMP, J. W., and FINNEGAN, J. J. (1957) A combination carbon-graphite electrode for dc arc analysis, *Appl. Spectry*, **11**, 158. [313, 314]

MELLICHAMP, J. W., and FINNEGAN, J. J. (1959) A comparison of carbon and graphite electrodes, *Appl. Spectry*, **13**, 126. [311, 313]

MINCZEWSKI, J. (1962a) Einige Bemerkungen über die Spektralspurenbestimmungsmethoden, *Acta Chim. Acad. Sci. Hung.*, **34**, 123. [265]

MINCZEWSKI, J. (1962b) Über die Methoden der Reinheitsprüfung der reinsten Metalle, *Chem. Anal. (Warsaw)*, **7**, 877. [265]

MINCZEWSKI, J. (1963) Über die analytischen Methoden zur Prüfung von reaktorreinem Uran. *Proc. 1st Intern. Symp. High-Purity Materials, Dresden 1961*, p. 69 (Akademie-Verlag, Berlin). [265]

MIRLIN, D. N., PIKUS, G. E., and YUR'EV, V. G. (1962) Determination of electron scattering cross-sections by the electrical conductivity of a weakly ionized gas, *Zh. Tekhn. Fiz.*, **32**, 766; *Soviet Phys. Tech. Phys.*, 1963, **7**, 559. [52]

MITCHELL, A. C. G., and ZEMANSKY, M. W. (1961) *Resonance Radiation and Excited Atoms*, 2nd ed. (Cambridge University Press). [89]

MITTELDORF, A. J. (1965) Emission spectrochemical methods. *Trace Analysis, Physical Methods*, p. 193 (Interscience Publishers, New York, London, Sydney). [265]

MITTELDORF, A. J. (1965a) Spectroscopic electrodes, *The Spex Speaker* **10**(1), 1 (Spex Industries, Inc., Metuchen, N.J.). [314]

MITTELDORF, A. J., and LANDON, D. O. (1963) The stabilized plasma jet fluid analyzer, *The Spex Speaker* **8**(1), 1 (Spex Industries, Inc., Metuchen, N.J.). [5]

MOCHALOV, K. N., and RAFF, E. L. (1956) The role played by the surrounding medium in the excitation of atomic spectra in an arc discharge, *Zh. Tekhn. Fiz.*, **26**, 505; *Soviet Phys. Techn. Phys.*, 1957, **1**, 487. [87, 231]

MOENKE-BLANKENBURG, L. (1964) Geochemische Superanalyse mit Gitterspektrographen. *Emissionsspektroskopie*, p. 351 (Akademie-Verlag, Berlin). [265]

MOHLER, F. L. (1941) Concepts of temperature in electric discharge phenomena. *Temperature, its Measurement and Control in Science and Industry, Symposium, New York 1939*, p. 734 (Reinhold Publ. Comp., New York, N.Y.). [78]

MOONEY, J. B., SCHODER, C. E., and GARBINI, L. J. (1964) Pyrolytic graphite spectrographic electrodes, *Anal. Chem.*, **36**, 703. [314]

MOORE, C. E. (1949, 1952, 1958) *Atomic Energy Levels*. Vol. I: 1 H–23 V (1949). Vol. II: 24 Cr–41 Nb (1952). Vol. III: 42 Mo–57 La; 72 Hf–89 Ac (1958) (N.B.S. Circular No. 467, Washington, D.C.). [23, 83, 94, 163, 175, 326]

MOORE, C. E. (1950, 1952a, 1962) *An Ultraviolet Multiplet Table*, Section 1: 1 H–23 V (1950). Section 2: 24 Cr–41 Nb (1952). Section 3: 42 Mo–57 La; 72 Hf–88 Ra (1962). Section 4: Finding List for 1 H–41 Nb (1962). Section 5: Finding List for 42 Mo–57 La; 72 Hf–88 Ra (1962) (N.B.S. Circular No. 488, Washington, D.C.). [175]

MORRISON, G. H., and RUPP, R. L. (1961) Ultratrace emission spectroscopy. Symposium on Extension of Sensitivity for Determining Various Constituents in Metals. *A.S.T.M. Spec. Tech. Publ. No. 308*, p. 44 (Am. Soc. Testing Materials, Philadelphia, Pa.). [87, 265]

MÜLLER, G. (1958) Über die Störung eines Lichtbogens durch Tauchsonden, *Z. Physik*, **151**, 460. [39]

MÜLLER, G., and FINKELNBURG, W. (1956). Über Potentialmessungen mit bewegten Sonden in Hochdrucklichtbögen, *Z. Angew. Phys.*, **8**, 282. [39]

NAGIBINA, I. M. (1958) Determination of the concentration of atoms in the plasma of an a.c. arc from line widths, *Opt. i Spektroskopiya*, **4**, 430 (in Russian). Translation in: *Optical Transition Probabilities, A Representative Collection of Russian Articles, 1932–1962.* (Israel Program for Scientific Translation, Jerusalem, 1963. Published for the U.S. Department of Commerce and the National Science Foundation, Washington, D.C., O.T.S.-63-11135.). [292]

NAGIBINA, I. M. (1959) Determination of atomic concentrations in arc discharges from the line widths and correlation between concentration in the solid and gaseous phases, *Izv. Akad. Nauk SSSR, Ser. Fiz.* **23**, 1056; *Bull. Acad. Sci. USSR, Phys. Ser.*, **23**, 1048. [292]

NAGIBINA, I. M. (1963) Some spectroscopic studies of the gas cloud of an arc discharge with powder samples blown in, *Zavodsk. Lab.* **29**, 680; *Ind. Lab. (USSR)*, **29**, 724. [292]

NALIMOV, V. V., NEDLER, V. V., and MEN'SHOVA, N. P. (1961) Metrological evaluations in the detection of small concentrations by the method of emission spectrum analysis, *Zavodsk. Lab.*, **27**, 861; *Ind. Lab. (USSR)*, **27**, 864. [264]

NAKANO, Makota (1962). Buffer effects in the spectrochemical analysis of niobium–tantalum mixtures, *Appl. Spectry*, **16**, 165. [7]

NEEB, K. H., and GEBAUHR, W. (1962) Verwendung eines 'Plasmabrenners' als spektrochemische Anregungsquelle, *Z. Anal. Chem.*, **190**, 92. [5]

NESTOR, O. H., and OLSEN, H. N. (1960) Numerical methods for reducing line and surface probe data, *S.I.A.M. Review*, **2**, 200. [140]

NICKEL, H. (1963) Beobachtungen über chemische Reaktionen in borhaltigen Graphitpresslingen im Gleichstrombogen, *Z. Anal. Chem.*, **198**, 55. [316]

NICKEL, H. (1965a) Die Verwendung von Eisen-59 zur Untersuchung von Vorgängen in Graphitelektroden bei Anregung im Gleichstrombogen, *Z. Anal. Chem.*, **209**, 243. [275, 295, 312, 316]

NICKEL, H. (1965b) Thermochemische Reaktionen in borhaltigen Graphitelektroden und ihre Einfluss auf spektroskopische Ergebnisse bei Anregung im Gleichstrombogen, *Spectrochim. Acta*, **21**, 363. [295, 309, 312, 316]

NICKEL, H. (1965c) Die Verwendung radioaktiver Nuklide zur Untersuchung physikalisch-chemischer Vorgänge in Graphitelektroden bei Lichtbogenanregung, *Spectrochim. Acta*, **21**, 2031. [312, 316]

NICKEL, H., and PFLUGMACHER, A. (1961a) Die quantitative spektrochemische Bestimmung der Hauptelemente in Flugstauben und Eisenerzen im Gleichstromkohlebogen, *Z. Anal. Chem.*, **180**, 401. [309]

NICKEL, H., and PFLUGMACHER, A. (1961b) Ein Beitrag zum spektrochemischen Problem der Beeinflussung der Linienintensität durch Unterschiede in der chemischen Bindung, *Z. Anal. Chem.*, **184**, 161. [309]

NIKONOVA, E. I., and PROKOF'EV, V. K. (1959) The distribution of neutral atoms in the plasma of a d.c. arc, *Opt. i Spektroskopiya*, **6**, 253; *Opt. Spectry (USSR)*, **6**, 162. [292]

NULL, M. R., and LOZIER, W. W. (1962) Carbon arc as a radiation standard, *J. Opt. Soc. Am.*, **52**, 1156. [15]

O'CONNELL, R. F. (1964) An apparatus for the vaporization-concentration method of spectrochemical analysis, *Appl. Spectry*, **18**, 179. [316]

ODINTSOV, A. I. (1963) Measurement of the oscillator strength of the resonance line of calcium by the absorption method in an atomic beam, *Opt. i Spektroskopiya*, **14**, 322; *Opt. Spectry (USSR)*, **14**, 172. [24, 25]

OLDENBERG, O. (1913) Spektroheliographische Untersuchungen am Lichtbogen, *Z. Wiss. Phot.*, **23**, 133. [126]

OLDENBERG, O. (1934) On abnormal rotation of molecules, *Phys. Rev.*, **46**, 210. [103]

OLSEN, H. N. (1963) The electric arc as a light source for quantitative spectroscopy, *J. Quant. Spectry Radiative Transfer*, **3**, 305. [30, 69]

ORDELMAN, J. E., SMIT, H. A., and TOLK, A. (1965) A general method for quantitative analysis with the d.c. arc performed with solid samples in briquetted form. *Rept. Reactor Centrum Nederland*, RCN-Int-65-040; *Spectrochim. Acta*, 1966, **22**, 313. [207]

ORNSTEIN, L. S., and BRINKMAN, H. (1931a) Temperature determination from band spectra. II, Rotational energy distribution in the CN and AlO bands, and temperature distribution in the arc, *Proc. Koninkl. Ned. Acad. Wetenschap.*, **34**, 498. [121, 123]

ORNSTEIN, L. S., and BRINKMAN, H. (1931b) Temperatur im Lichtbogen und Saha-Theorie, *Naturwissenschaften*, **19**, 462. [127]

ORNSTEIN, L. S., and BRINKMAN, H. (1934) Der thermische Mechanismus in der Säule des Lichtbogens, *Physica*, **1**, 797. [79, 84, 86, 87]

ORNSTEIN, L. S., BRINKMAN, H., and BEUNES, A. (1932) Prüfung der Comptonschen Bogentheorie durch Messung der Bogentemperatur als Funktion des Druckes, *Z. Physik*, **77**, 72. [67, 183]

ORNSTEIN, L. S., and VAN WIJK, W. R. (1930) Temperaturbestimmung im elektrischen Bogen aus dem Bandenspektrum, *Proc. Koninkl. Ned. Acad. Wetenschap.*, **33**, 44. [121]

OVECHKIN, G. V. (1964) The radiation intensity of the spectral lines emitted by the plasma of arc and spark discharges on taking reabsorption into account, *Zh. Analit. Khim.*, **19**, 43; *J. Anal. Chem. USSR*, **19**, 38. [91]

OWEN, L. E. (1961) Stable plasma jet for the excitation of solutions, *Appl. Spectry*, **15**, 150. [5]

OWEN, L. E. (1962) Spectral excitation with stabilized plasma jets. *Developments in Applied Spectroscopy* **1**, p. 143 (Plenum Press, New York, N.Y.). [5]

PAVLYUCHENKO, M. M., and DUBOVIK, K. V. (1963) The intensity of the spectral lines of microelements and the diagram of state of the system, *Zh. Analit. Khim.*, **18**, 1426; *J. Anal. Chem. USSR*, **18**, 1242. [309, 310]

PEARCE, W. J. (1958) Calculation of the radial distribution of photon emitters in symmetric sources. *Conference on Exremely High Temperatures, Boston, Mass.*, p. 123 (John Wiley and Sons, Inc., New York, Chapman and Hall Ltd, London). [139]

PEARCE, W. J. (1960) Plasma jet temperature study. *Optical Spectrometric Measurements of High Temperatures*, p. 125 (University of Chicago Press, Chicago, Ill.). [139]

PEARSE, R. W. B., and GAYDON, A. G. (1963) *The Identification of Molecular Spectra*, 3rd ed. (Chapman and Hall Ltd, London). [321–324]

PELZER, H. (1960) Thermal conductivity of gases in dissociation and first ionization. *Brit. Electr. All. Ind. Res. Ass. (E.R.A.)*, *Confid. Tech. Rept Ref.* G/T 324. [72]

PELZER, H. (1962) On the contribution of dissociation and ionization to the thermal conductivity of gases. *Proc. 5th Intern. Conf. Ionization Phenomena Gases, Munich 1961*, **1**, 846. (North-Holland Publ. Comp., Amsterdam). [72]

PENKIN, N. P. (1964) Determination of oscillator strengths in atomic spectra, *J. Quant. Spectry Radiative Transfer*, **4**, 41 (in Russian). [21]

PENKIN, N. P., and RED'KO, T. P. (1960) The relative oscillator strengths of certain lines of Zn I and Cd I, *Opt. i Spektroskopiya*, **9**, 680; *Opt. Spectry (USSR)*, **9**, 360. [20]

PENKIN, N. P., and SHABANOVA, L. N. (1963) Oscillator strength of spectral lines of aluminium, gallium, and indium atoms, *Opt. i Spektroskopiya*, **14**, 12; *Opt. Spectry (USSR)*, **14**, 5. [24, 25]

PENKIN, N. P., and SLAVENAS, I.-YU. YU. (1963) Absolute oscillator strengths of the resonance doublets of Ag I and Au I, *Opt. i Spektroskopiya*, **15**, 9; *Opt. Spectry (USSR)*, **15**, 3. [24, 25]

PEPPER, C. E., PARDI, A. J., and ATWELL, M. G. (1963) The effects of various grades of graphite electrodes on the carrier distillation of impurities in U_3O_8, *Appl. Spectry*, **17**, 114. [314]

PEREL, J., ENGLANDER, P., and BEDERSON, B. (1962) Measurement of total cross-sections for the scattering of low-energy electrons by lithium, sodium, and potassium, *Phys. Rev.*, **128**, 1148. [52]

PETERS, TH. (1956) Über den Zusammenhang des Steenbeckschen Minimumprinzips mit dem thermodynamischen Prinzip der minimalen Entropie-erzeugung, *Z. Physik*, **144**, 612. [77]

PETERS, TH. (1962) Bogenmodell und Steenbeck'sches Minimumprinzip. *Proc. 5th Intern. Conf. Ionization Phenomena Gases, Munich 1961*, **1**, 885 (North-Holland Publ. Comp., Amsterdam). [77]

PINGEL, H., and SICKMEIER, E. W. (1962) Die Bestimmung der Linienform aus der Linienkontour bei Linien mit Selbstumkehr. III, Bestätigung der Methodik durch direkte Messung, *Z. Physik*, **166**, 176. [89]

PLŠKO, E. (1964) Anwendung aufgelöster Spektren zur Untersuchung des Probentransports aus der Schüttelektrode in die Entladung. *Emissionsspektroskopie*, p. 225 (Akademie-Verlag, Berlin). [5]

PREOBRAZHENSKII, N. G. (1959) The functional description of spectrum line profiles in the Cowan-Dieke and Bartels theories, *Opt. i Spektroskopiya*, **7**, 274; *Opt. Spectry (USSR)*, **7**, 173. [89]

PREOBRAZHENSKII, N. G. (1962) Photometric width of spectrum lines as a measure of their intensity, *Izv. Akad. Nauk. SSSR, Ser. Fiz.*, **26**, 953; *Bull. Acad. Sci. USSR, Phys. Ser.*, **26**, 958. [9]

PREOBRAZHENSKII, N. G. (1963) An extension of the limits of applicability of the self-absorption method, *Opt. i Spektroskopiya*, **14**, 342; *Opt. Spectry (USSR)*, **14**, 183. [89]

PREUSS, E. (1956) Zur Spektralanalyse kleiner Substanzmengen, *Mikrochim. Acta*, 382. [308]

PRILEZHAEVA, N. A., and GORYACHEV, V. N. (1950) Determination of the concentration of sodium atoms in the positive column of an arc discharge, *Izv. Akad. Nauk SSSR, Ser. Fiz.*, **14**, 732; *Chem. Abstr.*, 1951, **45**, 4131b. [190]

PSZONICKI, L. (1960) Zur Wirkung der Träger bei der Verdampfung der Spurenverunreinigungen in der Spektralanalyse, *Proc. VIII*

Colloq. Spectros. Intern., Luzern 1959, p. 50 (Verlag H. R. Sauerländer und Co., Aarau, Switzerland). [316]

PSZONICKI, L. (1962) Studien über die Verdampfung der Elemente aus schwerflüchtigen spektrographischen Proben, *Acta Chim. Acad. Sci. Hung.*, **30**, 351. [316]

PSZONICKI, L., and MINCZEWSKI, J. (1962) The influence of AgCl and Ga₂O₃ as 'spectroscopic carrier' on distillation of trace impurities from spectrographic samples, *Spectrochim. Acta*, **18**, 1325. [316]

RAIKHBAUM, YA. D., and LUZHNOVA, M. A. (1959) Quantitative spectral analysis of ores by the introduction of samples into the arc with an air jet, *Zavodsk. Lab.*, **25**, 1449; *Ind. Lab. (USSR)*, **25**, 1519. [5]

RAIKHBAUM, YA. D., and MALYKH, V. D. (1960) A spectroscopic method for studying the diffusion of atoms in an electric arc, *Opt. i Spektroskopiya*, **9**, 425; *Opt. Spectry (USSR)*, **9**, 223. [57, 279, 281, 282]

RAIKHBAUM, YA. D., and MALYKH, V. D. (1961) On a possible cause of the 'carrier' effect in spectral analysis, *Opt. i Spektroskopiya*, **10**, 524; *Opt. Spectry (USSR)*, **10**, 269, [279, 281]

RAIKHBAUM, YA. D., MALYKH, V. D., and LUZHNOVA, M. A. (1963) A scintillation method for the spectral determination of tantalum and niobium in ores, *Zavodsk. Lab.*, **29**, 677; *Ind. Lab. (USSR)*, **29**, 721. [5]

RAISEN, E., CARRIGAN, R. A., RAZICUNAS, V., LOSEKE, W. A., and GROVE, E. L. (1964, 1965) Identification of chemical species in a plasma by emission spectroscopy. *Developments in Applied Spectroscopy* 3, p. 250. (Plenum Press, New York, N.Y.); *Appl. Spectry*, **19**, 41. [5]

RICARD, R. (1953a) L'effet carbone dans l'analyse spectrographique des matériaux réfractaires, *Spectrochim. Acta*, **5**, 417. [206]

RICARD, R. (1953b) Etude de l'effet carbone dans l'analyse spectrographique des matériaux réfractaires. *Proc. 16th Congr. G.A.M.S.*, p. 165 (Groupement pour l'Avancement des Méthodes Spectrographiques, Paris). [206]

RICHTMYER, F. K., and KENNARD, E. H. (1947). *Introduction to Modern Physics* (McGraw-Hill Book Comp., New York, N.Y.). [25]

RIEMANN, M. (1963) Ein stabilisierter Lichtbogen für die Lösungsspektralanalyse, *Monatsber. Deut. Akad. Wiss. Berlin*, **5**, 316. [52]

RIEMANN, M. (1964a) Die Messung von relativen und absoluten optischen Übergangswahrscheinlichkeiten des Cu I im wandstabilisierten Lichtbogen, *Z. Physik*, **179**, 38. [52]

RIEMANN, M. (1964b) Ein stabilisierter Lichtbogen für die Lösungsspektralanalyse. *Emissionsspektroskopie*, p. 173 (Akademie-Verlag, Berlin). [31, 52]

RIEMANN, M. (1965) Lösungsspektralanalyse mit Hilfe von Aerosolbeschickten wandstabilisierten Lichtbögen. *Proc. XII Colloq. Spectros. Intern., Exeter* 1965, p. 199 (Hilger and Watts Ltd, London). [31, 52]

RIGHINI, G. (1935) Température et pression relative des gaz dans un arc électric, *Physica*, **2**, 585. [154]

RINGLER, H. (1962) Radialer Potentialverlauf im Hochstromkohlebogen, *Z. Physik*, **169**, 273. [39]

RODER, O., and STAMPA, A. (1964) Photoelektrische Registrierung der Linienprofile thermisch angeregter He-I-Linien, *Z. Physik*, **178**, 348. [60]

ROES, R. L. (1962) The natural spectrum and the composition of the arc gas in the carbon arc in nitrogen and air. (In Dutch) Ph.D. Thesis, University of Utrecht. [63, 64, 69, 72–75, 144, 146, 317, 319–323]

ROES, R. L., and SMIT, J. A. (1957) Die Herkunft des Untergrundes im Kohlebogen-Spektrum (Proc. VI Colloq. Spectros. Intern., Amsterdam 1956), *Spectrochim. Acta*, **11**, 119. [322, 324]

ROLLWAGEN, W. (1939) Die physikalischen Erscheinungen der Bogenentladung in ihrer Bedeutung für spektralanalytische Untersuchungsmethoden, *Spectrochim. Acta*, **1**, 66. [67]

ROMPE, R., and STEENBECK, M. (1939) Der Plasmazustand der Gase, *Ergeb. Exakt. Naturw.*, **18**, 257. [59]

ROUBAULT, M., DE LA ROCHE, H., and GOVINDARAJU, K. (1962/63) L'analyse des roches silicatées par spectrométrie photo-électrique au Quantomètre A.R.L. et son contrôle par des roches étalons, *Sci. Terre*, **9**, 339. [5]

RUBEŠKA, I., and POLEJ, B. (1962) The shape of working curves using photometric line width, *Appl. Spectry*, **16**, 104. [9]

RUBEŠKA, I., and POLEJ, B. (1964a) Einige neue Gesichtspunkte im Linienbreitenverfahren. *Emissionsspektroskopie*, p. 148 (Akademie-Verlag, Berlin). [9]

RUBEŠKA, I., and POLEJ, B. (1964b) Spektrographische Indiumbestimmung mit Hilfe der photometrischen Linienbreite. *Emissionsspektroskopie*, p. 153 (Akademie-Verlag, Berlin). [9]

RUKOSNEVA, A. V. (1964) Determination of atomic concentrations and the degree of inhomogeneity in an arc discharge by the line absorption method, *Opt. i Spektroskopiya*, **17**, 340; *Opt. Spectry (USSR)*, **17**, 181. [292]

RUSANOV, A. K. (1941) Dependence of spectrum line intensities on the volatility of ore ingredients in a carbon arc, *Izv. Akad. Nauk. SSSR, Ser. Fiz.* **5**, 235 (in Russian); Cf. *Spectrochim. Acta*, 1947/49, **3**, 240. [315]

RUSANOV, A. K. (1943) Dependence of results of spectrographic analysis by the arc method on composition of minerals and ores *Trudy Vses. Konf. Anal. Khim.*, **2**, 211, (in, Russian); Cf. *Spectrochim. Acta*, 1947/49, **3**, 240, and *Index to the Literature on Spectrochemical Analysis*, Part II, 2184. [315]

RUSANOV, A. K. (1945) Influence of volatility of ore component on the temperature of the flame and electrodes of the carbon arc, *Izv. Akad. Nauk SSSR, Ser. Fiz.* **9**, 707 (in Russian). [45, 315]

RUSANOV, A. K., ALEKSEEVA, V. M., and IL'YASOVA, N. V. (1961) Spectrographic determination of germanium and other elements in ores with sulphidation of the latter during their evaporation, *Zh. Analit. Khim.*, **16**, 284; *J. Anal. Chem. USSR*, **16**, 300. [316]

RUSANOV, A. K., and BATOVA, N. T. (1961) Effect of composition of powder on the results of spectrographic analysis by the insufflation method, *Zavodsk. Lab.*, **27**, 299; *Ind. Lab. (USSR)*, **27**, 303. [5]

RUSANOV, A. K., and BATOVA, N. T. (1962) Characteristic of the method of blowing powders into the arc flame in the spectral analysis of ores, *Zh. Analit. Khim.*, **17**, 404; *J. Anal. Chem. USSR*, **17**, 405. [5]

RUSANOV, A. K., and KHITROV, V. G. (1958) Spectrographic analysis of ores by introducing the powder into the arc in a stream of air, *Spectrochim. Acta*, **10**, 404. [5]

RUSANOV, A. K., KHITROV, V. G., and BATOVA, N. T. (1959) A low temperature carbon arc as a spectrum excitation source for rubidium, cesium, thallium, and indium during spectrographic analysis of silicates, *Zh. Analit. Khim.*, **14**, 534; *J. Anal. Chem. USSR*, **14**, 581. [197]

RUSANOV, A. K., and VOROB'EV, V. S. (1964) Uniform blowing of high-dispersion powders into the arc flame in spectral analysis, *Zavodsk. Lab.*, **30**, 41; *Ind. Lab. (USSR)*, **30**, 45. [5]

RÜSSMANN, H. H. (1958) Methodische Untersuchungen an Spektralkohlen. Special Publication: *Aus dem Prüffeld der Ringsdorff-Werke* (Ringsdorff, Bad Godesberg-Mehlem, Western Germany). [311, 314]

SAHA, M. N. (1920) Ionization in the solar chromosphere, *Phil. Mag.*, **40**(6), 472. [157]

SAHA, M. N. (1921a) On a physical theory of stellar spectra, *Proc. Roy. Soc. (London)*, **99A**, 135. [157]

SAHA, M. N. (1921b) Versuch einer Theorie der physikalischen Erscheinungen bei hohen Temperaturen mit Anwendung auf die Astrophysik, *Z. Physik*, **6**, 40. [157]

SAMSONOVA, Z. N. (1962) On the mechanism of the 'carrier' effect on the intensity of spectral lines, *Opt. i Spektroskopiya*, **12**, 466; *Opt. Spectry (USSR)*, **12**, 257. [279, 284, 307]

SAWYER, R. A. (1951) *Experimental Spectroscopy* (Prentice Hall, Inc., New York, N.Y.). [8]

SCHLUGE, H., and FINKELNBURG, W. (1944) Über das Zischen des Homogenkohlebogens, *Z. Physik*, **122**, 714. [35]

SCHMICK, H., and SEELIGER, R. (1928) Studien über den Mechanismus des Lichtbogens. II, Glimmbogen und zischende Bogen, *Physikal. Z.*, **29**, 168. [35]

SCHMITZ, G. (1949) Zur Temperaturverteilung im freibrennenden Kohlelichtbogen, *Z. Physik*, **126**, 1. [144, 147]

SCHMITZ, G., and PATT, H. J. (1962) Eine Methode zur Bestimmung von Materialfunktionen von Gasen aus den Temperaturverteilungen zylindrischer Lichtbögen, *Z. Physik*, **167**, 163. [69]

SCHMITZ, G., and PATT, H. J. (1963) Die Bestimmung von Materialfunktionen eines Stickstoffplasmas bei Atmosphärendruck bis 15 000°K, *Z. Physik*, **171**, 449. [69]

SCHMITZ, G., PATT, H. J., and UHLENBUSCH, J. (1963) Eigenschaften und Parameterabhängigkeit der Temperaturverteilung und der Charakteristik eines zylinder-symmetrischen Stickstoffbogens, *Z. Physik*, **173**, 552. [69]

SCHNAUTZ, H. (1939) Linienabsorption und Gesamtabsorption der Kupferresonanzlinien sowie Bestimmung der Strahlungstemperatur in der Gassäule des Kupferlichtbogens, *Spectrochim. Acta*, **1**, 173. [128]

SCHOLZE, H., and GERBATSCH, R. (1964) Bemerkungen zur Reinheitsprüfung von Spektralkohlen, *Jenaer Rundschau* **9**, 77. [314]

SCHROLL, E. (1963) Über die Anwendung thermochemischer Reaktionen in der emissionsspektrographischen Spurenanalyse und ihre Bedeutung für den Carriereffekt, *Z. Anal. Chem.*, **198**, 40. [208, 315, 316]

SCHROLL, E., BRANDENSTEIN, M., JANDA, I., and ROCKENBAUER, W. (1960) Emissionsspektrographische Spurenanalyse mit der Doppelbogenmethode. *Proc. VIII Colloq. Spectros. Intern., Luzern* 1959, p. 145 (Verlag H.R. Sauerländer und Co., Aarau, Switzerland). [112, 208]

SCHROLL, E., and WENIGER, M. (1965) Eine empfindliche spektrochemische Analysenmethode zur Bestimmung von Germanium und

Zinn unter Verwendung sulfidierender thermo-chemischer Reagenzien, *Mikrochim. Ichnoanal. Acta*, 378. [316]

SCHULZ, P. (1947a) Der Einfluss der Positiven Ionen auf die Elektronenbeweglichkeit. I, Elektronenbeweglichkeit in Hochdruckplasma, *Ann. Physik*, **1**(6), 318. [53]

SCHULZ, P. (1947b) Zur Theorie der Hoch-strombogensäule, *Z. Naturforsch.*, **2a**, 662. [72]

SCHURER, K. (1964) Personal communication. [116, 117]

SCHUTTEVAER, J. W. (1942) Relative probab-ilities of transitions in the cadmium, zinc, and strontium atom. (In Dutch) Ph.D. Thesis, University of Utrecht. [20, 197]

SCHUTTEVAER, J. W., DE BONT, M. J., and VAN DEN BROEK, TH.H. (1943). Determination of the relative transition probabilities of some triplet lines in the atomic spectra of calcium and strontium, *Physica*, **10**, 544. [116]

SCHUTTEVAER, J. W., and SMIT, J. A. (1943) The relative probabilities of transitions in the zinc atom, *Physica*, **10**, 502. [20]

SCHWEITZER, E. (1927) Eine absolute Methode zur Ausführung der quantitativen Emissions-spektralanalyse, *Z. Anorg. Allg. Chem.*, **164**, 127. [4]

SCRIBNER, B. F., and MARGOSHES, M. (1962) Excitation of solutions in gas-stabilized arc sources. *Proc. IX Colloq. Spectros. Intern.*, Lyons 1961, **2**, 309 (Groupement pour l'Avance-ment des Méthodes Spectrographiques, Paris). [5]

SCRIBNER, B. F., and MULLIN, H. R. (1946) Carrier-distillation method for spectrographic analysis and its application to the analysis of uranium-base materials, *J. Res. Nat. Bur. Std.*, **37**, 379. [279]

SEMENOVA, O. P. (1945) On the question as to the mechanism of arc discharge, *Izv. Akad. Nauk SSSR, Ser. Fiz.*, **9**, 715 (in Russian). [228]

SEMENOVA, O. P. (1946) On the mechanism of arc discharge, *Doklady Akad. Nauk SSSR*, **51** 683; *Compt. Rend. Acad. Sci. URSS*, **51**, 683. [46, 228]

SEMENOVA, O. P., and DURKINA, A. V. (1957) Influence of gas composition on intensity of spectral lines at thermal excitation, *Opt. i Spek-troskopiya*, **2**, 34 (in Russian). [255]

SEMENOVA, O. P., and LEVCHENKO, M. A. (1962) Dependence of the effective ionization potential on the concentration of easily ionized impurities in the arc discharge, *Opt. i Spektros-kopiya*, **13**, 609; *Opt. Spectry (USSR)*, **13**, 347 [228, 230]

SERIN, P. A., and ASHTON, K. H. (1964) Spectrochemical analysis of aqueous solutions by the plasma jet, *Appl. Spectry*, **18**, 166. [5]

SHAW, D. M. (1962) Spectrographic methods using the Stallwood jet, *The Spex Speaker* **7**(3), 4 (Spex Industries, Inc., Metuchen, N.J.). [30]

SHAW, D. M., JOENSUU, O. I., and AHRENS, L. H. (1950) A double arc method for spectro-chemical analysis of geological materials, *Spectrochim. Acta*, **13**, 197. [112]

SHAW, D. M., WICKREMASINGHE, O. C., and WEBER, J. N. (1960) Spectrochemical determina-tion of lithium, sodium, potassium, and rubid-ium in rocks and minerals using the Stallwood jet, *Anal. Chim. Acta*, **22**, 398. [87]

SHUMAKER, J. B., and YOKLEY, C. R. (1964) The use of an analog computer in side-on arc spectroscopy, *Appl. Optics*, **3**, 83. [140]

SHVANGIRADZE, R. R., OGANEZOV, K. A., and CHIKHLADZE, B. YA. (1962) Temperature and thermal equilibrium of arcs burning in inert gases, *Opt. i Spektroskopiya*, **13**, 25, 613; *Opt. Spectry (USSR)*, **13**, 14, 350. [87]

SIROIS, E. H. (1964) Plasma arc solution spectrochemistry evaluation and optimization of operating parameters. Element calibrations by plasma jet spectrochemistry, *Anal. Chem.*, **36**, 2389, 2394. [5]

SLAVIN, M. (1938) Quantitative analysis based on spectral energy, *Ind. Eng. Chem., Anal. Ed.*, **10**, 407. [5]

SMIT, J. A. (1946) The determination of temperature from spectra, *Physica*, **12**, 683. [19, 79, 86, 102, 123]

SMIT, J. A. (1950) The production and measurement of constant high temperatures, up to 7000°K. (In Dutch) Ph.D. Thesis, University of Utrecht. [32, 59, 67, 86, 102, 117, 119, 122–124, 145, 151, 318]

SMIT-MIESSEN, M. M. and SPIER, J. L. (1942) Intensity profiles of nonresolved CN bands, *Physica*, **9**, 193. [123]

SNELLEMAN, W. (1965) A flame as a standard of temperature. Ph.D. Thesis, University of Utrecht. [58]

SNOOKS, E. C. (1962) Studies on the analysis of tungsten using the Stallwood jet, *The Spex Speaker*, **7** (3), 3 (Spex Industries, Inc., Metu-chen, N.J.). [30]

SOBOLEV, N. N. (1957) The shape and width of spectral lines emitted by a flame and by a direct-current arc (Proc. VI Colloq. Spectros Intern., Amsterdam 1956), *Spectrochim. Acta* **11**, 310. [89]

SOBOLEV, N. N., KOLESNIKOV, V. N., and KITAYEVA, V. F. (1960). Spectroscopic study of carbon arc burning in inert gases. *Proc. 4th Intern. Conf. Ionization Phenomena Gases, Uppsala 1959*, **1**, 371 (North-Holland Publ. Comp., Amster-dam). [87, 102]

Somers, P. J. (1954) Investigation of an arc discharge in nitrogen at increased pressure. (In Dutch) Ph.D. Thesis, University of Utrecht. [19, 43, 125]

Somers, P. J., and Smit, J. A. (1956) Measurements on arc discharges in nitrogen at 1 atm and higher pressure, Appl. Sci. Res., 6, Sect. B, 75. [43]

Sperling, J. (1950) Das Temperaturfeld im freien Kohlebogen, Z. Physik, 128, 269. (144, 147]

Spier, J. L., and Smit, J. A. (1942) Temperature determination from the CN bands 4216 and 4197 Å in the case of non-resolved rotational structure, Physica, 9, 597. [123]

Spier, J. L., and Smit-Miessen, M. M. (1942) On the determination of the temperature with the aid of non-resolved CN bands 3883 and 3871 Å, Physica, 9, 422. [123]

Spindler, D. C. (1961) D.C. arc analysis. Some effects of replacing graphite with carbon, Appl. Spectry, 15, 20. [206, 314]

Spitzer, L., and Härm, R. (1953) Transport phenomena in a completely ionized gas, Phys. Rev., 89, 977. [53]

Stallwood, B. J. (1954) Air-cooled electrodes for spectrochemical analysis of powders, J. Opt. Soc. Am., 44, 171. [30]

Steenbeck, M. (1932) Energetik der Gasentladungen. Physikal. Z., 33, 809. [76]

Steenbeck, M. (1960/61) Zur theoretischen Begründung des 'Minimumprinzips' für die Spannung einer Gasentladung und einige weitere Folgerungen, Beitr. Plasmaphysik, 1, 1. [77]

van Stekelenburg, L. H. M. (1943) Measurement of transition probabilities in the spectrum of titanium. (In Dutch) Ph.D. Thesis, University of Utrecht. [33, 34, 99, 105, 144, 145, 190, 294]

van Stekelenburg, L. H. M. (1946) Convection velocity in a direct current arc, Physica, 12, 289. [33]

van Stekelenburg, L. H. M., and Smit, J. A. (1948) Relative transition probabilities in the spectra of Ti I and Ti II, Physica, 4, 189. [19, 105, 144, 145]

Strock, L. W. (1953) Quantitative evaluation of a metal base method for determining major constituents in nonmetallic samples, Appl. Spectry, 7, 64. [87]

Strock, L. W. (1954) Some experimental evidence of collision processes in spectrochemical analysis, Appl. Spectry, 8, 105. [87]

Suits, G. C. (1934) Experiments with arcs at atmospheric pressure, Phys. Rev., 46, 252. [43]

Suits, G. C. (1939a) Convection currents in arcs in air, Phys. Rev., 55, 198. [34]

Suits, G. C. (1939b) High pressure arcs, Phys. Rev., 55, 561. [43]

Suits, G. C. (1941) High temperature gas measurements in arcs. Temperature, its Measurement and Control in Science and Industry. Symposium, New York, 1939 (Reinhold Publ. Comp., New York, N.Y.). [102]

Svoboda, V. (1962) Spectrochemical analysis of radioactive materials. II, Wall-stabilized arc discharge, Chem. Anal. (Warsaw), 7, 181. [31]

Symon, F. I. (1925) The diffusion of salt vapours in a Bunsen flame, Proc. Roy. Soc. Edinburgh, 46A, 15. [57]

Tappe, W., and van Calker, J. (1963) Quantitative spektrochemische Untersuchung mit hochfrequenten Plasmaflammen, Z. Anal. Chem., 198, 13. [5]

Terpstra, J. (1956) Measurement of the transition probabilities belonging to atomic lines in the spectrum of silver. (In Dutch) Ph.D. Thesis, University of Utrecht. [19, 135, 144]

Terpstra, J., and Smit, J. A. (1958) Measurement of 'optical' transition probabilities in the silver atom, Physica, 24, 937. [9, 19, 135, 144]

Theimer, O. (1957). Über die effektive Ionisierungsspannung eines Atoms im Inneren des Plasmas, Z. Naturforsch., 12a, 517. [93, 162]

Thiers, R. E., and Vallee, B. L. (1957) The effect of noble gases on the characteristics of the d.c. arc. (Proc. VI Colloq. Spectros. Intern, Amsterdam 1956), Spectrochim. Acta, 11, 179. [87, 313]

Treffzt, E. (1949) Wellenfunktionen und Übergangswahrscheinlichkeiten, Z. Astrophys., 26, 240. [174]

Treffzt, E. (1950) Wellenfunktionen und Übergangswahrscheinlichkeiten beim Atom Mg I, Z. Astrophys., 28, 67. [174]

Tymchuk, P., Russell, D. S., and Berman, S. S. (1965) The effect of halide carriers in the spectrographic examination of trace impurities in copper, Spectrochim. Acta, 21, 2051. [316]

Uhlenbusch, J. (1964) Berechnung der Materialfunktionen eines Stickstoff- und Argonplasmas aus gemessenen Bogendaten, Z. Physik, 179, 347. [69]

Unsöld, A. (1948) Zur Berechnung der Zustandssummen für Atome und Ionen in einem Teilweise ionisierten Gas, Z. Astrophys., 24, 355. [93]

Unsöld, A. (1955) Physik der Sternatmosphären (Springer-Verlag, Berlin, Göttingen, Heidelberg). [18, 19, 22, 23, 25, 47, 53, 60, 86, 89, 93, 150, 157, 202]

VAINSHTEIN, E. E. (1962). Nouveaux procédés pour les études des processus se déroulant dans le plasma des sources spectrales et recherches dans le but de trouver des voies de perfectionnements des méthodes de l'analyse spectrale des substances. *Proc. IX Colloq. Spectros. Intern.*, *Lyons* 1961, **1**, 105 (Groupement pour l'Avancement des Méthodes Spectrographiques, Paris). [5, 283, 284]

VAINSHTEIN, E. E., and BELYAEV, YU. I. (1958, 1959). Use of radioactive isotopes in studying the spatial distribution of elements in direct current arc plasma during spectrochemical determination of impurities in uranium, *Zh. Analit. Khim.*, 1958, **13**, 388; *J. Anal. Chem. USSR*, 1958, **13**, 441; *Intern. J. Appl. Radiation Isotopes*, 1959, **4**, 179. [283]

VAINSHTEIN, E. E., KOROLEV, V. V., and SAVINOVA, E. N. (1961). Excitation conditions of the spectra of elements in a plasma generator and the application of the latter for the spectrographic analysis of titanium alloys, *Zh. Analit. Khim.*, **16**, 532; *J. Anal. Chem. USSR*, **16**, 527. [5]

VALLEE, B. L., and ADELSTEIN, S. J. (1952) The effect of argon and argon-helium mixtures on the direct-current arc, *J. Opt. Soc. Am.*, **42**, 295. [87]

VALLEE, B. L., and BAKER, M. R. (1956) Anode temperatures and characteristics of the d.c. arc in noble gases, *J. Opt. Soc. Am.*, **46**, 77. [87]

VALLEE, B. L., and PEATTIE, R. W.' (1952) Volatilization rates of elements in the helium direct-current arc, *Anal. Chem.*, **24**, 434. [87]

VALLEE, B. L., REIMER, C. B., and LOOFBOUROW, J. R. (1950) The influence of argon, helium, oxygen, and carbon dioxide on emission spectra in the DC arc, *J. Opt. Soc. Am.*, **40**, 751. [87]

VALLEE, B. L., and THIERS, R. E. (1956) Effect of alkali and alkaline earth chlorides on the anode temperature of the d.c. arc, *J. Opt. Soc. Am.*, **46**, 83. [87]

VENDRIK, A. J. H. (1949) Optical investigation of alkali metal vapour in an acetylene flame. (In Dutch) Ph.D. Thesis, University of Utrecht. [57]

VOORHOEVE, P. G. (1946) Investigations in the spectrum of aluminium. (In Dutch) Ph.D. Thesis, University of Utrecht. [19, 121]

VOROB'EV, G. G. (1961) Quantitative spectrographic analysis of tektites and the silicate phase of meteorites, *Meteoritika, Moscow*, 185; Cf. *Anal Abstr.*, 1962, **9**, 4241. [207]

VUKANOVIĆ, V. M. (1960) Über den Einfluss des Elektronendruckes bei Spurenanalysen im Lichtbogen. *Proc. VIII Colloq. Spectros. Intern.*, *Luzern* 1959 (Verlag H. R. Sauerländer und Co., Aarau, Switzerland). [108, 110, 173, 279 285, 307]

VUKANOVIĆ, V. (1964) Erfahrungen über den Einfluss der Vorgänge im Plasma bei Spurenanalysen im Lichtbogen. *Emissionsspektroskopie*, p. 9 (Akademie-Verlag, Berlin). [108, 110, 144, 173, 279, 285, 307]

WALKER, R. E., and WESTENBERG, A. A. (1958a) Molecular diffusion studies in gases at high temperature. I, The 'point source' technique, *J. Chem. Phys.*, **29**, 1139. [56, 294]

WALKER, R. E., and WESTENBERG, A. A. (1958b) Molecular studies in gases at high temperature. II, Interpretation of results on the He-N_2 and CO_2-N_2 systems, *J. Chem. Phys.*, **29**, 1147. [56]

WALKER, R. E., and WESTENBERG, A. A. (1959) Molecular studies in gases at high temperature. III, Results and interpretation of He-A system, *J. Chem. Phys.*, **31**, 519. [56]

WALKER, R. E., and WESTENBERG, A. A. (1960) Molecular diffusion in gases at high temperature. IV, Results and interpretation of the CO_2-O_2, CH_4-O_2, H_2-O_2, and H_2O-O_2 systems, *J. Chem. Phys.*, **32**, 436. [56, 57]

WALKER, R. E., MONCHICK, L., WESTENBERG, A. A., and FAVIN, S. (1961) High temperature gaseous diffusion experiments and intermolecular potential energy functions. *Conf. Phys. Chem. Aerodynamics Space Flight*, p. 221 (Pergamon Press, New York, Oxford, London, Paris). [56 57]

WALLACE, L. (1962) Band-head wavelengths of C_2, CH, CN, CO, NH, NO, O_2, OH, and their ions, *Astrophys. J.*, *Supple. Ser.*, **7**, 165. [321]

WANG, M. S., and CAVE, W. T. (1964) Arc chamber for spectral excitation in controlled atmosphere, *Appl. Spectry.*, **18**, 189. [87]

WEBB, M. S. W., and WILDY, P. C. (1963) Determination of calcium in biological materials by means of a plasma jet, *Nature*, **198**, 1218. [5]

WEDEPOHL, K. H. (1953) Untersuchungen zur Geochemie des Zinks, *Geochim. Cosmochim. Acta*, **3**, 93. [112]

WEIZEL, W., and FASSBENDER, J. (1940) Über die Ursache des Zischens beim Homogenkohlebogen, *Z. Tech. Physik*, **21**, 391. [35]

WEIZEL, W., and FASSBENDER, J. (1943) Über die Vorgänge an der Anode eines zischenden Homogenkohlebogens, *Z. Physik*, **120**, 252. [35]

WEIZEL, W., and ROMPE, R. (1949) *Theorie Elektrischer Lichtbögen und Funken* (Barth-Verlag, Leipzig). [35, 41, 47, 68, 69, 80]

WENDT, R. H., and FASSEL, V. A. (1965) Induction-coupled plasma spectrometric excitation source, *Anal. Chem.*, **37**, 920. [5]

WEST, C. D., and HUME, D. N. (1964) Radio-frequency plasma emission spectrophotometer, *Anal. Chem.*, **36**, 412. [5]

WESTENBERG, A. A., and WALKER, R. E. (1959) Experiments in the molecular diffusion of gases at high temperatures. *Symp. Thermal Properties, Lafayette, Ind.* 1959, p. 314 (McGraw-Hill Book Comp., Inc., New York, Toronto, London). [56]

WHITE, H. E. (1934) *Introduction to Atomic Spectra* (McGraw-Hill Book Comp., Inc. New York, N.Y.). [25]

WHITTAKER, E. T., and WATSON, G. N. (1948) *A Course of Modern Analysis*, p. 229 (The Macmillan Company, New York, N.Y.). [135]

WIENECKE, R. (1956) Experimentelle und theoretische Bestimmung der Wärmeleitfähigkeit, *Z. Physik*, **146**, 39. [63, 72, 73]

WIESE, W. L. (1963) Present status of our knowledge of atomic transition probabilities. *Proc. X Colloq. Spectros. Intern., Maryland* 1962, p. 37 (Spartan Books, Washington, D.C.). [18, 19]

WIESE, W. L., PAQUETTE, D. R., and SOLARSKI, J. E. (1962) Experimental study of the Stark broadening of the Balmer line H_y. *Proc. 5th Intern. Conf. Ionization Phenomena Gases, Munich* 1961, **1**, 907 (North-Holland Publ. Comp., Amsterdam). [60]

WIESE, W. L., and SHUMAKER, J. B., Jr. (1961) Measurement of the transition probability of the O I multiplet at 6157 A, *J. Opt. Soc. Am.*, **51**, 937. [140]

VAN WIJK, W. R. (1932) Optische Untersuchung des Akkomodationskoeffizienten der Molekularrotationen eines verdünnten Gases, *Z. Physik*, **75**, 584. [103]

WILKINSON, P. G. (1963) Diatomic molecules of astrophysical interest: ionization potentials and dissociation energies, *Astrophys. J.*, **138**, 778. [321, 325, 326]

WILSON, H. A. (1912) The diffusion of alkali salt vapours in flames, *Phil. Mag.*, **24**(6), 118. [57, 294]

WITTE, H. (1934) Experimentelle Trennung von Temperaturanregung und Feldanregung im elektrischen Lichtbogen, *Z. Physik*, **88**, 415. [82, 87]

YOKLEY, C. R., and SHUMAKER, J. B. (1963) Computer for the Abel inversion, *Rev. Sci. Instr.*, **34**, 551. [140]

YOUNG, L. G. (1962) Emission spectroscopy of solutions, *Analyst*, **87**, 6. [5]

ZAIDEL, A. N., KALITEEVSKII, N. I., LIPIS, L. V., and CHAIKA, M. P. (1960) *Emission Spectrum Analysis of Atomic Materials*. Translated from a publication of the State Publishing House of Physicomathematical Literature, Moscow 1960 (U.S.A. Atomic Energy Commission, AEC-tr-5745, 1963). [38, 264, 265, 315, 316]

ZAIDEL, A. N., KALITEEVSKII, N. I., KUND, G. G., and FRATKIN, S. G. (1957) Spectral analysis by the evaporation method. III, On the role of 'carrier' in spectral analysis of uranium, *Opt. i Spektroskopiya*, **2**, 28 (in Russian). The main body of the subject has been summarized by ZAIDEL, KALITEEVSKII, LIPIS, and CHAIKA (1960). [279]

ZAIDEL, A. N., MALYSHEV, G. M., and SCHREIDER, E. YA. (1964) On the sensitivity of spectral analysis, *Opt. i Spektroskopiya*, **17**, 129; *Opt. Spectry (USSR)*, **17**, 65. [264]

ZAKHARIYA, N. F., TURULINA, O. P., KARPENKO, L. I., and VOLOSHCHENKO, I. A. (1963) The use of sulfiding agents in spectral analysis, *Zavodsk. Lab.*, **29**, 683, *Ind. Lab. (USSR)*, **29**, 728. [316]

ZEEMAN, P. B., and COETZER, F. J. (1961) Application of the tape technique to the spectrographic determination of Mg, Mn, P, and B in unashed plant material, *Appl. Spectry*, **15**, 161. [5]

ZWICKER, H. (1962) Die Bestimmung der Linienform aus der Linienkontour bei Linien mit Selbstumkehr. II, Anwendung auf Linien eines QuecksilberhöchstdruckBogens, *Z. Physik*, **166**, 163. [89]

SUBJECT INDEX